Impacts of Climate Change on Hydrology and Water Resources

Impacts of Climate Change on Hydrology and Water Resources

Editors

Sonia Raquel Gámiz-Fortis
Matilde García-Valdecasas Ojeda

 Basel • Beijing • Wuhan • Barcelona • Belgrade • Novi Sad • Cluj • Manchester

Editors
Sonia Raquel Gámiz-Fortis
Applied Physics Department
University of Granada
Granada
Spain

Matilde García-Valdecasas Ojeda
Applied Physics Department
University of Granada Granada
Spain

Editorial Office
MDPI
St. Alban-Anlage 66
4052 Basel, Switzerland

This is a reprint of articles from the Special Issue published online in the open access journal *Water* (ISSN 2073-4441) (available at: www.mdpi.com/journal/water/special_issues/8503G3XYV4).

For citation purposes, cite each article independently as indicated on the article page online and as indicated below:

Lastname, A.A.; Lastname, B.B. Article Title. *Journal Name* **Year**, *Volume Number*, Page Range.

ISBN 978-3-7258-1030-7 (Hbk)
ISBN 978-3-7258-1029-1 (PDF)
doi.org/10.3390/books978-3-7258-1029-1

© 2024 by the authors. Articles in this book are Open Access and distributed under the Creative Commons Attribution (CC BY) license. The book as a whole is distributed by MDPI under the terms and conditions of the Creative Commons Attribution-NonCommercial-NoDerivs (CC BY-NC-ND) license.

Contents

About the Editors vii

Preface ix

Ala A. M. Salameh, Matilde García-Valdecasas Ojeda, María Jesús Esteban-Parra, Yolanda Castro-Díez and Sonia R. Gámiz-Fortis
Extreme Rainfall Indices in Southern Levant and Related Large-Scale Atmospheric Circulation Patterns: A Spatial and Temporal Analysis
Reprinted from: *Water* 2022, 14, 3799, doi:10.3390/w14233799 1

Elisabeth Probst and Wolfram Mauser
Climate Change Impacts on Water Resources in the Danube River Basin: A Hydrological Modelling Study Using EURO-CORDEX Climate Scenarios
Reprinted from: *Water* 2022, 15, 8, doi:10.3390/w15010008 24

Ali Uzunlar and Muhammet Omer Dis
Novel Approaches for the Empirical Assessment of Evapotranspiration over the Mediterranean Region
Reprinted from: *Water* 2024, 16, 507, doi:10.3390/w16030507 53

Xuelei Zhang, Gaopeng Wang and Hejia Wang
Spatiotemporal Variations in Actual Evapotranspiration Based on LPJ Model and Its Driving Mechanism in the Three Gorges Reservoir Area
Reprinted from: *Water* 2023, 15, 4105, doi:10.3390/w15234105 81

Marco Possega, Matilde García-Valdecasas Ojeda and Sonia Raquel Gámiz-Fortis
Multi-Scale Analysis of Agricultural Drought Propagation on the Iberian Peninsula Using Non-Parametric Indices
Reprinted from: *Water* 2023, 15, 2032, doi:10.3390/w15112032 100

Citlalli Madrigal, Rama Bedri, Thomas Piechota, Wenzhao Li, Glenn Tootle and Hesham El-Askary
Water Whiplash in Mediterranean Regions of the World
Reprinted from: *Water* 2024, 16, 450, doi:10.3390/w16030450 122

Muhammad Imran, Jingming Hou, Tian Wang, Donglai Li, Xujun Gao and Rana Shahzad Noor et al.
Assessment of the Impacts of Rainfall Characteristics and Land Use Pattern on Runoff Accumulation in the Hulu River Basin, China
Reprinted from: *Water* 2024, 16, 239, doi:10.3390/w16020239 136

Samir Mainali and Suresh Sharma
Climate Change Effects on Rainfall Intensity–Duration– Frequency (IDF) Curves for the Lake Erie Coast Using Various Climate Models
Reprinted from: *Water* 2023, 15, 4063, doi:10.3390/w15234063 161

Vijendra Kumar, Naresh Kedam, Kul Vaibhav Sharma, Darshan J. Mehta and Tommaso Caloiero
Advanced Machine Learning Techniques to Improve Hydrological Prediction: A Comparative Analysis of Streamflow Prediction Models
Reprinted from: *Water* 2023, 15, 2572, doi:10.3390/w15142572 183

Shuang Hao, Anders Wörman, Joakim Riml and Andrea Bottacin-Busolin
A Model for Assessing the Importance of Runoff Forecasts in Periodic Climate on Hydropower Production
Reprinted from: *Water* **2023**, *15*, 1559, doi:10.3390/w15081559 207

George Varlas, Christina Papadaki, Konstantinos Stefanidis, Angeliki Mentzafou, Ilias Pechlivanidis and Anastasios Papadopoulos et al.
Increasing Trends in Discharge Maxima of a Mediterranean River during Early Autumn
Reprinted from: *Water* **2023**, *15*, 1022, doi:10.3390/w15061022 222

Demelash Ademe Malede, Tena Alamirew and Tesfa Gebrie Andualem
Integrated and Individual Impacts of Land Use Land Cover and Climate Changes on Hydrological Flows over Birr River Watershed, Abbay Basin, Ethiopia
Reprinted from: *Water* **2022**, *15*, 166, doi:10.3390/w15010166 246

Shees Ur Rehman, Afzal Ahmed, Gordon Gilja, Manousos Valyrakis, Abdul Razzaq Ghumman and Ghufran Ahmed Pasha et al.
A Laboratory Study of the Role of Nature-Based Solutions in Improving Flash Flooding Resilience in Hilly Terrains
Reprinted from: *Water* **2023**, *16*, 124, doi:10.3390/w16010124 268

About the Editors

Sonia Raquel Gámiz-Fortis

Dr Sonia Raquel Gámiz-Fortis is Full Professor at the Department of Applied Physics, the University of Granada. She graduated in Physical Sciences at this University in 1998. In 1999, she joined the Atmospheric Physics Group of this University, obtaining her Ph.D. in Physics in 2003.

She regularly engages in the teaching of different degrees (Physics, Biology, Environmental Sciences and Edification) at the University of Granada, and several subjects of the Master of Geophysical and Meteorology (GEOMET), such as Advance Physics Meteorology, and Climatology and Climate Change.

Her research focuses on the study and analysis of climate variability and climate change, particularly in the Earth physic area. She has participated in several research projects related to the seasonal forecast of the climate and impacts of climate change, and has led some projects mainly focused on the study of the projected changes in water resources in different regions of the world.

Matilde García-Valdecasas Ojeda

Dr. Matilde García-Valdecasas Ojeda is a postdoctoral researcher at the Atmospheric Physics group of the Department of Applied Physics at the University of Granada. She defended her PhD thesis in 2018, obtaining maximum qualification and Cum Laude and international mention, and was also awarded the 2017/2018 Extraordinary Doctorate Prizes in Sciences at this University. She was also awarded a postdoctoral contract at the OGS in Trieste (Italy), where she worked under the supervision of Dr. Coppola, a world expert in regional climate modeling.

Her scientific aims include assessing the impact of climate change using climate (the weather research and forecasting (WRF) model and ICTP regional climate model (RegCM)) and hydrological models (the variable infiltration capacity (VIC) model and CETEMPS hydrological meodel (CHyM)), with a special focus on droughts and floods. Her work has enabled her to collaborate with other Spanish and international groups, both in the field of climatology and other fields such as ecology or botany. She has been involved in several research projects (1 international, 4 nationals, and 3 regionals) and published numerous research papers in *JCR* journals.

Preface

Water is essential for many aspects of human life, including agriculture, industry and power generation. One of the major impacts of global warming is likely to affect hydrology and water resources, as climate change can alter the balance between the different components of the hydrological cycle. However, despite the developments in recent decades, research on the impact of climate change on hydrology and water resources necessitates improvement. The mechanisms underlying atmospheric circulation and the hydrological cycle, as well as the internal relationships between them, are not fully understood, and the effects of climate change on the hydrologic cycle are associated with significant uncertainty in both climate projections and hydrologic modeling approaches.

This volume of 13 chapters includes descriptions of a variety of studies for a variety of geographic regions by an international roster of authors.

Acknowledgments: Logistics were made possible by the dedicated efforts of Ms. Lvy Xiao and her team, and we gratefully acknowledge their assistance. Valuable feedback on several chapters was solicited and received from volunteer reviewers who lent their scientific expertise to suggest improvements in presentation.

Sonia Raquel Gámiz-Fortis and Matilde García-Valdecasas Ojeda
Editors

Article

Extreme Rainfall Indices in Southern Levant and Related Large-Scale Atmospheric Circulation Patterns: A Spatial and Temporal Analysis

Ala A. M. Salameh *, Matilde García-Valdecasas Ojeda, María Jesús Esteban-Parra, Yolanda Castro-Díez and Sonia R. Gámiz-Fortis

Departamento de Física Aplicada, Universidad de Granada, E18071 Granada, Spain
* Correspondence: alasalman84@correo.ugr.es

Abstract: This study aims to provide a comprehensive spatio-temporal analysis of the annual and seasonal extreme rainfall indices over the southern Levant from 1970 to 2020. For this, temporal and spatial trends of 15 climate extreme indices based on daily precipitation at 66 stations distributed across Israel and Palestine territories were annually and seasonally analyzed through the nonparametric Mann–Kendall test and the Sen's slope estimator. The annual averages for frequency-based extreme indices exhibited decreasing trends, significantly for the Consecutive Dry Days. In contrast, the percentiles- and intensity-based extreme indices showed increasing trends, significant for extremely wet days, Max 1- and 3-day precipitation amount indices. The study area had expanding periods of extreme dry spells for spring and correspondingly shortening extreme wet spells for spring, winter and the combined winter–spring. Moreover, most of spring indices showed negative trends. Conversely, most winter indices displayed positive trends. Regarding the influence of large-scale circulation patterns, the North Sea Caspian pattern, the Western Mediterranean Oscillation, and ENSO were the primary regulators of the winter, spring, and autumn extreme indices, respectively. These findings contribute to a better understanding of extreme rainfall variability in the Levant region and could be utilized in the management of water resources, drought monitoring, and flood control.

Keywords: extreme rainfall indices; Levant region; trend analysis; teleconnection indices

Citation: Salameh, A.A.M.; Ojeda, M.G.-V.; Esteban-Parra, M.J.; Castro-Díez, Y.; Gámiz-Fortis, S.R. Extreme Rainfall Indices in Southern Levant and Related Large-Scale Atmospheric Circulation Patterns: A Spatial and Temporal Analysis. *Water* 2022, 14, 3799. https://doi.org/10.3390/w14233799

Academic Editor: Davide Zanchettin

Received: 27 October 2022
Accepted: 20 November 2022
Published: 22 November 2022

Publisher's Note: MDPI stays neutral with regard to jurisdictional claims in published maps and institutional affiliations.

Copyright: © 2022 by the authors. Licensee MDPI, Basel, Switzerland. This article is an open access article distributed under the terms and conditions of the Creative Commons Attribution (CC BY) license (https://creativecommons.org/licenses/by/4.0/).

1. Introduction

Changes in extreme rainfall events must be assessed since they have extensive implications for human and environmental systems such as society, ecosystem, agriculture, water resources, and economic development [1–4]. Global warming has the potential to increase the frequency and intensity of extreme rainfall, where a warmer atmosphere with more water vapor creates a more active hydrological cycle [5–8]. Furthermore, small changes in mean precipitation due to global warming can cause significant changes in extreme precipitation [9,10]. On global and regional scales, many studies predict that under global warming, a greater increase is expected in extreme rainfall events as compared to the mean values [11–14]. In this context, numerous studies have reported increasing trends in extreme rainfall events in Saudi Arabia [15], Greece [16], India [17], the Mediterranean basin [18,19], and globally [20–22]. On the other hand, decreasing trends in extreme rainfall events were documented in many regions such as Turkey [23], Western Australia [24], northeast Bangladesh [25], Mongolia [26], and Ghana [27].

In the second half of the twentieth century, the Mediterranean region experienced a decrease in precipitation [28]. This trend is expected to continue, with total annual precipitation decreasing by up to 20% by 2050 [29]. In Turkey, Cyprus, Lebanon and Israel, the number of rainy days may decrease by 5–15 days at the mid-century and by 10–20 days per year at the end-of-century [30]. Additionally, in the framework of several paleo-hydrological and longer-term millennial-scale studies suggested a drying of regional climate that coincides with the decline of the Roman and Byzantine Empires in the Levant region [31–33]. The east Mediterranean area, including the Levant region, is considered one of the most vulnerable regions to climate change [11,34–36]. Climate variability in the Levant is accompanied by several environmental and developmental stresses such as frequent droughts, water shortages, population growth, political conflicts, weak infrastructure, and low adaptation capacity [37–40].

A literature review for the Levant indicates most climate studies have focused on long- and mid-term averages, with the majority of studies implemented in small geographical domains with a limited number of stations [41–45]. There have not been many studies on extreme temperature or precipitation indices, mostly because there are not much accessible daily data for the area. However, [46] examined the annual changes in extreme temperature and precipitation indices over Israel during 1950–2017. The authors observed a decline in the total amount of precipitation as well as a rise in the intensity of rainy days. They displayed a spatial coherence despite the fact that none of the regional patterns in the precipitation indices were statistically significant. [47,48] analyzed the changes of several extreme indices in the Middle East and Arab regions, but their studies included less than ten stations from the Levant region. The findings indicated that trends in precipitation indices, including the number of days with precipitation, the average precipitation intensity, and maximum daily precipitation events, are weak in general and do not show spatial coherence. Extreme temperature indices over Israel and Palestine, at annual and seasonal scales, using data of 28 stations from 1987–2017 were examined by [49]. They also examined their relationships with the large-scale atmospheric circulation patterns, but the study did not analyze any extreme rainfall indices.

Investigating the influence of the large-scale atmospheric circulation patterns on the extreme rainfall indices is vital to establish the basis for understanding the causes of rainfall variability and the causal mechanisms of these indices. For the Levant, most studies analyzed the influence of the large-scale circulation patterns on the mean precipitation values in Israel [50–53]. However, [54] looked at teleconnections regarding different rainfall daily intensities, including heavy precipitation.

Due to the fact that extreme indices have been found to be highly correlated with meteorological and hydrological disasters such as droughts, floods, and landslides [55,56], the necessity to investigate the variability of extreme rainfall indices in a region like the Levant is imperative. Until now, the extreme precipitation indices have not been analyzed at a seasonal scale in the southern Levant, where the study of precipitation changes using only the annual time scale may mask some considerable variations between seasons [57]. Moreover, the impact of the large-scale atmospheric circulation patterns on the extreme rainfall has not been investigated in southern Levant. Therefore, understanding the spatio-temporal variability of extreme precipitation and its related large-scale climate teleconnections mechanism in such a vulnerable region is essential to comprehend the extreme events response to global warming and finding better procedures to deal with water resources management.

The main objectives of this study are: (1) to provide a comprehensive spatio-temporal variability and trends analysis for the annual and seasonal extreme rainfall indices over the whole area of Israel and Palestine during the period 1970–2020; and (2) to investigate the relationships between the extreme rainfall indices in the southern Levant and the main large-scale atmospheric circulation patterns in the Northern Atlantic and Mediterranean Basin.

2. Materials and Methods
2.1. Study Area

The study area covers Israel and Palestine, which are located on the eastern edge of the Mediterranean Sea, roughly between 34°15′ E and 35°40′ E and 29°30′ N and 33°15′ N (Figure 1). It also conforms the western section of the southern Levant, with an area of about 27,000 km^2, and an elevation ranging from 392 m below sea level to 1208 m above sea level. According to the Köppen climate classification, the northern and central parts of the region have a Mediterranean climate (type Csa), while the southern and southeastern parts have semiarid (type BSh) and arid (type BWh) climates. The rainy season lasts from September to May, with 67% of annual precipitation falling in winter (December–February), 16% in spring (March–May), and 17% in autumn (September–November).

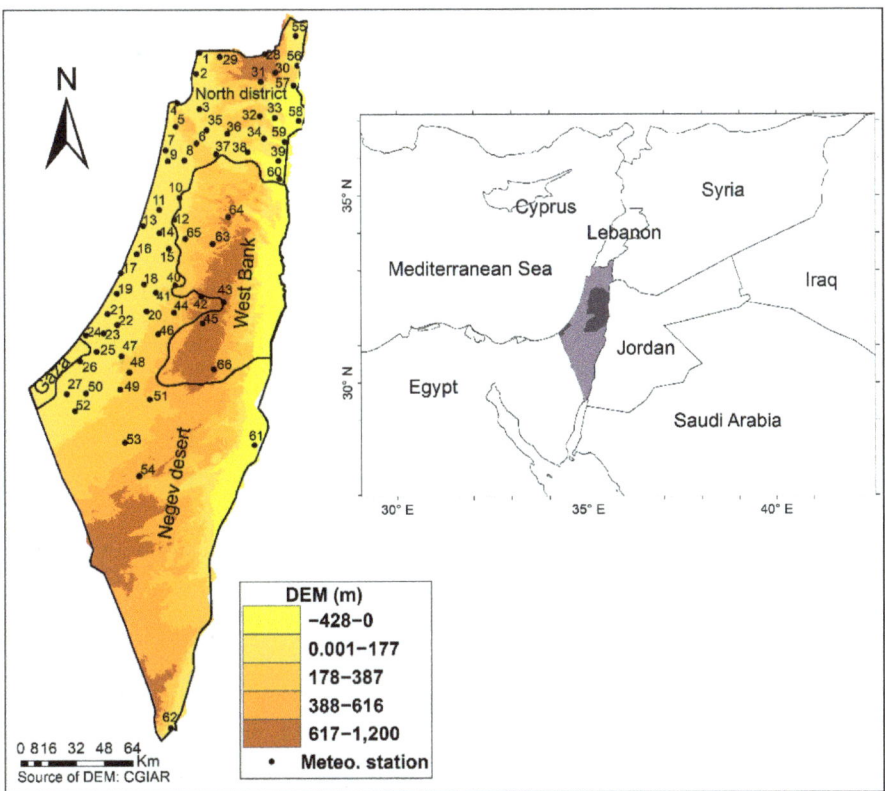

Figure 1. Study area location and the spatial distribution of the stations used in this study. Names of stations are mentioned in Table S2 (Supplementary Material).

2.2. Data and Quality Control

Observed daily precipitation data of an initial set of 75 stations distributed over Israel and Palestine were obtained from the Israel and Palestine Meteorological Departments (https://ims.data.gov.il, accessed on 1 May 2021 and http://www.pmd.ps/, accessed on 1 May 2021), respectively. Each station with a minimum record duration of 51 years (1970–2020), except for the Elqana and Karmel stations from the West Bank (Table S1), covered the period 1982–2020. Time series were subjected to a rigorous data quality control process to identify systematic errors (e.g., negative values or typing errors), missing data, and outliers [47,58].

A total of 66 stations with very few missing data (<0.42%) were considered in the analyses (Figure 1, Table S2). These missing days were handled using the spatial interpolation method based on nearby stations (distance < 8 km and correlation > 0.90) [49,59]. The software package RClimDex V1.3 allows for the detection of outliers on a daily timescale, with a range of thresholds for flagging unreliable data. The outliers were visually evaluated and compared with other nearby stations. In addition, the homogeneity for the selected daily time series was tested to avoid any false trends caused by any anthropogenic effects. The R-based 'RHtests_dlyPrcp,V4' software, based on the transPMFred algorithm [60], was used to detect multiple change points in the series, and adjust them using the 'quantile-matching' algorithm [61]. This technique is commonly used to detect change points in daily rainfall time series [62–64]. Finally, a total of 15 break points were detected in 15 out of 66 stations used in this study (Table S1 in Supplementary Material). Additional information about stations, including names, coordinates, and the period of record, as well as the missing values, is listed in the supplementary materials (Table S2 in Supplementary Material).

2.3. Methods

2.3.1. Indices of Extreme Precipitation

A total of 15 extreme precipitation indices were chosen based on 27 temperature and precipitation indices established by the Expert Team on Climate Change Detection Monitoring and Indices (ETCCDI) [12,65] and recommended by the World Meteorological Organization-Commission for Climatology (WMO-CCI). Table 1 contains a brief description of these indices, along with their acronyms. These indices were selected based on previous studies in the study area and Arab region [46–48,66], in order to evaluate the characteristics of extreme precipitation events, such as intensity, duration, and frequency. Following [20], a classification of extreme precipitation indices into five categories was used, with threshold indices (e.g., R1mm, R10mm, R20mm, and R50mm), absolute indices (e.g., Rx1day, Rx3day, and Rx5day), extreme percentiles (e.g., R95P and R99P), duration indices (CDD and CWD), and other indices (PRCPTOT, SDII, R95Ptot, and R99Ptot).

Table 1. Description of extreme precipitation indices used in this study.

No.	Index	Indicator Name	Definition	Unit
1	PRCPTOT	Annual total wet day precipitation	Annual total precipitation from days ≥1 mm	mm
2	R1mm	Number of wet days	Annual count of days when precipitation ≥1 mm	Days
3	R10mm	Number of heavy precipitation days	Annual count of days when precipitation ≥10 mm	Days
4	R20mm	Number of very heavy precipitation days	Annual count of days when precipitation ≥20 mm	Days
5	R50mm	Number of days above 50 mm	Annual count of days when precipitation ≥50 mm	Days
6	R95P	Very wet days	Annual total precipitation when daily precipitation amount >95th percentile	mm
7	R99P	Extremely wet days	Annual total precipitation when daily precipitation amount >99th percentile	mm
8	R95Ptot	Contribution from very wet days	100*R95P/PRCPTOT	%
9	R99Ptot	Contribution from extremely wet days	100*R99P/PRCPTOT	%
10	RX1day	Max 1-day precipitation amount	Monthly maximum 1-day precipitation	mm
11	RX3day	Max 3-day precipitation amount	Monthly maximum consecutive 3-day precipitation	mm
12	RX5day	Max 5-day precipitation amount	Monthly maximum consecutive 5-day precipitation	mm
13	SDII	Simple daily intensity index	Annual total precipitation divided by the number of wet days (defined as precipitation ≥1 mm) in the year	mm/day
14	CWD	Consecutive wet days	Maximum number of consecutive days when precipitation ≥1 mm	Days
15	CWD-DJF	Consecutive wet days in winter	Maximum number of consecutive days when precipitation ≥1 mm, between December to February	Days
16	CWD-MAM	Consecutive wet days in spring	Maximum number of consecutive days when precipitation ≥1 mm, between March to May	Days
17	CWD-DJFMAM	Consecutive wet days in winter and spring	Maximum number of consecutive days when precipitation ≥1 mm between December to March	Days
18	CDD	Consecutive dry days	Maximum number of consecutive days when precipitation <1 mm	Days
19	CDD-DJF	Consecutive dry days in winter	Maximum number of consecutive days when precipitation <1 mm, between December to February	Days
20	CDD-MAM	Consecutive dry days in spring	Maximum number of consecutive days when precipitation <1 mm, between March to May	Days
21	CDD-DJFMAM	Consecutive dry days in winter and spring	Maximum number of consecutive days when precipitation <1 mm between December to March	Days

The software package RClimDex v1.0 [67] developed by the Climate Research Branch of the Meteorological Service of Canada was used to calculate the extreme indices. The software and documentation are available at http://etccdi.pacificclimate.org, accessed on 1 August 2021. Such software performs the calculations using daily data and provides monthly and annual data for the indices. All indices were computed at annual time scale and at seasonal scale for PRCPTOT, R1mm, R10mm, R20mm, RX1day, RX3day, RX5day, and SDII indices. Additional index calculations such as consecutive dry days (CDD) and consecutive wet days (CWD) were performed for the wet months (e.g., CDD/CWD-DJF, CDD/CWD-MAM, and CDD/CWD-DJFMAM). Days from December to the end of February (DJF) were considered for winter, March to May (MAM) for spring and from September to November (SON) for autumn.

2.3.2. Trend Detection

The annual and seasonal trends of the various indices for each station were calculated for the period 1970–2020. The analysis was performed using the robust nonparametric Mann–Kendall test [68,69] with Sen's slope estimator [70], since it is a distribution-free test and less sensitive to outliers [71]. The Man–Kendall test has been widely used to assess the monotonic trend in extreme precipitation events and climatological time series globally and regionally [20,72–74]. All the time series were pre-whitened in order to correct the Mann–Kendall test for serial autocorrelation [71,75]. The statistical significance of the trends was assessed at 0.01 and 0.05 levels. This trend analysis was conducted using the R package "modifiedmk" [76].

2.3.3. Teleconnection Indices

Additionally, the monthly values of seven teleconnection indices, the North Atlantic Oscillation (NAO), the East Atlantic (EA) pattern, the EA/Western Russia (EA/WR) pattern, the Mediterranean Oscillation (MO), the Western Mediterranean Oscillation (WEMO), the North Sea-Caspian (NCP) pattern and El Niño-Southern Oscillation (ENSO), for the period 1970–2020 were collected from the Climate Prediction Center of the National Oceanic and Atmospheric Administration (http://www.cpc.ncep.noaa.gov/data/teledoc/telecontents.shtml, accessed on 1 September 2021), from the Climatic Research Unit of the University of Norwich (https://crudata.uea.ac.uk/cru/data/moi/, accessed on 1 September 2021) for MO and NCP, and from the Group of climatology of the University of Barcelona (http://www.ub.edu/gc/en/2016/06/08/wemo/, accessed on 1 September 2021) for WEMO. These monthly values were averaged to obtain seasonal and annual values. Afterward, their influence on the extreme precipitation indices was examined by using the Pearson correlation, as in other studies [77,78] based on detrended series for each station. The statistical significance of the correlations was assessed at the 5% level.

3. Results

3.1. Annual Trends of Extreme Precipitation Indices

Table 2 shows an overall view of the annual trend analysis through the total number of stations with increasing or decreasing trends, as well as the trends of the averaged time series over the study area from 1970 to 2020. The temporal behavior of some indices that exhibited significant increasing or decreasing trends is shown in Figure S1 in the Supplementary Material. More than 62% of the stations showed decreasing trends in the PRCPTOT, R1mm, R10mm, and CDD indices (Table 2). In contrast, the R95P, R95Ptot, RX1day, RX3day, RX5day, SDII, and CWD indices increased in more than 72% of the stations. For all extreme indices, the frequency of significant decreasing or increasing trends was less 16 stations, and between 38–86% of the stations did not exhibit trends in the R20mm, R50mm, R99P, and R99Ptot indices. The results showed significant increasing trends in the R99P (4.4 mm/decade), R99Ptot (0.78%/decade), RX1day (1.7 mm/decade), and RX3day (2.1 mm/decade) indices (Table 2, Figure S1a–d). On the other hand, a significant decreasing trend was observed for the CDD index (−2.7 day/decade) (Table 2, Figure S1c).

Table 2. Number of stations that showed increasing or decreasing trend along with the trend values for the annual extreme indices averaged over the study area, during the period 1970–2020. The number in brackets represents the counts of stations with statistically significant trends at the 95% confidence level. Asterisks indicate significance level: ** = ($p < 0.05$).

No.	Index	Total (+) Trends (Sig.)	Total (−) Trends (Sig.)	No Trend	Trend for Averaged Time Series
1	PRCPTOT	25 (0)	41 (1)	0	−2.9 (mm/decade)
2	R1mm	3 (0)	63 (11)	0	−1.1 (days/decade)
3	R10mm	15 (0)	51 (3)	2	−0.2 (days/decade)
4	R20mm	5 (0)	27 (2)	27	0.0 (days/decade)
5	R50mm	17 (0)	3 (0)	46	0.07 (days/decade)
6	R95P	48 (4)	14 (0)	4	5.3 (mm/decade)
7	R99P	9 (9)	0 (0)	57	4.4 ** (mm/decade)
8	R95Ptot	52 (2)	10 (0)	4	0.8 (%/decade)
9	R99Ptot	9 (6)	0 (0)	57	0.78 ** (%/decade)
10	RX1day	51 (3)	15 (0)	0	1.7 ** (mm/decade)
11	RX3day	48 (6)	18 (0)	0	2.1 ** (mm/decade)
12	RX5day	48 (1)	18 (0)	0	1.2 (mm/decade)
13	SDII	50 (4)	16 (0)	0	0.19 (mm/decade)
14	CWD	48 (0)	18 (0)	25	0.04 (days/decade)
15	CDD	9 (0)	57 (16)	0	−2.7 ** (days/decade)

Regarding the spatial distribution of the annual trends (Figure 2), the significant decreasing trends cover the west bank (with an average of −2.3 days/decade) and some northeastern locations of the study area for the R1mm index (Figure 2a). Significant increasing trends are observed in the northern regions for some intensity extreme indices, R99p, R99Ptot, and RX3day by averages of 2.1 mm/decade, 1.9%/decade, and 6.5 mm/decade, respectively (Figure 2b–d). For the CDD index (Figure 2e), a regional significant decreasing trend is grouped in the northern sites of the study area (with an average value around −6.5 days/decade) and the southern coastal locations (with value of −9.2 days/decade in average).

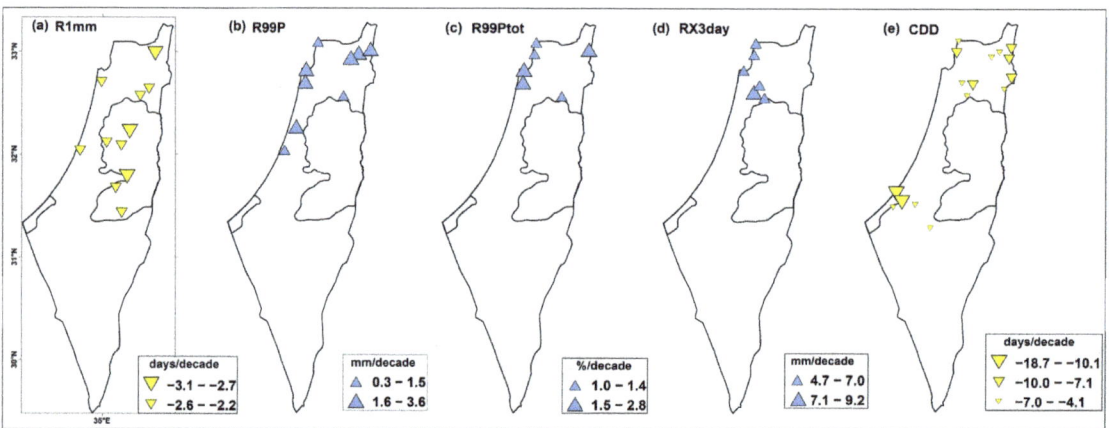

Figure 2. Spatial distribution of trends for the number of wet days (R1mm in days/decade), extremely wet days (R99P in mm/decade), contribution from extremely wet days (R99Ptot in %/decade), max 3-day precipitation amount (RX3day in mm/decade), and consecutive dry days (CDD in days/decade) indices exhibiting notably significant decreasing (yellow triangles) or increasing (blue triangles) trends at the 95% confidence level.

3.2. Seasonal Trends of the Extreme Precipitation Indices

In this section, the seasonal trends for the PRCPTOT, R1mm, R10mm, R20mm, RX1day, RX3day, RX5day, SDII, CWD and CDD indices were calculated for each station and for the entire study area on the basis of the averaged time series (Table 3). Figures 3–5 show the spatial distribution of the trends of some winter, spring, and autumn indices.

Table 3. Number of stations that showed increasing or decreasing trends along with the trend values for the averaged time series, 1970–2020. The number in brackets represents the counts of stations with statistically significant trends at the 95% confidence level. Asterisks indicate significance level: * = ($p < 0.1$), ** = ($p < 0.05$).

Index	Season	Tot. (+) Trends (Sig.)	Tot. (−) Trends (Sig.)	No Trend	Trend for Averaged Time Series
PRCPTOT	Winter	48 (0)	18 (0)	0	8.8 mm/decade
	Spring	2 (0)	64 (18)	0	−5.8 mm/decade
	Autumn	12 (0)	54 (0)	0	−1.9 mm/decade
R1mm	Winter	6 (0)	60 (3)	0	−0.6 days/decade
	Spring	7 (0)	59 (8)	0	−0.3 days/decade
	Autumn	20 (0)	37 (0)	9	−0.08 days/decade
R10mm	Winter	27 (0)	33 (0)	6	−0.05 days/decade
	Spring	4 (0)	62 (7)	0	−0.02 days/decade
	Autumn	12 (0)	28 (0)	26	0
R20mm	Winter	46 (0)	9 (0)	11	0.1 days/decade
	Spring	0 (0)	22 (5)	44	−0.09 days/decade
	Autumn	3 (0)	25 (4)	38	0.04 days/decade
RX1day	Winter	53 (6)	13 (0)	0	2.2 * mm/decade
	Spring	2 (0)	64 (21)	0	−2.1 * mm/decade
	Autumn	6 (0)	60 (9)	0	−1.8 mm/decade
RX3day	Winter	52 (7)	14 (0)	0	3.3 mm/decade
	Spring	1 (0)	65 (20)	0	−3.6 * mm/decade
	Autumn	7 (0)	59 (8)	0	−2.5 mm/decade
RX5day	Winter	46 (2)	20 (0)	0	1.7 mm/decade
	Spring	2 (0)	64 (15)	0	−3.7 mm/decade
	Autumn	11 (0)	55 (3)	0	−2.4 mm/decade
SDII	Winter	55 (11)	11 (0)	0	0.25 mm/decade
	Spring	6 (0)	60 (19)	0	−0.52 * mm/decade
	Autumn	3 (0)	63 (19)	0	−0.75 * mm/decade
CDD	Winter	22 (0)	44 (0)	0	−0.03 days/decade
	Spring	59 (21)	5 (0)	0	1.5 ** days/decade
	Winter-spring	59 (20)	7 (0)	0	1.7 * days/decade
CWD	Winter	15 (0)	25 (0)	26	−0.05 days/decade
	Spring	3 (0)	53 (16)	10	−0.2 days/decade
	Winter-spring	22 (0)	21 (0)	23	−0.01 days/decade

3.2.1. Winter Trends

Significant trends in the averaged time series were not found for all extreme indices, except for the RX1day index with an averaged value of 2.2 mm/decade (Table 3). The spatial distribution of some winter indices trends is shown in Figure 3. Overall, the extreme winter indices do not seem to change very quickly locally, as very few significant trends were observed (<11 stations) for all indices (Table 3).

The highest increasing trends for the PRCPTOT index were observed in the northern and northwestern locations by an average of 19.1 mm/decade, although they were not-significant (not shown). A percent of 91% of stations exhibited decreasing trends for the number of wet days index (Table 3), with significant values between −1.5 and −2.1 days/decade for some stations (Figure 3a). Additionally, these large declining trends in R1mm along with the increasing trends in the PRCPTOT index are reflected in the rising trends for the SDII index, as is mainly observed in the northern regions of the study area (Figure 3b). In this regard, eleven northern locations showed significant rising trends in the SDII index with an averaged value of 0.71 mm/decade.

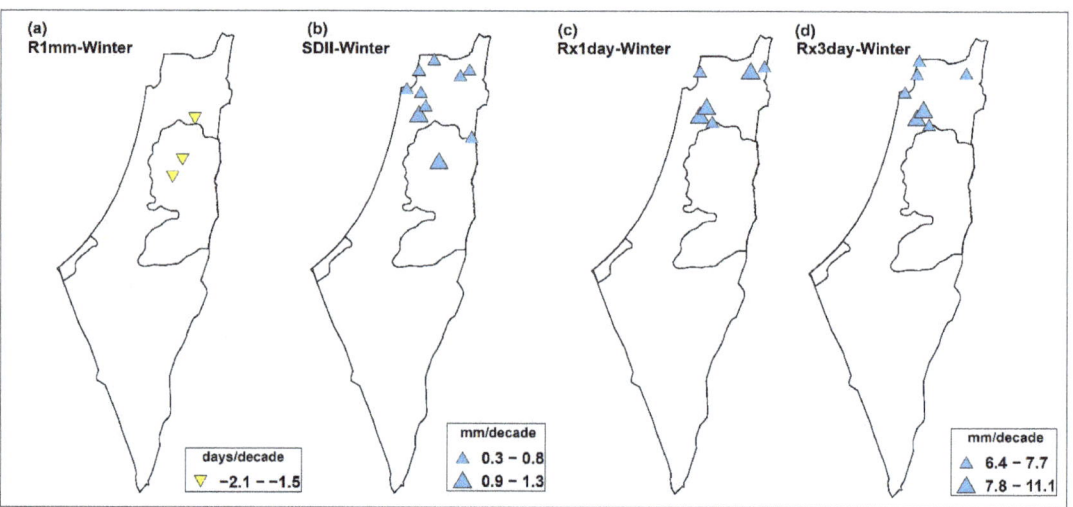

Figure 3. Spatial distribution of winter trends for some indices that exhibited notably significant decreasing (yellow triangles) or increasing (blue triangles) trends. Legend for RX1day and RX3day indices is common.

According to Figure 3c,d regarding the RX1day and RX3day indices, the central and northern regions vastly showed increasing trends, but only significant values were obtained for some northern locations with an averaged values around 4.7 and 8.1 mm/decade, respectively.

3.2.2. Spring Trends

Compared with other seasons, all spring extreme indices showed very rapid changes (Table 3, Figure 4). The results indicate that, in average for the whole area, the trends for the indices RX1day, RX3day, and SDII decreased significantly by −2.1, −3.6, and −0.52 mm/decade, respectively (Table 3). On the contrary, the CDD-MAM and CDD-DJFMAM indices showed significantly increasing trends of 1.5 and 1.7 days/decade, respectively (Table 3).

For the PRCPTOT index, a percentage of 97% of the stations (64 stations) showed a decreasing trend, with 28% (18 stations) showing a significant trend at the level of 0.05 (Table 3). Locally, a coherent and intense pattern of significantly decreasing trends can be seen (Figure 4a). The northern stations, Jerusalem Governorate, and the central region of the coastal areas had the highest values (around −8.9 mm/decade).

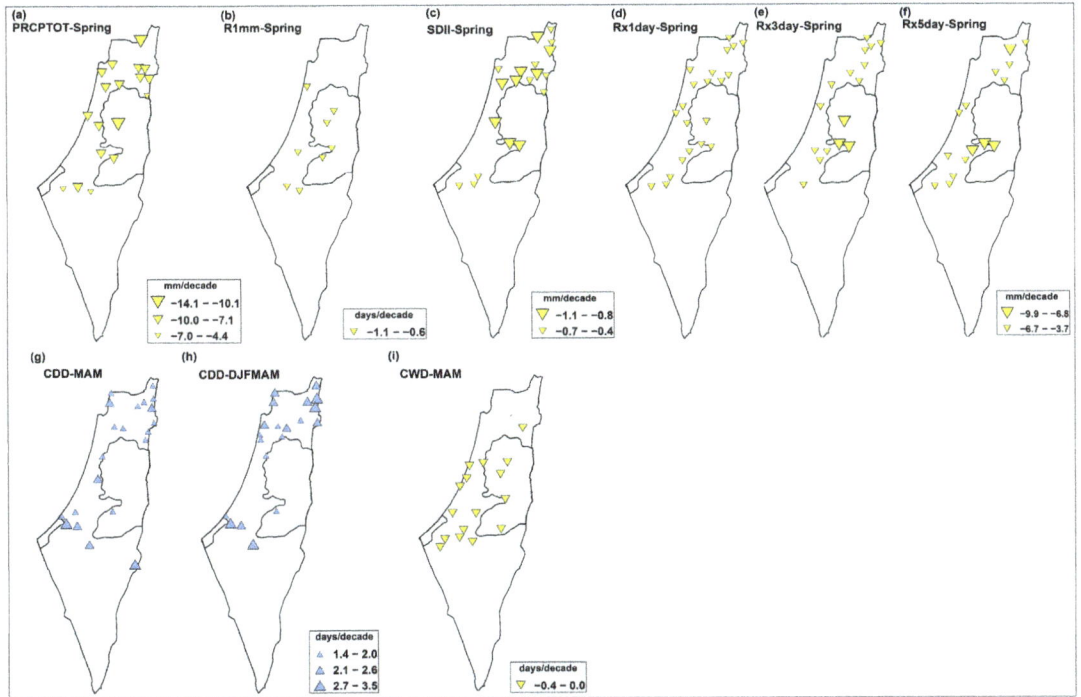

Figure 4. Spatial distribution of spring trends for some indices that exhibited notably decreasing (yellow triangles) or increasing (blue triangles) trends. Legends for RX1day, RX3day and RX5day, and for CDD-MAM and CDD-DJFMAM, are common, respectively.

For the R1mm index, 89% of the stations (Table 3) showed decreasing trends, and 14% (8 stations) of them reported significant trends. The R1mm index had a lower trend than the PRCPTOT for the most studied stations. It is also worth noting that the significant trends affected stations in the West Bank with an average of −0.89 days/decade. With respect to the SDII (Figure 4c), 91% of the stations showed decreasing trends, and 32% (19 stations) of them showed significant trends (Table 3). Although the R1mm index decreased at most sites, the SDII index also decreased due to the large decreases in the PRCPTOT index. Most stations that had a significant decreasing trend in the PRCPTOT index also showed a significant decreasing trend in the SDII index. The highest significant decreasing trends (with value from −0.9 to −1.1 mm/decade) were observed at ten locations in the northern regions.

For RX1day, RX3day, and RX5day indices, more than 94% (>62 stations) of the total stations showed decreasing trends, with significant trends for 33%, 31%, and 23% of the stations, respectively (Table 3, Figure 4e,g). The significant declining trends for these indices are concentrated in northern locations, the Jerusalem governorate, and east to the Gaza strip with an average value of −5.2 mm/decade for the RX3day and RX5day indices, and −2.9 mm/decade for the RX1day.

The CDD index in spring showed rising trends in 89% (59 stations) of the total stations, with significant trends in 36% (21 stations) of them (Figure 4g and Table 3). Very similar results were obtained for the CCD index for the combined winter and spring seasons, which also showed increasing trends in 88% (58 stations) of the stations, with significant trends in 20% (20 stations) of them (Figure 4h and Table 3). The significant increasing trends for CDD-MAM and CDD-DJFMAM (Figure 4g,h) covered many stations in the north of the study area and around the Gaza strip, with relatively higher values for the CDD-DJFMAM index (2.5 day/decade) than the CDD-MAM (2.1 day/decade).

The broad increasing trends for CDD-MAM index have led to the broad decreasing trends in CWD-MAM index (Table 3, Figure 4i). For this latter, 80% of stations had declining trends, with significant trends in 30% of stations (Table 3). These significant decreasing trends covered the central locations in the study area and many locations in the southern part of the coastal area, reaching an average value of -1.5 day/decade.

3.2.3. Autumn Trends

The results for the autumn trends indicated no significant decreasing or increasing trends observed for any of the indices, except for the SDII index, which had a significant decreasing trend of -0.75 mm/decade (Table 3). Furthermore, the frequency-based indices (R1mm, R10mm, and R20mm) did not reflect any remarkable changes in the area under investigation in 1970–2020. However, the PRCPTOT, RX1day, RX3day, and RX5day indices showed declining trends of -1.9, -1.8, -2.5, and -2.4 mm/decade, respectively. In terms of stations, 82%, 56%, 42%, and 38% of the stations showed declining trends in the PRCPTOT, R1mm, R10mm, and R20mm indices, respectively, with no notably significant trends. On the other hand, the intensity extreme indices RX1day, RX3day, RX5day, and SDII showed decreasing trends for more than 83% of the total stations, with 9, 8, 3, and 19 stations, respectively, showing significant decreasing trends.

The spatial distribution of trends for some indices is shown in Figure 5. Significant declining trends were found for the SDII index (Figure 5a) in the central area extended from 31.5° N to 32° N latitudes and in the northeastern locations of the study area with average values of -1.2 and -1 mm/decade, respectively. Spatial distributions of RX1day and RX3day indices can be observed in Figure 5b,c, with significant decreasing trends for many sites around the Jerusalem district with an average of -4.0 mm/decade.

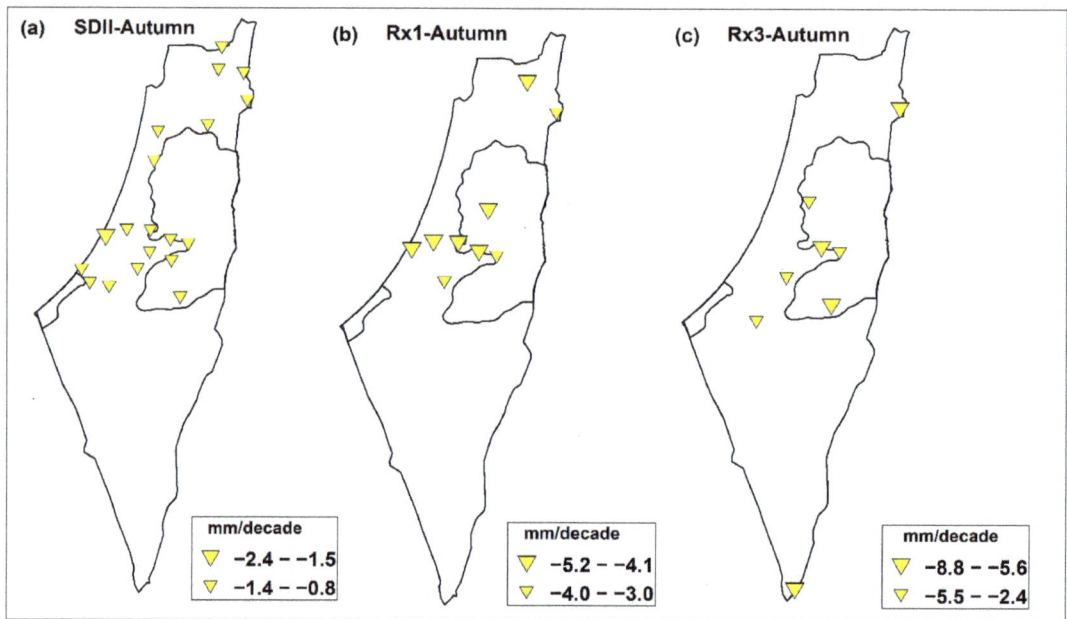

Figure 5. Spatial distribution of autumn trends for some indices that exhibited notably significant decreasing (yellow triangles) or increasing (blue triangles) trends.

3.3. Extreme Rainfall Indices and Teleconnection Patterns

In this section, the relationships between 15 extreme rainfall indices and seven large-scale circulation patterns (WEMO, EA/WR, NAO, EA, MO, NCP, and ENSO) were investigated to determine whether a particular circulation pattern could have some influence on the occurrence of precipitation extremes over the study area. Tables 4 and 5 summarize the number of stations with significant correlations between the extreme precipitation indices and the teleconnection indices at annual and seasonal time scales. Figure 6 shows the spatial distribution of the correlation coefficients for the most important relationships found between the circulation patterns and the extreme rainfall indices at an annual scale, while Figures 7–9 are for a seasonal scale.

3.3.1. Annual Scale

According to Table 4, large-scale circulation patterns had a more significant impact on the frequency-based indices than the intensity- and percentiles-based indices. For the intensity- and percentile-based indices, some influence was obtained for less than 30% of the stations and mainly related to the WEMO index. The results also revealed the MO index was the main driver for the R1mm, R10mm, R20mm, PRCPTOT, and R99Ptot indices. At the 95% confidence level, the threshold $|r| > 0.27$ for the Pearson correlation between extreme indices and teleconnection patterns results are significant.

Table 4. Number of stations with significant positive or negative correlations between extreme precipitation and teleconnection indices at an annual scale. Only significant results at the 95% confidence level are shown.

Index	WEMO +	WEMO −	EAWR +	EAWR −	NAO +	NAO −	EA +	EA −	MO +	MO −	NCP +	NCP −	ENSO +	ENSO −
PRCPTOT	3	0	0	2	10	0	0	0	39	0	3	0	0	33
R1mm	0	0	26	0	8	0	0	0	43	0	31	0	0	39
R10mm	3	0	5	0	6	0	2	0	40	0	6	0	0	28
R20mm	15	0	5	0	8	0	1	0	28	0	2	0	0	8
R50mm	14	0	2	0	2	3	0	0	3	0	4	0	0	8
R95P	19	0	1	0	1	0	1	1	3	0	2	0	0	1
R95Ptot	10	1	0	0	0	8	2	1	2	2	2	0	0	0
R99P	11	1	0	0	0	8	2	1	2	0	2	0	0	0
R99Ptot	2	0	5	0	0	8	0	0	1	12	5	0	0	0
RX1day	9	0	1	0	0	2	2	1	3	2	2	0	0	3
RX3day	6	5	2	0	6	5	1	0	0	2	0	0	0	1
RX5day	2	0	2	0	2	2	1	0	2	1	0	0	0	0
SDII	15	0	0	0	1	3	1	0	1	0	0	0	0	0
CDD	0	0	1	14	1	6	9	0	3	0	8	0	0	5
CWD	2	0	8	0	0	0	3	0	5	0	31	0	0	28

In detail, the index MO showed the highest frequency of significant correlation for the indices PRCPTOT, R1mm, R10mm, and R20mm with 59%, 65%, 61%, and 42% of the stations, respectively (Table 4). Its effect is concentrated between 31.4° N and 33.2° N, with correlation coefficients ranging from 0.27 to 0.38 for the indices PRCPTOT, R1mm, and R20mm (Figure 6a,b,d), and from 0.27 to 0.53 for the index R10mm (Figure 6c). The MO pattern was also the main driver for the R99Ptot index (Table 4, Figure 6e) with 20% of the stations correlating negatively with it (values between −0.27 and −0.38) and spatially covering some northern sites.

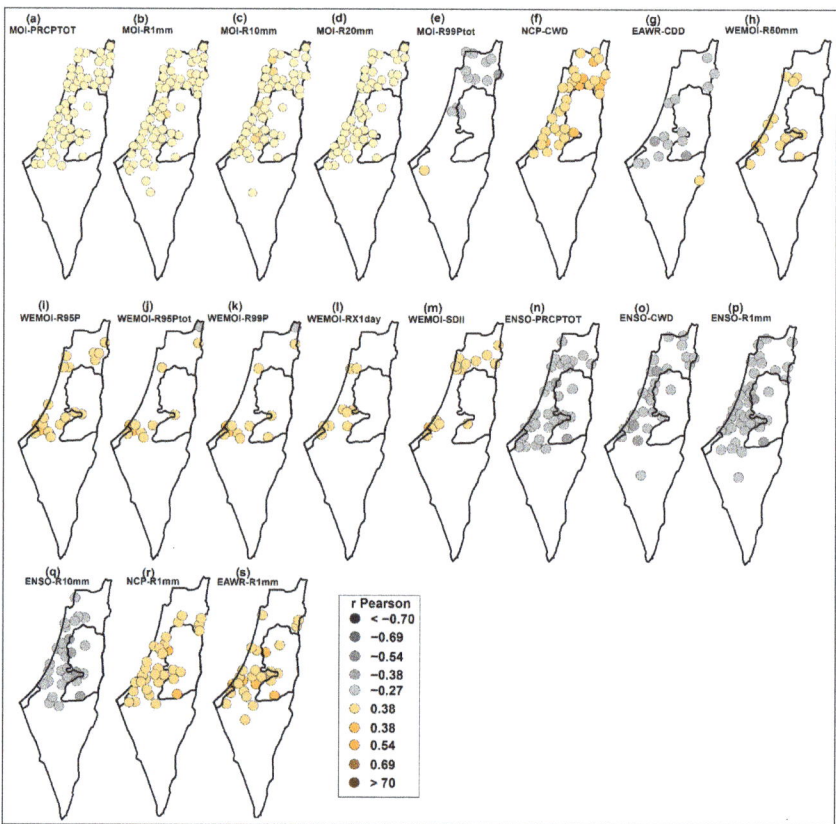

Figure 6. Spatial distribution of significant Pearson correlation coefficients between the teleconnection indices (MOI, NCP, EAWR, WEMOI, and ENSO) and the extreme precipitation indices (PRCPTOT, R1mm, R10mm, R20mm, R99Ptot, CWD, CDD, R95P, R95Ptot, R99P, RX1 day, and SDII), at annual scale.

In addition, the ENSO pattern was the second most important pattern affecting the PRCPTOT, CWD, R1mm, and R10mm indices (Table 4). A percentage of 50%, 42%, 59%, and 42% of the stations showed significant negative correlations, with values from −0.27 to −0.54, spatially speaking (Figure 6n–q).

For the CWD index, the NCP index was the main controller, showing a significant positive correlation with 47% of the stations (Table 4). Its effect extended over all stations at the northern boundaries of the West Bank (with correlation coefficients between 0.27 and 0.54), the northernmost locations, and the central and southern locations of the coastal area (Figure 6f). In addition, 47% of the stations showed remarkable positive significant correlations with the NCP for the R1mm index (Figure 6r). The EAWR pattern had a significant effect at 23% and 39% of the stations, respectively, with negative and positive significant correlations on the CDD and R1mm indices (Table 4). Spatially (Figure 6g,s), the EAWR effect mainly covered some central locations.

The WEMO index had a dominant significant influence on six extreme indices (R50mm, R95P, R95Ptot, R99P, RX1day, and SDII), with 21%, 29%, 17%, 18%, 14% and 23% of the stations showing a significant positive correlation with it (Table 4). The geographical domain of the WEMO effect was primarily concentrated at some locations in the Jerusalem district and north-eastern locations in the Gaza Strip for all six indices (Figure 6h–m). Finally, the EA and NAO indices are poorly correlated with the annual extreme precipitation indices.

3.3.2. Seasonal Scale

Table 5 summarizes the number of stations with significant correlations between the extreme precipitation indices and the teleconnection patterns at a seasonal scale. Figures 7–9 show the spatial distribution of significant Pearson correlation coefficients between the teleconnection indices and the extreme precipitation indices for winter, spring, and autumn, respectively. Based on the frequency of significant correlations, the results indicated no single dominant pattern on the seasonal precipitation extremes, as different patterns generally influence different seasons. In this context, the NCP pattern appeared as the dominant pattern on winter extreme precipitation indices, with the MO and EA/WR indices also showing a considerable frequency of significant correlations. On the other hand, the ENSO and WEMO indices showed high frequencies of significant correlations in autumn and spring, respectively.

Table 5. Number of stations with significant positive or negative correlations between extreme precipitation and teleconnection indices, at a seasonal scale. Only significant results at the 95% confidence level are shown.

Index	Season	WEMO +	WEMO −	EA/WR +	EA/WR −	NAO +	NAO −	EA +	EA −	MO +	MO −	NCP +	NCP −	ENSO +	ENSO −
PRCPTOT	Winter	0	0	20	0	0	0	0	0	43	0	51	0	0	2
	Spring	0	37	0	0	0	1	0	0	5	0	8	0	0	1
	Autumn	0	20	4	0	9	0	0	0	2	0	0	0	0	48
R1mm	Winter	0	7	41	0	0	0	0	0	53	0	58	0	0	1
	Spring	0	57	0	0	3	0	0	2	3	0	27	0	0	5
	Autumn	0	35	7	0	14	0	0	0	6	0	0	0	0	56
R10mm	Winter	0	0	30	0	1	0	1	0	58	0	55	0	0	5
	Spring	0	37	1	1	0	1	0	0	9	1	9	1	0	1
	Autumn	0	26	5	0	6	0	1	0	4	0	0	0	0	33
R20mm	Winter	0	0	6	0	2	0	0	0	32	0	30	0	0	6
	Spring	0	4	0	4	0	0	1	0	4	0	2	0	2	4
	Autumn	0	15	3	0	2	0	1	0	2	0	0	0	0	20
RX1day	Winter	1	2	3	1	1	2	1	0	2	1	4	0	0	5
	Spring	1	9	1	1	0	1	2	1	2	0	0	0	2	2
	Autumn	0	8	4	0	3	1	0	1	1	0	0	0	0	15
		+	−	+	−	+	−	+	−	+	−	+	−	+	−
RX3day	Winter	0	3	7	0	0	9	2	0	6	0	15	0	0	7
	Spring	0	8	0	0	0	0	0	1	4	0	3	0	0	2
	Autumn	0	9	0	0	13	0	0	0	0	0	1	0	0	19
RX5day	Winter	0	2	7	0	3	5	1	0	9	0	20	0	0	3
	Spring	0	16	0	0	0	0	0	0	1	0	16	0	0	1
	Autumn	0	5	0	0	28	0	0	0	0	0	1	0	0	38
SDII	Winter	0	0	0	0	0	1	3	0	1	0	1	0	0	4
	Spring	0	0	0	1	1	0	3	0	2	0	1	0	1	1
	Autumn	0	2	1	0	0	0	0	2	1	0	0	0	0	5
CDD	Winter	0	1	0	2	0	32	1	0	0	14	0	15	0	1
	Spring	7	0	1	3	0	12	12	0	1	0	0	9	1	3
CWD	Winter	0	1	50	0	0	0	1	0	1	0	48	0	1	25
	Spring	0	37	0	0	2	0	0	10	4	0	33	0	0	2

As shown in Table 5, the NCP index had a greater impact on seven winter extreme rainfall indices (PRCPTOT, R1mm, R10mm, R20mm, RX3day, RX5day, and CWD). It significantly correlated with them at 77, 88, 83, 45, 23, 30, and 73% of the stations, respectively. Figure 7a shows the spatial distribution of these significant correlation coefficients in winter, with the highest correlations (values between 0.54–0.69) for the R1mm index in the southern parts of the coastal area and around the Gaza Strip. Its effect was extensive for the PRCPTOT, R1mm, R10mm, R20mm, and CWD indices.

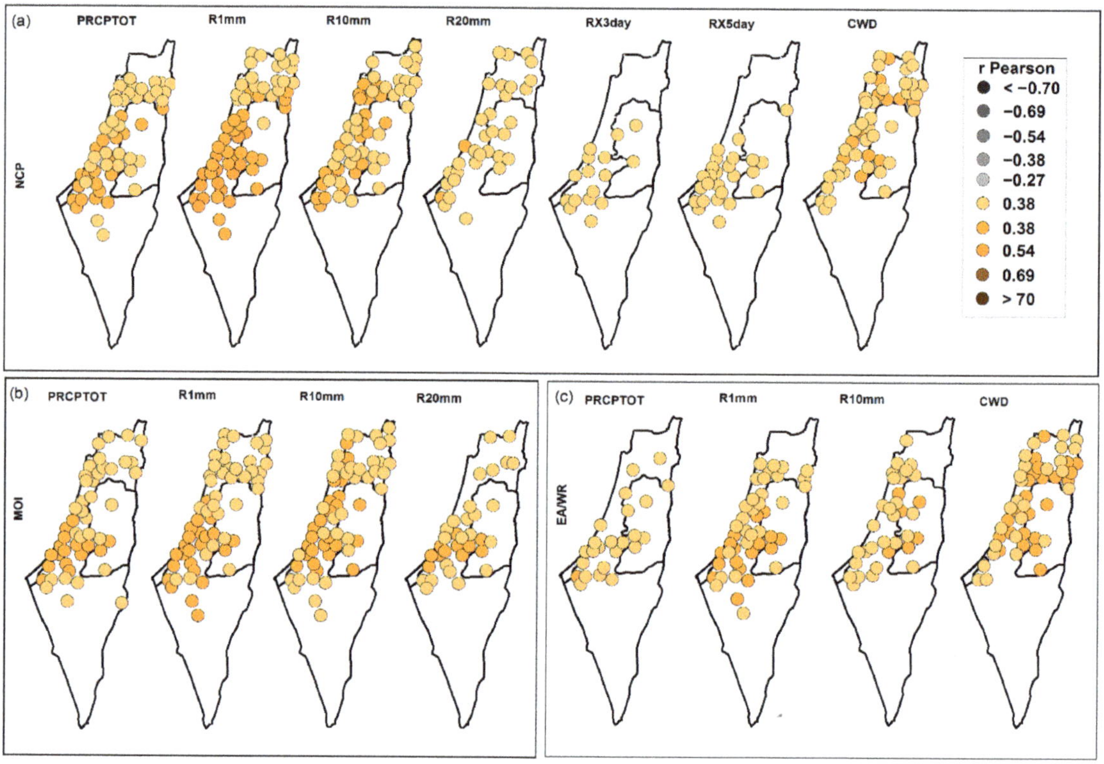

Figure 7. Spatial distribution of significant Pearson correlation coefficients between the teleconnection indices (NCP, MOI and EAWR) and the extreme precipitation indices (PRCPTOT, R1mm, R10mm, R20mm, RX3 day, RX5 day, and CWD), in winter.

The results also showed that the MO and EA/WR indices had remarkable effects on four winter extreme precipitation indices (Figure 7b,c). The winter indices PRCPTOT, R1mm, R10mm, and R20mm are positively correlated with the MO index at 65, 80, 88, and 48% of the total stations, respectively. For the R20mm index, MO covered more sites in the southern coastal region with a high band of significant correlations (values between 0.38–0.54) than the NCP index (Figure 7b). The EA/WR index (Figure 7c) showed a lower frequency of significant correlations for the PRCPTOT, R1mm, and R10mm indices, with 30, 62, and 45% of the stations affected, respectively, compared to the NCP and MO indices. In contrast, it showed a very high frequency of significant correlations with the CWD index for 76% of the total stations, with a spatial pattern similar to NCP.

The WEMO index was found to be the most influential pattern on six spring extreme precipitation indices (PRCPTOT, R1mm, R10mm, RX1day, RX5day, and CWD) with 56, 86, 56, 14, 24, and 56% of the total stations (Table 5), respectively, affected. The highest significant correlations (with values between -0.38 to -0.54) were with the R1mm index for most stations and with the PRCPTOT index at some central and northeast locations (Figure 8a). The WEMO-RX5day significant correlations are, however, concentrated in the coastal and northeastern locations. In addition to the WEMO index's effect, the NCP index exerted a certain positive influence on the R1mm, RX5day, and CWD indices, but with a lower frequency and magnitude (Figure 8b). As shown in Figure 8c, the effect of the EA index on the CDD index occurred only in spring at 13% of the stations, with significant correlations in the range of 0.27 to 0.38. The NAO pattern also had some negative impact on 13% of stations for the CDD index and was spatially distributed in the northern locations (Figure 8c).

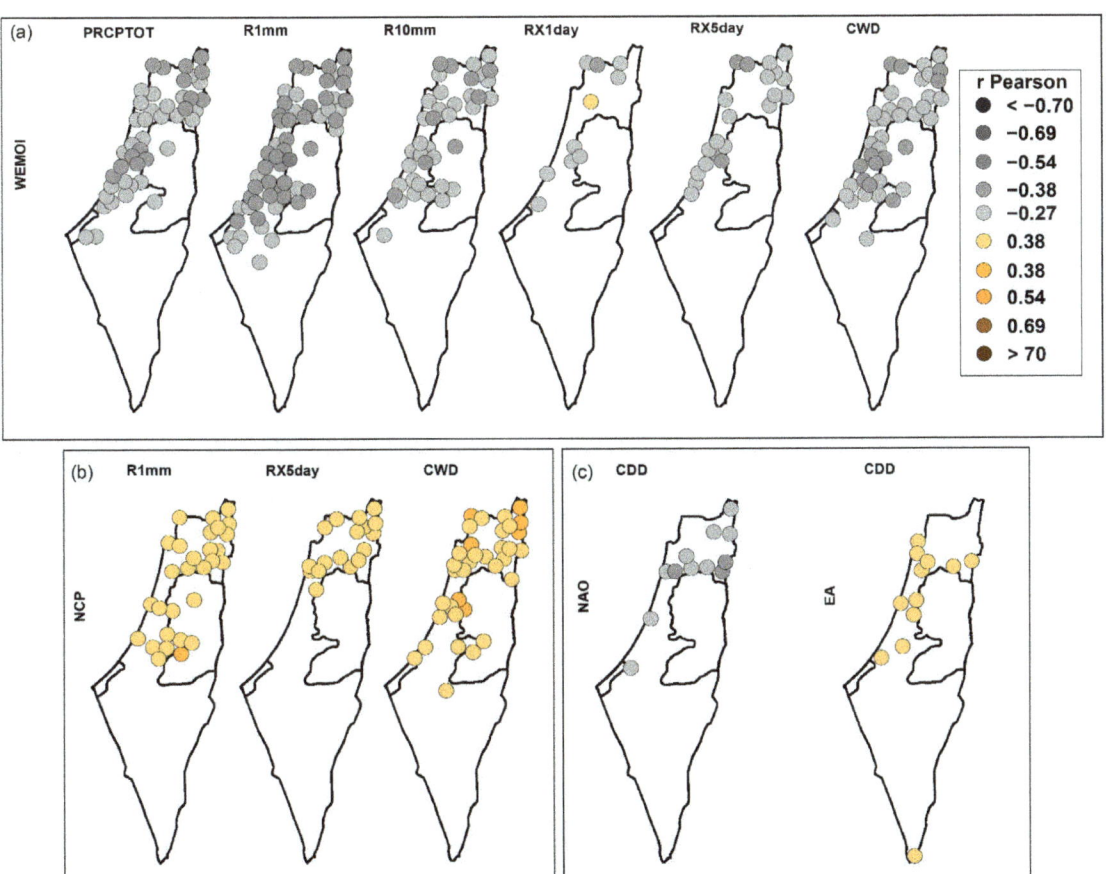

Figure 8. Spatial distribution of significant Pearson correlation coefficients between the teleconnection indices (WEMOI, NCP, NAO, and EA) and the extreme precipitation indices (PRCPTOT, R1mm, R10mm, RX1 day, RX5 day, CWD, and CDD), in spring.

For autumn, the results listed in Table 5 show that the ENSO pattern is the main regulator for seven extreme precipitation indices (PRCPTOT, R1mm, R10mm, R20mm, RX1day, RX3day, and RX5day). For these indices, 73, 58, 50, 30, 23, 29, and 58% of the total stations showed significant negative correlations. In addition, the indices PRCPTOT, R1mm, R10mm, and R20mm correlated negatively with the WEMO for 30, 53, 39, and 23% of the stations, respectively. The NAO pattern also showed some positive effects on the RX3day and RX5day indices at 20 and 42% of the stations, respectively, located in the central regions (Figure 9c). Figure 9 show the spatial distribution of the correlation values between these three circulation patterns and extreme precipitation indices. Most stations in the northern, coastal, and West Bank areas were significantly affected by the ENSO pattern, especially for the PRPTOT, R1mm, R10mm, and RX5day indices (Figure 9a). Other indices (R20mm, RX1day, and RX3day) showed spatially, almost isolated, patterns for the significant correlations, except for stations in the Jerusalem Governorate. The ENSO pattern affected more stations in all regions than the WEMO index.

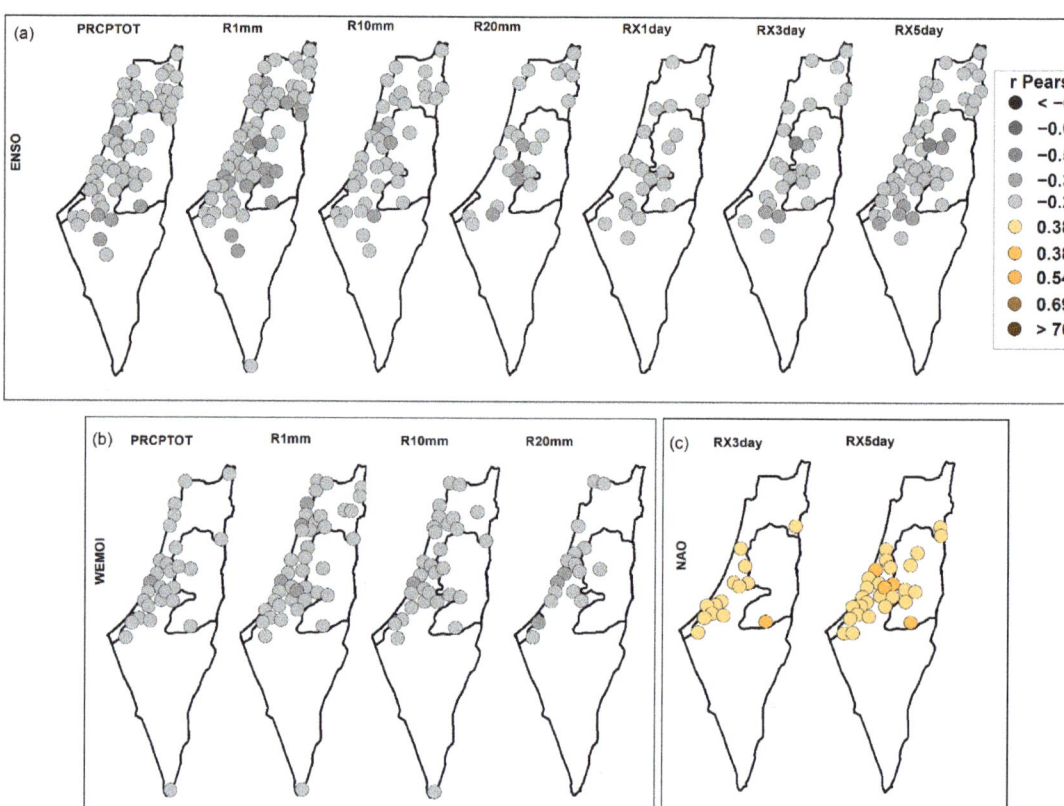

Figure 9. Spatial distribution of significant Pearson correlation coefficients between the teleconnection indices (ENSO, WEMOI, and NAO) and the extreme precipitation indices (PRCPTOT, R1mm, R10mm, R20mm, RX1 day, RX3 day, and RX5 day), in autumn.

4. Discussion

Results from this study show that the study area may be subjected to drought episodes in the future due to the vast decreasing trends in annual, spring, and autumn indices, mainly for the PRCPTOT, R1mm, R10mm, and R20mm indices, with more than 62% of the stations showing decreasing trends. This result is in agreement with other studies showing that the Eastern Mediterranean and the Middle East region are likely to be impacted by frequent and intense drought events [30,79–81]. A decreasing precipitation trend and a reduction in the annual number of precipitation days for the Mediterranean and Middle East regions have been also documented [16,18,45,82–84].

The trend analysis at annual scale showed that the study area tends to have more intense rainy days, where the R1mm index had a notably decreasing trend and the PRCP-TOT did not change much. This is also demonstrated by the increasing trends in all the heavy precipitation indices, with significant increases for R99p, R95ptot, RX1Day, and RX3Day. In [46], the authors showed the frequency-based extreme indices were affected by decreasing trends, whereas the percentiles- and intensity-based extreme indices were generally affected by increasing trends in Israel. However, they found a decreasing trend in the RX5Day during 1950–2017 while our study showed an increasing trend, which can be attributed to the different base periods, especially since they also found a positive trend in the RX5Day during 1988–2017. In [85], the authors found that extreme rainfall had been more intense, but less frequent over Jeddah, Saudi Arabia during 1979–2018.

The increase in the intensity of extreme rainfall events is a major impact of global warming [86,87]. In addition, the increase in heavy rainfall in spite of the decrease in the totals is associated with fewer rainy days and the increase in the frequency and persistence of sub-tropical anticyclones, particularly over the Mediterranean [18]. Decreasing trends in rainfall in the study area may be regarded as a manifestation of the increased influence of the subtropical high over the Mediterranean Basin. Such an evolution is implied by an expansion of the Hadley cell, attributed to the global warming [45,88].

At a global scale, [21] found the SDII, RX5day, R10mm, R20mm, and CDD indices are decreasing. However, SDII and RX5day indices are increasing over the study area. Similarly, CWD increases globally and in the study area [21]. The cause of the significant decreasing annual CDD may be connected to summer rainy days or rainy days during September/October when afterwards the dry season continues, sometimes until the end of November. Regarding spatial trends, the West Bank stations exhibited significant decreasing trends for the R1mm index, as well as the northern locations and several southern coastal locations for the CDD index.

Seasonally, the winter PRCPTOT, RX1day, RX3day, RX5day, and SDII indices showed increasing trends, significant for the northwestern locations in the SDII index due to the PRCPTOT increasing and the R1mm decreasing. It is also significant for several northern locations in the RX1day and RX3day indices. These increasing trends indicate the possibility of flooding in these areas; in particular, they affect many sites with an annual precipitation maximum of more than 1000 mm.

Most extreme indices showed decreasing trends in spring and autumn (more than 80% of the total stations); with central and northern locations, showing significant decreases in the spring RX1day, RX3day, and SDII indices. In addition, the central locations of the study area show significant decreasing trends in the autumn SDII, RX1day, and RX3day indices. Spring and autumn, according to these findings, are the seasons that contribute the most in those locations where annual declines in the PRCPTOT, RX1day, RX3day, RX5day, R1mm, R10mm, R20mm, and CDD indices are found; while winter is the season that contributes the most to annual increases in these indices in other sites. The authors of [45,89] also found that the declining trend in annual precipitation in Israel and the Mediterranean region is mainly due to the spring season.

The study area had longer periods of extreme dry spells (CDD) in spring and correspondingly shorter extreme wet spells (CWD) for winter, spring, and the combined winter–spring. The significant increase in the CDD for the winter–spring was caused by the spring and not by the winter, when most of the total precipitation occurred. In addition, this study found a negative non-significant trend for CDD index during winter, which means that there were no prolonged dry periods in the winter (DJF). The Mediterranean region has also witnessed significant increasing trends in the CDD spring index [89–91], that could affect crop growth and availability of water for irrigation in these areas. This trend was also documented in the southern Levant during the rainy season [45,92].

Regarding the influence of large-scale circulation patterns on extreme precipitations, the NCP, WEMO, and ENSO patterns seem to be the main regulators for the extreme rainfall indices in winter, spring, and autumn, respectively. The MO and EA/WR indices had a remarkable impact on four winter extreme rainfall indices, with the highest correlation values on the R10mm for the MO index, and on the R1mm for EA/WR. In this regard, no works in the literature establishes a direct link between the extreme precipitation indices and large-scale circulation patterns in Israel and Palestine. Most studies analyzed the influence of the large-scale circulation pattern on the mean values. However, it is important to note that some of these studies confirm some of the results obtained in this study.

Many studies have found that the NCP, MO, and EA/WR indices have a positive effect on winter precipitation in Israel, East Mediterranean and Europe [50,93–97]. The negative phase of the NCP refers to an increased counterclockwise anomaly around the western center of the NCP, i.e., the north of the Caspian Sea, and an increased clockwise anomaly circulation around the eastern pole. These anomalies imply increased westerly anomaly circulation towards central Europe and an increased easterly anomaly circulation towards Georgia, and eastern Turkey, which leads to an increase in southwesterly anomaly circulation towered the Balkans, western Turkey, and the Middle East, causing above normal temperatures and below normal rainfall in these regions. The opposite occurs in the positive phase of the NCP index [49]. The authors of [92] found that precipitation in Israel during the positive phase of the NCP is far greater than precipitation during the negative phase of the NCP, and that the influence of the NCP on the precipitation regime in Israel increases from the northern parts of the country to the south. The authors of [94] analyzed the relationship between the MO index and winter precipitation in southern Levant (Israel and Palestine) for the period 1960–1993, finding that winter precipitation is significantly associated with positive MO phases.

For the WEMO index, the positive phase corresponds to an anticyclone over the Azores enclosing the southwestern Iberian quadrant and low pressures in the Liguria Gulf, while its negative phase coincides with the Central European anticyclone located north of the Italian peninsula and a low-pressure center in the framework of the Iberian southwest [98]. The Levant rainfall in the negative phase is favored by maritime surface flow where the negative phase is associated with flows from the Mediterranean [98]. The relationship between the WEMO index and the rain regime in the Levant had never been detected before and more investigation is needed. On the other hand, Ref. [99] found that El Niño is associated with an increased rainfall over the north of Israel after 1970s due to the changes in the jet stream position, because if the jet stream shifts equatorward (during El Niño events) or poleward (during La Niña events) by a few degrees, significant changes in precipitation amounts can occur.

Note that the study of driving mechanism of extreme precipitation is highly complex due to many factors affecting regional precipitation variability such as regional environment characteristics or human activities must be taken into account [100]. However, although the analysis of the relationships between the atmospheric circulation patterns and the extreme rainfall indices presents limitations, it constitutes the starting point for the search of more complex relationships of a non-linear nature. In this sense, further research could involve a comprehensive physical mechanism analysis following the methodology applied for other regions [101,102] that consider the application of more sophisticated statistical techniques capable of analyzing the joint variability, using variables such as the sea surface temperature (SST) or the sea level pressure (SLP) in addition to extreme rainfall indices, could help in the search of potential predictors for the precipitation in this region.

5. Conclusions

This study presents a comprehensive annual and seasonal analysis of trends and variability for a set of 15 extreme precipitation indices using homogeneous and quality-controlled daily records for 66 stations distributed in Israel and Palestine, for the period 1970–2020. In addition, the relationships between these extreme indices and some large-scale circulation patterns covering the Atlantic Ocean and the Mediterranean Sea were examined at the annual and seasonal scales.

The main findings of this work can be summarized as follows:

- Substantial decreasing trends were found for extreme rainfall indices at an annual scale, and for spring and autumn seasons, mainly for the PRCPTOT, R1mm, R10mm, and R20mm indices.
- At an annual scale, southern Levant tends to have more intense rainy days, showing increased trends for all the heavy precipitation indices.

- Seasonally, winter PRCPTOT, RX1day, RX3day, RX5day, and SDII indices showed increasing trends, significant for SDII index in the northwestern locations in the area, related with the PRCPTOT increasing and the R1mm decreasing.
- In spring and autumn, most extreme indices showed decreasing trends, these being the seasons mostly contributing to the annual declines in the PRCPTOT, RX1day, RX3day, RX5day, R1mm, R10mm, R20mm, and CDD indices.
- Southern Levant had experienced longer periods of extreme dry spells (CDD) in spring and consistently shorter extreme wet spells (CWD) for winter, spring, and the combined winter–spring season.
- The NCP, WEMO, and ENSO atmospheric circulation patterns are the main regulators for the extreme rainfall indices in winter, spring, and autumn, respectively.

Overall, the results obtained from this study serve as a strong warning in the southern Levant because the trends detected toward drier conditions and more intense rainy days, will increases the risk of flooding, food security, emigration, life loss, and property damages in a region, which is already suffering from several developmental stresses such as poor infrastructure, rapid population growth, and scarcity of water resources. In this sense, the findings obtained in this work are very important and constitutes the baseline for decision makers on the findings of potential solutions and for implementing efficient mitigation and adaptation strategies in the Levant.

Supplementary Materials: The following supporting information can be downloaded at: https://www.mdpi.com/article/10.3390/w14233799/s1, Table S1. List of stations where change points in daily rainfall were detected. Table S2. The final list of the meteorological stations used in this study. Figure S1. Temporal evolution for some averaged extreme rainfall indices over all stations, period 1970–2020. Red line is indicating a LOWESS smoothing.

Author Contributions: Conceptualization, M.J.E.-P., S.R.G.-F., A.A.M.S. and Y.C.-D.; methodology, M.J.E.-P., A.A.M.S. and S.R.G.-F.; software, A.A.M.S.; validation, A.A.M.S., M.J.E.-P. and S.R.G.-F.; formal analysis, A.A.M.S.; investigation, A.A.M.S. and S.R.G.-F.; resources, M.J.E.-P., S.R.G.-F. and Y.C.-D.; data curation, A.A.M.S., M.J.E.-P., S.R.G.-F. and Y.C.-D.; writing—original draft preparation, A.A.M.S.; writing—review and editing, M.J.E.-P., S.R.G.-F., A.A.M.S., M.G.-V.O. and Y.C.-D.; visualization, A.A.M.S. and M.J.E.-P.; supervision, M.J.E.-P., S.R.G.-F. and Y.C.-D.; project administration, M.J.E.-P. and S.R.G.-F.; funding acquisition, M.J.E.-P. and S.R.G.-F. All authors have read and agreed to the published version of the manuscript.

Funding: This research has been carried out in the framework of the projects P20_00035 funded by FEDER/Junta de Andalucía-Consejería de Transformación Económica, Industria, Conocimiento y Universidades, and LifeWatch-2019-10-UGR-01 co-funded by the Ministry of Science and Innovation through the FEDER funds from the Spanish Pluriregional Operational Program 2014–2020 (POPE), LifeWatch-ERIC action line.

Data Availability Statement: The data presented in this study are available on request from the first author.

Conflicts of Interest: The authors declare no conflict of interest.

References

1. Najafi, M.R.; Moradkhani, H. A Hierarchical Bayesian Approach for the Analysis of Climate Change Impact on Runoff Extremes. *Hydrol. Process.* **2014**, *28*, 6292–6308. [CrossRef]
2. Abiodun, B.J.; Mogebisa, T.O.; Petja, B.; Abatan, A.A.; Roland, T.R. Potential Impacts of Specific Global Warming Levels on Extreme Rainfall Events over Southern Africa in CORDEX and NEX-GDDP Ensembles. *Int. J. Climatol.* **2020**, *40*, 3118–3141. [CrossRef]
3. Ellwanger, J.H.; Kulmann-Leal, B.; Kaminski, V.L.; Valverde-Villegas, J.M.; da VEIGA, A.B.G.; Spilki, F.R.; Fearnside, P.M.; Caesar, L.; Giatti, L.L.; Wallau, G.L.; et al. Beyond Diversity Loss and Climate Change: Impacts of Amazon Deforestation on Infectious Diseases and Public Health. *An Acad Bras Cienc.* **2020**, *92*, e20191375. [CrossRef] [PubMed]
4. Lima, A.O.; Lyra, G.B.; Abreu, M.C.; Oliveira-Júnior, J.F.; Zeri, M.; Cunha-Zeri, G. Extreme Rainfall Events over Rio de Janeiro State, Brazil: Characterization Using Probability Distribution Functions and Clustering Analysis. *Atmos. Res.* **2021**, *247*, 105221. [CrossRef]

5. Easterling, D.R.; Evans, J.L.; Groisman, P.Y.; Karl, T.R.; Kunkel, K.E.; Ambenje, P. Observed Variability and Trends in Extreme Climate Events: A Brief Review. *Bull. Am. Meteorol. Soc.* **2000**, *81*, 417–425. [CrossRef]
6. Trenberth, K.E. Changes in Precipitation with Climate Change. *Clim. Res.* **2011**, *47*, 123–138. [CrossRef]
7. Zhu, Q.; Yang, X.; Ji, F.; Liu, D.L.; Yu, Q. Extreme Rainfall, Rainfall Erosivity, and Hillslope Erosion in Australian Alpine Region and Their Future Changes. *Int. J. Climatol.* **2020**, *40*, 1213–1227. [CrossRef]
8. Towfiqul Islam, A.R.M.; Rahman, M.S.; Khatun, R.; Hu, Z. Spatiotemporal Trends in the Frequency of Daily Rainfall in Bangladesh during 1975–2017. *Theor. Appl. Climatol.* **2020**, *141*, 869–887. [CrossRef]
9. Katz, R.W.; Brown, B.G. Extreme Events in a Changing Climate: Variability Is More Important than Averages. *Clim. Chang.* **1992**, *21*, 289–302. [CrossRef]
10. de Lima, M.I.P.; Santo, F.E.; Ramos, A.M.; de Lima, J.L.M.P. Recent Changes in Daily Precipitation and Surface Air Temperature Extremes in Mainland Portugal, in the Period 1941-2007. *Atmos. Res.* **2013**, *127*, 195–209. [CrossRef]
11. Solomon, S.D.; Qin, M.; Manning, Z.; Chen, M.; Marquis, K.B.; Averyt, M.T.; Miller, H.L.; Solomon, S.; Qin, D.; Manning, M.; et al. Summary for Policymakers. In *Climate Change 2007: The Physical Science Basis. Contribution of Working Group I to the Fourth Assessment Report of the Intergovernmental Panel on Climate Change*; Qin, D., Manning, M., Chen, Z., Marquis, M., Averyt, K., Tignor, M., Miller, H.L., Eds.; Cambridge University Press: New York, NY, USA; Geneva, Switzerland, 2007. [CrossRef]
12. Zhang, X.; Alexander, L.; Hegerl, G.C.; Jones, P.; Tank, A.K.; Peterson, T.C.; Trewin, B.; Zwiers, F.W. Indices for Monitoring Changes in Extremes Based on Daily Temperature and Precipitation Data. *Wiley Interdiscip. Rev. Clim. Chang.* **2011**, *2*, 851–870. [CrossRef]
13. Şensoy, S.; Türkoğlu, N.; Akçakaya, A.; Ekici, M.; Demircan, M.; Ulupınar, Y.; Atay, H.; Tüvan, A.; Demirbaş, H. Trends in Turkey Climate Indices from 1960 To 2010. *Meteoroloji. Gov. Tr.* **2013**. Available online: https://www.researchgate.net/publication/289520845_Trends_in_Turkey_Climate_Indices_from_1960_to_2010 (accessed on 26 October 2022).
14. Yilmaz, A.G. The Effects of Climate Change on Historical and Future Extreme Rainfall in Antalya, Turkey. *Hydrol. Sci. J.* **2015**, *60*, 2148–2162. [CrossRef]
15. Almazroui, M. Rainfall Trends and Extremes in Saudi Arabia in Recent Decades. *Atmosphere* **2020**, *11*, 964. [CrossRef]
16. Nastos, P.T.; Zerefos, C.S. On Extreme Daily Precipitation Totals at Athens, Greece. *Adv. Geosci.* **2007**, *10*, 59–66. [CrossRef]
17. Deshpande, N.R.; Kulkarni, A.; Krishna Kumar, K. Characteristic Features of Hourly Rainfall in India. *Int. J. Climatol.* **2012**, *32*, 1730–1744. [CrossRef]
18. Alpert, P.; Ben-Gai, T.; Baharad, A.; Benjamini, Y.; Yekutieli, D.; Colacino, M.; Diodato, L.; Ramis, C.; Homar, V.; Romero, R.; et al. The Paradoxical Increase of Mediterranean Extreme Daily Rainfall in Spite of Decrease in Total Values. *Geophys. Res. Lett.* **2002**, *29*, 31-1–31-4. [CrossRef]
19. Goodess, C.M.; Jones, P.D. Links between Circulation and Changes in the Characteristics of Iberian Rainfall. *Int. J. Climatol.* **2002**, *22*, 1593–1615. [CrossRef]
20. Alexander, L.v.; Zhang, X.; Peterson, T.C.; Caesar, J.; Gleason, B.; Klein Tank, A.M.G.; Haylock, M.; Collins, D.; Trewin, B.; Rahimzadeh, F.; et al. Global Observed Changes in Daily Climate Extremes of Temperature and Precipitation. *J. Geophys. Res. Atmos.* **2006**, *111*, 1042–1063. [CrossRef]
21. Donat, M.G.; Alexander, L.V.; Yang, H.; Durre, I.; Vose, R.; Caesar, J. Global Land-Based Datasets for Monitoring Climatic Extremes. *Bull. Am. Meteorol. Soc.* **2013**, *94*, 997–1006. [CrossRef]
22. Westra, S.; Alexander, L.V.; Zwiers, F.W. Global Increasing Trends in Annual Maximum Daily Precipitation. *J. Clim.* **2013**, *26*, 3904–3918. [CrossRef]
23. Acar, Z.; Gönençgil, B. Investigation of Extreme Precipitation Indices in Turkey. *Theor. Appl. Climatol.* **2022**, *148*, 679–691. [CrossRef]
24. Haylock, M.; Nicholls, N. Trends in Extreme Rainfall Indices for an Updated High Quality Data Set for Australia, 1910–1998. *Int. J. Climatol.* **2000**, *20*, 1533–1541. [CrossRef]
25. Basher, M.A.; Stiller-Reeve, M.A.; Saiful Islam, A.K.M.; Bremer, S. Assessing Climatic Trends of Extreme Rainfall Indices over Northeast Bangladesh. *Theor. Appl. Climatol.* **2018**, *134*, 441–452. [CrossRef]
26. Tong, S.; Li, X.; Zhang, J.; Bao, Y.; Bao, Y.; Na, L.; Si, A. Spatial and temporal variability in extreme temperature and precipitation events in Inner Mongolia (China) during 1960–2017. *Sci. Total. Environ.* **2019**, *649*, 75–89. [CrossRef] [PubMed]
27. Larbi, I.; Hountondji, F.C.C.; Annor, T.; Agyare, W.A.; Gathenya, J.M.; Amuzu, J. Spatio-Temporal Trend Analysis of Rainfall and Temperature Extremes in the Vea Catchment, Ghana. *Climate* **2018**, *6*, 87. [CrossRef]
28. *Xoplaki Climate Variability over the Mediterranean*; University of Bern: Bern, Switzerland, 2002.
29. Black, E. The impact of climate change on daily precipitation statistics in Jordan and Israel. *Atmospheric Sci. Lett.* **2009**, *10*, 192–200. [CrossRef]
30. Lelieveld, J.; Hadjinicolaou, P.; Kostopoulou, E.; Chenoweth, J.; El Maayar, M.; Giannakopoulos, C.; Hannides, C.; Lange, M.A.; Tanarhte, M.; Tyrlis, E.; et al. Climate change and impacts in the Eastern Mediterranean and the Middle East. *Clim. Chang.* **2012**, *114*, 667–687. [CrossRef]
31. Mayewski, P.A.; Rohling, E.; Stager, J.C.; Karlén, W.; Maasch, K.A.; Meeker, L.D.; Meyerson, E.A.; Gasse, F.; Van Kreveld, S.; Holmgren, K.; et al. Holocene climate variability. *Quat. Res.* **2004**, *62*, 243–255. [CrossRef]

32. Orland, I.J.; Bar-Matthews, M.; Kita, N.T.; Ayalon, A.; Matthews, A.; Valley, J.W. Climate deterioration in the Eastern Mediterranean as revealed by ion microprobe analysis of a speleothem that grew from 2.2 to 0.9 ka in Soreq Cave, Israel. *Quat. Res.* **2009**, *71*, 27–35. [CrossRef]
33. Frumkin, A.; Magaritz, M.; Carmi, I.; Zak, I. The Holocene climatic record of the salt caves of Mount Sedom Israel. *Holocene* **1991**, *1*, 191–200. [CrossRef]
34. Giorgi, F. Climate Change Hot-Spots. *Geophys. Res. Lett.* **2006**, *33*, 101029. [CrossRef]
35. Lu, J.; Vecchi, G.A.; Reichler, T. Expansion of the Hadley cell under global warming. *Geophys. Res. Lett.* **2007**, *34*. [CrossRef]
36. Zittis, G.; Almazroui, M.; Alpert, P.; Ciais, P.; Cramer, W.; Dahdal, Y.; Fnais, M.; Francis, D.; Hadjinicolaou, P.; Howari, F.; et al. Climate Change and Weather Extremes in the Eastern Mediterranean and Middle East. *Rev. Geophys.* **2022**, *60*, e2021RG000762. [CrossRef]
37. Al-Qinna, M.I.; Hammouri, N.A.; Obeidat, M.M.; Ahmad, F.Y. Drought analysis in Jordan under current and future climates. *Clim. Chang.* **2011**, *106*, 421–440. [CrossRef]
38. Shadeed, S. Spatio-temporal Drought Analysis in Arid and Semi-arid Regions: A Case Study from Palestine. *Arab. J. Sci. Eng.* **2013**, *38*, 2303–2313. [CrossRef]
39. Mathbout, S.; Lopez-Bustins, J.A.; Martin-Vide, J.; Bech, J.; Rodrigo, F.S. Spatial and temporal analysis of drought variability at several time scales in Syria during 1961–2012. *Atmospheric Res.* **2018**, *200*, 153–168. [CrossRef]
40. Lange, M.A. Impacts of Climate Change on the Eastern Mediterranean and the Middle East and North Africa Region and the Water–Energy Nexus. *Atmosphere* **2019**, *10*, 455. [CrossRef]
41. Ben-Gai, T.; Bitan, A.; Manes, A.; Alpert, P. Long-term changes in annual rainfall patterns in southern Israel. *Theor. Appl. Clim.* **1994**, *49*, 59–67. [CrossRef]
42. Ben-Gai, T.; Bitan, A.; Manes, A.; Alpert, P.; Rubin, S. Spatial and Temporal Changes in Rainfall Frequency Distribution Patterns in Israel. *Arch. Meteorol. Geophys. Bioclimatol. Ser. B* **1998**, *61*, 177–190. [CrossRef]
43. Freiwan, M.; Kadioglu, M. Spatial and temporal analysis of climatological data in Jordan. *Int. J. Clim.* **2008**, *28*, 521–535. [CrossRef]
44. Ghanem, A.A. Climatology of the areal precipitation in Amman/Jordan. *Int. J. Clim.* **2011**, *31*, 1328–1333. [CrossRef]
45. Ziv, B.; Saaroni, H.; Pargament, R.; Harpaz, T.; Alpert, P. Trends in rainfall regime over Israel, 1975–2010, and their relationship to large-scale variability. *Reg. Environ. Chang.* **2014**, *14*, 1751–1764. [CrossRef]
46. Yosef, Y.; Aguilar, E.; Alpert, P. Changes in extreme temperature and precipitation indices: Using an innovative daily homogenized database in Israel. *Int. J. Clim.* **2019**, *39*, 5022–5045. [CrossRef]
47. Zhang, X.; Aguilar, E.; Sensoy, S.; Melkonyan, H.; Tagiyeva, U.; Ahmed, N.; Kutaladze, N.; Rahimzadeh, F.; Taghipour, A.; Hantosh, T.H.; et al. Trends in Middle East climate extreme indices from 1950 to 2003. *J. Geophys. Res. Earth Surf.* **2005**, *110*. [CrossRef]
48. Donat, M.G.; Peterson, T.C.; Brunet, M.; King, A.D.; Almazroui, M.; Kolli, R.K.; Boucherf, D.; Al-Mulla, A.Y.; Nour, A.Y.; Aly, A.A.; et al. Changes in extreme temperature and precipitation in the Arab region: Long-term trends and variability related to ENSO and NAO. *Int. J. Climatol.* **2014**, *34*, 581–592. [CrossRef]
49. Salameh, A.A.M.; Gámiz-Fortis, S.R.; Castro-Díez, Y.; Abu Hammad, A.; Esteban-Parra, M.J. Spatio-temporal analysis for extreme temperature indices over the Levant region. *Int. J. Clim.* **2019**, *39*, 5556–5582. [CrossRef]
50. Kutiel, H.; Benaroch, Y. North Sea-Caspian Pattern (NCP)—An upper level atmospheric teleconnection affecting the Eastern Mediterranean: Identification and definition. *Theor. Appl. Clim.* **2002**, *71*, 17–28. [CrossRef]
51. Price, C.; Stone, L.; Huppert, A.; Rajagopalan, B.; Alpert, P. A possible link between El Niño and precipitation in Israel. *Geophys. Res. Lett.* **1998**, *25*, 3963–3966. [CrossRef]
52. Kelley, C.; Ting, M.; Seager, R.; Kushnir, Y. Mediterranean precipitation climatology, seasonal cycle, and trend as simulated by CMIP5. *Geophys. Res. Lett.* **2012**, *39*. [CrossRef]
53. Ziv, B.; Dayan, U.; Kushnir, Y.; Roth, C.; Enzel, Y. Regional and global atmospheric patterns governing rainfall in the southern Levant. *Int. J. Clim.* **2006**, *26*, 55–73. [CrossRef]
54. Yosef, Y.; Saaroni, H.; Alpert, P. Trends in Daily Rainfall Intensity Over Israel 1950/1-2003/4. *Open Atmos. Sci. J.* **2009**, *3*, 196–203. [CrossRef]
55. Ávila, A.; Justino, F.; Wilson, A.; Bromwich, D.; Amorim, M. Recent precipitation trends, flash floods and landslides in southern Brazil. *Environ. Res. Lett.* **2016**, *11*, 114029. [CrossRef]
56. Spinoni, J.; Barbosa, P.; Bucchignani, E.; Cassano, J.; Cavazos, T.; Christensen, J.H.; Christensen, O.B.; Coppola, E.; Evans, J.; Geyer, B.; et al. Future Global Meteorological Drought Hot Spots: A Study Based on CORDEX Data. *J. Clim.* **2020**, *33*, 3635–3661. [CrossRef]
57. Garnaut, R. *The Garnaut Climate Change Review: Final Report*; Cambridge University Press: Cambridge, UK, 2008.
58. Trewin, B. A daily homogenized temperature data set for Australia. *Int. J. Clim.* **2013**, *33*, 1510–1529. [CrossRef]
59. Sibson, R. A Brief Description of Natural Neighbour Interpolation. In *Interpreting Multivariate Data*; John Wiley & Sons: New York, NY, USA, 1981.
60. Wang, X.L.; Feng, Y. *RHtestsV4 User Manual*; Climate Research Division, Atmospheric Science and Technology Directorate, Science and Technology Branch, Environment Canada: Toronto, ON, Canada, 2013.
61. Wang, X.L.; Chen, H.; Wu, Y.; Feng, Y.; Pu, Q. New Techniques for the Detection and Adjustment of Shifts in Daily Precipitation Data Series. *J. Appl. Meteorol. Clim.* **2010**, *49*, 2416–2436. [CrossRef]

62. Villafuerte, M.Q.; Matsumoto, J.; Kubota, H. Changes in extreme rainfall in the Philippines (1911-2010) linked to global mean temperature and ENSO. *Int. J. Clim.* **2015**, *35*, 2033–2044. [CrossRef]
63. Wang, S.; Jiang, F.; Ding, Y. Spatial coherence of variations in seasonal extreme precipitation events over Northwest Arid Region, China. *Int. J. Clim.* **2015**, *35*, 4642–4654. [CrossRef]
64. Wu, C.; Huang, G. Projection of climate extremes in the Zhujiang River basin using a regional climate model. *Int. J. Clim.* **2016**, *36*, 1184–1196. [CrossRef]
65. Klein Tank, A.B.G.; Zwiers, F.W. *Guidelines on Analysis of Extremes in a Changing Climate in Support of Informed Decisions for Adaptation*; World Meteorological Organization (WMO): Geneva, Switzerland, 2009.
66. Ajjur, S.B.; Riffi, M.I. Analysis of the observed trends in daily extreme precipitation indices in Gaza Strip during 1974–2016. *Int. J. Clim.* **2020**, *40*, 6189–6200. [CrossRef]
67. Zhang, X.B.F.Y. *R ClimDex 1.0 User Mannual*; Climate research branch environment Canada: Downsview, ON, Canada, 2004.
68. Mann, H.B. Non-Parametric Test Against Trend. *Econometrica* **1945**, *13*, 245–259. [CrossRef]
69. Kendall, M.G. Rank Correlation Methods. *Biometrika* **1957**, *44*, 298. [CrossRef]
70. Sen, P.K. Estimates of the regression coefficient based on Kendall's Tau. *J. Am. Stat. Assoc.* **1968**, *63*, 1379–1389. [CrossRef]
71. Zhang, X.; Vincent, L.A.; Hogg, W.D.; Niitsoo, A. Temperature and precipitation trends in Canada during the 20th century. *Atmos.-Ocean* **2000**, *38*, 395–429. [CrossRef]
72. New, M.; Hewitson, B.; Stephenson, D.B.; Tsiga, A.; Kruger, A.; Manhique, A.; Gomez, B.; Coelho, C.A.S.; Masisi, D.N.; Kululanga, E.; et al. Evidence of trends in daily climate extremes over southern and west Africa. *J. Geophys. Res. Atmos.* **2006**, *111*, D14102. [CrossRef]
73. Demir, V.; Keskin, A. Water level change of lakes and sinkholes in Central Turkey under anthropogenic effects. *Arch. Meteorol. Geophys. Bioclimatol. Ser. B* **2020**, *142*, 929–943. [CrossRef]
74. Citakoglu, H.; Minarecioglu, N. Trend analysis and change point determination for hydro-meteorological and groundwater data of Kizilirmak basin. *Arch. Meteorol. Geophys. Bioclimatol. Ser. B* **2021**, *145*, 1275–1292. [CrossRef]
75. Wang, X.L.; Swail, V.R. Changes of Extreme Wave Heights in Northern Hemisphere Oceans and Related Atmospheric Cir-culation Regimes. *J. Clim.* **2001**, *14*. [CrossRef]
76. Patakamuri, S.K.; O'Brien, N. Package 'Modifiedmk' (Version 1.4.0): Modified Versions of Mann Kendall and Spearman's Rho Trend Tests. CRAN 2021. Available online: https://cran.r-project.org/web/packages/modifiedmk/modifiedmk.pdf (accessed on 1 September 2021).
77. Efthymiadis, D.; Goodess, C.; Jones, P. Trends in Mediterranean gridded temperature extremes and large-scale circulation influences. *Nat. Hazards Earth Syst. Sci.* **2011**, *11*, 2199–2214. [CrossRef]
78. Popov, T.; Gnjato, S.; Trbić, G.; Ivanišević, M. Recent Trends in Extreme Temperature Indices in Bosnia and Herzegovina. *Carpathian J. Earth Environ. Sci.* **2018**, *13*, 211–224. [CrossRef]
79. Yenigun, K.; Ibrahim, W.A. Investigation of drought in the northern Iraq region. *Meteorol. Appl.* **2019**, *26*, 490–499. [CrossRef]
80. Nouri, N.; Homaee, M. Drought trend, frequency and extremity across a wide range of climates over Iran. *Meteorol. Appl.* **2020**, *27*, e1899. [CrossRef]
81. Hameed, M.; Ahmadalipour, A.; Moradkhani, H. Drought and food security in the middle east: An analytical framework. *Agric. For. Meteorol.* **2020**, *281*, 107816. [CrossRef]
82. AlSarmi, S.; Washington, R. Recent observed climate change over the Arabian Peninsula. *J. Geophys. Res. Earth Surf.* **2011**, *116*. [CrossRef]
83. Shohami, D.; Dayan, U.; Morin, E. Warming and drying of the eastern Mediterranean: Additional evidence from trend analysis. *J. Geophys. Res. Earth Surf.* **2011**, *116*. [CrossRef]
84. Hochman, A.; Mercogliano, P.; Alpert, P.; Saaroni, H.; Bucchignani, E. High-resolution projection of climate change and extremity over Israel using COSMO-CLM. *Int. J. Clim.* **2018**, *38*, 5095–5106. [CrossRef]
85. Luong, T.M.; Dasari, H.P.; Hoteit, I. Extreme precipitation events are becoming less frequent but more intense over Jeddah, Saudi Arabia. Are shifting weather regimes the cause? *Atmos. Sci. Lett.* **2020**, *21*, e981. [CrossRef]
86. Cheng, C.S.; Auld, H.; Li, Q.; Li, G. Possible impacts of climate change on extreme weather events at local scale in south–central Canada. *Clim. Change* **2012**, *112*, 963–979. [CrossRef]
87. Papalexiou, S.M.; Montanari, A. Global and Regional Increase of Precipitation Extremes Under Global Warming. *Water Resour. Res.* **2019**, *55*, 4901–4914. [CrossRef]
88. Seidel, D.J.; Fu, Q.; Randel, W.J.; Reichler, T.J. Widening of the tropical belt in a changing climate. *Nat. Geosci.* **2008**, *1*, 21–24. [CrossRef]
89. Hertig, E.; Seubert, S.; Paxian, A.; Vogt, G.; Paeth, H.; Jacobeit, J. Changes of total versus extreme precipitation and dry periods until the end of the twenty-first century: Statistical assessments for the Mediterranean area. *Arch. Meteorol. Geophys. Bioclimatol. Ser. B* **2013**, *111*, 1–20. [CrossRef]
90. Sillmann, J.; Roeckner, E. Indices for extreme events in projections of anthropogenic climate change. *Clim. Change* **2008**, *86*, 83–104. [CrossRef]
91. Tebaldi, C.; Hayhoe, K.; Arblaster, J.M.; Meehl, G.A. Going to the Extremes: An Intercomparison of Model-Simulated Historical and Future Changes in Extreme Events. *Clim. Change* **2006**, *79*, 185–211. [CrossRef]

92. Saaroni, H.; Ziv, B.; Lempert, J.; Gazit, Y.; Morin, E. Prolonged dry spells in the Levant region: Climatologic-synoptic analysis. *Int. J. Clim.* **2015**, *35*, 2223–2236. [CrossRef]
93. Kutiel, H.; Paz, S. Sea Level Pressure Departures in the Mediterranean and their Relationship with Monthly Rainfall Conditions in Israel. *Arch. Meteorol. Geophys. Bioclimatol. Ser. B* **1998**, *60*, 93–109. [CrossRef]
94. Törnros, T. On the relationship between the Mediterranean Oscillation and winter precipitation in the Southern Levant. *Atmos. Sci. Lett.* **2013**, *14*, 287–293. [CrossRef]
95. Baltacı, H.; Akkoyunlu, B.O.; Tayanç, M. Relationships between teleconnection patterns and Turkish climatic extremes. *Arch. Meteorol. Geophys. Bioclimatol. Ser. B* **2018**, *134*, 1365–1386. [CrossRef]
96. Redolat, D.; Monjo, R.; Lopez-Bustins, J.A.; Martin-Vide, J. Upper-Level Mediterranean Oscillation index and seasonal variability of rainfall and temperature. *Theor. Appl. Clim.* **2019**, *135*, 1059–1077. [CrossRef]
97. Müller-Plath, G.; Lüdecke, H.-J.; Lüning, S. Long-distance air pressure differences correlate with European rain. *Sci. Rep.* **2022**, *12*, 10191. [CrossRef]
98. Martin-Vide, J.; Lopez-Bustins, J.-A. The Western Mediterranean Oscillation and rainfall in the Iberian Peninsula. *Int. J. Clim.* **2006**, *26*, 1455–1475. [CrossRef]
99. Krichak, S.O.; Alpert, P. Decadal trends in the east Atlantic-west Russia pattern and Mediterranean precipitation. *Int. J. Clim.* **2005**, *25*, 183–192. [CrossRef]
100. Zhang, X.; Chen, Y.; Fang, G.; Li, Y.; Li, Z.; Wang, F.; Xia, Z. Observed changes in extreme precipitation over the Tienshan Mountains and associated large-scale climate teleconnections. *J. Hydrol.* **2022**, *606*, 127457. [CrossRef]
101. Zhu, J.; Zhou, H.; Xiao, H.; Wang, X. Singular value decomposition (SVD) based correlation analysis of climatic factors and extreme precipitation in Hunan Province, China, during 1960–2009. *J. Water Clim. Change* **2021**, *12*, 3602–3616. [CrossRef]
102. Quishpe-Vásquez, C.; Gámiz-Fortis, S.R.; García-Valdecasas-Ojeda, M.; Castro-Díez, Y.; Esteban-Parra, M.J. Tropical Pacific sea surface temperature influence on seasonal streamflow variability in Ecuador. *Int. J. Clim.* **2019**, *39*, 3895–3914. [CrossRef]

Article

Climate Change Impacts on Water Resources in the Danube River Basin: A Hydrological Modelling Study Using EURO-CORDEX Climate Scenarios

Elisabeth Probst *[ ] and Wolfram Mauser

Department of Geography, Ludwig-Maximilians-Universität München (LMU), Luisenstraße 37, 80333 Munich, Germany
* Correspondence: elisabeth.probst@iggf.geo.uni-muenchen.de

Abstract: Climate change affects the hydrological cycle of river basins and strongly impacts water resource availability. The mechanistic hydrological model PROMET was driven with an ensemble of EURO-CORDEX regional climate model projections under the emission scenarios RCP2.6 and RCP8.5 to analyze changes in temperature, precipitation, soil water content, plant water stress, snow water equivalent (SWE) and runoff dynamics in the Danube River Basin (DRB) in the near (2031–2060) and far future (2071–2100) compared to the historical reference (1971–2000). Climate change impacts remain moderate for RCP2.6 and become severe for RCP8.5, exhibiting strong year-round warming trends in the far future with wetter winters in the Upper Danube and drier summers in the Lower Danube, leading to decreasing summer soil water contents, increasing plant water stress and decreasing SWE. Discharge seasonality of the Danube River shifts toward increasing winter runoff and decreasing summer runoff, while the risk of high flows increases along the entire Danube mainstream and the risk of low flows increases along the Lower Danube River. Our results reveal increasing climate change-induced discrepancies between water surplus and demand in space and time, likely leading to intensified upstream–downstream and inter-sectoral water competition in the DRB under climate change.

Keywords: climate change; hydrology; water resources; precipitation; runoff; soil moisture; regional climate model; hydrological model; Danube; PROMET

Citation: Probst, E.; Mauser, W. Climate Change Impacts on Water Resources in the Danube River Basin: A Hydrological Modelling Study Using EURO-CORDEX Climate Scenarios. *Water* **2023**, *15*, 8. https://doi.org/10.3390/w15010008

Academic Editors: Sonia Raquel Gámiz-Fortis and Matilde García-Valdecasas Ojeda

Received: 19 November 2022
Revised: 9 December 2022
Accepted: 13 December 2022
Published: 21 December 2022

Copyright: © 2022 by the authors. Licensee MDPI, Basel, Switzerland. This article is an open access article distributed under the terms and conditions of the Creative Commons Attribution (CC BY) license (https://creativecommons.org/licenses/by/4.0/).

1. Introduction

River basins are complex hydrological systems, in which hydrological processes are highly interconnected and highly susceptible to climate change. The recently published Sixth Assessment Report (AR6) of the Intergovernmental Panel on Climate Change (IPCC) stated that the global surface temperature has already increased by +1.09 °C (+0.95 °C to +1.20 °C) in 2011–2020 compared to 1850–1900 [1]. This recent development is already quite rapidly reducing the remaining scope for further greenhouse gas emissions in light of the Paris Agreement targets of limiting the global surface temperature increase to well below +2 °C and preferably below +1.5 °C above pre-industrial levels [2].

A global average temperature increase is not equivalent to a uniform temperature increase across the globe, but rather to a heterogeneous spatial distribution with hot and cold spots of temperature increase [3–5]. For the Danube River Basin (DRB), for example, the results from the IPCC AR6 suggested that a global average temperature increase of +2 °C compared to the pre-industrial level would translate into a median temperature increase of +2.7 °C on spatial average across the basin [6,7].

Large and densely inhabited transboundary watersheds, such as the DRB, are especially susceptible to climate change, which directly and indirectly affects water resource availability [8]. With the International Commission for the Protection of the Danube River

(ICPDR) founded in 1998, the DRB has a long tradition of transboundary water management with the main goal of achieving sustainable water resource management in line with the EU Water Framework Directive (WFD). Moreover, the ICPDR has developed and pursues a climate change adaptation strategy [9] based on recent results in climate change impact research.

Based on an extensive review of numerous climate change case studies in the DRB [10], the ICPDR anticipates an increase in the mean annual temperature of 4.0 °C to 5.0 °C until 2100 in the DRB, under the Representative Concentration Pathway (RCP) emission scenario RCP8.5 [9]. Moreover, the ICPDR identified overarching spatial and seasonal precipitation trends, according to which wet regions tend to become wetter, dry regions tend to become drier and precipitation seasonality tends to shift toward wetter winters and drier summers [9].

In the literature, few hydrological climate change impact modelling studies exist which consider the whole DRB. They have mostly relied on general circulation models (GCMs) of coarser resolution rather than regional climate models (RCMs), or on GCM-RCM combinations of former generations. Stagl and Hattermann [11] investigated river discharge in the DRB by using the process-based watershed model SWIM [12] driven with ENSEMBLES GCM-RCM combinations under SRES (Special Report on Emissions Scenarios) [13] based on the Coupled Model Intercomparison Project Phase 3 (CMIP3) [14]. Subsequently, Stagl and Hattermann [15] investigated discharge in the DRB by driving SWIM with a GCM ensemble of CMIP5 [16] under different RCPs. Here, coarse-resolution GCMs were directly used as the meteorological driver instead of high-resolution RCMs. However, hydrological climate change impact assessment requires higher-resolution meteorological drivers such as RCMs, since GCMs cannot resolve circulation patterns down to the finer scales on which hydrological processes occur [17]. Bisselink et al. [18] investigated water scarcity in the DRB by using the process-based hydrological model LISFLOOD [19,20] driven by a EURO-CORDEX [21] GCM-RCM ensemble under a 2 °C global mean temperature increase according to the RCP8.5 scenario. The analyzed future periods were not uniformly selected for the entire ensemble, but were centered on the year of exceeding 2 °C global mean temperature for each single GCM [18].

To the best of our knowledge, a systematic and comparative hydrological modelling study using the (to date) latest-generation high-resolution EURO-CORDEX [21] GCM-RCM simulations under different RCPs and uniformly selected future periods for the entire ensemble—enabling to directly compare the development of different components of the water cycle (e.g., precipitation, soil and snow water, discharge) in a near and far future time frame—is not yet available for the whole DRB.

We present a hydrological climate change impact modelling study for the DRB, using the process-based hydrological model PROMET [22], which interlinks dynamic vegetation and hydrological modelling and was chosen due to its strong physical basis and predictive power. Probst and Mauser [23] successfully applied PROMET in the DRB, driven with meteorological reanalysis data to analyze the influence of different climatologies used for bias correction on the quality of the simulated spatial discharge. They fully validated the PROMET model in the DRB using the observed daily discharge at major sub-basin gauges [23].

In this study, we evaluated future projections of temperature and of different components of the hydrological cycle, namely precipitation, soil water content, snow water equivalent, the resulting river runoff dynamics as well as plant water stress as a water-related vegetation variable and their respective interdependencies in the DRB. These variables are of direct relevance for water resource availability, and they address the dynamic response of vegetation development to climate change in the DRB. For this, we analyzed simulations of the PROMET hydrological model (30″ spatial resolution) driven by a selected high-resolution GCM-RCM ensemble of the (to date) most recent and most high-resolution regional climate change projections for Europe from the EURO-CORDEX initiative [21]. Within EURO-CORDEX, GCM simulations from the CMIP5 initiative were dynamically regionalized by RCMs. For each variable, we compared the results of the ensemble mean of the GCM-RCM-driven simulations for the historical period (1971–2000) with the results of the ensemble means of the GCM-RCM-driven simulations under the two emission scenarios RCP2.6 and RCP8.5 for both the near future (2031–2060) and the far future (2071–2100). By comparing RCP2.6 and RCP8.5, we highlighted the differences between two very different future projections for the DRB: (i) a scenario aiming to keep global warming likely below 2 °C above pre-industrial levels (RCP2.6); (ii) a high-end scenario with very high greenhouse gas emissions (RCP8.5) [24].

2. Materials and Methods

2.1. The Danube River Basin

The DRB is the second largest river basin in Europe, covering an area of ~817,000 km² [25]. From its source in the Black Forest to its mouth in the Black Sea, the Danube's river length amounts to 2857 km [26]. The DRB is a complex watershed, featuring heterogeneous natural characteristics in terms of topography and climate, which strongly influences basin-wide hydrology. The watershed comprises snow-and ice-covered high mountain ranges, forest-covered low mountain ranges, sparsely vegetated karst regions, plateaus with river valleys and wide agricultural plains [26] (Figure 1).

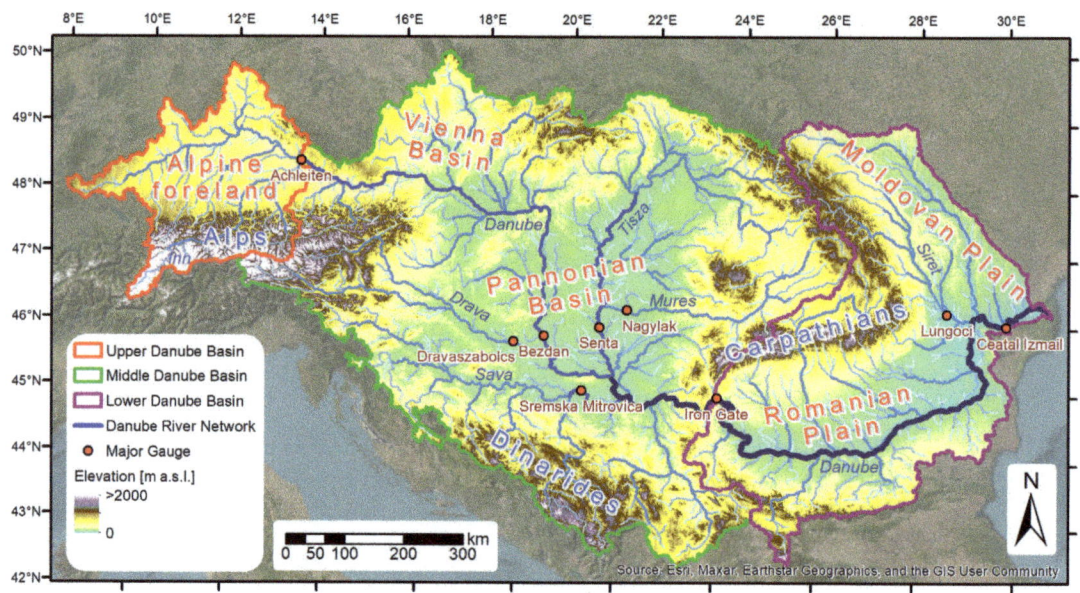

Figure 1. Map of the Danube River Basin with its river network, major gauges and the subdivision into Upper, Middle and Lower Danube. Major sub-basins (gauges in parentheses): Upper Danube (Achleiten), Middle Danube (Bezdan and Iron Gate), Drava (Dravaszabolcs), Sava (Sremska Mitrovica), Mures (Nagylak), Tisza (Senta), Siret (Lungoci) and the Lower Danube (Ceatal Izmail). Major gauges along the Danube main course: Achleiten, Bezdan, Iron Gate, Ceatal Izmail. Data sources: Farr et al. [27], Lehner et al. [28], Global Runoff Data Centre (GRDC) [29], International Commission for the Protection of the Danube River (ICPDR) [30] and Esri [31].

The DRB is located in a transition zone between three Köppen–Geiger climate types. It mainly lies in the humid continental climate (Dfb, Dfa) zone, with influences of the humid subtropical climate (Cfa) around the Dinarides and the semi-arid steppe climate (BSk) at the Danube delta [32]. Since the DRB closely borders the oceanic climate (Cfb) at the very western Upper Danube and the Mediterranean climate (Csa, Csb) in the southern DRB [32], influences of these two climate zones are found along the peripheries of the basin [33]. The current complex interference of the different climate zones in the DRB roughly results in a gradient of decreasing mean annual precipitation and increasing mean annual temperature from the north-west to the south-east.

The DRB is divided into three parts: the Upper Danube, extending from the Danube's source to the Devín Gate in the Little Carpathians near the Austrian–Slovakian border [25] (in some sources, the Upper Danube extends down to gauge Achleiten [34]), the subsequent Middle Danube, extending to the Iron Gate at the Serbian–Romanian border and the following Lower Danube, extending down to the mouth in the Black Sea [25]. Hence, the DRB comprises sub-basins with the full range of hydrological characteristics from mountain to lowland watersheds. In Table 1, we show the main morphological and hydrographic characteristics of the Upper, Middle and Lower Danube.

Table 1. Main morphological and hydrographic characteristics of the Upper, Middle and Lower Danube (data sources: Schiller et al. [26], Regionale Zusammenarbeit der Donauländer (RZD) [35], Farr et al. [27] and Global Runoff Data Centre (GRDC) [29]).

Characteristics	Upper Danube	Middle Danube	Lower Danube
Major geomorphological units (not exhaustive)	Swabian/Franconian Alb, Bavarian Forest, Bohemian–Moravian Highland, Alpine Foreland, Northern Calcareous and Central Alps	Carpathians, Carnic Alps, Karawanks, Julian Alps, Dinarides, Pannonian Basin, Transylvanian Plateau	Carpathians, Balkans, Romanian/Bulgarian Plain, Dobrogea Hills, Moldavian Plateau
Terrain height range [m a.s.l.]	303–3676	35–3449	−2–2683
Sub region area [km^2]	76,653	576,232	807,000
Major tributary rivers (>20,000 km^2 basin area)	Inn	Morava, Drava, Tisza, Sava, Velika Morava,	Olt, Siret, Prut
Outlet gauge	Achleiten	Iron Gate/Orsova	Ceatal Izmail
- Altitude [m a.s.l.]	287.7	44.0	0.6
- Distance from estuary [km]	2223	955	72
- MQ [m^3/s] (1971–2000) *,**	1417	5430	6401
- MNQ [m^3/s] (1971–2000) *,**	659	2075	3045
- MHQ [m^3/s] (1971–2000) *,**	3821	10,636	11,104

Notes: * MQ: mean flow; MNQ: mean low flow; MHQ: mean high flow. ** Orsova: MQ, MNQ, MHQ are calculated for 1971–1990, due to end of records in 1990.

2.2. The Mechanistic Hydrological Model PROMET

The mechanistic hydrological model PROMET (Processes of Radiation, Mass and Energy Transfer) is a physically-based land surface process model developed by Mauser and Bach [22], interlinking dynamic vegetation and hydrological modelling. It simulates spatially distributed fluxes of water, carbon and energy on hourly time steps while strictly adhering to the law of conservation of mass and energy in space and time [22]. A detailed description of the model theory can be found in Mauser and Bach [22], Hank et al. [36], Mauser et al. [37] and Zabel et al. [38]. PROMET is structured in modules (e.g., meteorology, land surface, vegetation, soil, groundwater, runoff formation and routing), with the individual modules steadily exchanging information.

The dynamic vegetation module follows the approaches of Farquhar et al. [39], Chen et al. [40] and Ball et al. [41] to simulate the net primary production and canopy conductivity. Stomatal conductance of the canopy is primarily regulated by water supply, which is determined from the water content in the rooted soil zone [36]. According to the approach of Jarvis and Morison [42], the inhibition of the stomatal conductance is calculated from the soil suction in the rooted zone [36]. In PROMET, this inhibition can be interpreted as a plant-specific water stress variable, given as a percentage reduction of potential transpiration. It ranges from 0% to 100% inhibition of transpiration and primary production. A water stress value of 0% thereby stands for completely absent water stress, meaning that the plant's stomata are fully open. In this case, the actual transpiration equals the potential transpiration. In contrast, a water stress value of 100% stands for the maximum water stress, meaning that the plant's stomata remain fully closed. The dynamic process description in the vegetation module allows for a full consideration of the alterations in plant growth resulting from a changing climate.

The routing module, transferring water through the basin's channel network, follows the approach of Cunge [43] and Todini [44]. Due to PROMET's strong physical basis and predictive power, a classical calibration of the PROMET parameters to fit observed discharge is not appropriate [22,23]. Instead, literature sources and measurements are used for the parametrization of the initial values within PROMET, maintaining the model's physical consistency [22].

PROMET is driven with hourly time series of spatially distributed meteorological variables, which are downscaled and disaggregated within the PROMET meteorology module in space and time (30" and 1 h in this study) according to Marke et al. [45]. Simultaneously, temperature and precipitation are linearly (and spatially distributed) bias-corrected using long-term climatologies as a reference [23].

The fully validated PROMET model setup of Probst and Mauser [23] (30" spatial resolution) served as the basis for this study. Here, topographical information was taken from the SRTM (Shuttle Radar Topography Mission) digital elevation model [27], soil property data were derived from the Harmonized World Soil Database (HWSD) [46] and watershed topography information was taken from the HydroSHEDS database [28]. For land use information, we used the first existing consolidated land use map for the DRB of Probst and Mauser [23] including the spatial distribution of agricultural crops, which is based on a mosaic of the CORINE Land Cover 2012 [47] and the ESA CCI Land Cover 2015 [48] augmented by cultivation statistics [49]. The described input datasets are harmonized datasets covering the entire DRB, which allows for a physically consistent and uniform parametrization of the PROMET input parameters for soil, vegetation and hydrology all over the DRB. This comprehensive parametrization had successfully been validated within the DRB [23]. All gridded input data come with a spatial resolution of 30" (~0.00833333°).

2.3. Regional Climate Models as Meteorological Drivers

2.3.1. Selection of Appropriate Climate Models

In this study, we drove PROMET with EUR-11 GCM-RCM simulations from the ensemble for Europe within the World Climate Research Program Coordinated Regional Downscaling Experiment (EURO-CORDEX) [21]. From the EURO-CORDEX simulations, we selected an appropriate ensemble of different combinations of RCMs driven by GCMs.

For this, we consulted the literature, such as the ensemble audit from the Bavarian State Office for Environment authored by Zier et al. [50], in which the EURO-CORDEX ensemble was checked for plausibility concerning circulation patterns and climate processes within the hydrological region of Bavaria (which covers the entire Upper Danube basin). Zier et al. [50] assessed a set of four indicators for temperature and precipitation to identify possible biases or shifts between the GCM-RCM simulation results and the KliRef2014 temperature and precipitation reference climatologies, which were created by the authors based on interpolated observations. The indicators assessed by Zier et al. [50] were (i) the deviation of the mean annual values ("quantity indicator"); (ii) the deviation of the mean monthly values ("quantity indicator with seasonality reference"); (iii) the deviation of the seasonality of the mean monthly values ("seasonality indicator"); (iv) the deviation of the spatial distribution of the mean annual values ("pattern indicator") from the climatologies. Zier et al. [50] rated models as implausible, particularly when a spatial offset of precipitation due to drift effects or insufficient and inverse annual dynamics had been encountered. The authors considered it unlikely that these biases can be reliably removed by bias correction, which makes the respective models inappropriate for hydrological impact modelling in Bavarian watersheds and downstream [50]. GCM-RCM simulations with a uniform bias were considered appropriate, since this indicates that the climatic processes are represented correctly in the climate model, but merely with an incorrect order of magnitude that can be compensated with a bias correction [50].

Furthermore, the selected GCM-RCM ensemble had to meet a set of technical requirements. These were (i) 0.11° spatial resolution (EUR-11); (ii) at least 3 h temporal resolution;

(iii) availability of the "historical", "RCP2.6" and "RCP8.5" experiments; (iv) availability of the required PROMET meteorological input variables (total cloud cover [%], wind speed (10 m height) [m/s], air temperature (2 m height) [K], dew point temperature (2 m height) [K], surface pressure [Pa], total precipitation [mm], downward surface solar radiation [W/m^2] and downward surface thermal radiation [W/m^2]).

Based on these considerations, we selected three suitable EURO-CORDEX GCM-RCM combinations for this study:

- RACMO22Ev1 (RCM) driven with ICHEC-EC-EARTH (r12) (GCM) (denoted as ICHEC-RACMO in this study);
- RCA4v1 (RCM) driven with ICHEC-EC-EARTH (r12) (GCM) (denoted as ICHEC-RCA4 in this study);
- RCA4v1a (RCM) driven with MPI-M-MPI-ESM-LR (r1) (GCM) (denoted as MPI-RCA4 in this study).

These three GCM-RCM combinations have also been considered plausible and without serious biases in other study regions within the DRB, e.g., in the Alps [51] or in the Pannonian Basin [52]. Moreover, the RCMs involved have been found to be amongst the better-performing in the Carpathian region [53].

For each of the three GCM-RCM combinations, we used the historical experiment for the reference period (1971–2000) (the time frame was selected due to EURO-CORDEX data availability) as well as the RCP2.6 and RCP8.5 experiments for both the near future (2031–2060) and the far future (2071–2100) as drivers for the PROMET hydrological simulations.

2.3.2. Bias Correction of Climate Model Simulations

Climate model simulations typically exhibit systematic biases in climate variables [54], which introduce additional uncertainties to future climate projections. Hence, we applied a spatially distributed linear bias correction of temperature and precipitation based on reference climatologies to the downscaled (30″ spatial resolution) and disaggregated (1 h temporal resolution) climate model forcing data. As reference climatologies, we used the compiled climatologies (30″ spatial resolution) from Probst and Mauser [23], which had already been used for the generation of the historical ERA5-GPW forcing dataset [23] and had been successfully validated within the PROMET setup for the DRB in the mentioned study. The compiled climatologies are: (i) the global WorldClim 2 temperature climatology [55] for the bias correction of temperature; (ii) a mosaic of the global WorldClim 2 precipitation climatology [55] and the Alpine precipitation climatologies GLOWA (Globaler Wandel; engl. Global Change) [56] and PRISM (Parameter-elevation Regression on Independent Slope Model) [57] for the bias correction of precipitation. These climatologies are based on interpolated observations, whereby the two Alpine precipitation climatologies, GLOWA and PRISM, are derived from a particularly high-resolution station density [23]. The other meteorological variables apart from temperature and precipitation were not bias-corrected.

The hindcast in Figure 2 shows the performance of the bias correction. Here, we show the long-term mean monthly temperature and precipitation (1980–2000) on spatial average over the DRB and the resulting long-term mean monthly discharge (1980–2000) at the outlet gauge Ceatal Izmail. The simulations driven by the three bias-corrected GCM-RCM ensemble members (ICHEC-RACMO, ICHEC-RCA4 and MPI-RCA4) were compared with the simulation driven by the ERA5-GPW forcing dataset of Probst and Mauser [23]. As can be seen, the correspondence between ERA5-GPW long-term temperature and precipitation seasonalities with their counterparts from the bias-corrected GCM-RCM simulations is very high. The same holds true for discharge: the long-term discharge seasonalities of the simulations driven by the bias-corrected GCM-RCM ensemble members are in good agreement to the long-term discharge seasonality of the ERA5-GPW-driven simulation.

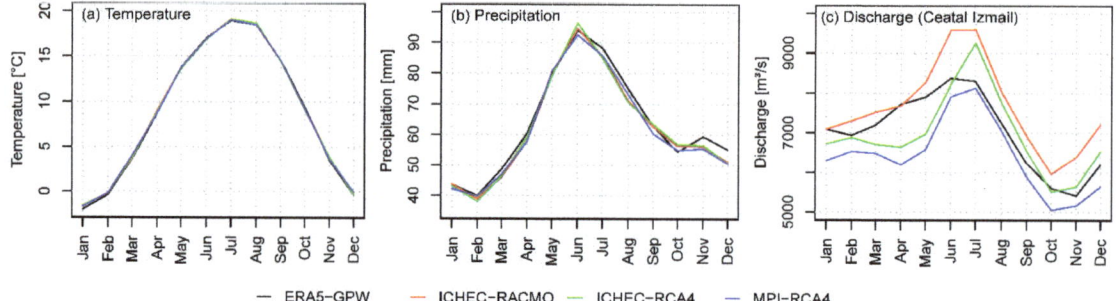

Figure 2. Hindcast of long-term mean monthly temperature [°C] (**a**) and precipitation [mm] (**b**) (1980–2000) on spatial average over the Danube River Basin as well as long-term mean monthly discharge [m³/s] (**c**) at outlet gauge Ceatal Izmail (1980–2000). The simulations driven by the bias-corrected GCM-RCM ensemble members ICHEC-RACMO, ICHEC-RCA4 and MPI-RCA4 were compared with the ERA5-GPW-driven simulation of Probst and Mauser [23].

3. Results

3.1. Temperature and Precipitation

3.1.1. Basin-Wide and Regional Trends

The interplay between temperature and precipitation is one of the main determinants of the hydrological cycle in a watershed. Climate change-induced alterations in temperature and precipitation are major drivers for changes in the hydrological cycle and water resource availability in both space and time. In Figure 3, we show the development of the mean annual temperature and precipitation on spatial average over the whole DRB as a result of the following simulations: the simulation driven with the historical ERA5-GPW forcing dataset [23], the historical GCM-RCM-driven simulations (ensemble mean) and the GCM-RCM-driven simulations (ensemble mean) under RCP2.6 and RCP8.5 for the near and far future.

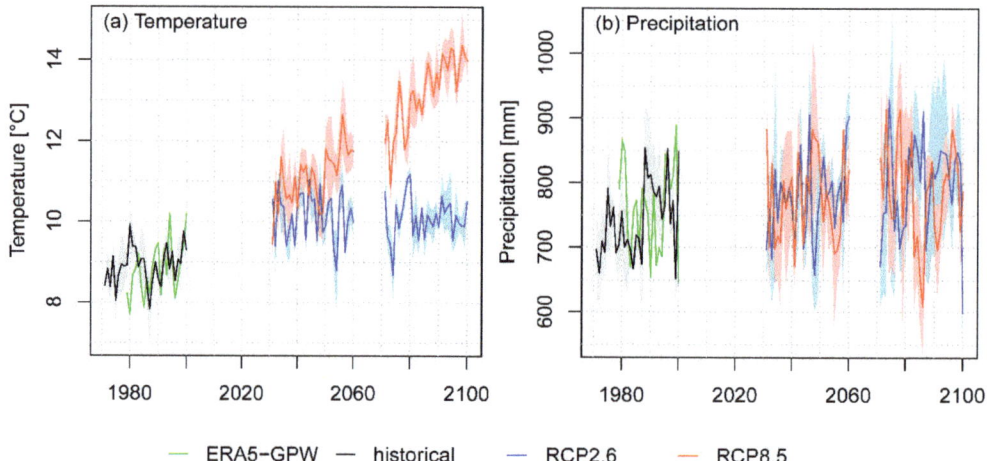

Figure 3. Mean annual temperature [°C] (**a**) and precipitation [mm] (**b**) according to the historical ERA5-GPW-driven simulation (1980–2000) [23] (green) and to the historical GCM-RCM-driven simulations (ensemble mean; 1971–2000) (black) as well as according to the GCM-RCM-driven simulations (ensemble mean) under RCP2.6 (blue) and RCP8.5 (red) for the near future (2031–2060) and the far future (2071–2100) on spatial average over the whole Danube River Basin. Lines: ensemble mean; shadings: standard deviation of the ensemble.

In Table 2, we additionally show the changes in the long-term mean annual and seasonal temperature and precipitation on spatial average for the Upper, Middle and Lower Danube as well as for the Danube overall according to the GCM-RCM-driven simulations (ensemble mean) under RCP2.6 and RCP8.5 for the near and far future compared to the historical GCM-RCM-driven simulations (ensemble mean).

Table 2. Changes in the long-term mean annual and seasonal temperature [°C] and precipitation [%] on spatial average for the Upper, Middle and Lower Danube as well as for the Danube overall according to the GCM-RCM-driven simulations (ensemble mean) under RCP2.6 and RCP8.5 for the near future (2031–2060) and the far future (2071–2100) compared to the historical GCM-RCM-driven simulations (ensemble mean, 1971–2000).

	Temperature [°C]				Precipitation [%]			
Emission Scenario	Upper Danube	Middle Danube	Lower Danube	Danube Overall	Upper Danube	Middle Danube	Lower Danube	Danube Overall
RCP2.6 (2031–2060)								
Annual	+1.3	+1.2	+1.3	+1.2	+4.6%	+3.7%	+6.5%	+4.5%
DJF *	+1.5	+1.4	+1.6	+1.5	+13.4%	+11.7%	+15.0%	+12.6%
MAM *	+1.2	+1.2	+1.4	+1.3	+6.4%	+7.2%	+5.6%	+6.7%
JJA *	+1.2	+1.1	+1.1	+1.1	+1.1%	+0.6%	+6.6%	+2.1%
SON *	+1.3	+1.1	+1.0	+1.1	+1.4%	−1.6%	+1.1%	−0.7%
RCP2.6 (2071–2100)								
Annual	+1.2	+1.2	+1.3	+1.2	+5.5%	+7.1%	+8.6%	+7.2%
DJF	+1.3	+1.4	+1.5	+1.4	+16.7%	+12.9%	+11.8%	+13.2%
MAM	+1.2	+1.4	+1.6	+1.4	+5.4%	+8.6%	+4.7%	+7.3%
JJA	+1.1	+1.0	+1.0	+1.0	−0.1%	+3.3%	+7.8%	+3.8%
SON	+1.1	+0.9	+0.9	+0.9	+5.3%	+6.6%	+12.5%	+7.6%
RCP8.5 (2031–2060)								
Annual	+2.1	+2.2	+2.3	+2.2	+7.1%	+5.1%	+2.8%	+4.9%
DJF	+2.6	+2.5	+2.5	+2.5	+26.6%	+13.3%	+16.9%	+15.9%
MAM	+1.9	+2.2	+2.4	+2.2	+5.5%	+8.6%	+8.2%	+8.1%
JJA	+2.0	+2.1	+2.3	+2.2	+0.2%	−1.9%	−6.5%	−2.6%
SON	+2.0	+2.0	+2.0	+2.0	+3.8%	+4.7%	+1.0%	+3.8%
RCP8.5 (2071–2100)								
Annual	+4.2	+4.2	+4.4	+4.3	+9.8%	+7.0%	+0.4%	+5.9%
DJF	+4.6	+4.7	+4.8	+4.7	+23.8%	+27.3%	+19.8%	+25.3%
MAM	+3.8	+3.9	+4.3	+4.0	+16.2%	+12.0%	+3.9%	+10.7%
JJA	+4.3	+4.4	+4.8	+4.5	−2.4%	−7.4%	−12.6%	−7.9%
SON	+4.1	+3.8	+3.9	+3.9	+11.1%	+6.0%	+2.9%	+6.1%

Notes: * DJF: December, January, February; MAM: March, April, May; JJA: June, July, August; SON: September, October, November.

While Figure 3 shows that the historical mean annual temperature and precipitation developments from the ERA5-GPW-driven simulation and from the historical GCM-RCM-driven simulations are in good agreement, the future developments under RCP2.6 and RCP8.5 reveal very different trends. In particular, basin-wide temperatures differ significantly between RCP2.6 and RCP8.5. While temperature rise remains on a relatively low and constant level under RCP2.6 (+1.2 °C on long-term basin-wide average for both the near and far future compared to the historical period; Table 2), the RCP8.5 temperature rise features a significant increasing trend especially for the far future (+2.2 °C and +4.3 °C on long-term basin-wide average for the near and far future compared to the historical period; Table 2), culminating in a temperature rise of +5.1 °C by 2100. Concerning annual precipitation, we identified no such clear trends as for temperature. For both RCP2.6 and RCP8.5, a slight increase in annual precipitation can be observed (RCP2.6: +4.5% and +7.2% on long-term basin-wide average for the near and far future compared to the historical period; RCP8.5: +4.9% and +5.9% on long-term basin-wide average for the near and far future compared to the historical period; Table 2). In addition to the long-term average increase in precipitation, strong variability is visible in Figure 3.

In Figure 4, we show regional differences in temperature and precipitation. Here, we show boxplots of annual temperature and precipitation change for the Upper, Middle and Lower Danube according to the simulations driven by the three individual GCM-RCM ensemble members under RCP2.6 and RCP8.5 for the near and far future compared to the historical reference.

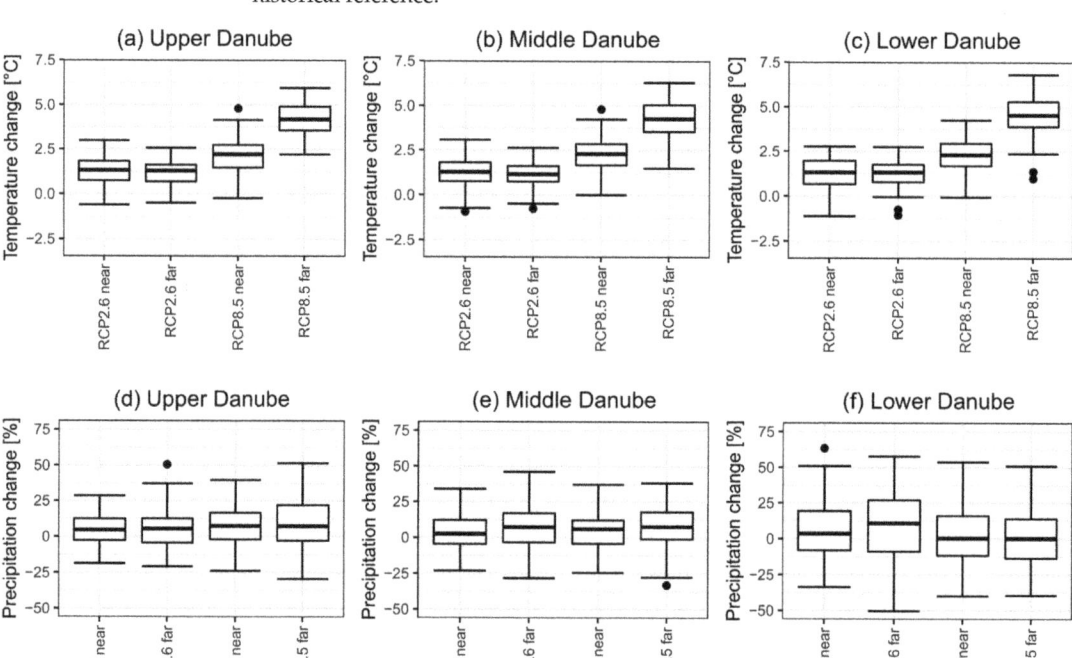

Figure 4. Boxplots of the annual temperature change [°C] (**a**–**c**) and the annual precipitation change [%] (**d**–**f**) on spatial average for the Upper, Middle and Lower Danube according to the simulations driven by the three individual GCM-RCM ensemble members under RCP2.6 and RCP8.5 for the near future (2031–2060) and the far future (2071–2100) compared to the long-term mean of the GCM-RCM-driven simulations for the historical period (1971–2000). The boxplots show yearly simulation results of the three individual GCM-RCM ensemble members. Whiskers: 1.5-fold interquartile range; black dots: minimum and maximum values.

According to Figure 4, the future temperature development uniformly points to an increase in median temperature for both emission scenarios, both future periods, and for all three Danube sub regions. Especially the RCP8.5 median temperature increase (far future) points to high values of +4.2 °C to +4.5 °C across the sub regions. While for RCP2.6 (near and far future) and RCP8.5 (near future), single colder years still occur in which annual temperature is lower than the long-term mean annual temperature of the historical period, no single colder year occurs for RCP8.5 (far future). The future precipitation development points to a slight to negligible increase in median annual precipitation for both emission scenarios and both future periods in all sub regions, with a range that extends from +0.2% (RCP8.5 far future, Lower Danube) to +11.0% (RCP2.6 far future, Lower Danube). In the Lower Danube, median annual precipitation under RCP8.5 remains nearly unchanged for both the near and far future. At the same time, the variability of the annual precipitation is especially large for the Lower Danube compared to the Upper and Middle Danube.

3.1.2. Changes in Seasonality

To thoroughly assess the impacts of climate change, it is also necessary to take a closer look at possible seasonal changes. In Figure 5, we show seasonality plots of the long-term mean monthly developments of temperature and precipitation on spatial average for the

Upper, Middle and Lower Danube. Here, we compared the ensemble mean results from the historical GCM-RCM-driven simulations and the GCM-RCM-driven simulations under RCP2.6 and RCP8.5 for the near and far future.

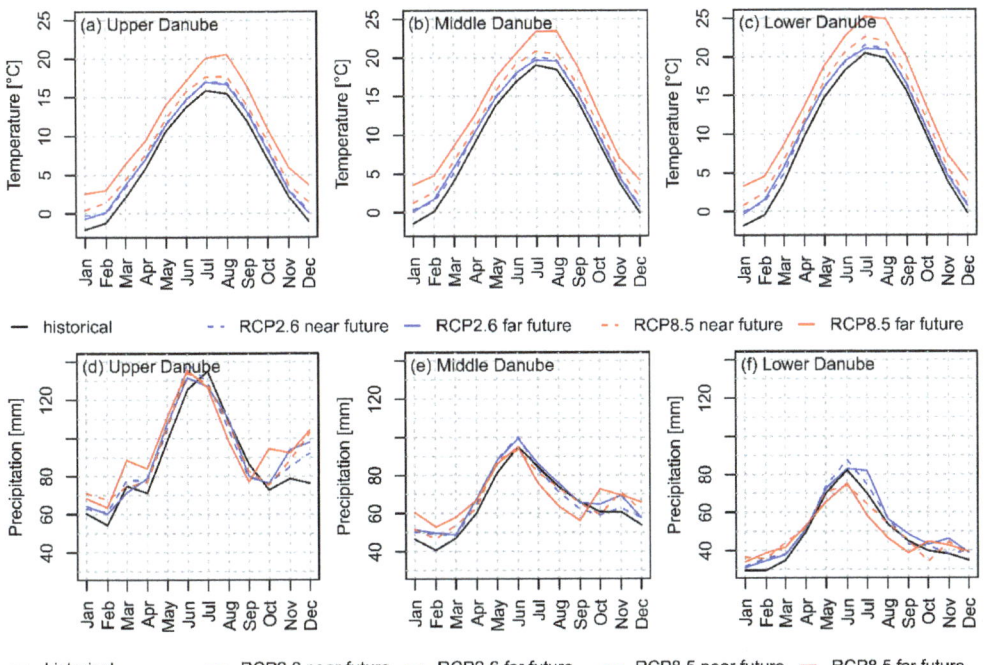

Figure 5. Long-term mean monthly developments of temperature [°C] (**a–c**) and precipitation [mm] (**d–f**) according to the historical GCM-RCM-driven simulations (ensemble mean, 1971–2000) and the GCM-RCM-driven simulations (ensemble mean) under RCP2.6 and RCP8.5 for the near future (2031–2060) and the far future (2071–2100) on spatial average for the Upper, Middle and Lower Danube.

Figure 5 shows that the future temperature increases more or less evenly throughout the year in the Upper, Middle and Lower Danube without shifts in seasonality. In all sub regions, however, a greater warming trend occurs in the winter season than during the rest of the year in all scenarios (e.g., RCP8.5, near future: +2.5 °C to +2.6 °C in the winter season versus +1.9 °C to +2.4 °C during the other seasons) (Table 2). Especially the warming trend according to RCP8.5 (far future) is stronger in the winter season (+4.6 °C to +4.8 °C) and the summer season (+4.3 °C to +4.8 °C) in all sub regions than during the other seasons (+3.8 °C to +4.3 °C) (Table 2).

For precipitation, the plots reveal more distinct shifts in seasonality: for the Upper Danube, the increase in the mean annual precipitation under RCP2.6 and RCP8.5 (near and far future; Figure 5 and Table 2) is mainly attributable to an above-average increase in the winter and spring season. For RCP8.5 (far future), for example, the long-term mean seasonal precipitation is projected to change by +23.8% and +16.2% in the winter and spring season, versus −2.4% and +11.1% in the summer and autumn season in the Upper Danube. In addition, a shift of peak precipitation in the Upper Danube from July to June is projected in both RCP2.6 and RCP8.5 (near and far future), which is associated with a shift of the summer precipitation regime forward by one month. For the Lower Danube, there is an increase in precipitation visible in the summer season for RCP2.6 (+6.6% and +7.8% for the near and far future), but a moderate to strong decrease throughout summer (May–September) for RCP8.5 (e.g., −6.5% and −12.6% in the summer season for the near and far future). For the winter season, the simulations point to an increase in precipitation in the Lower Danube for all scenarios and all future periods, ranging from +11.8% (RCP2.6 far future) to +19.8% (RCP8.5 far future). For the Middle Danube, the major trends from the Upper Danube (especially the increase in winter precipitation) and the Lower Danube (especially the decrease in summer precipitation) blend, particularly under RCP8.5 (far future).

3.1.3. Changes in Spatial Patterns

Given the natural heterogeneity of the DRB, it is expected that climate change effects will also be spatially distributed in a very heterogeneous manner. In Figures 6 and 7, we show maps of the long-term mean seasonal temperature and precipitation (ensemble mean) in the DRB according to the historical GCM-RCM-driven simulations as well as the mean seasonal (absolute or relative) changes (ensemble mean) according to the GCM-RCM-driven simulations under RCP2.6 and RCP8.5 for the near and far future.

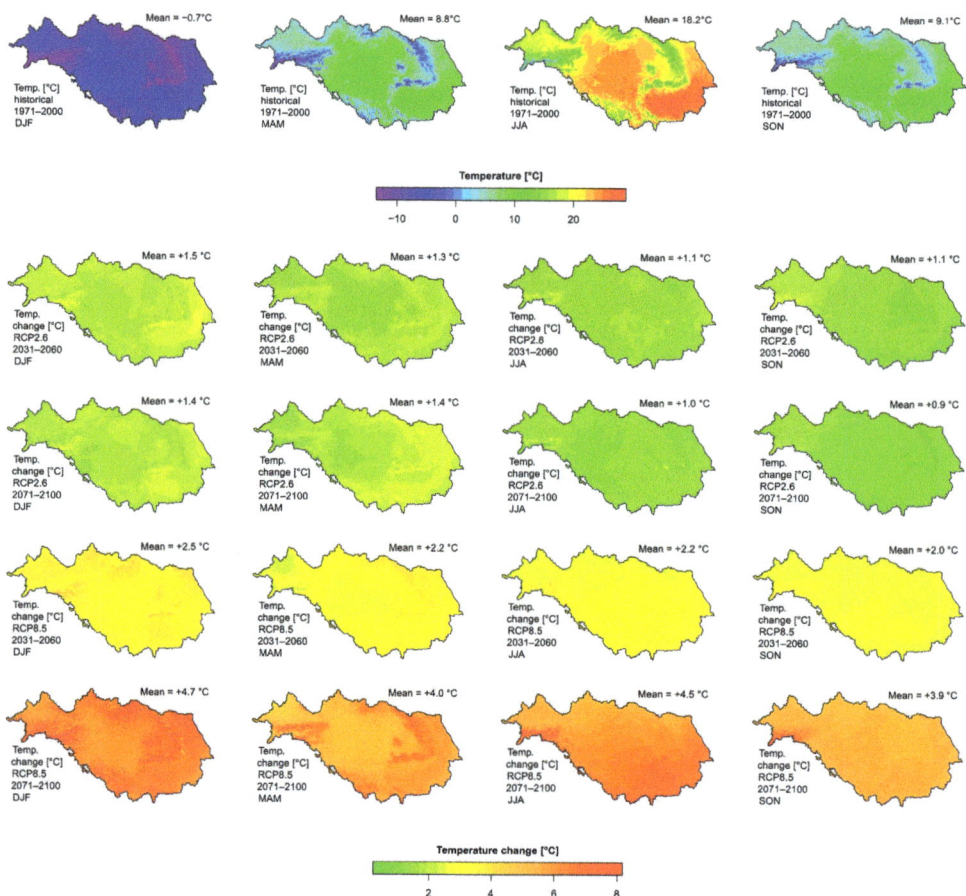

Figure 6. Maps of the long-term mean seasonal temperature [°C] for the historical GCM-RCM-driven simulations (ensemble mean, 1971–2000) as well as the long-term mean seasonal temperature change [°C] according to the GCM-RCM-driven simulations (ensemble mean) under RCP2.6 and RCP8.5 for the near future (2031–2060) and the far future (2071–2100) in the Danube River Basin.

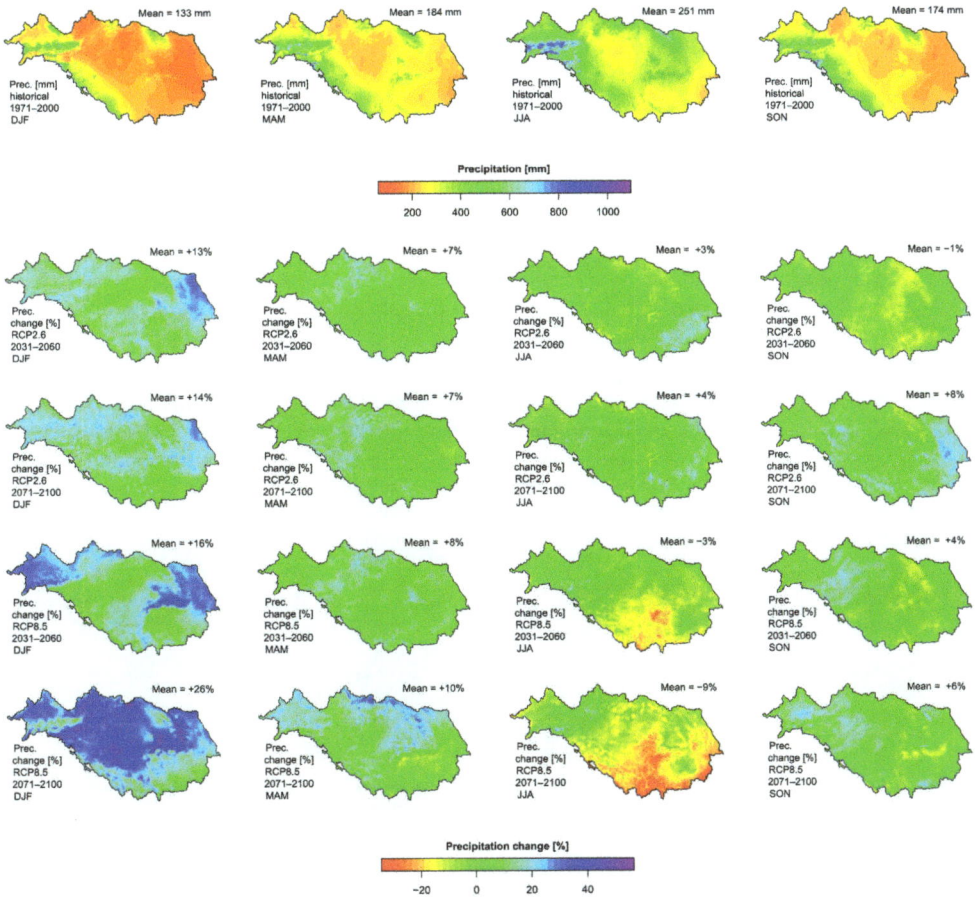

Figure 7. Maps of the long-term mean seasonal precipitation [mm] for the historical GCM-RCM-driven simulations (ensemble mean, 1971–2000) as well as the long-term mean seasonal precipitation change [%] according to the GCM-RCM-driven simulations (ensemble mean) under RCP2.6 and RCP8.5 for the near future (2031–2060) and the far future (2071–2100) in the Danube River Basin.

Figure 6 shows a relatively uniform spatial distribution of the mean seasonal temperature rise under RCP2.6 and RCP8.5 for both the near and far future. Especially for RCP8.5 (far future), however, the spatial patterns hint at a relatively stronger temperature increase in the mountains (e.g., +6 °C up to +8 °C during the winter season) compared to the basin overall (e.g., +4.7 °C on spatial average during the winter season). The historical simulations show that the highest long-term mean temperatures in the summer season occur in the basins of the Middle and Lower Danube (e.g., the Romanian Plain with an average summer temperature of >22 °C). For RCP8.5 (far future), a particular hot spot of summer temperature rise of +5 °C and beyond lies in the Lower Danube, additionally amplifying hot summer temperatures.

Figure 7 shows a gradient of decreasing mean seasonal precipitation, roughly from the north-west to the south-east of the DRB. The change maps show different spatial hot spots of precipitation increase or decrease depending on the scenario. For RCP2.6 (near and far future), precipitation increases in almost every season nearly throughout the basin, ranging from −1% (RCP2.6 near future, autumn season) to +14% (RCP2.6 far future, winter season) on spatial average. For RCP8.5 (near and far future), precipitation generally increases in the winter season and decreases in the summer season in large parts of the basin, ranging from −9% (RCP8.5 far future, summer season) to +26% (RCP8.5 far future, winter season) on spatial average. For RCP8.5 (near future), a tendency of increasing winter precipitation in the Upper Danube basin, across the northern Carpathian ridge and in the Moldovan Plain of +20% up to +40% and decreasing summer precipitation in the southern Lower Danube basin of −20% down to −26% is visible. For RCP8.5 (far future), the winter precipitation increase extends almost over the entire DRB (with hot spots of up to +57%) and the summer precipitation decrease intensifies and extends over large southern parts of the Middle and Lower Danube basin with −20% down to −36%, additionally amplifying drier summers in this region.

3.2. Soil Water Content, Plant Water Stress and Snow Water Equivalent

3.2.1. Changes in Seasonality

In Figure 8, we show seasonality plots of the long-term mean monthly developments of available soil water content in the rooted zone, plant water stress and snow water equivalent (SWE) on spatial average for the Upper, Middle and Lower Danube. The developments result from the historical GCM-RCM-driven simulations (ensemble mean) and the GCM-RCM-driven simulations (ensemble mean) under RCP2.6 and RCP8.5 for the near and far future.

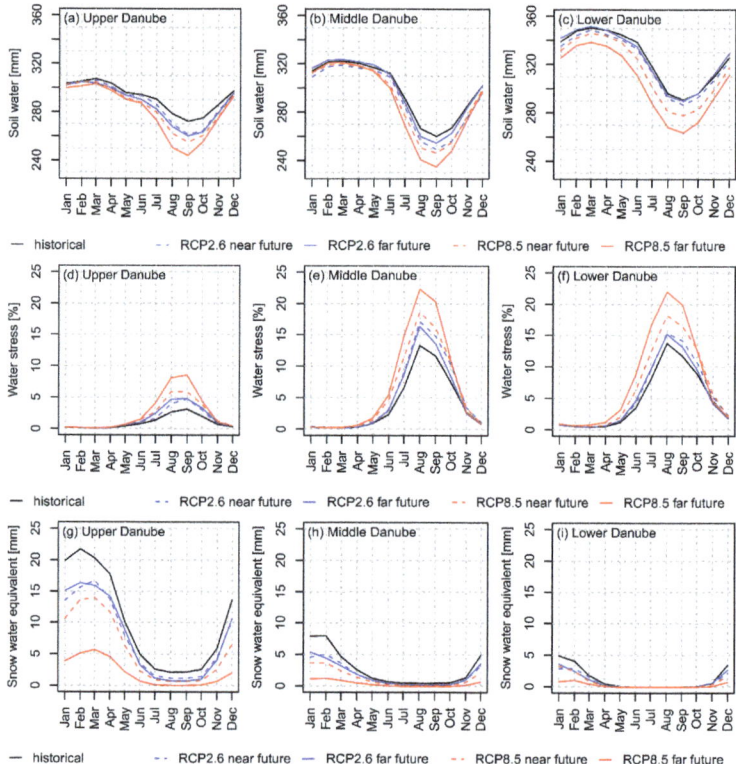

Figure 8. Long-term mean monthly developments of soil water content in the rooted zone [mm] (**a–c**), plant water stress [% reduction of potential transpiration] (**d–f**) and snow water equivalent [mm] (**g–i**) according to the historical GCM-RCM-driven simulations (ensemble mean, 1971–2000) and the GCM-RCM-driven simulations (ensemble mean) under RCP2.6 and RCP8.5 for the near future (2031–2060) and the far future (2071–2100) on spatial average over the Upper, Middle and Lower Danube.

Figure 8 shows that the long-term mean monthly soil water content decreases in late summer and early autumn in both the Upper and Middle Danube for all scenarios and particularly strong for RCP8.5 (far future), with the most severe percentage soil water reduction occurring in September. For RCP8.5 (far future), for example, soil water reduction in September amounts to −28 mm (−10.2%) in the Upper Danube and −25 mm (−9.6%) in the Middle Danube. In the Lower Danube, soil water decreases more or less uniformly throughout the year with a likewise considerable reduction for RCP8.5 (far future) and the comparatively strongest percentage reduction in July with −30 mm (−9.4%) for RCP8.5 (far future).

The depleting soil water storages directly translate into rising water stress experienced by vegetation (especially agricultural crops, grasslands and forests) in the DRB. Our simulations show that the long-term mean monthly plant water stress intensifies in the Upper, Middle and Lower Danube, especially during the summer and autumn months of diminished soil water toward the end of the period of maximum forest transpiration and toward the end of the vegetation period of dominant agricultural summer crops such as maize, sunflower, soybean, potatoes and sugar beet. The historical simulations show the maximum monthly plant water stress values in late summer for all the sub regions, resulting in 3% for the Upper Danube in September, 13% for the Middle Danube in August and 14% for the Lower Danube in August. For RCP8.5 (far future), plant water stress intensifies to maximum values of 9% for the Upper Danube, 22% for the Middle Danube and 22% for the Lower Danube in the same months.

Furthermore, the simulations show a reduction in the long-term mean monthly SWE, which indicates that the amount of water stored in snow decreases for all scenarios and especially for RCP8.5 (far future). In the winter season (DJF), simulated future SWE decreases all over the DRB, with SWE reductions of −14.7 mm (−80%) in the Upper Danube, −5.9 mm (−85%) in the Middle Danube and −3.2 mm (−78%) in the Lower Danube on spatial average for RCP8.5 (far future). Summer snow cover in the high mountain ranges of the Alps in the Upper Danube vanishes almost completely for RCP8.5 (far future).

3.2.2. Changes in Spatial Patterns

In Figures 9–11, we show maps of the long-term mean seasonal soil water content in the rooted zone, plant water stress and SWE in the DRB according to the historical GCM-RCM-driven simulations (ensemble mean) as well as the mean seasonal values or changes according to the GCM-RCM-driven simulations (ensemble mean) under RCP2.6 and RCP8.5 for the near and far future.

Figure 9 reveals that almost all scenarios and future periods (except for RCP2.6 far future in the winter and spring season) show negative changes in soil water on spatial average for the four seasons. For RCP2.6 (near and far future), the seasonal soil water change ranges between −8 mm (RCP2.6, near future) in the autumn season and +2 mm (RCP2.6, far future) in the winter season. Here, the change maps only show local alterations of soil water, with a concentration of refilling soil water storages around the Pannonian Basin in winter and spring and depleting soil water storages in some forested areas of Transylvania in summer and autumn. For RCP8.5 (near and far future), the seasonal soil water change ranges between −3 mm (RCP8.5, near future) in the winter and spring season and −22 mm (RCP8.5, far future) in the summer season. For RCP8.5, hot spots of more significant changes can be encountered, in which the soil water particularly increases in the Pannonian Basin during winter and spring and decreases in the forested areas of Transylvania, the Carpathian foothills and the southern Dinarides during summer and autumn (near and far future).

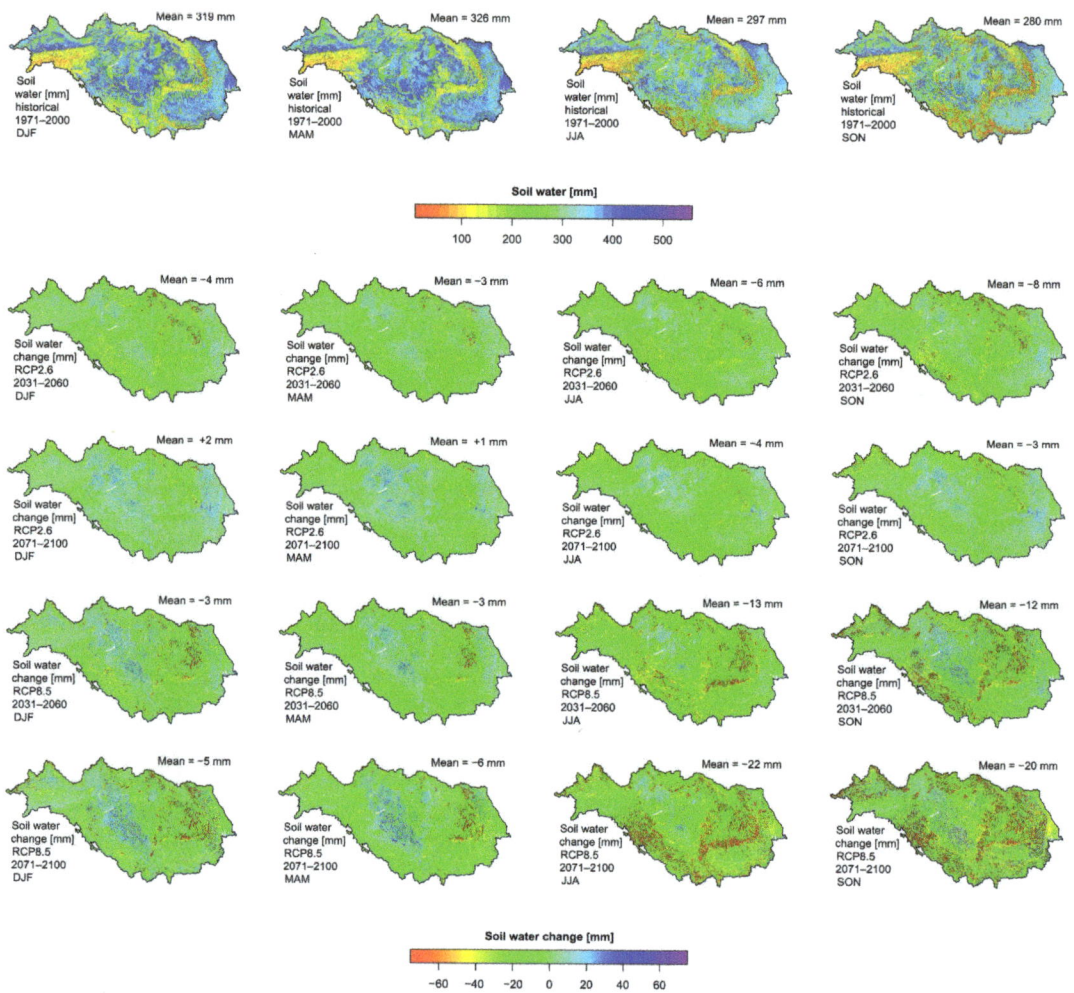

Figure 9. Maps of the long-term mean seasonal soil water content in the rooted zone [mm] for the historical GCM-RCM-driven simulations (ensemble mean, 1971–2000) as well as the long-term mean seasonal soil water change [mm] according to the GCM-RCM-driven simulations (ensemble mean) under RCP2.6 and RCP8.5 for the near future (2031–2060) and the far future (2071–2100) in the Danube River Basin.

Figure 10 shows that the patterns of increasing plant water stress follow the patterns of depleting soil water storages. Long-term mean seasonal plant water stress is largely restricted to the summer and autumn season. In the historical simulations, the mean plant water stress amounts to 7.1% on spatial average during both summer and autumn. During summer, the extensive croplands in the lowlands of the Middle and Lower Danube are especially affected by water stress values of >25%, which can be interpreted as serious water shortages reducing agricultural yields. During autumn, hot spots of the plant water stress shift to areas of deciduous forests, with values of >30%. For all scenarios and future periods, summer and autumn water stress intensifies on spatial average. Especially for RCP8.5 (far future), the maps point to a more widespread and more intense plant water stress of 13.8% during summer and 11.0% during autumn on spatial average. Distinct hot spots occur in the Pannonian Basin, the Romanian Plain (water stress values of >40% for agricultural crops in the summer season) and around the Carpathian foothills (water stress values of >50% for deciduous forests in the autumn season).

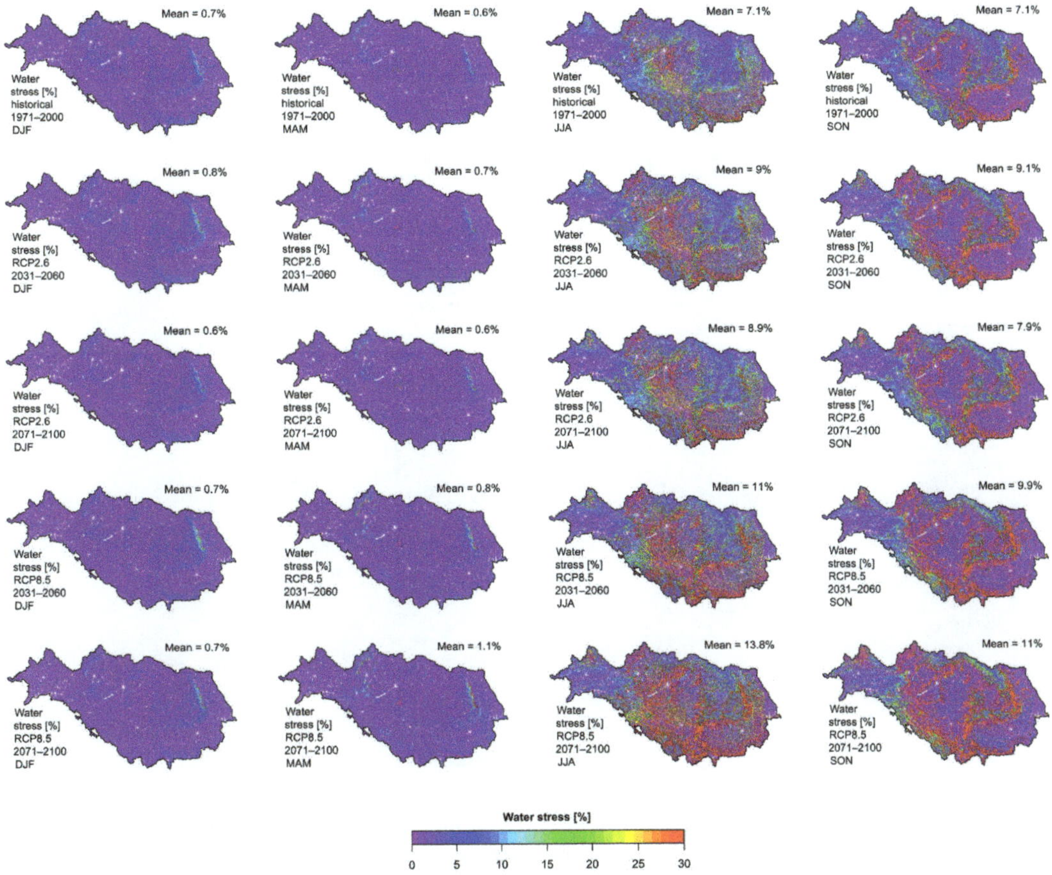

Figure 10. Maps of the long-term mean seasonal plant water stress [% reduction of potential transpiration] for the historical GCM-RCM-driven simulations (ensemble mean, 1971–2000) as well as for the GCM-RCM-driven simulations (ensemble mean) under RCP2.6 and RCP8.5 for the near future (2031–2060) and the far future (2071–2100) in the Danube River Basin.

Figure 11 shows that the long-term mean seasonal SWE concentrates in the mountain regions during winter and spring. For the historical period, the long-term mean SWE in the winter season amounts to 7.35 mm on spatial average, with SWE values of >100 mm in the Alpine mountain ranges. The change maps reveal significant reductions in the seasonal SWE for all scenarios all over the DRB. The strongest reduction is again projected for RCP8.5 (far future), for which the SWE decreases considerably in the Alps, Carpathians and Dinarides during autumn (from 0.89 mm to 0.08 mm), winter (from 7.35 mm to 1.28 mm) and spring (from 3.62 mm to 0.83 mm), and almost completely vanishes during summer (from 0.65 mm to 0.05 mm).

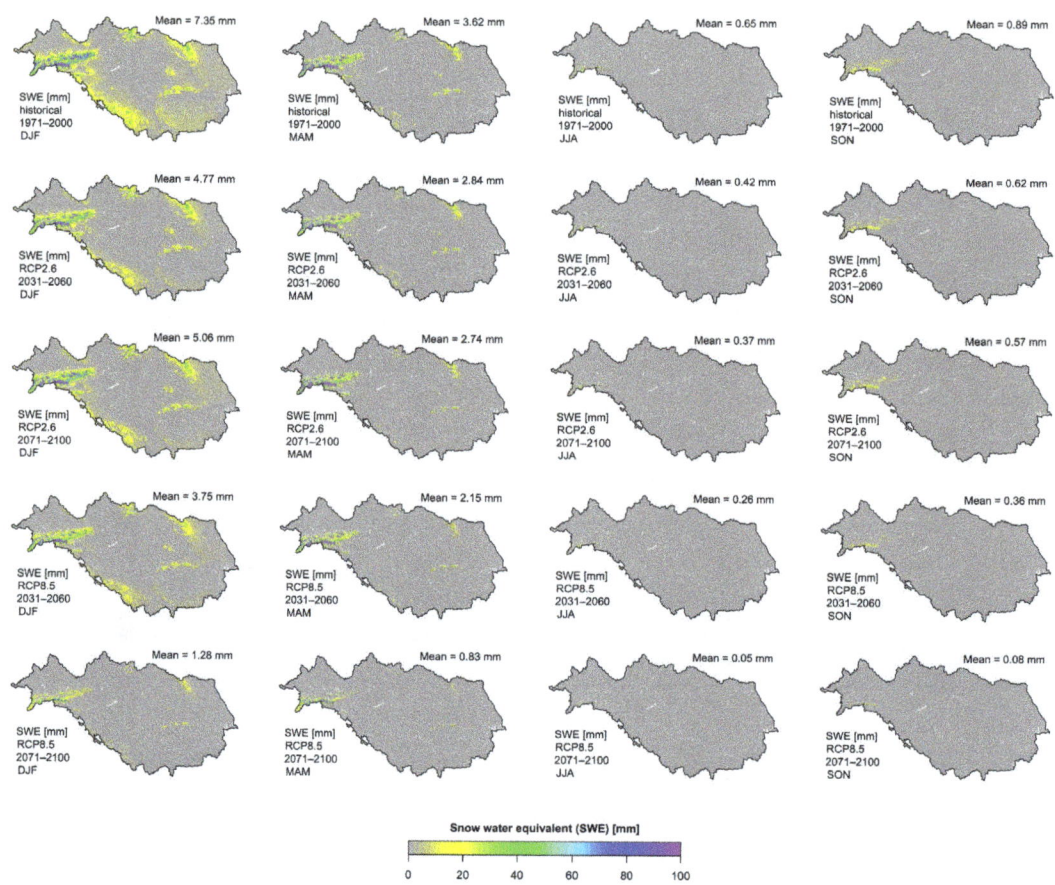

Figure 11. Maps of the long-term mean seasonal snow water equivalent (SWE) [mm] for the historical GCM-RCM-driven simulations (ensemble mean, 1971–2000) as well as for the GCM-RCM-driven simulations (ensemble mean) under RCP2.6 and RCP8.5 for the near future (2031–2060) and the far future (2071–2100) in the Danube River Basin.

3.3. Discharge

3.3.1. Annual and Seasonal Trends in Danube Sub-Basins

The future projections of the temperature and the components of the hydrological cycle, such as precipitation, soil water and snow water, hold direct consequences for the runoff dynamics in the DRB. In Table 3, we show the changes in the long-term mean annual and seasonal mean discharge (MQ) in the entire DRB and its major sub-basins (Figure 1) according to the GCM-RCM-driven simulations (ensemble mean) under RCP2.6 and RCP8.5 for the near and far future compared to the historical simulations (ensemble mean).

Table 3 shows that the trends of the future MQ changes (increasing or decreasing) generally follow the precipitation development, although some time lag is visible. Surprisingly, despite the decreasing soil water content in the root zone and the increasing plant water stress, the long-term mean annual MQ increases for all scenarios and all future periods at the Danube outlet gauge in Ceatal Izmail (+2.8% and +10.7% for RCP2.6 in the near and far future; +3.1% and +5.2% for RCP8.5 in the near and far future). For RCP2.6 (near future), the MQ tends to increase in winter and spring, and to decrease in summer and autumn across a majority of sub-basins. Interestingly, for RCP2.6 (far future), in contrast, the MQ increases in all seasons for almost all sub-basins. For RCP8.5 (near and far future), Table 3 hints at a clear dichotomy: on the one hand, the MQ increases for all seasons in the Upper Danube, Middle Danube (Bezdan) and Drava basin (e.g., +22.5% and +26.4% for winter MQ in the Upper Danube for RCP8.5 in the near and far future). On the other hand, the MQ increases for the winter and spring season (+0.5% to +19.1%) and decreases in the summer and autumn season (−1.5% to −17.1%) in the downstream sub-basins (Sava, Mures, Tisza, Siret and the Lower Danube). Hereby, the reduction in the seasonal MQ according to RCP8.5 is more severe for the Sava, Tisza and Lower Danube basin for the far future. Concerning the annual MQ, the increase in the winter and spring MQ over-compensates for the decrease in the summer and autumn MQ in most sub-basins. At the Danube outlet, the winter and spring surplus outweighs the summer and autumn deficit for all scenarios.

Table 3. Changes in the long-term mean annual and seasonal mean discharge (MQ) [%] in the Danube River Basin and its major sub-basins (at respective gauges in parentheses) according to the GCM-RCM-driven simulations (ensemble mean) under RCP2.6 and RCP8.5 for the near future (2031–2060) and the far future (2071–2100) compared to the historical GCM-RCM-driven simulations (ensemble mean, 1971–2000).

Emission Scenario	Upper Danube (Achleiten)	Middle Danube (Bezdan)	Drava (Dravaszabolcs)	Sava (Sremska Mitrovica)	Mures (Nagylak)	Tisza (Senta)	Siret (Lungoci)	Lower Danube (Ceatal Izmail)
RCP2.6 (2031–2060)								
Annual	+4.6%	+4.8%	+6.4%	−0.9%	−0.8%	−4.0%	+4.4%	+2.8%
DJF *	+9.2%	+7.8%	+11.6%	+2.1%	−1.8%	−8.3%	+16.3%	+3.4%
MAM *	+4.6%	+5.5%	+7.4%	+0.6%	+5.0%	+2.5%	+9.5%	+4.4%
JJA *	+5.9%	+7.9%	+8.9%	−0.3%	−2.8%	−2.5%	−0.8%	+5.1%
SON *	−1.8%	−3.1%	−1.9%	−8.0%	−3.3%	−9.3%	+0.5%	−3.1%
RCP2.6 (2071–2100)								
Annual	+6.9%	+9.5%	+12.7%	+9.6%	+7.4%	+5.6%	+20.0%	+10.7%
DJF	+20.4%	+20.2%	+16.6%	+13.5%	+14.7%	+7.8%	+37.6%	+16.9%
MAM	+3.0%	+6.8%	+10.7%	+8.0%	+7.1%	+8.1%	+11.6%	+8.8%
JJA	+3.8%	+7.8%	+13.0%	+8.5%	−0.2%	+3.5%	+16.0%	+9.5%
SON	+0.6%	+2.8%	+10.9%	+7.4%	+14.9%	+3.3%	+23.9%	+7.4%
RCP8.5 (2031–2060)								
Annual	+7.2%	+8.1%	+9.0%	−2.0%	−3.9%	+0.1%	−4.3%	+3.1%
DJF	+22.5%	+21.1%	+17.3%	+3.1%	+9.8%	+5.7%	+11.5%	+12.1%
MAM	+4.4%	+6.2%	+6.1%	+0.5%	+7.2%	+5.4%	+13.0%	+6.0%
JJA	+2.8%	+4.7%	+5.9%	−7.1%	−13.5%	−2.7%	−16.5%	−1.5%
SON	−0.2%	+0.0%	+8.1%	−6.7%	−14.1%	−9.2%	−10.6%	−4.4%
RCP8.5 (2071–2100)								
Annual	+10.5%	+11.5%	+9.4%	−0.2%	+0.6%	+4.5%	−5.2%	+5.2%
DJF	+26.4%	+27.7%	+33.3%	+14.7%	+13.2%	+13.1%	+9.5%	+19.1%
MAM	+9.2%	+10.0%	+8.5%	+7.4%	+17.3%	+18.3%	+13.0%	+11.8%
JJA	+3.3%	+6.2%	+0.8%	−16.7%	−11.1%	−3.5%	−14.7%	−2.9%
SON	+4.1%	+2.0%	+0.6%	−11.9%	−11.5%	−10.8%	−17.1%	−7.2%

Notes: * DJF: December, January, February; MAM: March, April, May; JJA: June, July, August; SON: September, October, November.

In Figure 12, we show seasonality plots of the long-term mean monthly developments of discharge for the Upper Danube (gauge Achleiten), Middle Danube (gauge Iron Gate) and Lower Danube (gauge Ceatal Izmail). The developments result from the historical GCM-RCM-driven simulations (ensemble mean) and the GCM-RCM-driven simulations (ensemble mean) under RCP2.6 and RCP8.5 for the near and far future.

The changes in the future discharge regimes are a direct response to the changes in the precipitation regimes. In the Upper Danube, the increase in winter precipitation, the slight decrease in summer precipitation (especially for RCP8.5, far future) and the forward shift of the precipitation peak directly translates into corresponding changes in seasonal discharge regimes at gauge Achleiten. Here, the long-term mean monthly MQ at gauge Achleiten increases by +26.4% on average in the winter season and decreases by −7.1% during August and September (RCP8.5, far future). The same tendencies can be seen for the Middle Danube (gauge Iron Gate) and Lower Danube (gauge Ceatal Izmail), where especially the strong decrease in summer precipitation (RCP8.5, far future) translates into a decrease in the summer discharge. Here, the long-term mean monthly MQ increases by

+21.9% at gauge Iron Gate and +19.1% at gauge Ceatal Izmail in the winter season and decreases by −12.1% at gauge Iron Gate and −11.3% at gauge Ceatal Izmail during August and September. However, the decrease in summer discharge blends with the discharge changes of the Upper Danube (especially with the increase in winter discharge) that are transferred to downstream reaches.

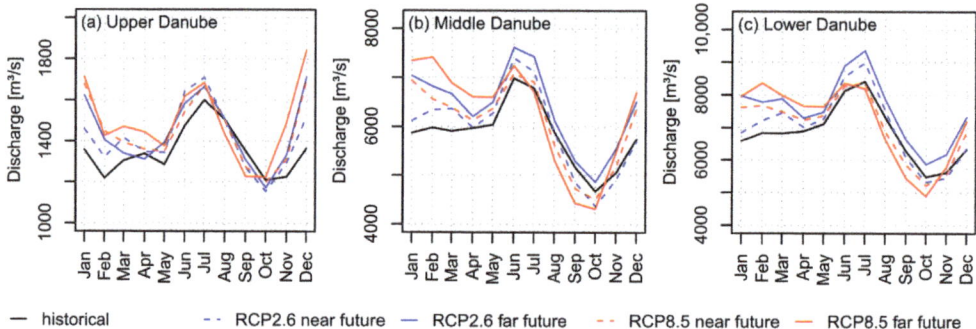

Figure 12. Long-term mean monthly developments of discharge [m^3/s] according to the historical GCM-RCM-driven simulations (ensemble mean, 1971–2000) and the GCM-RCM-driven simulations (ensemble mean) under RCP2.6 and RCP8.5 for the near future (2031–2060) and the far future (2071–2100) for the Upper Danube at gauge Achleiten, the Middle Danube at gauge Iron Gate and the Lower Danube at gauge Ceatal Izmail.

3.3.2. Changes in Spatial Patterns

In terms of discharge, smaller tributary rivers in drier regions are expected to be particularly affected by climate change. In Figure 13, we show maps of the changes in the long-term mean seasonal MQ in the DRB according to the GCM-RCM-driven simulations (ensemble mean) under RCP2.6 and RCP8.5 for the near and far future compared to the historical GCM-RCM-driven simulations (ensemble mean).

Figure 13 shows that the changes in the long-term mean seasonal MQ under different scenarios and future periods are very heterogeneously distributed across the DRB. On basin-wide average, the range of seasonal MQ change varies between −8% (RCP8.5, far future, summer season) and +19% (RCP8.5, far future, winter season). An interesting finding for the MQ changes under RCP2.6 is that negative MQ changes occur much more frequently for the near future than for the far future. Shortages in water resource availability in rivers under the RCP2.6 scenario are more frequently projected in the near future and are compensated by higher precipitation in the far future. For RCP2.6 (near future), the mean seasonal MQ change ranges from −3% (autumn) to +5% (winter) on basin-wide average. Here, local hot spots of runoff reductions are especially found in the Pannonian Basin and the Tisza basin in winter, summer and autumn. Cold spots (regions with runoff increase) are particularly found in the Alps in the winter season and the Danube delta region in autumn and winter. For RCP2.6 (far future), the mean seasonal MQ change ranges from +8% (summer) to +18% (winter) on basin-wide average. Here, cold spots of an increasing MQ predominate the map, especially in the Pannonian Basin (all year), the Upper Danube (winter season) and the Danube delta region with the Romanian and Moldovan Plain (autumn and winter).

For the RCP8.5 scenario though, significantly greater reductions in the MQ occur throughout the basin. For RCP8.5 (near future), the mean seasonal MQ change ranges from −5% (summer and autumn) to +12% (winter) on basin-wide average. Here, widespread hot spots of an MQ decrease can be found in the entire Lower Danube and the Sava basin in summer and autumn. Simultaneously, winter precipitation and thus, MQ is increasing in the Upper Danube, especially in the Alpine region. For RCP8.5 (far future), the mean seasonal MQ change ranges from −8% (summer) to +19% (winter) on basin-wide average.

Here, severe decreases in the MQ (changes of up to −50%) can be observed for smaller tributary rivers scattered almost all over the basin in summer and autumn. Around the Carpathian Mountains, the MQ decreases all over the year. Likewise, winter precipitation and thus, MQ increases in the Upper Danube (especially in the Alps) and the Pannonian Basin, and spring precipitation and thus, MQ increases in the Pannonian Basin.

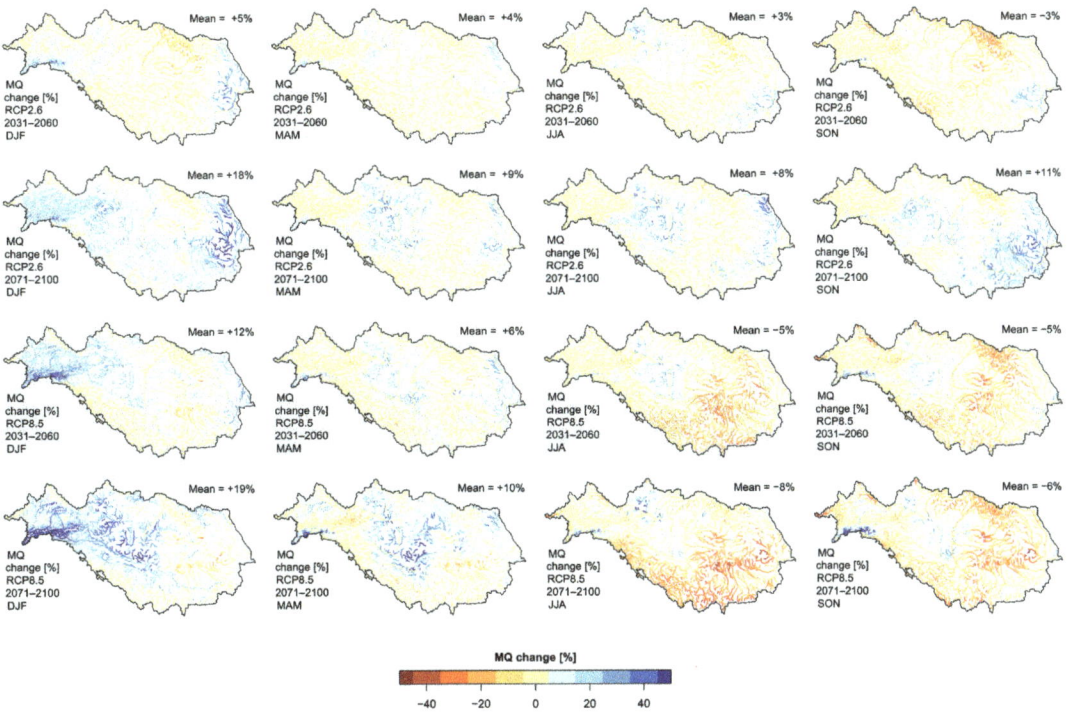

Figure 13. Maps of the long-term mean seasonal MQ change [%] in the Danube River Basin according to the GCM-RCM-driven simulations (ensemble mean) under RCP2.6 and RCP8.5 for the near future (2031–2060) and the far future (2071–2100) compared to the historical GCM-RCM-driven simulations (ensemble mean, 1971–2000).

3.3.3. Changes in the Risk of High and Low Flows

To thoroughly analyze the impacts of climate change on runoff dynamics, it is also necessary to take a closer look on the development of high and low flows. In Figure 14, we show flow duration curves for the Upper Danube (gauge Achleiten), the Middle Danube (gauges Bezdan and Iron Gate) and the Lower Danube (gauge Ceatal Izmail) (sub panels a–d). The flow duration curves show the average number of days in a year, in which the discharge values are exceeded. The colored curves in Figure 14 refer to the ensemble mean of the historical GCM-RCM-driven simulations as well as to the ensemble mean of the GCM-RCM-driven simulations under RCP2.6 and RCP8.5 for the near and far future. Figure 14 additionally zooms into the flow duration curves of the respective gauges in two intervals: (i) the interval of the flow duration curves near the historical long-term mean annual low flow (MNQ) values (thereby showing the number of undershoot days; sub panels e–h); (ii) the interval of the flow duration curves near the historical long-term mean annual high flow (MHQ) values (thereby showing the number of exceedance days; sub panels i–l) at the respective gauges.

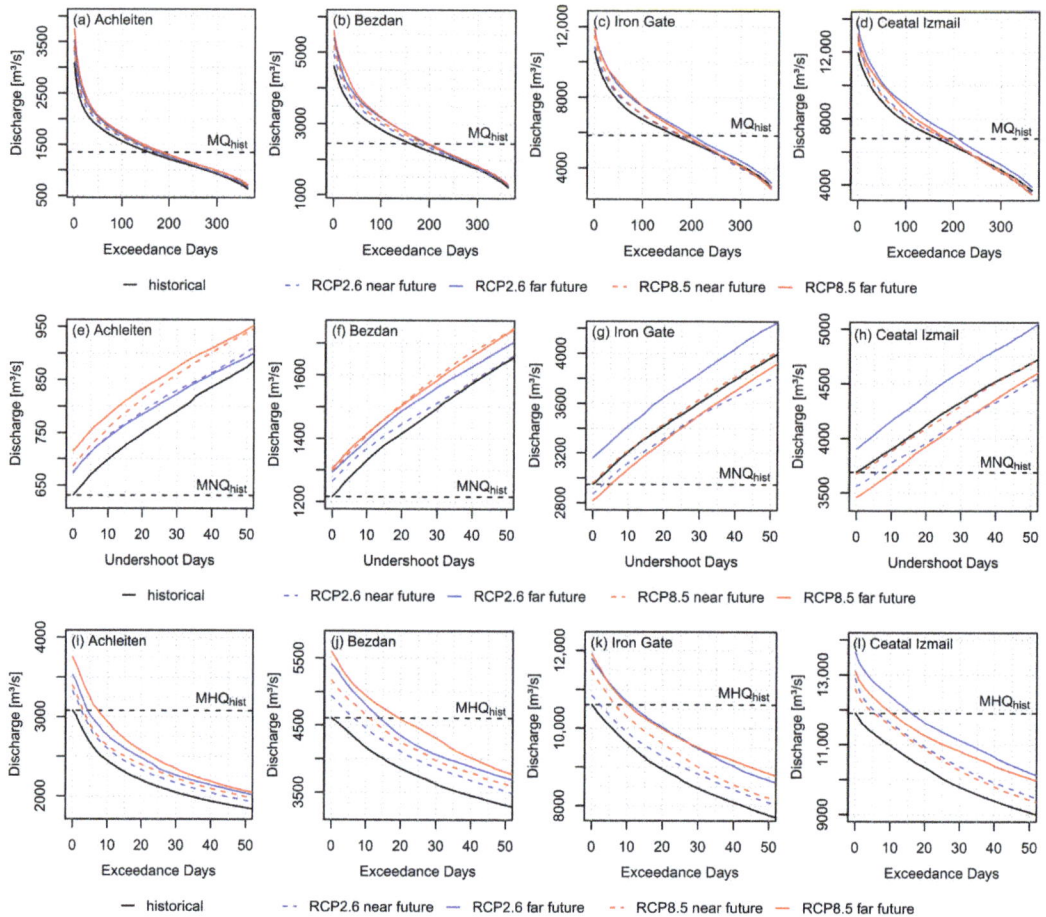

Figure 14. Flow duration curves for different gauges along the main course of the Danube (gauges Achleiten, Bezdan, Iron Gate and Ceatal Izmail) for the historical GCM-RCM-driven simulations (ensemble mean, 1971–2000) as well as for the GCM-RCM-driven simulations (ensemble mean) under RCP2.6 and RCP8.5 for the near future (2031–2060) and the far future (2071–2100). The graphs in subplots (**a–d**) show the average number of days in a year in which the daily discharge values are exceeded. The graphs in subplots (**e–h**) are zoomed-in near the historical long-term mean annual low flow (MNQ) values and the graphs in subplots (**i–l**) are zoomed-in near the historical long-term mean annual high flow (MHQ) values at the respective gauges.

According to Figure 14, the parts of the flow duration curves that lie beyond the historical long-term mean annual MQ values show a distinct trend: the curves from the future GCM-RCM-driven simulations generally lie above the curves from the historical simulations. This means that on average, a selected discharge value is exceeded on more days in a year. This points to a development, in which a greater water volume is in the river system on an increasing number of days in a year. The zoomed-in plots near the historical MNQ values and near the historical MHQ values at the respective gauges provide insight into the development of low and high flows. For the Upper Danube (gauge Achleiten) and the Middle Danube (gauge Bezdan), the historical MNQ values (631 m³/s and 1216 m³/s) are on average not undershot in an average year according to all scenarios and future periods (RCP2.6 and RCP8.5, near and far future). For the Middle Danube (gauge Iron

Gate) and the Lower Danube (gauge Ceatal Izmail), the historical MNQ is on average undershot on some days of the year according to most scenarios and future periods. For example, the historical MNQ values at gauges Iron Gate and Ceatal Izmail (2948 m^3/s and 3689 m^3/s) are undershot on 6 and 11 days (RCP8.5, far future). This means that low flow conditions along the Middle and Lower Danube River are projected to become more frequent in the future.

For all considered gauges of the Danube River (gauges Achleiten, Bezdan, Iron Gate and Ceatal Izmail), the historical MHQ is on average overshot on some days of the year according to all scenarios and future periods. For example, the historical MHQ values at the gauges Achleiten, Bezdan, Iron Gate and Ceatal Izmail (3081 m^3/s, 4611 m^3/s, 10,612 m^3/s and 11,907 m^3/s) are overshot on 8, 20, 11 and 11 days (RCP8.5, far future). This means that high flow conditions all along the Danube main course are projected to become more frequent in the future.

4. Discussion

4.1. The Big Picture

4.1.1. Trends of Temperature and Precipitation

On basin-wide average, we find a very significant and continuous warming trend with rising mean annual temperatures for RCP8.5, which begins in the near future and continues to strengthen in the far future (Figure 3 and Table 2). For RCP2.6, the temperature slightly increases from the historical baseline but soon saturates at a constant level in both the near and far future (Figure 3 and Table 2). The maps of the temperature increase under RCP2.6 and RCP8.5 in the near and far future (Figure 6), as simulated by the GCM-RCM ensemble, are very similar in their magnitude to the corresponding maps of Stagl and Hattermann [15], as simulated by a GCM ensemble. However, our maps show a much stronger spatial differentiation, especially in mountainous terrain, due to the higher spatial resolution of the RCMs used in this study. Especially for RCP8.5 (far future), we found a relatively stronger temperature increase in the mountains compared to the remaining basin. These findings are in agreement with a widely recognized phenomenon of elevation-dependent warming, according to which mountain areas are subject to (in most cases) stronger or weaker warming trends than the surrounding lowlands [58]. This is due to the varying responses of different land covers to climate change and to the snow-albedo feedback, which accelerates warming where snow cover retreats, and thus exposes surfaces of lower albedo [58].

Concerning precipitation, our results show an increase in the basin-wide, mean annual precipitation for both RCP2.6 and RCP8.5 in the near and far future (Figure 3 and Table 2), which may be surprising at first glance. This finding is very much in line with the results of Bisselink et al. [59], who found a considerable increase in precipitation under RCP8.5 in the far future over large parts of the Upper and Middle Danube, and a large uncertainty of precipitation development in the Lower Danube, as simulated by a GCM-RCM ensemble. In fact, we find that the variability of future annual precipitation is considerable, especially for the Lower Danube (Figures 3 and 4). This may add some degree of uncertainty to the precipitation projections.

Especially interesting is the future development of precipitation seasonality in the DRB. We find a general trend of winter precipitation increase, which is most pronounced in the Upper and Middle Danube, and a general trend of summer precipitation decrease, which is most pronounced in the Middle and Lower Danube (Figure 5). This is in line with the findings of Stagl and Hattermann [11] (ENSEMBLES near future) and Bisselink et al. [59] (RCP8.5, far future), who also identified more winter rainfall and less summer rainfall in a roughly north-western to south-eastern gradient in their future simulations in the DRB. Our findings are also in accordance with the key trends summarized by the International Commission for the Protection of the Danube River (ICPDR) [9], according to which wet regions will become wetter, dry regions will become drier and a seasonality shift will occur toward wetter winters and drier summers. In our simulations, the increase in winter precipitation outweighs the decrease in summer precipitation in the annual budget so that the mean annual precipitation on spatial average increases both basin-wide and over each of the different sub regions, namely the Upper, Middle and Lower Danube (Table 2).

This development can be traced back to the geographical location of the DRB in the transition zone between the humid continental climate dominated by the westerlies in the northern part, and the Mediterranean-influenced climate in the southern part, for each of which diverging climate change effects and impacts are predicted. According to IPCC-AR5 and AR6, (winter) rainfall is projected to increase for the DRB's northern part of humid continental climate, whereas (summer) rainfall is projected to decrease in the DRB's southern part of Mediterranean-influenced climate, leading to more frequent and intense summer droughts [6]. Case studies for the Mediterranean region find a very clear warming trend with, at the same time, equivocal trends for precipitation [60–62]. Between these two climate zones, and thus in large parts of the DRB, there is a broad transition zone, where even the sign of future precipitation change (positive or negative) is of high uncertainty [6].

4.1.2. Trends of Soil Water Content, Plant Water Stress and Snow Water Equivalent

Our simulation results for the soil water content are a direct consequence of the temperature and precipitation developments in the basin and hold direct implications for plant water stress. Although our results show an increase in the annual precipitation for both scenarios and future periods, the soil water content tends to decrease for both scenarios and future periods during the summer and autumn season in the Upper and Middle Danube, and during the whole year in the Lower Danube, especially for RCP8.5 (near and far future, Figure 8). Our projected strong increase in the winter MQ for the three gauges draining the Upper, Middle and Lower Danube (Figure 12) suggests that soil water storage does not benefit much from the increase in (especially winter) rainfall. In fact, much of this surplus water seems to be directly passed on to runoff formation without effectively refilling soil water storage (Figure 9). Furthermore, multiple interacting processes lead to the decrease in the soil water content, particularly in the late summer and early autumn: decreasing precipitation in the Middle and Lower Danube, which goes in line with decreasing cloud cover and increasing shortwave radiation input, the highest temperature (which increases particularly strongly for RCP8.5, far future), and thus, the highest saturation deficit, amplifying evapotranspiration. Collectively, these processes cause the soil water storage to continuously be drained by vegetation during the summer. By the late summer and early autumn, the soil water content drops to its seasonally lowest level due to high evapotranspiration rates, which are particularly observed for agricultural crops and forests during their most active phases in the vegetation period. Therefore, the projected decreasing soil water contents are particularly noticeable in agricultural or forested areas (Figure 9).

Directly related to the decreasing soil water content, we find an increasing tendency for plant water stress in the summer and early autumn months, which becomes most pronounced for RCP8.5 (far future) (Figures 8 and 10). During summer, agricultural crops are particularly affected by water stress due to high (potential) transpiration rates. This development will most likely increase the pressure on expanding irrigation of staple crops throughout the Middle and Lower Danube, with adverse consequences on the volume of water available for runoff. During the late summer and autumn, deciduous and coniferous forests are affected by heavy water stress after they have drained the soil water storages through strong transpiration over the summer. This will most likely increase the risk of forest fires in the future.

Moreover, we find a severe basin-wide reduction in the SWE for all scenarios and future periods, but most significantly for RCP8.5 (far future) (Figures 8 and 11). Most likely, this is attributable to the strong increase in (winter) temperature for all scenarios, especially for RCP8.5 (far future). Here, we find a stronger relative increase in winter temperatures in the mountains than in the remaining basin, leading to a reduced and delayed snow cover formation of likely shallower snow thicknesses and covers, as well as to an earlier and accelerated snowmelt. Consequently, the contribution of water from snowmelt might become increasingly insignificant for the runoff dynamics in mountain watersheds and for the water supply within the DRB, leading to a shift of river regimes from snow- to precipitation-dominated.

4.1.3. Trends of Discharge

Our projections of precipitation are directly reflected in the projections of discharge. Our results show a tendential increase in the mean annual MQ in the Upper Danube and the mountain watersheds of the Middle Danube in all scenarios and a decrease in the mean annual MQ in the lowland watersheds of the Middle and Lower Danube for RCP2.6 (near future) and RCP8.5 (near and far future) (Table 3). Following the precipitation trends with some time lag, we see a tendency of an increasing winter MQ (especially in the Upper and Middle Danube) and a decreasing summer and autumn MQ (especially in the Middle and Lower Danube) (Table 3 and Figures 12 and 13). Our results show less severe decreases in the mean annual and seasonal MQ across the sub-basins than the results of Stagl and Hattermann [15], who simulated stronger decreases in the MQ across the sub-basins and seasons (except for the Upper Danube and the mountain watersheds of the Middle Danube, where the winter MQ increases). Our more moderate MQ decreases might be traced back to a generally wetter precipitation development, especially in winter. Our results are more in line with the findings of Bisselink et al. [59], who showed a basin-wide winter MQ increase for a 2 °C warmed-up climate and widespread uncertainties for the development in the other seasons. However, the spatial patterns of the MQ development in our maps for RCP2.6 and RCP8.5 in the near and far future (Figure 13) are quite similar to the corresponding maps of Stagl and Hattermann [15] and Bisselink et al. [59].

Our findings on low flows along the Danube main course, which become less frequent along the Upper Danube River and more frequent along the Middle and Lower Danube River (Figure 14), are only partly in line with the general tendency of the results of Stagl and Hattermann [15]. Instead, they projected decreasing low flow levels in the whole Danube for RCP8.5 (far future). In contrast, Bisselink et al. [59] projected rising low flow levels in the winter season for the whole basin, but indicated major uncertainties for the rest of the year (RCP8.5 far future; although here, the authors included the effects of changing land use and water demand in addition to climate change).

Our findings on more frequent high flows all along the Danube main course (Figure 14) confirm a general conclusion in hydrological climate change impact research that extremes are likely to become more frequent and more extreme with future climate change. However, our results contradict the results of Stagl and Hattermann [15], who projected decreasing high flow levels for the whole Danube for all scenarios. In contrast, Bisselink et al. [59] projected rising high flow levels in the winter half for the whole basin, with again, major uncertainties for the rest of the year (RCP8.5 far future).

Our results on more frequent low flows along the Middle and Lower Danube River have direct implications on the future seasonal navigability of rivers. This is of special interest for sectors such as the shipping industry, which are particularly susceptible to low flows. However, the increasing winter discharges expected from our results, together with the large installed hydropower plants, e.g., at the Iron Gate, might have the potential to stabilize future renewable energy systems in the region by increasing hydropower production in winter, when solar potentials are and will remain low.

4.2. Sources of Uncertainties

As models provide a simplified representation of reality, modelling studies are inevitably subject to uncertainties. Uncertainties in hydrological modelling studies on climate change impacts on water resources typically arise from (amongst others) the GCMs, the RCMs and the hydrological model itself [63]. Uncertainties due to the hydrological model are related to the model's structure, the parametrization and the input data used [63].

Uncertainties due to the climate models are related to the many degrees of freedom in the modelling of feedbacks between the atmosphere and the hydrosphere [64]. In climate change impact studies, the influence of the sources of uncertainties varies according to the future period considered. In the near future (roughly until the middle of the 21st century), uncertainties due to the climate models may overweigh, whereas in the far future (roughly until the end of the 21st century), uncertainties due to the selected RCP scenarios may be more influential [15,63]. In general, uncertainties arising from the climate models are viewed to override uncertainties arising from the hydrological model [64].

Strategies to reduce uncertainties arising from the hydrological model imply a proper validation of the model, which is a prerequisite for hydrological climate change impact modelling [63]. We used the fully validated PROMET model setup for the DRB described in Probst and Mauser [23]. Here, we took full advantage of the physical consistency of PROMET, which—assuming that hydrological processes will remain unchanged under climate change—should equally predict both the present and future states of hydrological systems [22]. Therefore, we did not calibrate the hydrological model using historical discharges to avoid distortions of the model's predictive power, but rather we used a comprehensive and proven parametrization based on measurements and values within the literature [22,23].

Strategies to reduce climate model uncertainties imply a wise selection of appropriate GCM-RCM drivers and their bias correction. We chose a selected ensemble of EURO-CORDEX climate change projections, consisting of three GCM-RCM combinations, which meet the criterion of realistic model dynamics in the mountain watersheds of Bavaria [50], in the entire Alpine region [51], in the Carpathians [53] and in the Pannonian Basin [52]. Especially the Bavarian watersheds completely cover the Upper Danube, with a significant part located in the Alps. We argue that it is of particular importance to represent the hydrological processes and runoff dynamics in these mountain (head) watersheds as correctly as feasible. However, correctly representing the atmospheric dynamics and small-scale circulation patterns in mountains such as the Alps is challenging, and mountain hydrology is especially sensitive to biases introduced by the meteorological inputs such as temperature or precipitation.

Among the GCM-RCM combinations that were approved as plausible and used in this study, no GCM-RCM combination is completely bias-free. Kotlarski et al. [54] found a cold bias of temperature for the reanalysis-driven RCMs RCA4 and RACMO across Europe during winter and summer (at least, winter cold biases are quite common among RCMs) and a cold bias exceeding $-3\ °C$ in the winter temperature across the Alpine ridge for RCA4 and RACMO. Additionally, RCA4 typically features wet precipitation biases across Europe during the winter season [54]. However, since these biases are comparatively small and occur uniformly throughout the year without compromising the seasonal dynamics, this can be compensated with a bias correction [50]. The hindcast of temperature, precipitation and discharge in the DRB (Figure 2) demonstrates the effectiveness of the applied bias correction.

5. Conclusions

In this study, we performed a detailed hydrological climate change impact modelling study on different components of the hydrological cycle in the DRB, for which the future developments of temperature and precipitation are the main drivers. For this, we analyzed the future projections of temperature and different components of the water cycle, namely precipitation, soil water content, snow water equivalent and river discharge, as well as plant water stress, using the mechanistic hydrological model PROMET. The PROMET model was driven by a selected ensemble of EURO-CORDEX GCM-RCM combinations under the emission scenarios RCP2.6 and RCP8.5 for the near future (2031–2060) and the far future (2071–2100). The results were compared to the GCM-RCM-driven simulation results of the historical reference period (1971–2000).

The basin-wide mean annual temperature is projected to rise considerably in the DRB (RCP2.6: +1.2 °C in near and far future; RCP8.5: +2.2 °C and +4.3 °C in near and far future) with the Lower Danube and the mountain regions experiencing the strongest temperature rise. The strongest seasonal temperature rise is projected in the winter and summer season. The basin-wide mean annual precipitation is projected to slightly increase (RCP2.6: +4.5% and +7.2% in near and far future; RCP8.5: +4.9% and +5.9% in near and far future), although strong variability can be observed. The rise of the mean annual precipitation is the combined effect of a temperature rise and a significant shift in precipitation seasonality with rising winter rainfall and declining summer rainfall, with the winter surplus slightly over-compensating for the summer loss. This seasonality shift follows a north-western to south-eastern gradient, with the highest winter rainfall increase occurring in the Upper Danube and the highest summer rainfall decrease occurring in the Lower Danube. For RCP8.5, this development intensifies in comparison to RCP2.6.

However, increasing precipitation is not capable to refill the soil water storage in winter, as much of the water goes directly into runoff. Soil water content is projected to decrease particularly in summer and autumn, due to the combined effect of decreasing precipitation, increasing temperature and thus, increasing saturation deficits. Therefore, soil water storage is depleted by the amplified evapotranspiration, particularly by forests, grasslands and crops, which increases the plant water stress risk. The role of snow water storage for water resources in the DRB, however, is declining sharply, due to the rising winter temperatures, especially in the mountains.

The mean annual MQ at the Danube outlet is projected to slightly rise (RCP2.6: +2.8% and +10.7% in near and far future; RCP8.5: +3.1% and +5.2% in near and far future), but the picture varies for different sub-basins and seasons. Generally, the winter MQ tends to rise in the Upper Danube and the mountain watersheds of the Middle Danube (e.g., Drava, Sava), whereas the summer and autumn MQ tends to fall in the lowland and low mountain watersheds of the Middle Danube (e.g., Mures, Tisza) and the Lower Danube (e.g., Siret). This holds direct consequences for the risk of low flows in the DRB, which is of special interest along the main course of the Danube for navigability reasons. Along the Upper and Middle Danube River (gauges Achleiten and Bezdan), low flows become less frequent in an average year, whereas along the Middle and Lower Danube River (gauges Iron Gate and Ceatal Izmail), low flows become more frequent in an average year. In the future, this development is expected to hamper shipping along the Lower Danube River on more days of the year, presumably during dry episodes in the summer and autumn seasons. In contrast, the risk of high flows is projected to rise, as high flows become more frequent in an average year, all along the Danube main course.

Our results show that the hydrological impacts of climate change will not follow national borders, but are of a transboundary nature, connecting the downstream with the upstream regions of the Danube. Hence, they are highly relevant for formulating and updating state-of-the-art basin-wide climate change adaptation strategies for the DRB under the leadership of a coordinating water management entity, such as the ICPDR. Climate change trends in the DRB can indeed be broken down to a simplified tendency, according to which temperature increases strongly, wet regions become wetter and dry regions become drier [9]. More winter rainfall in the Upper Danube and less summer rainfall in the Middle and Lower Danube will lead to an increasing discrepancy between the largest water surplus and the largest water demand—which is mainly due to agriculture—in space and time. Thus, the present water competition between upstream and downstream countries and between different sectors is very likely to increase with intensifying water shortages.

Further research needs to be undertaken to assess the dynamics of the upstream–downstream and inter-sectoral water competition in the present, and under climate change scenarios, to provide important insights for future-oriented basin-wide water resource management in the DRB.

Author Contributions: Conceptualization, E.P. and W.M.; Methodology, E.P. and W.M.; Software, E.P. and W.M.; Validation, E.P.; Formal analysis, E.P.; Investigation, E.P.; Resources, W.M.; Data curation, E.P.; Writing—original draft preparation, E.P.; Writing—review and editing, E.P. and W.M.; Visualization, E.P.; Supervision, W.M.; Project administration, W.M.; Funding acquisition, W.M. All authors have read and agreed to the published version of the manuscript.

Funding: This research was funded by the Federal Ministry of Education and Research (BMBF) in the frame of the ViWA project within the research initiative "Global Resource Water (GRoW)", grant number 02WGR1423A.

Data Availability Statement: The data will be provided upon request.

Conflicts of Interest: The authors declare no conflict of interest.

References

1. Intergovernmental Panel on Climate Change (IPCC). Summary for Policymakers. In *Climate Change 2021: The Physical Science Basis. Contribution of Working Group I to the Sixth Assessment Report of the Intergovernmental Panel on Climate Change*; Masson-Delmotte, V., Zhai, P., Pirani, A., Connors, S.L., Péan, C., Berger, S., Caud, N., Chen, Y., Goldfarb, L., Gomis, M.I., et al., Eds.; Cambridge University Press: Cambridge, UK; New York, NY, USA, 2021; pp. 3–32.
2. United Nations Framework Convention on Climate Change (UNFCCC). *Adoption of the Paris Agreement, 21st Conference of the Parties*; United Nations: Paris, France, 2015.
3. Arias, P.A.; Bellouin, N.; Coppola, E.; Jones, R.G.; Krinner, G.; Marotzke, J.; Naik, V.; Palmer, M.D.; Plattner, G.-K.; Rogelj, J.; et al. Technical Summary. In *Climate Change 2021: The Physical Science Basis. Contribution of Working Group I to the Sixth Assessment Report of the Intergovernmental Panel on Climate Change*; Masson-Delmotte, V., Zhai, P., Pirani, A., Connors, S.L., Péan, C., Berger, S., Caud, N., Chen, Y., Goldfarb, L., Gomis, M.I., et al., Eds.; Cambridge University Press: Cambridge, UK; New York, NY, USA, 2021; pp. 33–144.
4. Doblas-Reyes, F.J.; Sörensson, A.A.; Almazroui, M.; Dosio, A.; Gutowski, W.J.; Haarsma, R.; Hamdi, R.; Hewitson, B.; Kwon, W.-T.; Lamptey, B.L.; et al. Linking Global to Regional Climate Change. In *Climate Change 2021: The Physical Science Basis. Contribution of Working Group I to the Sixth Assessment Report of the Intergovernmental Panel on Climate Change*; Masson-Delmotte, V., Zhai, P., Pirani, A., Connors, S.L., Péan, C., Berger, S., Caud, N., Chen, Y., Goldfarb, L., Gomis, M.I., et al., Eds.; Cambridge University Press: Cambridge, UK; New York, NY, USA, 2021; pp. 1363–1512.
5. Kjellström, E.; Nikulin, G.; Strandberg, G.; Christensen, O.B.; Jacob, D.; Keuler, K.; Lenderink, G.; van Meijgaard, E.; Schär, C.; Somot, S.; et al. European climate change at global mean temperature increases of 1.5 and 2 °C above pre-industrial conditions as simulated by the EURO-CORDEX regional climate models. *Earth Syst. Dynam.* **2018**, *9*, 459–478. [CrossRef]
6. Iturbide, M.; Fernández, J.; Gutiérrez, J.M.; Bedia, J.; Cimadevilla, E.; Díez-Sierra, J.; Manzanas, R.; Casanueva, A.; Baño-Medina, J.; Milovac, J.; et al. Repository Supporting the Implementation of FAIR Principles in the IPCC-WG1 Atlas. 2021. Available online: https://github.com/IPCC-WG1/Atlas (accessed on 14 November 2022).
7. Gutiérrez, J.M.; Jones, R.G.; Narisma, G.T.; Alves, L.M.; Amjad, M.; Gorodetskaya, I.V.; Grose, M.; Klutse, N.A.B.; Krakovska, S.; Li, J.; et al. *Atlas*; Cambridge University Press: Cambridge, UK; New York, NY, USA, 2021.
8. Dogaru, D.; Mauser, W.; Balteanu, D.; Krimly, T.; Lippert, C.; Sima, M.; Szolgay, J.; Kohnova, S.; Hanel, M.; Nikolova, M.; et al. Irrigation Water Use in the Danube Basin: Facts, Governance and Approach to Sustainability. *J. Environ. Geogr.* **2019**, *12*, 1–12. [CrossRef]
9. International Commission for the Protection of the Danube River (ICPDR). *Climate Change Adaptation Strategy*; ICPDR: Vienna, Austria, 2019.
10. ICPDR/LMU. *Revision and Update of the Danube Study. Integrating and Editing New Scientific Results in Climate Change Research and the Resulting Impacts on Water Availability to Revise the Existing Adaptation Strategies in the Danube River Basin*; Ludwig-Maximilians-Universität: Munich, Germany, 2018.
11. Stagl, J.C.; Hattermann, F.F. Impacts of Climate Change on the Hydrological Regime of the Danube River and Its Tributaries Using an Ensemble of Climate Scenarios. *Water* **2015**, *7*, 6139–6172. [CrossRef]
12. Krysanova, V.; Hattermann, F.; Wechsung, F. Development of the ecohydrological model SWIM for regional impact studies and vulnerability assessment. *Hydrol. Process.* **2005**, *19*, 763–783. [CrossRef]
13. Van der Linden, P.; Mitchell, J.F.B. *ENSEMBLES: Climate Change and Its Impacts: Summary of Research and Results from the ENSEMBLES Project*; Met Office Hadley Centre: Exeter, UK, 2009; p. 160.
14. Meehl, G.A.; Covey, C.; Delworth, T.; Latif, M.; McAvaney, B.; Mitchell, J.F.B.; Stouffer, R.J.; Taylor, K.E. THE WCRP CMIP3 Multimodel Dataset: A New Era in Climate Change Research. *Bull. Am. Meteorol. Soc.* **2007**, *88*, 1383–1394. [CrossRef]
15. Stagl, J.C.; Hattermann, F.F. Impacts of Climate Change on Riverine Ecosystems: Alterations of Ecologically Relevant Flow Dynamics in the Danube River and Its Major Tributaries. *Water* **2016**, *8*, 566. [CrossRef]
16. Taylor, K.E.; Stouffer, R.J.; Meehl, G.A. An Overview of CMIP5 and the Experiment Design. *Bull. Am. Meteorol. Soc.* **2012**, *93*, 485–498. [CrossRef]

17. Maraun, D.; Wetterhall, F.; Ireson, A.M.; Chandler, R.E.; Kendon, E.J.; Widmann, M.; Brienen, S.; Rust, H.W.; Sauter, T.; Themeßl, M.; et al. Precipitation downscaling under climate change: Recent developments to bridge the gap between dynamical models and the end user. *Rev. Geophys.* **2010**, *48*, 1–34. [CrossRef]
18. Bisselink, B.; Roo, A.d.; Bernhard, J.; Gelati, E. Future projections of water scarcity in the Danube River Basin due to land use, water demand and climate change. *J. Environ. Geogr.* **2018**, *11*, 25–36. [CrossRef]
19. De Roo, A.P.J.; Wesseling, C.G.; Van Deursen, W.P.A. Physically based river basin modelling within a GIS: The LISFLOOD model. *Hydrol. Process.* **2000**, *14*, 1981–1992. [CrossRef]
20. Van Der Knijff, J.M.; Younis, J.; De Roo, A.P.J. LISFLOOD: A GIS-based distributed model for river basin scale water balance and flood simulation. *Int. J. Geogr. Inf. Sci.* **2010**, *24*, 189–212. [CrossRef]
21. Jacob, D.; Petersen, J.; Eggert, B.; Alias, A.; Christensen, O.B.; Bouwer, L.M.; Braun, A.; Colette, A.; Déqué, M.; Georgievski, G.; et al. EURO-CORDEX: New high-resolution climate change projections for European impact research. *Reg. Environ. Change* **2014**, *14*, 563–578. [CrossRef]
22. Mauser, W.; Bach, H. PROMET—Large scale distributed hydrological modelling to study the impact of climate change on the water flows of mountain watersheds. *J. Hydrol.* **2009**, *376*, 362–377. [CrossRef]
23. Probst, E.; Mauser, W. Evaluation of ERA5 and WFDE5 forcing data for hydrological modelling and the impact of bias correction with regional climatologies: A case study in the Danube River Basin. *J. Hydrol. Reg. Stud.* **2022**, *40*, 101023. [CrossRef]
24. Intergovernmental Panel on Climate Change (IPCC). Summary for Policymakers. In *Climate Change 2014: Synthesis Report. Contribution of Working Groups I, II and III to the Fifth Assessment Report of the Intergovernmental Panel on Climate Change*; Core Writing Team, Pachauri, R.K., Meyer, L.A.E., Eds.; IPCC: Geneva, Switzerland, 2014; p. 151.
25. Jungwirth, M.; Haidvogl, G.; Hohensinner, S.; Waidbacher, H.; Zauner, G. *Österreichs Donau. Landschaft—Fisch—Geschichte*; Institut für Hydrobiologie u. Gewässermanagement, BOKU: Vienna, Austria, 2014; p. 420.
26. Schiller, H.; Miklós, D.; Sass, J. Chapter 2. The Danube River and its Basin Physical Characteristics, Water Regime and Water Balance. In *Hydrological Processes of the Danube River Basin*; Brilly, M., Ed.; Springer: Dordrecht, The Netherlands; Heidelberg, Germany; London, UK; New York, NY, USA, 2010; pp. 25–77.
27. Farr, T.G.; Rosen, P.A.; Caro, E.; Crippen, R.; Duren, R.; Hensley, S.; Kobrick, M.; Paller, M.; Rodriguez, E.; Roth, L.; et al. The Shuttle Radar Topography Mission. *Rev. Geophys.* **2007**, *45*, 1–33. [CrossRef]
28. Lehner, B.; Verdin, K.; Jarvis, A. New Global Hydrography Derived from Spaceborne Elevation Data. *Eos* **2008**, *89*, 93–94. [CrossRef]
29. Global Runoff Data Centre (GRDC). The Global Runoff Data Centre. 2019. Available online: https://grdc.bafg.de (accessed on 14 November 2022).
30. International Commission for the Protection of the Danube River (ICPDR). Danube River Basin Water Quality Database. 2021. Available online: https://www.icpdr.org/wq-db/ (accessed on 14 November 2022).
31. Esri. "World Imagery" [basemap]. Scale Not Given. "World Imagery", 8 November 2022. 2009. Available online: https://www.arcgis.com/home/item.html?id=10df2279f9684e4a9f6a7f08febac2a9 (accessed on 18 November 2022).
32. Beck, H.E.; Zimmermann, N.E.; McVicar, T.R.; Vergopolan, N.; Berg, A.; Wood, E.F. Present and future Köppen-Geiger climate classification maps at 1-km resolution. *Sci. Data* **2018**, *5*, 180214. [CrossRef]
33. Kovács, P. Chapter 5. Characterization of the Runoff Regime and Its Stability in the Danube Catchment. In *Hydrological Processes of the Danube River Basin*; Brilly, M., Ed.; Springer: Dordrecht, The Netherlands; Heidelberg, Germany; London, UK; New York, NY, USA, 2010; pp. 143–173.
34. Mauser, W.; Prasch, M.; Weidinger, R.; Stöber, S. GLOWA-Danube. In *Regional Assessment of Global Change Impacts: The Project GLOWA-Danube*; Mauser, W., Prasch, M., Eds.; Springer International Publishing: Cham, Germany, 2016; pp. 3–18.
35. Regionale Zusammenarbeit der Donauländer (RZD). *Die Donau und ihr Einzugsgebiet. Eine hydrologische Monographie. Teil 1: Texte*; Bayerisches Landesamt für Wasserwirtschaft: Munich, Germany, 1986; p. 377.
36. Hank, T.B.; Bach, H.; Mauser, W. Using a Remote Sensing-Supported Hydro-Agroecological Model for Field-Scale Simulation of Heterogeneous Crop Growth and Yield: Application for Wheat in Central Europe. *Remote Sens.* **2015**, *7*, 3934–3965. [CrossRef]
37. Mauser, W.; Klepper, G.; Zabel, F.; Delzeit, R.; Hank, T.; Putzenlechner, B.; Calzadilla, A. Global biomass production potentials exceed expected future demand without the need for cropland expansion. *Nat. Commun.* **2015**, *6*, 8946. [CrossRef]
38. Zabel, F.; Putzenlechner, B.; Mauser, W. Global Agricultural Land Resources—A High Resolution Suitability Evaluation and Its Perspectives until 2100 under Climate Change Conditions. *PLoS ONE* **2014**, *9*, e107522. [CrossRef]
39. Farquhar, G.D.; von Caemmerer, S.; Berry, J.A. A biochemical model of photosynthetic CO_2 assimilation in leaves of C_3 species. *Planta* **1980**, *149*, 78–90. [CrossRef]
40. Chen, D.-X.; Coughenour, M.B.; Knapp, A.K.; Owensby, C.E. Mathematical simulation of C_4 grass photosynthesis in ambient and elevated CO_2. *Ecol. Modell.* **1994**, *73*, 63–80. [CrossRef]
41. Ball, J.T.; Woodrow, I.E.; Berry, J.A. A model predicting stomatal conductance and its contribution to the control of photosynthesis under different environmental conditions. In *Progress in Photosynthesis Research. Proceedings of the VIIth International Congress on Photosynthesis, Providence, RI, USA, 10–15 August 1986*; Biggins, J., Ed.; Springer Science+Business Media: Dordrecht, The Netherlands, 1987; pp. 221–224.
42. Jarvis, P.G.; Morison, J.I.L. The control of transpiration and photosynthesis by stomata. In *Stomatal Physiology*; Jarvis, P.G., Mansfield, T.A., Eds.; Cambridge University Press: Cambridge, UK, 1981; pp. 247–279.

43. Cunge, J.A. On the Subject of a Flood Propagation Computation Method (Muskingum Method). *J. Hydraul. Res.* **1969**, *7*, 205–230. [CrossRef]
44. Todini, E. A mass conservative and water storage consistent variable parameter Muskingum-Cunge approach. *Hydrol. Earth Syst. Sci.* **2007**, *11*, 1645–1659. [CrossRef]
45. Marke, T.; Mauser, W.; Pfeiffer, A.; Zängl, G.; Jacob, D.; Strasser, U. Application of a hydrometeorological model chain to investigate the effect of global boundaries and downscaling on simulated river discharge. *Environ. Earth Sci.* **2014**, *71*, 4849–4868. [CrossRef]
46. FAO/IIASA/ISRIC/ISSCAS/JRC. Harmonized World Soil Database (Version 1.2). 2012. Available online: http://www.fao.org/soils-portal/soil-survey/soil-maps-and-databases/harmonized-world-soil-database-v12/en/ (accessed on 14 November 2022).
47. European Environmental Agency (EEA). CORINE Land Cover (CLC2012). 2012. Available online: https://land.copernicus.eu/pan-european/corine-land-cover/clc-2012 (accessed on 14 November 2022).
48. European Space Agency (ESA). Land Cover CCI Product User Guide Version 2 Technical Report 2015. Available online: http://maps.elie.ucl.ac.be/CCI/viewer/download/ESACCI-LC-Ph2-PUGv2_2.0.pdf (accessed on 14 November 2022).
49. EUROSTAT. Crops by Classes of Utilised Agricultural Area in Number of Farms and Hectare by NUTS 2 Regions (ef_lus_allcrops). 2013. Available online: https://ec.europa.eu/eurostat/web/agriculture/data/database (accessed on 14 November 2022).
50. Zier, C.; Müller, C.; Komischke, H.; Steinbauer, A.; Bäse, F. *Das Bayerische Klimaprojektionsensemble—Audit und Ensemblebildung*; Bayerisches Landesamt für Umwelt (LfU): Augsburg, Germany, 2020; p. 55.
51. Smiatek, G.; Kunstmann, H.; Senatore, A. EURO-CORDEX regional climate model analysis for the Greater Alpine Region: Performance and expected future change. *J. Geophys. Res. Atmos.* **2016**, *121*, 7710–7728. [CrossRef]
52. Lazić, I.; Tošić, M.; Djurdjević, V. Verification of the EURO-CORDEX RCM Historical Run Results over the Pannonian Basin for the Summer Season. *Atmosphere* **2021**, *12*, 714. [CrossRef]
53. Torma, C.Z. Detailed validation of EURO-CORDEX and Med-CORDEX regional climate model ensembles over the Carpathian Region. *Idojaras* **2019**, *123*, 217–240. [CrossRef]
54. Kotlarski, S.; Keuler, K.; Christensen, O.B.; Colette, A.; Déqué, M.; Gobiet, A.; Goergen, K.; Jacob, D.; Lüthi, D.; van Meijgaard, E.; et al. Regional climate modeling on European scales: A joint standard evaluation of the EURO-CORDEX RCM ensemble. *Geosci. Model Dev.* **2014**, *7*, 1297–1333. [CrossRef]
55. Fick, S.E.; Hijmans, R.J. WorldClim 2: New 1-km spatial resolution climate surfaces for global land areas. *Int. J. Climatol.* **2017**, *37*, 4302–4315. [CrossRef]
56. Früh, B.; Schipper, H.; Pfeiffer, A.; Wirth, V. A pragmatic approach for downscaling precipitation in alpine-scale complex terrain. *Meteorol. Z.* **2006**, *15*, 631–646. [CrossRef]
57. Frei, C.; Schär, C. A precipitation climatology of the Alps from high-resolution rain-gauge observations. *Int. J. Climatol.* **1998**, *18*, 873–900. [CrossRef]
58. Pepin, N.C.; Arnone, E.; Gobiet, A.; Haslinger, K.; Kotlarski, S.; Notarnicola, C.; Palazzi, E.; Seibert, P.; Serafin, S.; Schöner, W.; et al. Climate Changes and Their Elevational Patterns in the Mountains of the World. *Rev. Geophys.* **2022**, *60*, e2020RG000730. [CrossRef]
59. Bisselink, B.; Bernhard, J.; Gelati, E.; Jacobs, C.; Adamovic, M.; Mentaschi, L.; Lavalle, C.; de Roo, A. *Impact of a Changing Climate, Land Use, and Water Usage on Water Resources in the Danube River Basin*; EUR 29228 EN; Publications Office of the European Union: Luxembourg, Luxembourg, 2018. [CrossRef]
60. Todaro, V.; D'Oria, M.; Secci, D.; Zanini, A.; Tanda, M.G. Climate Change over the Mediterranean Region: Local Temperature and Precipitation Variations at Five Pilot Sites. *Water* **2022**, *14*, 2499. [CrossRef]
61. Mersin, D.; Tayfur, G.; Vaheddoost, B.; Safari, M.J.S. Historical Trends Associated with Annual Temperature and Precipitation in Aegean Turkey, Where Are We Heading? *Sustainability* **2022**, *14*, 13380. [CrossRef]
62. Kastridis, A.; Kamperidou, V.; Stathis, D. Dendroclimatological Analysis of Fir (A. borisii-regis) in Greece in the frame of Climate Change Investigation. *Forests* **2022**, *13*, 879. [CrossRef]
63. Kundzewicz, Z.W.; Krysanova, V.; Benestad, R.E.; Hov, Ø.; Piniewski, M.; Otto, I.M. Uncertainty in climate change impacts on water resources. *Environ. Sci. Policy* **2018**, *79*, 1–8. [CrossRef]
64. Vetter, T.; Huang, S.; Aich, V.; Yang, T.; Wang, X.; Krysanova, V.; Hattermann, F. Multi-model climate impact assessment and intercomparison for three large-scale river basins on three continents. *Earth Syst. Dynam.* **2015**, *6*, 17–43. [CrossRef]

Disclaimer/Publisher's Note: The statements, opinions and data contained in all publications are solely those of the individual author(s) and contributor(s) and not of MDPI and/or the editor(s). MDPI and/or the editor(s) disclaim responsibility for any injury to people or property resulting from any ideas, methods, instructions or products referred to in the content.

Article

Novel Approaches for the Empirical Assessment of Evapotranspiration over the Mediterranean Region

Ali Uzunlar and Muhammet Omer Dis *

Department of Civil Engineering, Faculty of Engineering and Architecture, Kahramanmaras Sutcu Imam University, Onikisubat, Kahramanmaras 46050, Türkiye
* Correspondence: momerdis@ksu.edu.tr; Tel.: +90-344-300-1664

Abstract: The hydrological cycle should be scrutinized and investigated under recent climate change scenarios to ensure global water management and to increase its utilization. Although the FAO proposed the use of the Penman–Monteith (PM) equation worldwide to predict evapotranspiration (ET), which is one of the most crucial components of the hydrological cycle, its complexity and time-consuming nature, have led researchers to examine alternative methods. In this study, the performances of numerous temperature-driven ET methods were examined relative to the PM using daily climatic parameters from central stations in 11 districts of the Kahramanmaras province. Owing to its geographical location and other influencing factors, the city has a degraded Mediterranean climate with varying elevation gradients, while its meteorological patterns (i.e., temperature and precipitation) deviate from those of the main Mediterranean climate. A separate evaluation was performed via ten different statistical metrics, and spatiotemporal ET variability was reported for the districts. This study revealed that factors such as altitude, terrain features, slope, aspect geography, solar radiation, and climatic conditions significantly impact capturing reference values, in addition to temperature. Moreover, an assessment was conducted in the region to evaluate the effect of modified ET formulae on simulations. It can be drawn as a general conclusion that the Hargreaves–Samani and modified Blaney–Criddle techniques can be utilized as alternatives to PM in estimating ET, while the Schendel method exhibited the lowest performance throughout Kahramanmaras.

Keywords: empirical equations; evapotranspiration; modification; Penman–Monteith

Citation: Uzunlar, A.; Dis, M.O. Novel Approaches for the Empirical Assessment of Evapotranspiration over the Mediterranean Region. *Water* **2024**, *16*, 507. https://doi.org/10.3390/w16030507

Academic Editors: Renato Morbidelli, Sonia Raquel Gámiz-Fortis and Matilde García-Valdecasas Ojeda

Received: 13 December 2023
Revised: 30 January 2024
Accepted: 1 February 2024
Published: 5 February 2024

Copyright: © 2024 by the authors. Licensee MDPI, Basel, Switzerland. This article is an open access article distributed under the terms and conditions of the Creative Commons Attribution (CC BY) license (https://creativecommons.org/licenses/by/4.0/).

1. Introduction

Evapotranspiration (ET), a term used in environmental science, hydrology, and agriculture to describe the process by which water is transferred from the Earth's surface to the atmosphere, has become one of the most complex components of the hydrological cycle as it depends on multi-climatological parameters and their interactions [1–3]. This important phenomenon is a component of the Earth's water cycle, as it influences the movement and distribution of water in the atmosphere, the availability of water resources, and the climate. It is necessary to understand the connections between evapotranspiration and ecosystem type in response to climate change [4,5]. In the long term, the water that is directly available for human consumption and control is what separates evapotranspiration from continental precipitation. Therefore, quantitative ET knowledge is necessary for quantitative assessments of water resources and the impacts of climate and land-use change on these resources [6–8].

To better comprehend this complex formation and determine the amount of water lost through ET, definitions have been made based on a variety of assumptions [9–13]. ET approaches can be divided into three categories: hydrological and water balance methods, analytical methods based on climate variables, and empirical methods [14]. The water balance technique and basin hydrology for measuring or indirectly determining the ET value consists of sampling soil water change and lysimeter testing. Since this is primarily a physically based method, its use in climate change assessment is limited to the laboratory [15,16]. The second approach, known as the micrometeorological method, uses a scientific understanding of the physics of evapotranspiration. Mathematical relationships have been developed to describe these processes through two fundamental climatological components: energy balance and mass transport [10,17–19]. The third method centers on developing empirical relationships that are often site-specific and based on regional climatological conditions and often used in regression analysis. These techniques are often calibrated by correlating experimental predictions with observed data [20–22].

ET can be computed by the aerodynamic approach when energy is unlimited and by the energy balance method when vapor movement (mass transfer) is limitless. Normally, however, both of these factors are limiting; therefore, a combination of the two methods is required [17]. The Penman–Monteith (PM) technique, recommended worldwide by the Food and Agriculture Organization (FAO), is one of the most universally used energy balance and mass transfer-based methods to compute evaporation on terrestrial surfaces [14]. The equation developed by Penman [23] to calculate evaporation from an open water surface was modified by Monteith [24] by including canopy conductance to represent the ET rate from a vegetative surface. In the 1980s, the method was developed and presented in detail by combining the canopy and soil evaporation [25]. Allen et al. [1] updated the PM approach by adapting the ET values to the reference grass plant with constant albedo and surface resistance and proposed it as the FAO-56 Penman–Monteith equation. The PM method, recommended by the FAO for applications worldwide, has been analyzed in various regions for decades and has yielded sufficiently accurate results [26–30].

The fact that the FAO-56 PM equation requires a large number of meteorological data (i.e., temperature, humidity, radiation, and wind speed) makes the solution of the equation difficult, and obtaining such data is not always possible. Although Allen et al. [1] expressed the solvability of the equation using auxiliary formulae based on temperature data for the PM approach, many alternative empirical methods (i.e., temperature-based, radiation-based, and a combination of them) have been investigated for use in cases where sufficient data are not available [31–35]. For instance, Tabari et al. [34] evaluated the performance of thirty-one alternative empirical methods using meteorological data obtained under humid conditions in Northern Iran and the PM method as a reference. Analysis of the data revealed that, in general, the best results were attained with the Blaney–Criddle and Hargreaves methods compared to the PM equation; mass transfer-based approaches underestimated ET, whereas overestimations were more dominant in temperature-based and radiation-based methods. Sarlak and Bagcaci [35] evaluated the performances of six empirical ET approaches, namely Blaney–Criddle, Jensen–Haise, Makking, Turc, Priestley–Taylor, and Hargreaves–Samani, compared to the PM method using daily meteorological data from five stations in Konya Closed Basin. They concluded that in the absence of daily observation data, the Turc, Hargreaves–Samani, and Priestley–Taylor techniques, which require less data, can be used as an alternative to PM. Similarly, Song et al. [36] applied twelve different ET estimation methods relative to the PM method recommended by the FAO in northeast China, which they divided into eight sub-regions according to the climate and land types. In the study, temperature-based (Blaney–Criddle, Thornthwaite, Romanenko-1, and Romanenko-2), radiation-based (Hargreaves–Samani, Turc, Makking, and H-Makking), and combination methods (Linacre, simplified Penman–Monteith, Valiantzas-1, and Valiantzas-2) were employed using data over 126 stations for more than half a century. H-Makking and Valiantzas-2 approaches can be considerable as alternative methods in agricultural areas in northern regions of China, while Valiantzas-2, Romanenko-2, and H-Makking methods are more suitable during crop growth periods. Although the results varied regionally and seasonally, the temperature was the most sensitive parameter for estimating ET values. Furthermore, it is claimed that the Turc approach constantly tends to deviate negatively whereas the Hargreaves–Samani method produces notable biases.

Researchers have performed local calibrations as well as modifications of these empirical methods for use in various regions under differing conditions, even though many of the methods developed for a particular location are widely utilized worldwide. For instance, Cobaner et al. [37] compared the reference evapotranspiration values determined using the PM technique in the Mediterranean region with the values derived using calibrated Hargreaves–Samani equations, which require less data. According to this study, the ET values produced by the Hargreaves–Samani equations calibrated with minimum and maximum temperatures, as well as minimum and maximum humidity data, were close to the values calculated using the PM formula. They concluded that the Hargreaves–Samani equation calibrated with the maximum temperature is better than those calibrated with other meteorological data. Similarly, monthly ET values were computed using the modified Blaney–Criddle and Hargreaves–Samani equations, obtaining data from three stations located in the semi-arid climate regions of Pakistan and compared with PM-driven ET values [38]. Overall, it was stated in the study that both equations overestimated PM-driven ET values, although the findings of the Hargreaves–Samani method were superior to those of the modified Blaney–Criddle method. In another example, the effect of modified approaches on ET was investigated in Eastern Turkiye [14]. Study results showed that the modified Hargreaves–Samani approach formed by the constant values in the Hargreaves–Samani equation revealed better results than the Hargreaves–Samani equation over the region, while altitude-based modified Hargreaves–Samani technique has the lowest correlation results among the other methods in the study. Additionally, they concluded that the modification of the Blaney–Criddle formula increased the performance relative to the Blaney–Criddle equation.

It is necessary to conduct studies to ascertain the suitability and accuracy of empirical techniques for different regions that have emerged from investigations conducted in local areas with soil structure as well as certain climatic and environmental characteristics. Several empirical approaches can be used with varying degrees of success, owing to the unique characteristics of each place and the availability of a limited number of measurable climatic and environmental factors. Hydrological studies are significant for the effective management of water resources in Kahramanmaras, given their considerable water potential. Unfortunately, ET measurements are not available for Turkiye, and these values vary spatiotemporally. However, no comprehensive ET study has been conducted in the city, and a district-based ET study was implemented for the first time in the region. Additionally, the city has complex climate zones, varying vegetation cover, and uneven distribution of elevation gradient.

In this study, the quantitative determination of ET, one of the most significant water losses of the hydrological cycle, can contribute to studies in this field and will also play an important role in various hydrological planning studies, such as agricultural irrigation projects and basin management, to be carried out in the study area. Data-scarce ET time series were assessed utilizing ten climatological-driven ET approaches relative to the PM method over eleven districts of Kahramanmaras region, Turkiye. In this study, the performances of ten empirical evapotranspiration approaches, namely Blaney–Criddle (BC), modified Blaney–Criddle (BC_M), Hamon (HM), modified Hamon (HM_M), Hargreaves–Samani (HS), Kharrufa (KH), Romanenko (RM), Schendel (SC), Thornthwaite (TH), and Penman–Monteith at 0.5 m ($PM_{0.5}$), were evaluated at daily and monthly temporal resolution in Kahramanmaras province and its eleven districts and compared to the reference PM method. In addition, the effect of modified techniques on evapotranspiration estimates was investigated over the region.

2. Materials and Methods

The flow diagram used in the methodology is shown in Figure 1. In this study, ET values were simulated using Penman–Monteith at 0.5 m ($PM_{0.5}$), Blaney–Criddle (BC), modified Blaney–Criddle (BC_M), Hamon (HM), modified Hamon (HM_M), Hargreaves–Samani (HS), Kharrufa (KH), Schendel (SC) methods, and the reference Penman–Monteith (PM) at daily temporal resolutions, whereas monthly and annual ET values were derived by taking the averages. The Thornthwaite (TH) and Romanenko (RM) techniques were utilized to compute monthly ET values and produce yearly ET values. Days with missing data and ET values equal to zero were excluded from computation.

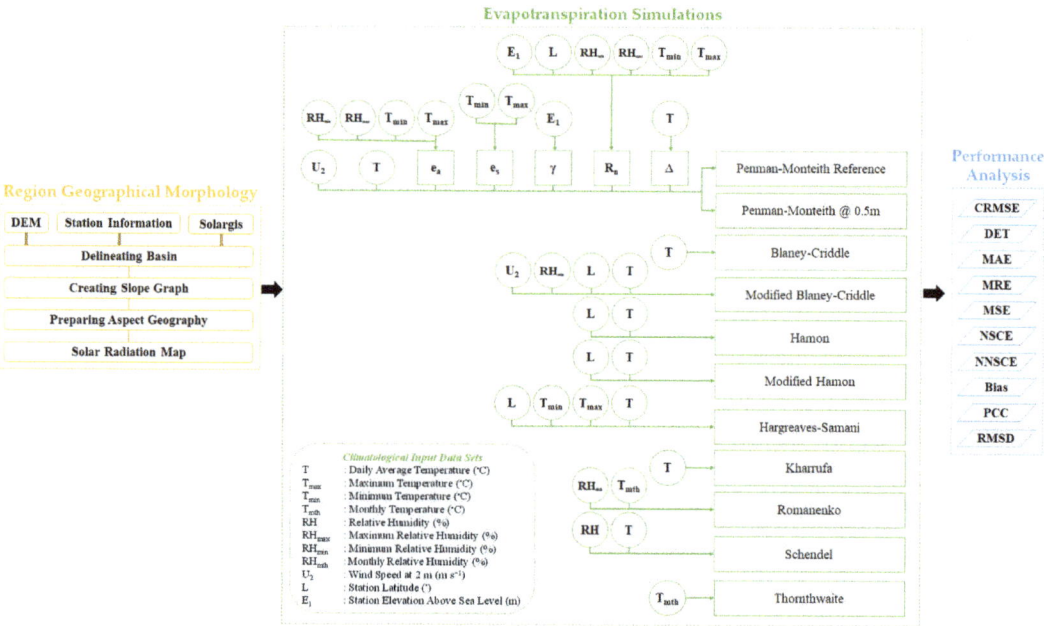

Figure 1. Flow diagram of the ET analysis methodology over Kahramanmaras.

The following statistical indices were used to assess the performance of the methods: centered root mean square error (CRMSE), determination coefficient (DET), mean absolute error (MAE), mean relative error (MRE), mean squared error (MSE), Nash–Sutcliffe efficiency coefficient (NSCE), normalized Nash–Sutcliffe efficiency (NNSCE), percentage error (Bias), Pearson's correlation coefficient (PCC), and root mean square error (RMSD).

2.1. Study Area and Data Sets

Kahramanmaras is the eleventh largest city in Turkiye, with a surface area of 14,346 km^2, and is situated between 37–39 northern parallels and 36–38 eastern meridians. The province's northern regions are quite mountainous, with landforms mostly consisting of mountains that are extensions of the Taurus Mountains in the southeast and the depressions that separate them (Figure 2). Digital elevation model (DEM) data for the study area were obtained from the US Geological Survey (USGS) website [39]. After adjusting the required projections and coordinate system as well as delineating the area, the acquired data were examined using the Arc-GIS program, a scalable integrated geographic information system software developed by ESRI. Figure 2 was obtained by processing the coordinate information of the relevant stations shown in Table 1 into the program. As can be seen from the figure, the altitude of the study area ranges from 130 to 3075 m, and the regions where stations S3 and S10 are located have the highest altitude, while the areas where stations S2 and S11 are located have the lowest elevation. Although the altitude values in the territories containing stations S4 and S9 are relatively close to one another, the high-elevation difference between the regions where stations S1 and S3 are located can be seen in Figure 2.

Kahramanmaras is located at the junction of Eastern Anatolia, Southeastern Anatolia, Central Anatolia, and geographical Mediterranean regions. The city has climate characteristics closer to the "Degraded Mediterranean Climate" due to its geographical location and the influence of other factors. Nevertheless, its temperature and precipitation patterns deviate from those of the main Mediterranean climate. In general, summers in the province are hot and dry, and winters are warm and rainy. The fact that the provincial territory is located in the transition area of the Mediterranean and Southeastern Anatolia regions has caused the climatic conditions in the city to differ. For instance, Andirin district is under the influence of maritime climate, and the Mediterranean climate is more dominant in the southern part of the city, while the northern part of the province experiences a more continental climate [21,40–42].

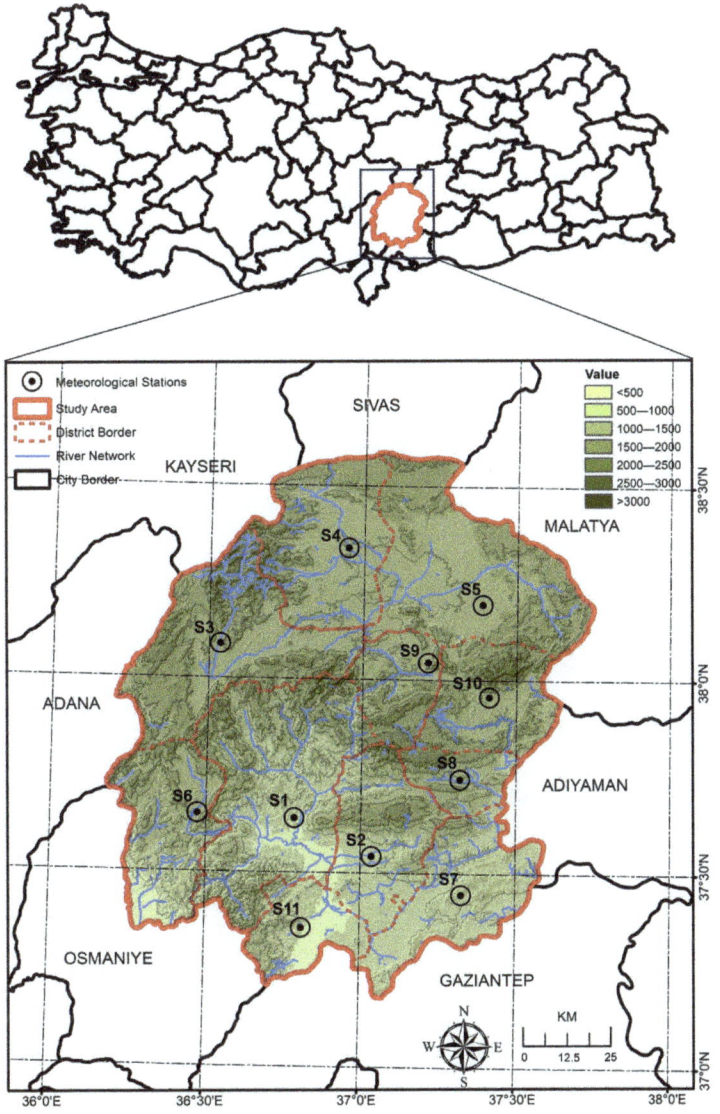

Figure 2. Map of Kahramanmaras and the location of the MGM meteorological stations.

Table 1. Details of the weather stations that were utilized in the research.

Station Number	Station Name	District	Coordinates	Elevation (m)	Available Time Period
S1	17255-Kahramanmaras	Onikisubat	37.58 36.92	572	01.01.2000–31.12.2021
S2	17256-Kahramanmaras Airport	Dulkadiroglu	37.54 36.97	525	01.01.2017–31.12.2021
S3	17866-Goksun	Goksun	38.02 36.48	1344	01.01.2000–31.12.2021
S4	17868-Afsin	Afsin	38.24 36.92	1230	01.01.2000–31.12.2021
S5	17870-Elbistan	Elbistan	38.20 37.20	1137	01.01.2000–31.12.2021
S6	18156-Andirin	Andirin	37.59 36.36	1108	01.01.2013–31.12.2021
S7	18157-Pazarcik	Pazarcik	37.47 37.24	787	01.01.2013–31.12.2021
S8	18279-Caglayancerit	Caglayancerit	37.75 37.37	1001	01.03.2014–31.12.2021
S9	18280-Ekinozu	Ekinozu	38.05 37.1872	1246	01.03.2014–31.12.2021
S10	18281-Nurhak	Nurhak	37.96 37.45	1368	01.03.2014–31.12.2021
S11	18282-Turkoglu	Turkoglu	37.38 36.84	535	01.03.2014–31.12.2021

Table 1 displays the details of the stations with data sets from various time periods that were generated using data obtained from the Turkish State Meteorological Service's (MGM) Meteorological Data Information Presentation and Sales System (MEVBIS). MEVBIS system is a project designed to archive and quickly present all meteorological observation data produced by the MGM after quality control and format conversion [43]. There are a total of 32 meteorological stations within the Kahramanmaras city borders, and some stations measure limited climatological data sets such as only temperature, wind speed, precipitation, or snow. Meteorological measurements have not been made at some stations since 1990 due to the changing city borders; some stations have a history of only a few years, with some regions gaining district status in 2012 with Kahramanmaras becoming a metropolitan city, and some do not have an uninterrupted continuous data set [21,44]. However, ET simulations, especially the Penman–Monteith equation, require numerous data sets, and obtaining such data is not always possible. Within the scope of this study, meteorological stations where the longest-term data were measured in each 11 districts were preferred. Meteorological stations' properties, such as their names, districts in which they are located, coordinates, and altitudes, are given in detail in Table 1. Using the PM simulations as a reference, the meteorological-driven ET predictions are evaluated for a period of 22 years (2000 through 2021) over the Onikisubat, Goksun, Afsin, and Elbistan regions; 9 years (2013 through 2021) over Andirin and Pazarcik counties; 8 years (2014 through 2021) over Caglayancerit, Ekinozu, Nurhak, and Turkoglu districts; and 5 years (2017 through 2021) over Dulkadiroglu (Table 1). Required data sets, namely average temperature (T, °C), maximum temperature (T_{max}, °C), minimum temperature (T_{min}, °C), average relative humidity (RH, %), maximum relative humidity (RH_{max}, %), minimum relative humidity (RH_{min}, %), average wind speed at 2 m height (u_2, m s^{-1}), and sunshine duration (n, hr) for each station, were acquired from the MEVBIS module at daily temporal resolution within the scope of this study. In this study, multi-paradigm numerical calculation software developed by MathWorks r2006a MATLAB program was utilized in a holistic sense with Microsoft 365 Excel program for manipulations on the obtained data sets, preparation of required time series, ET calculations, and graphing.

2.2. Evapotranspiration Estimation Methods

2.2.1. FAO Penman–Monteith Method (PM)

This study examined the applicability of various empirical approaches in the Kahramanmaras, where the widely known PM method was utilized as a reference in ET quantitative calculations. The method recommended by FAO for use globally is as in Equation (1) in its revised form by adapting the ET values with the reference grass plant with a constant albedo of 0.23, a surface resistance of 70 s m^{-1}, and adequately irrigated at a height of 0.12 m [1,23,24]. The ET_{PM} in the equation expresses the reference evapotranspiration in mm d^{-1}, and Δ (kPa °C^{-1}) denotes the slope of the vapor pressure curve at average air temperature and is calculated using Equations (2) and (3). In these equations, G symbolizes the soil heat flux density (MJ m^{-2} d^{-1}) and can be considered zero for daily calculations, T is the daily average air temperature (°C), u_2 denotes the wind speed at 2 m height (m s^{-1}), e_s emblematizes the saturated vapor pressure (kPa), e_a stands for actual vapor pressure (kPa), γ represents the psychrometric constant (kPa °C^{-1}), and e°(T) indicates the saturated vapor pressure (kPa) at air temperature T (°C).

$$ET_{PM} = \frac{0.408\Delta(R_n - G) + \gamma \frac{900}{T+273} u_2 (e_s - e_a)}{\Delta + \gamma(1 + 0.34 u_2)} \quad (1)$$

$$\Delta = \frac{4099 \, e°(T)}{(T + 237.3)^2} \quad (2)$$

$$e°(T) = 0.6108 \, exp\left(\frac{17.27 \, T}{T + 237.3}\right) \quad (3)$$

On the other hand, R_n (MJ m^{-2} d^{-1}) refers to the net radiation at the crop surface and is equal to the difference between the incoming net shortwave radiation (R_{ns}) and the outgoing net longwave radiation (R_{nl}). While R_{ns} is also known as net solar radiation and can be calculated using Equation (4), R_s is the part of solar radiation that is not reflected from the surface. The value 0.23 in the equation indicates the albedo coefficient for green grass surfaces. R_s (Solar radiation) can be derived as suggested by Hargreaves–Samani [45] and is shown in Equation (5) when there is no measured data of solar radiation. In the equation, K_{rs} symbolizes the calibration coefficient and can be taken as 0.16, whereas R_a indicates extraterrestrial radiation (MJ m^{-2} d^{-1}). T_{max} and T_{min} are the maximum and minimum absolute temperatures over 24 h, respectively. Net outgoing longwave radiation R_{nl}, the difference between the outgoing and incoming longwave radiation, is computed according to Equation (6).

$$R_{ns} = (1 - 0.23) R_s \quad (4)$$

$$R_s = K_{rs} R_a \sqrt{T_{max} - T_{min}} \quad (5)$$

$$R_{nl} = \acute{\varepsilon} \sigma \left(\frac{(T_{max} + 273.15)^4 + (T_{min} + 273.15)^4}{2}\right)\left(1.35 \frac{R_s}{R_{so}} - 0.35\right) \quad (6)$$

where σ is the Steffan–Boltzmann constant and has a value of 4.90×10^{-9} MJ m^{-2} d^{-1} K^{-4}. $\acute{\varepsilon}$ represents the air humidity correction factor and is defined as in Equation (7), while R_{so} expresses the clear-sky solar radiation (MJ m^{-2} d^{-1}) and is calculated as shown in Equation (8). In the following equation, E_1 represents the station elevation above sea level (m).

$$\acute{\varepsilon} = 0.34 - 0.14 \sqrt{e_a} \quad (7)$$

$$R_{so} = R_a \left(0.75 + 2 \times 10^{-5} E_1\right) \quad (8)$$

In Equation (9), where extraterrestrial radiation is calculated, G_{sc} is the solar constant and has a value of 0.082 MJ m^{-2} min^{-1}. d_r is the inverse relative distance factor for the Earth–Sun and is unitless (Equation (10)), w_s indicates sunset hour angle in radians (Equation (11)), δ refers to solar declination in radians (Equation (12)), φ symbolizes station latitude (L) in radians (Equation (13)), and i stands for the Julian day of the year.

$$R_a = \left(\frac{24(60)}{\pi}\right) G_{sc} d_r [w_s \, sin(\delta) \, sin(\phi) + cos(\phi) \, cos(\delta) \, sin(w_s)] \quad (9)$$

$$d_r = 1 + 0.033 cos\left(\frac{2\pi}{365} i\right) \quad (10)$$

$$w_s = \cos^{-1}\left[-tan(\phi)\, tan(\delta)\right] \tag{11}$$

$$\delta = 0.409 sin\left(\frac{2\pi}{365}i - 1.39\right) \tag{12}$$

$$\phi = \frac{\pi}{180} L \tag{13}$$

The psychrometric constant, γ, which is proportional to the mean atmospheric pressure, can be computed using Equations (14) and (15). While β (kPa) used in the equations indicates atmospheric pressure as a function of altitude, λ refers to the latent heat of vaporization, and its value is 2.45 MJ kg^{-1}. On the other hand, in cases where the u_2 data required in Equation (1) are unavailable, the wind speed data measured at various altitudes can be converted to the wind speed at 2 m height using Equation (16). In the equation, z_w (m) denotes the height of the measurement location from the ground, while u_z (m s^{-1}) indicates the wind speed at height z_w. Lastly, the saturated (e_s) and actual vapor pressures (e_a), can be calculated using daily e°(T_{min}), e°(T_{max}), minimum relative humidity (RH$_{min}$), and maximum relative humidity (RH$_{max}$) data by Equations (17) and (18), respectively. e°(T_{min}) and e°(T_{max}) represent the saturation vapor pressure (kPa) at the daily minimum and maximum temperatures, respectively.

$$\gamma = 0.00163 \left(\frac{\beta}{\lambda}\right) \tag{14}$$

$$\beta = 101.3 \left(\frac{293 - 0.0065\, E_1}{293}\right)^{5.26} \tag{15}$$

$$u_2 = u_z \left(\frac{4.87}{\ln(67.8 z_w - 5.42)}\right) \tag{16}$$

$$e_s = \frac{e°(T_{max}) - e°(T_{min})}{2} \tag{17}$$

$$e_a = \frac{e°(T_{max})\frac{RH_{min}}{100} - e°(T_{min})\frac{RH_{max}}{100}}{2} \tag{18}$$

To calculate ET$_{PM}$ in cases where the measured climatic parameters are insufficient or missing in the Penman–Monteith method (Equation (1)), which requires various climatic parameters, Hargreaves–Samani [45] and Allen et al. [1] proposed auxiliary equations from Equations (2)–(18) in the FAO Irrigation and Drainage Paper 56. Alternative approaches are being investigated because PM-driven ET and all of these supplementary equations are laborious and time-consuming.

2.2.2. FAO Penman–Monteith 0.5 m Method (PM$_{0.5}$)

In the standardized reference evapotranspiration method report prepared by the Environmental and Water Resources Institute of the American Society of Civil Engineers [46], the standardized PM$_{0.5}$ method was recommended for ET$_{PM_{0.5}}$ in mm d^{-1}, which will occur on a tall plant-covered surface with a height of approximately 50 cm, as given in Equation (19).

$$ET_{PM_{0.5}} = \frac{0.408\Delta(R_n - G) + \gamma \frac{1600}{T+273} u_2 (e_s - e_a)}{\Delta + \gamma(1 + 0.38 u_2)} \tag{19}$$

2.2.3. Blaney–Criddle Method (BC)

The BC equation, developed by Blaney and Morin [47] for use in evapotranspiration estimations, was modified by Blaney and Criddle, revised in 1950, and presented as

Equation (20) [48,49]. The symbol ET_{BC} denotes Blaney–Criddle-driven ET values in mm d^{-1} in the equation that uses the daily average air temperature parameter, whereas the symbol p can be used to define the mean daily percentage of annual daylight hours, which varies based on latitude. In the given equation, the seasonal crop coefficient, k, was considered to be 0.85.

$$ET_{BC} = k\, p\, (0.457\, T + 8.13) \tag{20}$$

2.2.4. Modified Blaney–Criddle Method (BC$_M$)

Climatic conditions would also affect ET, according to Doorenbos and Pruitt's [31] study, in addition to the crop coefficient in the original Blaney–Criddle equation. Equation (21) was generated by including the adjustment variables a and b, which considered humidity, sunshine duration, and daytime wind speed. Here, ET_{BC_M} stands for evapotranspiration values obtained using the modified Blaney–Criddle technique. Variable a, which can be computed using Equation (22), is a function of RH$_{min}$, n (actual sunshine duration), and N (maximum possible sunshine duration), while the b value varies depending on the daily average daytime wind speed (U) (m s^{-1}), in addition to RH$_{min}$ and n/N ratio (Equation (23)). In cases where daytime wind speed data are unavailable, 1.33 times the average wind speed can be considered for the u value. For the coefficients e_0, e_1, e_2, e_3, e_4, and e_5 employed in Equation (23), the values of 0.81917, −0.0040922, 1.0705, 0.065649, −0.0059684, and −0.0005967 were used, respectively [1,31,50,51].

$$ET_{BC_M} = a + b\,[\,p(0.457\, T + 8.13)\,] \tag{21}$$

$$a = 0.0043\, RH_{min} - \frac{n}{N} - 1.41 \tag{22}$$

$$b = e_0 + e_1 RH_{min} + e_2 \left(\frac{n}{N}\right) + e_3 U + e_4 RH_{min}\left(\frac{n}{N}\right) + e_5 RH_{min} U \tag{23}$$

2.2.5. Hamon Method (HM)

The method for predicting evapotranspiration on a daily scale suggested by Hamon [52] is given in Equation (24), and ET_{HM} symbolizes Hamon-driven ET estimations. The constant C in the equation has a value of 0.0055 and was converted from inches to mm and utilized as 0.1397 in the computations. While D, which indicates the 12 h possible sunshine duration (N/12), is estimated as in Equation (25), P_t is the saturated water vapor density at the daily average temperature and can be calculated using Equation (26).

$$ET_{HM} = C\, D^2\, P_t \tag{24}$$

$$D = \frac{w_s}{12 \times 24\pi} \tag{25}$$

$$P_t = 4.95\, exp(0.062\, T) \tag{26}$$

2.2.6. Modified Hamon Method (HM$_M$)

The approach proposed by Hamon [53] is given by Equation (27). ET_{HM_M} symbol denotes values obtained from modified Hamon-based evapotranspiration simulations. The constant H in the equation, which was converted from inches to millimeters and used as 0.1651 in the calculations, had a value of 0.0065. Equation (28) was used to obtain the P_t values using the modified Hamon technique. After the daily evapotranspiration estimations were determined using the Hamon and modified Hamon techniques, in this study, a local calibration coefficient of 1.2 was set based on the suggestions of previous studies [54–56].

$$ET_{HM_M} = H\, D\, P_t \tag{27}$$

$$P_t = \frac{e^0(T)}{T + 273.3} \tag{28}$$

2.2.7. Hargreaves–Samani Method (HS)

The technique was developed using 8-year daily lysimeter data representing 8–15 cm grass clipping evapotranspiration in California by Hargreaves [57] and was modified by Hargreaves and Samani [58] (Equation (29)). ET_{HS} denotes evapotranspiration predictions based on the Hargreaves–Samani approach, and the coefficient λ^{-1} (0.408) was used to convert evapotranspiration values into mm d^{-1}.

$$ET_{HS} = \frac{1}{\lambda} 0.0023 R_a \sqrt{T_{max} - T_{min}} (T + 17.8) \tag{29}$$

2.2.8. Kharrufa Method (KH)

The nonlinear equation based on the relationship between temperature and the average daily percentage of annual daylight hours in the year (p) proposed by Kharrufa [59] is given by Equation (30). The symbol ET_{KH} in the equation represents Kharrufa-driven daily evapotranspiration predictions.

$$ET_{KH} = 0.34 \, p \, T^{1.3} \tag{30}$$

2.2.9. Schendel Method (SC)

While the Schendel [60] method, which uses daily average temperature and relative humidity climatic data, is determined as shown in Equation (31), ET_{SC} symbolizes the evapotranspiration values obtained according to the Schendel technique.

$$ET_{SC} = 16 \left(\frac{T}{RH} \right) \tag{31}$$

2.2.10. Romenenko Method (RM)

According to Equation (32) developed by Romanenko [61], the ET_{RM} values were derived monthly using temperature (T_{mth}, °C) and relative humidity (RH_{mth}, %) monthly average climatic data.

$$ET_{RM} = 0.0018 \, (T_{mth} + 25)^2 (100 - RH_{mth}) \tag{32}$$

2.2.11. Thornthwaite Method (TH)

Thornthwaite [13] defined "potential evapotranspiration" and proposed an equation in his study on climate classification, emphasizing the significance of evapotranspiration value for climate classification (Equation (33)). In Equation (33), derived from the relationship between monthly average temperature and evapotranspiration (ET_{TH}), the ij symbol denotes the monthly temperature index (Equation (34)); the letter I symbolizes the annual temperature index, which is the sum of monthly temperature indices; the coefficient α is determined by the annual temperature index (Equation (35)).

$$ET_{TH} = 1.6 \left(\frac{10 T_{mth}}{I} \right)^{\alpha} \tag{33}$$

$$ij = \left(\frac{T_{mth}}{5} \right)^{1.514} \tag{34}$$

$$\alpha = 6.751 \times 10^{-7} \times I^3 - 7.711 \times 10^{-5} \times I^2 + 1.791 \times 10^{-2} \times I + 0.4924 \tag{35}$$

2.3. Statistical Metrics

Eleven distinct evapotranspiration techniques were applied to derive ET estimations for eleven different stations in Kahramanmaras. The performance of eight of these methods—which are computed on a daily time scale—was evaluated using numerous statistical performance assessment indicators compared to the reference PM technique. Performance evaluation indices were computed based on ET values in a daily timeframe to obtain more precise results.

Statistical metrics are widely used to assess model performance in hydrological analyses and applications [62–65]. BC, BC_M, HM, HM_M, HS, KH, $PM_{0.5}$, and SC-driven ET performances were evaluated using CRMSE, PCC, DET, MAE, MRE, MSE, NSCE, NNSCE, Bias, and RMSD statistical indices at a daily temporal resolution. The random error between the simulations and the reference values to the mean reference value was assessed using CRMSE. It can be calculated using Equation (36), and its values vary from 0 to $+\infty$. A lower CRMSE indicates better consistency, whereas a value of zero signifies no random error between the time series [65,66]. PCC is the covariance of the two variables divided by the product of their standard deviations and can be calculated using Equation (37). It measures the strength and direction of the linear relationship between two variables. If PCC is close to 1 (-1), it suggests a strong positive (negative) correlation, while converging to 0 indicates no systematic linear relationship between the estimations and references. However, it is important to note that a zero correlation does not necessarily imply the absence of any relationship between the variables; it simply means that there is no linear relationship [67]. DET quantifies the model's goodness of fit and can be expressed as shown in Equation (38). It is a statistical metric that determines the proportion of the variance in the dependent variable that is explained by the independent variables in a regression model. Its value ranges from 0 to 1; convergence to 1 indicates that the simulation explains a greater proportion of the variability in the dependent variable [68]. The absolute value of the variation between the simulated ET magnitude and PM was defined as the MAE and calculated using Equation (39). The MAE values vary between 0 and $+\infty$, with a value of 0 denoting perfect predictions. The MRE was utilized to evaluate the average relative difference between the estimated ET values relative to PM (Equation (40)). It is particularly useful for evaluating the accuracy of predictions in simulations in terms of over/under estimations and varies from $-\infty$ to $+\infty$ [64,69]. The MSE was preferred to measure the average squared difference between the estimated ET relative to the PM simulation. Equation (41) can be used to obtain this accuracy metric, and its values range from 0 to $+\infty$. A zero MSE implies that the alternative ET methods perfectly match PM-driven ET values. MAE treats all errors equally, unlike MSE, which squares the errors and may assign more weight to large errors [63,70,71]. The NSCE is a widely used metric for assessing the performance of hydrological applications and can be expressed as shown in Equation (42). NSCE ranges from $-\infty$ to 1, with higher values indicating better model performance and 1 being the ideal simulation [72]. The NNSCE error metric can be obtained by dividing the NSCE by a normalization factor to ensure that it remains between 0 and 1 (Equation (43)). The normalizing process makes the NSCE more interpretable, and the NNSCE is less affected by the scale of the data and is confined to a consistent range. The average tendency of the ET simulations to be larger or smaller than the reference ET can be measured by Bias (Equation (44)), and its values vary from $-\infty$ (underestimation) to $+\infty$ (overestimation) [73]. RMSD describes the difference between model simulations and reference ET in the units of the variable. Its values, which range from 0 to $+\infty$, close to zero imply a perfect fit, while increases indicate an increment in the error in ET predictions [62,64].

As for the terms used in the statistical metrics formulae between Equations (36) and (45), NT indicates the total number of data, $ET_{r,ti}$ denotes the evapotranspiration values obtained based on the PM reference method at ti^{th} day, $ET_{p,ti}$ symbolizes the evapotranspiration estimates obtained according to the alternative estimation method at ti^{th} day, while $\overline{ET_p}$ and

$\overline{ET_r}$ express the mean evapotranspiration values of the prediction and reference methods, respectively.

$$CRMSE\ (\%) = \frac{100}{\overline{ET_r}} \sqrt{\frac{\sum \left(ET_{p,ti} - ET_{r,ti} - \left(\overline{ET_p} - \overline{ET_r}\right)\right)^2}{NT}} \tag{36}$$

$$PCC\ (-) = \frac{\sum \left(ET_{r,ti} - \overline{ET_r}\right)\left(ET_{p,ti} - \overline{ET_p}\right)}{\sqrt{\sum \left(ET_{r,ti} - \overline{ET_r}\right)^2 \left(ET_{p,ti} - \overline{ET_p}\right)^2}} \tag{37}$$

$$DET = PCC^2 \tag{38}$$

$$MAE\ (\text{mm d}^{-1}) = \frac{\sum |ET_{p,ti} - ET_{r,ti}|}{NT} \tag{39}$$

$$MRE\ (-) = \frac{\sum (ET_{p,ti} - ET_{r,ti})}{\sum ET_{r,ti}} \tag{40}$$

$$MSE\ (\text{mm}^2\ \text{d}^{-2}) = \frac{\sum (ET_{p,ti} - ET_{r,ti})^2}{NT} \tag{41}$$

$$NSCE\ (-) = 1 - \frac{\sum (ET_{r,ti} - ET_{p,ti})^2}{\sum (ET_{r,ti} - \overline{ET_r})^2} \tag{42}$$

$$NNSCE\ (-) = \frac{1}{2 - NSCE} \tag{43}$$

$$Bias\ (\%) = 100 \times \frac{\sum (ET_{p,ti} - ET_{r,ti})}{\sum ET_{r,ti}} \tag{44}$$

$$RMSD\ (\text{mm d}^{-1}) = \sqrt{\frac{\sum (ET_{p,ti} - ET_{r,ti})^2}{NT}} \tag{45}$$

3. Results

The daily ET values for each station were estimated using the PM, PM$_{0.5}$, BC, BC$_M$, HM, HM$_M$, HS, KH, and SC techniques, resulting in the box plot shown in Figure 3. A box plot visualizes the five-number summary of the dataset: the minimum, first quartile, median, third quartile, and maximum. The box plot was produced using the fourth-generation programming language MATLAB. Alternative methods are shown on the horizontal axis of the box plot in Figure 3, whereas evapotranspiration values in mm d^{-1} predicted using these approaches are displayed logarithmically on the vertical axis. As shown in the figure, days with missing data and negative ET values were not evaluated, while the extreme ET values and estimates produced by each approach were presented with values from 0.001 to 100 mm d^{-1}. As can be seen from the figure, PM-driven simulations were overestimated by the PM$_{0.5}$-based approach over all districts, but the HM method, which displays a symmetrical box plot distribution, consistently underestimates the references in the study area. The positive results of the modifications to the BC method are shown in the graph. For example, the BC$_M$ approach yielded more effective outcomes than the BC method, which gave ET values within a narrow range when compared to the reference method over the region. Additionally, at numerous stations, the results acquired by the HM$_M$ technique were comparable to those obtained by the reference ET$_{PM}$ values, with higher performance than the HM approach, revealing the importance of modifications made to the HM equation.

The ET values obtained were clustered in the range of 1 to 6 mm d^{-1} (between the lower and upper quartiles) at Kahramanmaras station (S1), where 22 years of comprehensive data are available, which means that half of all ET values are in this range, considering the PM technique used as reference. The PM$_{0.5}$ approach results reveal that it slightly overestimates the ET values with respect to the reference PM method. The BC-driven ET values, shown in black, were aggregated in a narrower range at the S1 station relative to the reference method. The BC method simulated the minimum (maximum) ET values with overestimation (underestimation). It can be seen that the methods that give results close to the reference method at the S1 station are the HM$_M$ and HS approaches, although there are slight differences in the values of the whiskers. It is clear from the box plot of the HM approach with the best performance, shown in blue, that its quarters are distributed uniformly, and there is no skewness in the data to the PM technique. The upper outlier and interquartile range of the ET simulations produced by the BC$_M$ and KH methods—which are represented in gray and orange, respectively—produced similar findings to those of the reference method, but they had a longer minimum outlier owing to the underestimation of the small ET values. Overestimation is dominant in the maxima of SC-based ET predictions, whereas the opposite is true for smaller values less than one.

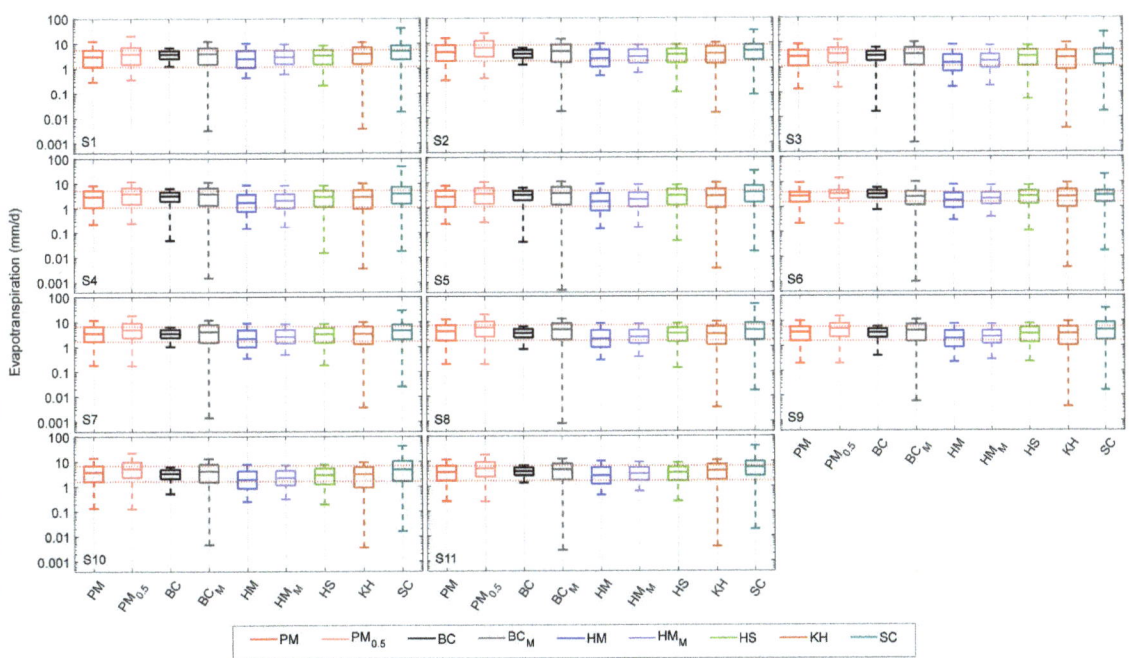

Figure 3. Box plots of evapotranspiration from the various approaches for the 11 stations at the daily temporal resolution.

Additionally, the reference ET values for the S2 station, which had the least amount of accessible climatic data, were concentrated between 2 and 9.5 mm d^{-1}. The BC$_M$ and SC techniques yield ET values within a similar range; nevertheless, the values obtained by both methods simulate the minimal ET values with strong underestimation. It can be seen that while the PM$_{0.5}$ approach produces ET values with a comparable distribution to the reference method, as in the S1 station, it continually produces a slight overestimation. The findings can be made more accurate by multiplying the values by the calibration coefficients to minimize this bias. The BC-based ET values generated for the Dulkadiroglu district were clustered in a narrower range, similar to the results at the S1 station. Although different distributions are attained in the HM, HM_M, HS, and KH approaches, where the maximum ET values are estimated to be close to each other, it has been observed that the underestimations are predominant compared with the PM reference method.

Moreover, as can be seen in the box plot of the S3 station, it can be seen that the HS method, shown in green, produces results substantially closer to the reference approach, although it slightly underestimates the minimum outlier. Although the ET values generated by the SC approach remained in the interquartile range compared to the reference method, the absolute values of the extreme ETs were higher. When the $PM_{0.5}$ technique was examined in terms of distribution, it was observed that it was quite similar to the reference method, although it tended to yield slightly higher ET values compared to the reference method. The convergence of the BC-driven ET time series in the first and third quartiles clearly shows a concentration in a narrower range, similar to those from other stations. It has been noted that the BC, BC_M, and KH approaches underestimated the minimum values for quartiles smaller than 0.05 at the S3 station. However, the BC_M and KH approaches performed well in the interquartile range and higher whisker values. Although the HM and HM_M methods capture ET values with very small variations in a negative way from the reference values, their distributions are similar to those in the PM.

In addition, when looking at the box plot of stations S4 and S5, the extreme ETs were close to one another, and the ET values computed using the reference technique were concentrated in a similar range. It was observed that the ET values acquired by the HS and KH techniques at both stations were concentrated in a similar range and captured the majority of the reference values, although the ET values smaller than the lower quartile were predicted with underestimation. Although the extreme ET values in whiskers simulated with the HM and HM_M techniques were close to the reference values, it was observed that both stations estimated ET values with a slight constant underestimation in the interquartile range compared to the reference values. The BC_M and SC methods produce overestimation (underestimation) ET values in the upper (lower) whisker compared to the reference method at both stations, while $PM_{0.5}$-driven ET simulations overestimated ET values in all quartiles.

Additionally, ET_{PM} values, varying between 1.8 and 4 mm d^{-1} for the interquartile range, were clustered in a narrower range relative to other stations, and they were captured using the HM_M, HS, and SC approaches at the S6 station. While the $PM^{0.5}$ and BC methods produce higher ET values for values between the 25th and 75th percentiles with respect to the PM technique, ET_{BC} values are concentrated in a narrower range. Although it was discovered that the ET_{HM} results were close to the reference for extreme values, these estimations clustered with underestimation until the median. Although BC_M-based ET estimates produce values close to the PM, it has been monitored that this method underestimates ET values smaller than the 25th percentile. As can be seen from the figure, the box plot produced by the KH approach has a wider spread with a higher standard deviation, and ET_{KH} values are simulated with a significant degree of bias in comparison to the reference method.

In the box plot of stations S7 and S8, where ET_{PM} values are close to each other, it is seen that the ET_{BC} simulations are similar to the other stations with the clustered in a narrower range. The ET values estimated by these two approaches are close to one another, and their maximum (minimum) values are simulated less (more) than those of the reference method. The $PM_{0.5}$ model overestimated the reference ET values with a slight difference, similar to the previous six stations. In the Pazarcik and Caglayancerit districts, the formula that yielded the closest results to the reference method was the HS approach. The BC_M, KH, and SC approaches, which simulated the minimum ET values up to the first quartiles with a high deviation compared to the reference method, captured the ET_{PM} values at both stations for the values interquartile range. It can also be seen from Figure 3 that the evapotranspiration values obtained via the HM and HM_M techniques had a more symmetrical distribution, even though the ET values at both stations underestimated the reference values.

At the S9 station, the interquartile range of reference ET values varies between 1.8 and 6 mm d^{-1}, and the best performance was acquired with the ET$_{HS}$ formula. In the Ekinozu district, although the BC$_M$, KH, and SC approaches underestimated the minimum ET values in the lower whisker with high deviation compared with the reference method, ET$_{PM}$ values were captured between the 25th and 75th percentiles. Underestimation is dominant in HM and HM$_M$-driven predictions, and BC-driven simulations are concentrated in a narrower range with a small standard deviation, whereas KH-based values are spread over a wider range with a high standard deviation. In addition, while the ET$_{BC}$ values are in a narrower range with a small standard deviation in Nurhak and Turkoglu, the PM$_{0.5}$ technique tends to overestimate evapotranspiration time series compared to the reference method, as in the other stations in general.

When the box plot of the S10 station was examined, while ET$_{KH}$ and ET$_{SC}$ values had high variance compared to ET$_{PM}$, both methods underestimated evapotranspiration smaller than the median value, and strong overestimation was dominant in ET$_{SC}$ values greater than the median. The BC$_M$ approach revealed the most accurate results relative to the reference method at the S10 station (except for the lower whisker), while the best performance was obtained with HS at the S11 station, with insignificant underestimations in the upper whisker. Another result obtained from the figure is that HS, HM, and HM$_M$-based estimates in the Nurhak district have a symmetric distribution and simulate ET values slightly less than the reference values. In the Turkoglu district, ET values in the interquartile range were captured by the BC$_M$, HM, HM$_M$, and KH approaches in addition to the ET$_{HS}$ formula. The graph also shows that the values in the lower than 25th percentile are underestimated by the BC$_M$, KH, and SC techniques.

To investigate the impact of slope on evapotranspiration, its map was prepared using the DEM model via the Arc-GIS program, as seen in Figure 4a. When the slopes of Nurhak and Turkoglu were compared, it was observed that the slope of the district where the S10 station was located was steeper than S11. On the other hand, while low ET values were detected in Nurhak (1368 m) at high altitudes, a positive effect of the slope on ET was observed. Similarly, while the S8 station in Caglayancerit (1001 m) is expected to have lower ET values than the S7 station in Pazarcik (787 m) because of the significant altitude differences, it is believed that the higher slope of Caglayancerit contributes to an increase in ET. In addition, the altitude of the S5 station in Elbistan is 1137 m, and Nurhak ET values are higher than those of S5 and S7 despite the higher altitude. Although the district has a high altitude, its steep slope is predicted to have a directly proportional effect on ET.

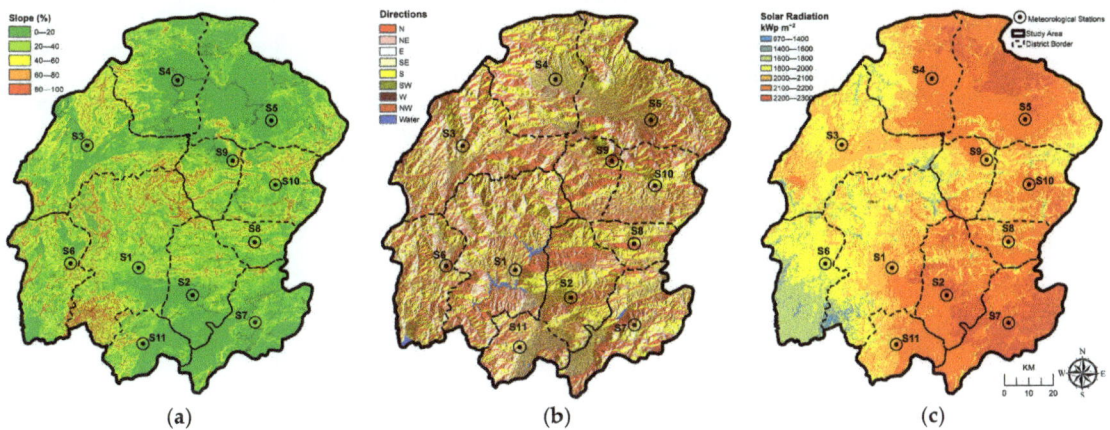

Figure 4. (**a**) Kahramanmaras slope map; (**b**) Kahramanmaras aspect geography map; (**c**) Kahramanmaras solar radiation map.

While examining the effect of the slope on ET, an aspect geography map over the study area given in Figure 4b was prepared to take into account the direction in which the slope was formed and the angle of receiving sunlight to make more accurate evaluations. To further support and elucidate the aspect map assessments and identify any subtle variations on a station basis, the solar radiation map shown in Figure 4c was generated. The data connected to the prepared map via Arc-GIS were obtained from the internet portal developed by Solargis and financed by the Energy Sector Management Assistance Program (ESMAP) [74]. Larger ET_{PM} values are found in S4, located in Afsin, even though the slopes of the S4 and S5 stations are close to one another when Figures 3 and 4a are examined together. However, it becomes clear from the aspect map in Figure 4b that Afsin has more southern slopes and that these slopes have a positive impact on ET.

Additionally, examining the solar radiation maps of the two stations in Figure 4c, it can be concluded that Elbistan has significantly stronger solar radiation than Afsin; on the other hand, a slight discrepancy occurred in ET values of these stations due to the positive effect of solar radiation on ET. Upon analyzing the aspect (Figure 4b) and solar radiation (Figure 4c) maps of the districts containing the S2 (525 m) and S11 (535 m) stations with similar altitudes (Figure 2) and slopes (Figure 4a), it was observed that Dulkadiroglu had larger ET_{PM} values (Figure 3) than Turkoglu because of the dominance of the southern slopes and high solar radiation. Moreover, when the slopes in Figure 4a of S1 (572 m) and S2 stations with similar heights were examined, it was revealed that the S1 station had a greater slope but exhibited lower ET values. This is because Dulkadiroglu has stronger solar radiation and more southern slopes, as shown in Figure 4b,c, and these two factors have a linear effect on ET.

When the ET values of the S8 (1001 m) and S9 (1246 m) stations, which have close slopes, were compared (Figures 2 and 4a), higher ETs were reached in Caglayancerit because of their lower altitudes (Figure 3). It is also estimated that this may have been triggered because Caglayancerit has more southern slopes (Figure 4b) and higher solar radiation (Figure 4c) than Ekinozu. As another example, as can be seen in Figure 3, ET values in a similar range for S10 (1368 m) and S11 (535 m) stations (Figure 2), which have a high-altitude difference, reveal the importance of the slope (Figure 4a) and the abundance of southern slopes (Figure 4b) in evapotranspiration estimations. On the other hand, even though S5 (1137 m) and S6 (1108 m), whose station altitudes are close to one another, show no discernible differences in their aspect maps (Figure 4b), it is obvious that Andirin has a steeper slope (Figure 4a) and less solar radiation than Elbistan (Figure 4c). Thus, although Elbistan has a flatter land structure, high solar radiation resulted in larger evapotranspiration values in S5.

For the monthly variation analysis of alternative ET methods across Kahramanmaras in a general manner, Thornthwaite and Romanenko methods were added, and the combined ET values of all stations are presented with the scatter plot in Figure 5. The X- and Y-axes of the graph represent the monthly evapotranspiration values in mm d^{-1} for the reference PM and alternative approaches, respectively. In addition, the calculated monthly correlation value and linear trend line with its formula between the reference PM and other methods are displayed in the plots.

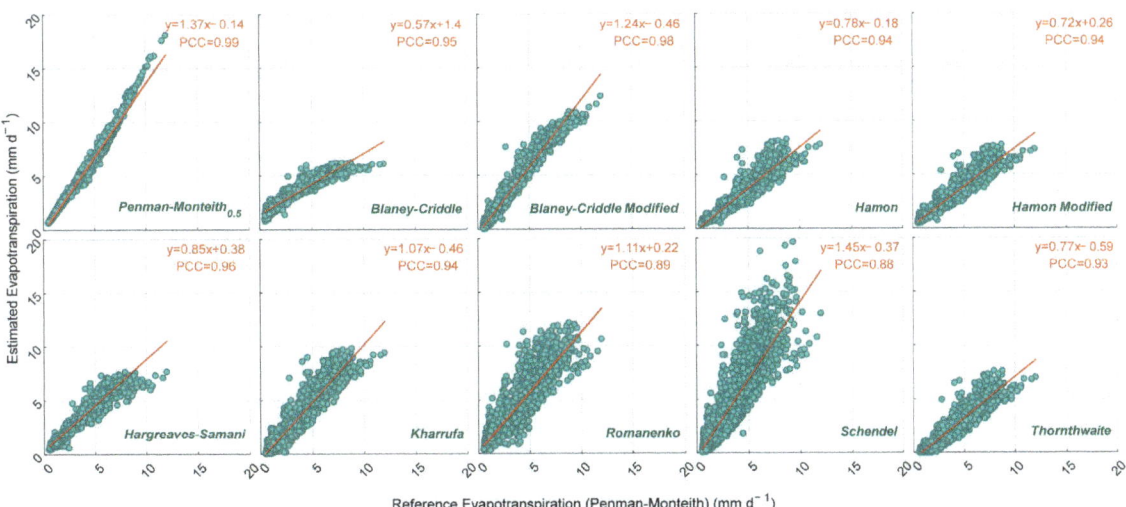

Figure 5. Scatter plots of evapotranspiration from the various approaches and reference (PM) at the monthly average.

As can be seen from the scatter plot shown in Figure 5, the $PM_{0.5}$ method has a strong positive correlation with the reference PM method with a PCC value of 0.99, while it estimates high ET values on a monthly basis more so than the references. While the BC-driven simulation revealed a correlation of 0.95, it overestimated (underestimated) low (high) ET_{PM} values. This discrepancy became wider for values greater than 6 mm d^{-1}. On the other hand, a significant improvement is observed in the BC_M technique compared to the original method, and it estimates the reference values slightly higher with a strong correlation (0.98). Additionally, the scatter plots of the HM and HM_M methods reveal similar results, with a correlation of 0.94 in both methods. Unlike other methods, the two equations, which include water vapor density, tend to underestimate ET_{PM} values.

The HS approach has a PCC value of 0.96, indicating a strong positive relationship with the PM method, whereas it underestimates the maximum extremes in ET estimations relative to the reference values. As can be seen in the box plot in Figure 3, although KH estimated the minimum ET values lower than the references, it is observed that the trend line in the scatter plot in Figure 5, obtained with monthly values with high standard deviation, produces similar results compared to the reference method with 0.94 correlation and yields an approximate slope of 1:1. As can be seen from Figure 5, the RM and SC methods show the highest variance and have correlation values of 0.89 and 0.88, respectively. In general, overestimation is dominant in the RM and SC formulae, and the SC method performs the poorest due to the significantly higher inconsistency in the maximum ET values. The results show that the monthly average temperature-driven ET_{TH} simulation underestimates the reference ET_{PM} values, while it has a lower standard deviation in its distribution compared to the other monthly methods, RM, and a strong PCC value of 0.93.

Figure 6a was obtained to evaluate the CRMSE index of eight different ET estimation methods, which vary between 10.63 and 91.69%, at 11 stations in Kahramanmaras. The BC_M-based simulations at the S2 station produced the lowest CRMSE of 10.63% among all values, whereas the HS technique at stations S1, S3, S4, S5, and S9 reached the lowest value. The best results in terms of CRMSE performance are seen to be the KH method, with values of 19.73% and 19.16% at stations S7 and S8, respectively. Moreover, the $PM_{0.5}$, BC_M, and HM methods produced lower CRMSE values than the other methods at stations S6, S10, and S11, respectively. It was also noted that the SC technique, which ranged from 34.94 to 91.69%, showed the worst performance compared with other approaches at all stations.

Examining Figure 6b, which demonstrates the variation of the determination coefficient, reveals that the $PM_{0.5}$ approach is the most consistent with the reference method, having the highest DET values (0.94–1) at all stations. On the other hand, the obtained results detect that the SC technique has the lowest DET values ranging from 0.61 to 0.82. The highest DET values, after $PM_{0.5}$ simulations, were produced with the BC method at the S1 station; the HS method at stations S3, S4, and S5; and the BC_M method at the other seven stations.

Additionally, the distribution of the MAE index values, which had the same unit (mm d^{-1}) as evapotranspiration, is shown in Figure 6c. Even though the BC_M has the lowest MAE values at S2, S8, and S10, the HS technique is the most successful approach since it shows the least absolute difference at the other eight stations. Although the methods with the highest MAE differ depending on the station, the SC technique has the worst MAE performance at the six stations.

When the results were evaluated according to the MRE index, one of the indices frequently used in performance assessments, BC_M showed the best overall result (Figure 6d). The HM_M method has the highest accuracy, with an MRE value of 0.0 at station S1, whereas negative MRE values indicate an underestimation tendency at other stations. In general, the $PM_{0.5}$ and SC (HM and KH) techniques with positive (negative) MRE values overestimated (underestimated) ET compared to the reference method.

The MSE error metric results for the eleven stations are presented in Figure 6e. While the lowest performance is achieved with SC, surprisingly, the $PM_{0.5}$ values are the highest MSEs after SC. Upon examining the results per station, it is seen that the BC_M approach at stations S2, S8, and S10; the KH method at station S7; and the HS technique at the other seven stations have the least error with MSE values closest to zero.

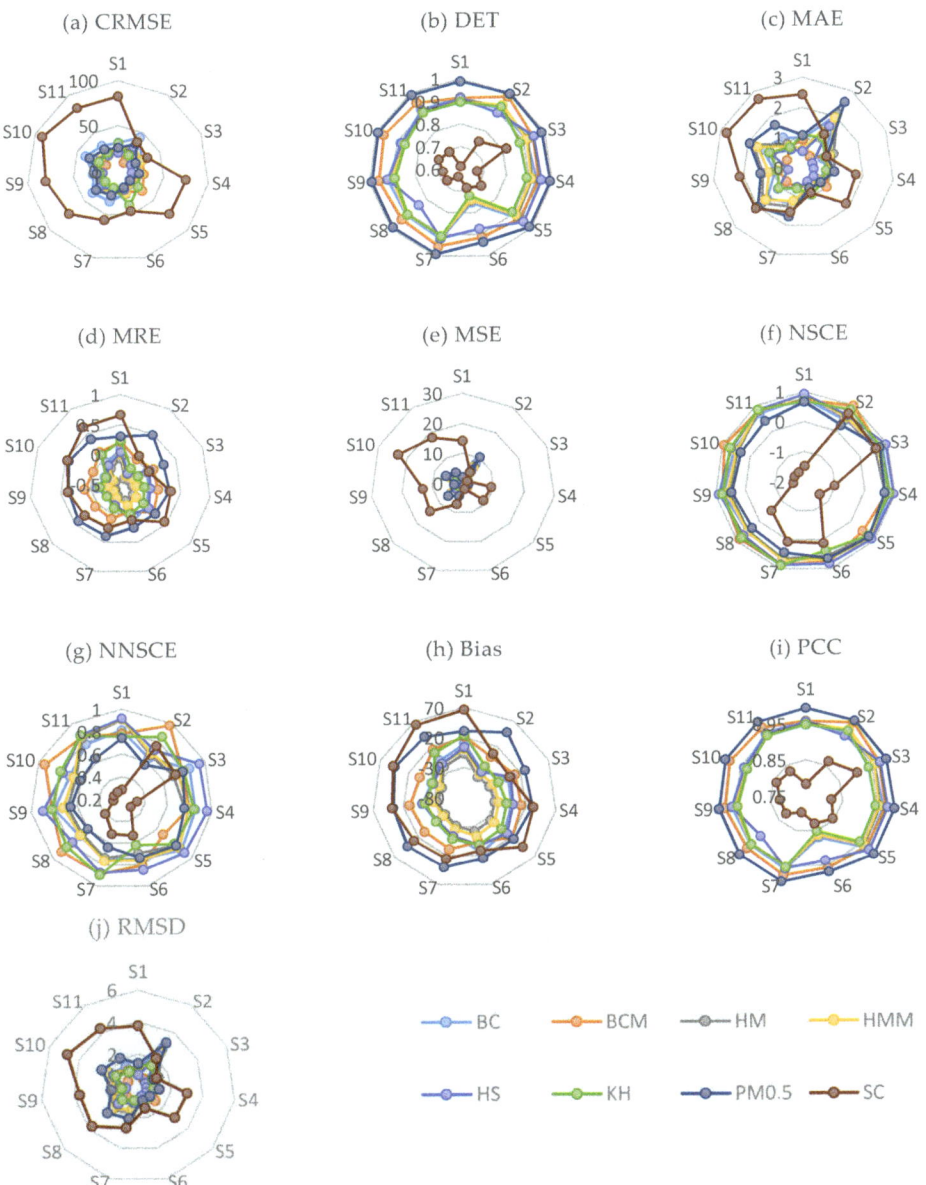

Figure 6. Variation of statistical metrics. (**a**) centered root mean square error; (**b**) determination coefficient; (**c**) mean absolute error; (**d**) mean relative error; (**e**) mean squared error; (**f**) Nash–Sutcliffe coefficient of efficiency; (**g**) normalized Nash–Sutcliffe coefficient of efficiency; (**h**) percent bias; (**i**) Pearson's correlation coefficient; (**j**) root mean square deviation.

In Figure 6f, the NSCE statistical index is used to evaluate the predictive accuracy of evapotranspiration algorithms. As can be seen from the graph, NSCE values indicated performance higher than 0.5 across stations and methods except for the SC technique, while negative NSCE values were captured for SC methods (except S2 and S3). While the BC_M method exhibited the best performance at stations S2, S7, S8, and S10, this method reached the highest NSCE value at station S2 with a value of 0.98. The HS method produced the closest result to the reference values for the other seven stations.

NNSCE is an extension of NSCE and is used as a benchmark metric, allowing for easier comparison and interpretation (Figure 6g). NNSCE performed furthest from the reference method were HM in S3, $PM_{0.5}$ in S2, and the SC method in the other nine stations. At stations S2, S7, S8, and S10, the BC_M method yields accurate results by giving the NNSCE value closest to one, and the KH method converges to 0.9 at station S7. As can be observed from the NNSCE index, the HS technique worked well for the remaining seven stations.

It is observed from Figure 6h that while the BC_M, $PM_{0.5}$, and SC methods, overall, take positive Bias values, indicating a tendency to overestimate, underestimation is more dominant in KH, HM, and HM_M approaches with general negative Bias values. Another noteworthy point when the results are examined in terms of the Bias error metric is that the BC_M and KH methods have performances closest to zero at the majority of the stations.

The PCC results are given in Figure 6i, and it can be seen that the strongest positive correlation is the $PM_{0.5}$ method, ranging between 0.97 and 1.00, while the lowest is the SC technique, varying between 0.78 and 0.91.

Figure 6j displays the variation in the RMSD metric at 11 stations, which has the same unit as evapotranspiration as the MAE performance assessment index. As can be seen from the figure, while the BC_M method performed better than the other empirical formulae with RMSD values closest to zero at stations S2, S8, and S10, the KH approach showed a successful performance at station S7. The figure indicates that for the other seven stations, the HS method produced RMSD values with a discrepancy of only a maximum of 1 mm d^{-1}. The HS (SC) method achieved the best (worst) performance among all the methods at station S3 (S10) with an RMSD value of 0.49 (4.80) mm d^{-1}.

4. Discussion

As can be seen from the box plot in four equal quartiles, information was obtained regarding the extremes of ET values, the range in which they were clustered, and their distribution. The BC_M, KH, and SC-based ET simulations underestimated minimum values smaller than the 25th percentile relative to the ET_{PM}. Saud et al. [75] analyzed the spatiotemporal variation of several methods over Al-Anbar province, western Iraq. They found that the Kharrufa equation tended to underestimate ET values, similar to the above-mentioned finding. On the other hand, the SC method tends to predict maximum extreme ET values larger than the third quartile than the reference method. The performance of the $PM_{0.5}$ approach was higher for the minimum ET values, which were especially smaller than the lower quartile. However, it produces more ET values, albeit with a slight difference, at larger ET values compared with the reference technique. Additionally, the ET_{BC} values were clustered in a narrower range than the reference ET values with smaller standard deviations. It was concluded that the BC_M method produced more successful results than the BC method, showing values close to those of the PM method for values greater than the lower quartile. As a matter of fact, in this sense, the positive effect of the modifications on the BC-based ET equation was observed, as in the previous studies [14,76]. It is also understood that the ET values obtained with the HM and HM_M approaches have a more symmetrical distribution; however, the HM method underestimates ET_{PM} at all stations, whereas the modified HM method achieves more successful results than the original HM method. The results of these two modified equations support the importance of adjustment in the original formulae [77–79]. For example, Proutsos et al. [19] evaluated 127 ET approaches in Mediterranean urban green sites and concluded that the adjusted models performed more accurate ET simulations overall compared to the original equations. Compared to other techniques, the HS method produces results that are closest to the reference ET_{PM} when examining the box plot of the method at all stations.

The graphs and maps were analyzed to see how altitude, slope, aspect geography, and solar radiation affected the evapotranspiration values. The results showed that evapotranspiration varies proportionally with the impact of altitude on Kahramanmaras. The results support the findings of Lin et al. [80] regarding ET correlation with topography in the Xiliao River Plain, while the outcomes are different from those reported by Ablikim et al. [81], in which ET values rise with the increment in altitude in the Urumqi River Basin. For instance, the reason for the high ET values in S2, which has the lowest altitude of 525 m, is thought to be the adverse effect of altitude on ET. The observation of lower ET values at the S3 station, which has a higher elevation (1344 m) than the S2 station, shows that the same effect occurs. The fact that the S9 station, at an altitude of 1246 m, had lower ET values relative to the S8 station at an altitude of 1001 m supports this effect. As S7 (787 m) exhibited smaller ET values than S2 and S11, which have altitudes of 535 m, it is another example that elevation has an adverse effect on ET. While station S1 (572 m) had smaller ET values relative to the S2 station, indicating the impact of altitude, a larger discrepancy in ET values was observed in response to the minor elevation differences. On the other hand, despite the high-altitude variation between stations S1 and S3 (572 m/1344 m), the difference between the ET values of both stations was lower. Similar results were observed at stations between S10 (1368 m) and S11 (535 m) and between S7 (787 m) and S8 (1001 m). These results reveal that the inverse proportion of altitude on ET may vary depending on altitude and that factors other than elevation may also be effective for ET. For this aim, considering previous studies showing that distributions of slope and aspect may have an effect on evapotranspiration, slope, aspect geography, and solar radiation maps, these were also investigated by correlating with ET [82–84]. A strong correlation with ET has been observed on the southern slopes, and solar radiation is another factor controlling ET. Among all the stations, for example, S6, which has a narrow ET distribution with a minor standard deviation and stands out with its low ET values compared to other stations, appears to have the lowest solar radiation. As another example, it is recognized that the S2 station at the lowest altitude exhibits the highest ET values as a result of all these factors, such as the fact that it has a more southern slope compared to other stations and intense solar radiation, in addition to its steeper slope.

Additionally, within the scope of this study, the combined performances of all methods across the city were examined on a monthly basis using a scatter plot. For instance, the Schendel approach produced ET values at some stations that were similar to the reference method, as can be seen from the box plot; on the other hand, the method yielded the lowest performance of all the alternative methods over the entire city when the monthly based scatter plot was examined. Both the BC and TH techniques underestimated ET_{PM} reference values, while an overestimation tendency was observed for the $PM_{0.5}$ method, although the highest correlation was achieved with it. As stated in the methodology section, the modified Blaney–Criddle approach applies parameters computed using the minimum relative humidity, ratio of actual sunshine to the maximum possible sunshine duration, and daytime wind speed instead of the seasonal crop coefficient used in the original method. This resulted in a significant improvement in the evapotranspiration estimations, as seen from the scatter plot. In the Hamon methods, which tend to underestimate ET, the modified version produced better ET values than the original equation. This graph reveals the importance of modification analysis more clearly and supports the results of previous studies [14,76–78]. Another important result obtained from the scatter plot is that the KH method, which tends to predict lower daily ET minimum extremes, showed a high performance in monthly simulations across the city. Additionally, the HS method produced the best result by providing a high correlation, although it underestimated the maximum extreme values.

The performances of the ET approaches on a daily scale were computed and graphed using ten various statistical criteria to enhance the evaluation of the methods to be more detailed and sensitive for each station. As can be seen from the NSCE results, a value higher than 0.5 demonstrates that alternative approaches (except SC) can sufficiently replicate the variability of the ET_{PM}. Conversely, negative NSCE values indicate that for SC-driven simulations at the stations, the mean of the ET_{PM} values is a better predictor than the SC empirical method. The main difference for NNCSE lies in the normalization of the NSCE value, which helps NNSCE to be less sensitive to the variability in the reference data and allows for better comparison across other empirical approaches, regardless of its variance. The highest NSCE and NNSCE values were obtained with BC_M-driven simulations in S2, S8, and S10, whereas the best results were yielded with ET_{HS} estimations in S7, and it is the HS method at the other seven stations. The obtained findings were observed more clearly in the NNSCE graph. It is essential to note that while the NSCE is a crucial error metric for measuring predictive accuracy, it has some limitations. For instance, it gives equal weight to both overestimation and underestimation errors, which might not always reflect the true importance of such errors in the direction of discrepancy. In comparison to ET_{PM}, in general, the underestimation is more dominant in Hargreaves–Samani, Kharrufa, and both Hamon equations with negative MRE values, whereas the BC_M, $PM_{0.5}$, and SC techniques overestimated the reference values with positive MRE indices. These results are seen more clearly in the Bias metric. Additionally, overall, HS, BC_M, and KH-driven simulations yielded the best MSE, CRMSE, and RMSD results, while the $PM_{0.5}$ (SC) method had the highest (lowest) value at all stations according to the DET metric. All approaches, excluding SC, have strong positive correlations greater than the PCC value of 0.9 at all stations except for S6. The BC (HS) technique exhibits the highest PCC values in S1 (S3, S4, and S5) after the $PM_{0.5}$ approach, while the BC_M method, in S2 and the six stations between S6 and S11, is secondary in terms of PCC performance.

However, this study has some limitations; for instance, in addition to the aforementioned factors, ET may also be affected by elements such as vegetation cover, soil map, land use, and land cover belonging to the relevant study area, and they were not evaluated within the scope of this study. Despite these limitations, the results of this study can help future studies mitigate the effects of drought and the prejudice of hydrological modeling results over the study area. Furthermore, the findings motivate future studies to analyze how well alternative empirical approaches are performed in other areas with features comparable to the analyzed region.

5. Conclusions

Empirical approaches for calculating ET values hold a significant place in the literature because of their advantages, such as simplicity of use when utilizing meteorological data and obtaining results in a short time. Evapotranspiration may be quantitatively assessed using a variety of techniques. Numerous studies have been conducted to find the most straightforward and accurate method that can be applied in various study regions, and these have been updated in response to evolving circumstances. In this study, evapotranspiration simulations carried out various techniques such as Penman–Monteith, Penman–Monteith at 0.5 m, Blaney–Criddle, modified Blaney–Criddle, Hamon, modified Hamon, Hargreaves–Samani, Kharrufa, Schendel, Romanenko, and Thornthwaite using daily meteorological data from stations in 11 districts of Kahramanmaras province. In evaluating the alternative methods at daily and monthly temporal resolutions, the Penman–Monteith method, recommended by the Food and Agriculture Organization for use worldwide, was considered as a reference.

A box plot was generated using ET values derived from daily scale estimations utilizing the PM, $PM_{0.5}$, BC, BC_M, HM, HM_M, HS, and KH, along with SC methods, and assessments were conducted on a station basis. The HM method, which shows a symmetrical box plot distribution, underestimated the ET_{PM} over the region, while the $PM_{0.5}$ method overestimated the reference ET_{PM} values at all stations. In contrast to the BC method, which produced ET values in a narrow range compared with the reference method in the 11 districts, the BC_M method produced more successful results. The HM_M method, which has a symmetrical box plot distribution, produced results close to those of the reference method at some of the stations, indicating that the modification of the HM method was positively reflected. Additionally, underestimation is dominant in minimum whisker ET values obtained from BC_M, KH, and SC-driven simulations. In the Onikisubat district, the HM, HM_M, and HS methods yielded the highest performances among the other methods. Although the BC_M, KH, and SC techniques underestimated the minimum extremes, they generally overestimated ET values compared to the reference method. In Dulkadiroglu, where the highest ET values were produced among all stations, the approaches that gave the closest results to the reference method for the interquartile range were the BC_M, KH, and SC. In the Goksun district, BC_M, HS, KH, and SC-based ET simulations captured ET_{PM} variations greater than the 25th percentile, whereas they predicted the minimum ET values to be lower. It was concluded that the HS and KH approaches, which underestimated the minimum outliers, provided the closest results to the reference method in the districts of Afsin and Elbistan, where similar ET values were achieved. For Andirin, where ET fluctuations were in the lowest range among all stations, the HS and SC methods produced the most accurate results for ET_{PM} values in the interquartile range. The HS method, which slightly underestimated the reference ET values, exhibited the highest accuracy in Pazarcik, Caglayancerit, and Ekinozu districts. In Nurhak, the BC_M method was the most successful, slightly underestimating the minimum ET_{PM} outliers, while the second-highest performance belonged to the HS simulations with underestimation relative to the reference method. Finally, in the Turkoglu district, the HM_M and HS methods produced results similar to those of the reference method.

To evaluate the statistical performance of the methods on a daily scale, the central square mean error, determination coefficient, mean absolute error, mean relative error, mean squared error, Nash–Sutcliffe efficiency coefficient, normalized NSCE, percentage error, Pearson's correlation coefficient, and root mean square error metrics were applied. According to the CRMSE index, the HS method had the lowest values in Onikisubat, Goksun, Afsin, Elbistan, and Ekinozu, whereas the BC_M approach achieved the highest performance in Dulkadiroglu and Nurhak. Additionally, KH resulted in the smallest CRMSE in Pazarcik and Caglayancerit, whereas $PM_{0.5}$ and HM were the best in Andirin and Turkoglu, respectively. The $PM_{0.5}$ approach performed well at all stations based on the DET and PCC metrics because of its similarity with the reference method, although SC-based simulations produced the lowest values. After the $PM_{0.5}$-driven performance, the methods showing the highest correlations are the HS method at Onikisubat, Goksun, Afsin, and Elbistan, similar to the CRMSE metric, whereas the BC_M approach has the highest at other stations. Moreover, the most successful results were obtained via the BC_M approach in Dulkadiroglu, Caglayancerit, as well as Nurhak, and the HS method yielded MAE values less than 0.5 mm d^{-1} at other stations, while the SC and $PM_{0.5}$ formulae produced strong discrepancy in terms of MAE. An underestimation tendency is observed in the HM, HM_M, and KH methods with negative MRE and Bias indices, while the $PM_{0.5}$ and SC methods overestimated the ET_{PM} values. Additionally, the BC_M and HS techniques are generally the ones that are closest to zero, while the methods with the least error vary based on the stations in terms of MRE and Bias metrics. MSE and MRE produced comparable outcomes, and SC-based ET simulations performed the poorest in terms of both statistical indices. The RMSD values more clearly displayed inconsistencies and corroborated those derived from the MSE index. Additionally, the lowest performances were obtained with SC and $PM_{0.5}$ formulae, while other methods generally received NSCE values greater than 0.7 and BC_M, HS as well as KH-driven ET predictions exhibited the best NSCE values. The negative NSCE values in the SC-driven simulations indicated that the model did not capture the variability and patterns present in the reference value. This finding typically means that SC predictions perform poorly and might be less accurate than simply using the mean of the ET_{PM} data as a prediction. NNSCE performances support these results and reveal the variation in accuracy/discrepancy on a station basis more clearly.

In this study, the monthly averages of ET values generated by the TH and RM techniques—which can compute ET on a monthly scale—as well as the daily ET values produced by other methods, were utilized to build a scatter plot. An overestimation tendency is observed for the $PM_{0.5}$ approach, with the strongest correlation value of 0.99 for the reference among the alternative methodologies. The approach that generates the ET value at a height of 50 cm is likely to have been overestimated as a result of the standardized as a result of coefficient modifications in the original PM equation. The ET_{BC} values were clustered, ranging from 1.82 to 6.15 mm d^{-1}, and the BC model overestimated low ET values, whereas it underestimated high ET values relative to ET_{PM} values. On the other hand, the modified BC version used the "a" and "b" coefficients computed depending on various climatic parameters (i.e., relative humidity, wind speed, and sunshine duration) instead of the "k" seasonal crop coefficient in the formula and yielded significant improvement in the results. After this adjustment, it was concluded that the BC_M approach, which has the second highest correlation with a PCC value of 0.98, can be used as an alternative to the PM method over many districts in the region. Although the HM and HM_M techniques involving water vapor density underestimated the ET_{PM} values with an identical PCC (0.94), unlike other alternative methods, the modified version produced better results than the original Hamon formula. Furthermore, even though the 1.2 local calibration coefficient improved the results in the equations of both techniques, it is anticipated that regional and seasonal modifications to the included coefficient will improve the accuracy of the ET estimations. In addition, the HS approach produced ET values that were similar to the reference method throughout Kahramanmaraş, demonstrating a successful performance with a low bias between ET_{PM} and ET_{HS} and a high correlation of 0.96 PCC. The KH technique, in which the linear trend line is close to that of the PM method with a PCC value of 0.94, produced accurate ET_{PM} at some stations, although it had higher noise in overall ET estimates. Moreover, the TH method, which underestimates ET values compared to the reference method, showed a high correlation with a PCC value of 0.93. Among all alternative empirical approaches, SC and RM methods generated the highest deviation in the simulations relative to the ET_{PM} values and smallest PCC values of 0.88 and 0.89, respectively. Additionally, both methods tended to overestimate the evapotranspiration time series compared with the reference method. Examining the equations for both methods revealed that ET values were derived only from the average temperature and relative humidity data. However, the majority of other alternative formulae are functions of sunshine duration in addition to the aforementioned parameters, and coefficients derived depending on sunshine duration are also enhanced in the correlation.

Within the scope of this study, where the impact of terrain characteristics and altitude on ET was also assessed, a slope, aspect geography, and solar radiation map of the study area were prepared. It was concluded that altitude has an adverse effect on ET, although the evaluation of altitude alone might not be comprehensive except in rare circumstances. Upon evaluating the aspect, slope, and solar radiation maps in this manner on a station basis, it was observed that the slope positively impacted ET. While the southern slopes of the slope are another factor that increases ET, it has been detected that interconnected solar radiation raises ET. Along with these elements in terrain characteristics, examining the effects of land use and vegetation on ET can motivate future studies.

In light of all examinations and evaluations, it can be concluded that the BC_M and HS approaches can be utilized as alternatives to the PM method in estimating evapotranspiration values over Kahramanmaras province. Additionally, the KH technique, which only employs temperature data, can be listed as an alternative for accurately capturing ETs. While the worst results in the region were obtained with SC-driven ET simulations, the $PM_{0.5}$ method consistently overestimated the ET_{PM} values despite having a high correlation. Investigating the effectiveness of the alternative empirical methodologies assessed in this study in other locations with features comparable to the region is another matter that can attract attention. The obtained ET results will play a significant role in the planning of areas for agriculture and forestry, in determining the usable water potential of dams, in the accurate estimation of water losses in rainfall-runoff simulations, and in hydrometeorological applications such as forecasting drought or flood predictions.

Author Contributions: A.U.: material preparation, data collection, analysis, and manuscript writing. M.O.D.: collecting meteorological data, evaluating analysis, interpretation of the findings, manuscript writing, supervising, editing, and submission. All authors have read and agreed to the published version of the manuscript.

Funding: This research received no external funding.

Data Availability Statement: The data of this study are available from the authors upon request.

Acknowledgments: The authors want to thank Kahramanmaras Sutcu Imam University and the Turkish State Meteorological Service for their support and data collection.

Conflicts of Interest: The authors declare no conflicts of interest.

References

1. Allen, R.G.; Pereira, L.S.; Raes, D.; Smith, M. *Crop Evapotranspiration-Guidelines for Computing Crop Water Requirements*; FAO Irrigation and Drainage Paper 56; Food and Agriculture Organization of the UN: Rome, Italy, 1998. Available online: http://www.climasouth.eu/sites/default/files/FAO%2056.pdf (accessed on 23 June 2023).
2. Srdic, S.; Srdevic, Z.; Stricevic, R.; Cerekovic, N.; Benka, P.; Rudan, N.; Rajic, M.; Todorovic, M. Assessment of Empirical Methods for Estimating Reference Evapotranspiration in Different Climatic Zones of Bosnia and Herzegovina. *Water* **2023**, *15*, 3065. [CrossRef]
3. Su, Q.; Dai, C.; Zhang, Q.; Zhou, Y. Analysis of Potential Evapotranspiration in Heilongjiang Province. *Sustainability* **2023**, *15*, 15374. [CrossRef]
4. Yang, H.; Luo, P.; Wang, J.; Mou, C.; Mo, L.; Wang, Z.; Fu, Y.; Lin, H.; Yang, Y.; Bhatta, L.D. Ecosystem Evapotranspiration as a Response to Climate and Vegetation Coverage Changes in Northwest Yunnan, China. *PLoS ONE* **2015**, *10*, e0134795. [CrossRef]
5. Yue, P.; Zhang, Q.; Ren, X.; Yang, Z.; Li, H.; Yang, Y. Environmental and biophysical effects of evapotranspiration in semiarid grassland and maize cropland ecosystems over the summer monsoon transition zone of China. *Agric. Water Manag.* **2022**, *264*, 107462. [CrossRef]
6. Dingman, S.L. *Physical Hydrology*, 2nd ed.; Prentice Hall: Upper Saddle River, NJ, USA, 2002; 646p.
7. Liu, W.; Zhang, B.; Han, S. Quantitative Analysis of the Impact of Meteorological Factors on Reference Evapotranspiration Changes in Beijing, 1958–2017. *Water* **2020**, *12*, 2263. [CrossRef]
8. Qiu, L.; Wu, Y.; Shi, Z.; Chen, Y.; Zhao, F. Quantifying the Responses of Evapotranspiration and Its Components to Vegetation Restoration and Climate Change on the Loess Plateau of China. *Remote Sens.* **2021**, *13*, 2358. [CrossRef]
9. Penman, H.L. Evaporation: An introductory survey. *Neth. J. Agric. Sci.* **1956**, *4*, 9–29. [CrossRef]
10. Yates, D.; Strzepek, Z. *Potential Evapotranspiration Methods and their Impact on the Assessment of River Basin Runoff Under Climate Change*; IIASA Working Paper; IIASA: Laxenburg, Austria, 1994; pp. 46–94. Available online: https://pure.iiasa.ac.at/id/eprint/4163/1/WP-94-046.pdf (accessed on 16 June 2023).
11. Verstraeten, W.W.; Veroustraete, F.; Feyen, J. Assessment of Evapotranspiration and Soil Moisture Content Across Different Scales of Observation. *Sensors* **2008**, *8*, 70–117. [CrossRef]
12. McMahon, T.A.; Peel, M.C.; Lowe, L.; Srikanthan, R.; McVicar, T.R. Estimating actual, potential, reference crop and pan evaporation using standard meteorological data: A pragmatic synthesis. *Hydrol. Earth Syst. Sci.* **2013**, *17*, 1331–1363. [CrossRef]
13. Thornthwaite, C.W. An Approach toward a Rational Classification of Climate. *Geogr. Rev.* **1948**, *38*, 55–94. [CrossRef]
14. Uzunlar, A.; Oz, A.; Dis, M.O. The Effect of Modified Approaches on Evapotranspiration Estimates: Case Study over Van. *Cukurova UMFD* **2022**, *37*, 973–988. [CrossRef]
15. Meissner, R.; Rupp, H.; Haselow, L. Chapter 7—Use of lysimeters for monitoring soil water balance parameters and nutrient leaching. In *Climate Change and Soil Interactions*; Elsevier: Amsterdam, The Netherlands, 2020; pp. 171–205. [CrossRef]
16. Brown, S.; Wagner-Riddle, C.; Debruyn, Z.; Jordan, S.; Berg, A.; Ambadan, J.T.; Congreves, K.A.; Machado, P.V.F. Assessing variability of soil water balance components measured at a new lysimeter facility dedicated to the study of soil ecosystem services. *J. Hydrol.* **2021**, *603 Pt C*, 127037. [CrossRef]
17. Chow, V.T.; Maidment, D.R.; Mays, L.W. *Applied Hydrology*; McGraw Hill: Singapore, 1988; 572p.
18. López-Olivari, R.; Fuentes, S.; Poblete-Echeverría, C.; Quintulen-Ancapi, V.; Medina, L. Site-Specific Evaluation of Canopy Resistance Models for Estimating Evapotranspiration over a Drip-Irrigated Potato Crop in Southern Chile under Water-Limited Conditions. *Water* **2022**, *14*, 2041. [CrossRef]
19. Proutsos, N.; Tigkas, D.; Tsevreni, I.; Alexandris, S.G.; Solomou, A.D.; Bourletsikas, A.; Stefanidis, S.; Nwokolo, S.C. A Thorough Evaluation of 127 Potential Evapotranspiration Models in Two Mediterranean Urban Green Sites. *Remote Sens.* **2023**, *15*, 3680. [CrossRef]
20. Ashour, M.A.; Abdel Nasser, M.S.; Abu-Zaid, T.S. Field Study to Evaluate Water Loss in the Irrigation Canals of Middle Egypt: A Case Study of the Al Maanna Canal and Its Branches, Assiut Governorate. *Limnol. Rev.* **2023**, *23*, 70–92. [CrossRef]
21. Dis, M.O. A New Approach for Completing Missing Data Series in Pan Evaporation Using Multi-Meteorologic Phenomena. *Sustainability* **2023**, *15*, 15542. [CrossRef]
22. Gourgouletis, N.; Gkavrou, M.; Baltas, E. Comparison of Empirical Eto Relationships with ERA5-Land and In Situ Data in Greece. *Geographies* **2023**, *3*, 499–521. [CrossRef]
23. Penman, H.L. Natural Evaporation from Open Water, Bare Soil and Grass. *Proc. Math. Phys. Eng. Sci. P Roy Soc. A-Math. Phys.* **1948**, *193*, 120–145. [CrossRef]
24. Monteith, J.L. Evaporation and environment. *Symp. Soc. Exp. Biol.* **1965**, *19*, 205–234.

25. Shuttleworth, W.J.; Wallace, J.S. Evaporation from sparse crops-an energy combination theory. *Quart. I. R. Met. Soc.* **1985**, *111*, 839–855. [CrossRef]
26. Calder, I.R. Transpiration observations from a spruce forest and comparisons with predictions from an evaporation model. *J. Hydrol.* **1978**, *38*, 33–47. [CrossRef]
27. Jensen, M.E.; Burman, R.D.; Allen, R.G. *Evapotranspiration and Irrigation Water Requirements*; ASCE: New York, NY, USA, 1990; 332p.
28. Lemeur, R.; Zhang, L. Evaluation of three evapotranspiration models in terms of their applicability for an arid region. *J. Hydrol.* **1990**, *114*, 395–411. [CrossRef]
29. Althoff, D.; Santos, R.A.d.; Bazame, H.C.; Cunha, F.F.d.; Filgueiras, R. Improvement of Hargreaves–Samani Reference Evapotranspiration Estimates with Local Calibration. *Water* **2019**, *11*, 2272. [CrossRef]
30. Lin, E.; Qiu, R.; Chen, M.; Xie, H.; Khurshid, B.; Ma, X.; Quzhen, S.; Zheng, S.; Cui, Y.; Luo, Y. Assessing forecasting performance of daily reference evapotranspiration: A comparative analysis of updated temperature penman-monteith and penman-monteith forecast models. *J. Hydrol.* **2023**, *626*, 130317. [CrossRef]
31. Doorenbos, J.; Pruitt, W.O. *Crop Water Requirements, FAO Irrigation and Drainage Paper 24*; Food and Agriculture Organization of the United Nations, Viale delle Terme di Caracalla: Rome, Italy, 1977; 144p. Available online: https://www.fao.org/3/s8376e/s8376e.pdf (accessed on 23 January 2023).
32. Brouwer, C.; Heibloem, H. *Irrigation Water Management: Irrigation Water Needs, Irrigation Water Management Training Manual No. 3*; Land and Water Development Division FAO Via delle Terme di Caracalla: Rome, Italy, 1986. Available online: https://www.fao.org/3/S2022E/s2022e00.htm (accessed on 23 January 2023).
33. Singh, V.; Xu, C. Evaluation and generalization of 13 mass-transfer equations for determining free water Evaporation. *Hydrol. Process.* **1997**, *11*, 311–323. [CrossRef]
34. Tabari, H.; Grismer, M.E.; Trajkovic, S. Comparative analysis of 31 reference evapotranspiration methods under humid conditions. *Irrig. Sci.* **2011**, *31*, 107–117. [CrossRef]
35. Sarlak, N.; Bagcaci, S.C. The Assessment of Empirical Potential Evapotranspiration Methods: A Case Study of Konya Closed Basin. *Teknik Dergi* **2020**, *565*, 9755–9772. [CrossRef]
36. Song, X.; Lu, F.; Xiao, W.; Zhu, K.; Zhou, Y.; Xie, Z. Performance of 12 reference ET estimation method compared with the Penman-Monteith method and the potential influences in northeast China. *Meteorol. Appl.* **2018**, *26*, 83–96. [CrossRef]
37. Cobaner, M.; Citakoglu, H.; Haktanir, T.; Yelkara, F. Determination of optimum Hargreaves-Samani equation for Mediterranean region. *DUMF* **2016**, *7*, 181–190.
38. Hafeez, M.; Chatha, Z.A.; Khan, A.A.; Gulshan, A.B.; Basit, A.; Tahira, F. Comparative Analysis of Reference Evapotranspiration by Hargreaves and Blaney-Criddle Equations in Semi-Arid Climatic Conditions. *Curr. Res. Agric. Sci.* **2020**, *7*, 525–557. [CrossRef]
39. USGS Geographic Information Systems, Digital Elevation Model Data Website. Available online: https://earthexplorer.usgs.gov/ (accessed on 3 May 2023).
40. Sarigul, O.; Turoglu, H. Flashflood and Flood Geographical Analysis and Foresight in Kahramanmaraş City. *J. Geogr.* **2020**, *40*, 275–293. [CrossRef]
41. Directorate General of Environmental Impact Assessment, Permit and Inspection. Kahramanmaras City 2021 Year Environmental Status Report. Republic of Turkiye Ministry of Environment, Urbanization and Climate Change. 2021. Available online: https://webdosya.csb.gov.tr/db/ced/icerikler/k.-maras-ilcdr-2021-20220706104818.pdf (accessed on 23 January 2023).
42. Cinar, M. Investigation of Mechanical and Physical Features of Cementitious Jet Grout Applications for Various Soil Types. *Buildings* **2023**, *13*, 2833. [CrossRef]
43. MEVBIS, The Meteorological Data Information Presentation and Sales System User's manuel, the Turkish State Meteorological Service, TC Orman ve Su Isleri Bakanligi, Ankara, Turkiye, 49, March 2017. Available online: https://mevbis.mgm.gov.tr/mevbis/ui/HelpMenu/MEVBIS_Kullanim_Kilavuzu.pdf (accessed on 22 July 2022).
44. Ceyhan, Z.N.; Dis, M.O. Determining the Future Population of Kahramanmaras via Multiple Projection Methods. *Cukurova UMFD* **2022**, *37*, 1155–1164. [CrossRef]
45. Hargreaves, G.H.; Samani, Z.A. Estimating Potential Evapotranspiration. *J. Irrig. Drain. Div.* **1982**, *108*, 225–230. [CrossRef]
46. ASCE-EWRI. *The ASCE Standardized Reference Evapotranspiration Equation*; Technical Committee Report to the Environmental and Water Resources Institute of the American Society of Civil Engineers from the Task Committee on Standardization of Reference Evapotranspiration; ASCE-EWRI: Reston, VA, USA, 2005; 173p.
47. Blaney, H.F.; Morin, K.V. Evaporation and Consumptive Use of Water Empirical Formulas. *Trans. Am. Geophys. Union* **1942**, *23*, 76–83. [CrossRef]
48. Blaney, H.F.; Criddle, W.D. *Determining Water Requirements in Irrigation Areas from Climatological and Irrigation Data*; United States Department of Agriculture, Soil Conservation Service: Washington, DC, USA, 1950; 48p. Available online: https://ia800300.us.archive.org/4/items/determiningwater96blan/determiningwater96blan.pdf (accessed on 22 July 2022).
49. Blaney, H.F.; Criddle, W.D. *Determining Consumptive Use and Irrigation Water Requirements*; Technical Bulletin No. 1275; United States Department of Agriculture in Cooperation with the Office of Utah State Engineer: Washington, DC, USA, 1962; 59p.
50. Frevert, D.K.; Hill, R.W.; Braaten, B.C. Estimation of FAO evapotranspiration coefficients. *J. Irrig. Drain. Eng.* **1983**, *109*, 265–270. [CrossRef]
51. Allen, R.G.; Pruitt, W.O. FAO-24 Reference Evapotranspiration Factors. *J. Irrig. Drain. Eng.* **1991**, *117*, 758–772. [CrossRef]

52. Hamon, W.R. Estimating Potential Evapotranspiration. *J. Hydraul. Div.* **1961**, *87*, 107–120. [CrossRef]
53. Hamon, W.R. *Computation of Direct Runoff Amounts from Storm Rainfall*; Research Cooperative with the University of Mississippi and Mississippi State University: Starkville, MS, USA, 1963; pp. 52–62. Available online: https://iahs.info/uploads/dms/063006.pdf (accessed on 29 November 2023).
54. Lu, J.; Sun, G.; McNulty, S.G.; Amatya, D.M. A comparison of six potential evapotranspiration methods for regional use in the Southeastern United States. *JAWRA J. Am. Water Resour. Assoc.* **2005**, *41*, 621–633. [CrossRef]
55. Xystrakis, F.; Matzarakis, A. Evaluation of 13 Empirical Reference Potential Evapotranspiration Equations on the Island of Crete in Southern Greece. *J. Irrig. Drain. Eng.* **2011**, *137*, 211–222. [CrossRef]
56. Xie, E.; Wang, A. Comparison of Ten Potential Evapotranspiration Models and Their Attribution Analyses for Ten Chinese Drainage Basins. *Adv. Atmos. Sci.* **2020**, *37*, 959–974. [CrossRef]
57. Hargreaves, G.H. Moisture Availability and Crop Production. *Trans. ASAE* **1975**, *18*, 980–984. [CrossRef]
58. Hargreaves, G.H.; Samani, Z.A. Reference Crop Evapotranspiration from Ambient Air Temperature. *Appl. Eng. Agric.* **1985**, *1*, 96–99. [CrossRef]
59. Kharrufa, N. Simplified equation for evapotranspiration in arid regions. *Beiträge Zur Hydrol. Sonderh.* **1985**, *5*, 39–47.
60. Schendel, U. Vegetations Wasserverbrauch Und -Wasserbedarf. *Habilit. Kiel* **1967**, *137*, 1–11.
61. Romanenko, V.A. Computation of the autumn soil moisture using a universal relationship for a large area. *Proc. Ukr. Hydrometeorol. Res. Inst. Kiev* **1961**, *3*, 12–25.
62. Boyle, D.P.; Gupta, H.V.; Sorooshian, S. Toward improved calibration of hydrologic models: Combining the strengths of manual and automatic methods. *Water Resour. Res. AGU* **2000**, *36*, 3663–3674. [CrossRef]
63. Krause, P.; Boyle, D.P.; Bäse, F. Comparison of different efficiency criteria for hydrological model assessment. *Adv. Geosci.* **2005**, *5*, 89–97. [CrossRef]
64. Moriasi, D.N.; Arnold, J.G.; Van Liew, M.W.; Bingner, R.L.; Harmel, R.D.; Veith, T.L. Model evaluation guidelines for systematic quantification of accuracy in watershed simulations. *Trans. ASABE* **2007**, *50*, 885–900. [CrossRef]
65. Dis, M.O.; Anagnostou, E.; Mei, Y. Using high-resolution satellite precipitation for flood frequency analysis: Case study over the Connecticut River Basin. *J. Flood Risk Manag.* **2018**, *11*, S514–S526. [CrossRef]
66. Alpaslan, H.; Akgonen, A.I. Investigation of double tee moment connections under monotonic loading. *J. Fac. Eng. Archit. Gazi Univ.* **2021**, *36*, 2271–2286. [CrossRef]
67. Schober, P.; Boer, C.; Schwarte, L.A. Correlation Coefficients: Appropriate Use and Interpretation. *Anesth. Analg.* **2018**, *126*, 1763–1768. [CrossRef] [PubMed]
68. Ozer, D.J. Correlation and the coefficient of determination. *Psychol. Bul.* **1985**, *97*, 307–315. [CrossRef]
69. Kat, C.J.; Els, P.S. Validation metric based on relative error. *Math. Comput. Model. Dyn. Syst.* **2012**, *18*, 487–520. [CrossRef]
70. Willmot, C.J. On the evaluation of model performance in physical geography. In *Spatial Statistics and Models*; Theory and Decision Library, Gaile, G.L., Willmott, C.J., Eds.; Springer: Dordrecht, The Netherlands, 1984; Volume 40, pp. 443–460. [CrossRef]
71. Legates, D.R.; McCabe, G.J. Evaluating the use of "goodness-of-fit" measures in hydrologic and hydroclimatic model validation. *Water Resour. Res.* **1999**, *35*, 233–241. [CrossRef]
72. Nash, J.E.; Sutcliffe, J.V. River flow forecasting through conceptual models: Part 1. A discussion of principles. *J. Hydrol.* **1970**, *10*, 282–290. [CrossRef]
73. Gupta, H.V.; Sorooshian, S.; Yapo, P.O. Status of automatic calibration for hydrologic models: Comparison with multilevel expert calibration. *J. Hydrol. Eng.* **1999**, *4*, 135–143. [CrossRef]
74. Global Solar Atlas, Direct Normal Irradiation (DNI) (GeoTIFF), Turkiye Data Website. Available online: https://globalsolaratlas.info/download/turkey (accessed on 16 June 2023).
75. Saud, A.; Said, M.A.M.; Abdullah, R.; Hatem, A. Temporal and spatial variability of potential evapotranspiration in semi-Arid Region: Case study the Valleys of Western Region of Iraq. *Int. J. Eng. Sci. Technol.* **2014**, *6*, 653.
76. Heydari, M.M.; Tajamoli, A.; Ghoreishi, S.H.; Darbe-Esfahani, M.K.; Gilasi, H. Evaluation and calibration of Blaney–Criddle equation for estimating reference evapotranspiration in semiarid and arid regions. *Environ. Earth. Sci.* **2015**, *74*, 4053–4063. [CrossRef]
77. McCabe, G.J.; Hay, L.E.; Bock, A.; Markstrom, S.L.; Atkinson, R.D. Inter-annual and spatial variability of Hamon potential evapotranspiration model coefficients. *J. Hydrol.* **2015**, *521*, 389–394. [CrossRef]
78. Okkan, U.; Kiymaz, H. Questioning of Empirically Derived and Locally Calibrated Potential Evapotranspiration Equations for a Lumped Water Balance Model. *Water Supply* **2020**, *20*, 1141–1156. [CrossRef]
79. Usta, S. Development of Jensen Haise Method Reference Evapotranspiration Estimation Equations Suitable for Black Sea Region Climatic Conditions. *EJOSAT* **2022**, *38*, 415–427. [CrossRef]
80. Lin, N.; Jiang, R.Z.; Liu, Q.; Yang, H.; Liu, H.L.; Yang, Q. Quantifying the Spatiotemporal Variation of Evapotranspiration of Different Land Cover Types and the Contribution of Its Associated Factors in the Xiliao River Plain. *Remote Sens.* **2022**, *14*, 252. [CrossRef]
81. Ablikim, K.; Yang, H.; Mamattursun, A. Spatiotemporal Variation of Evapotranspiration and Its Driving Factors in the Urumqi River Basin. *Sustainability* **2023**, *15*, 13904. [CrossRef]
82. Bennie, J.; Huntley, B.; Wiltshire, A.; Hill, M.O.; Baxter, R. Slope, aspect and climate: Spatially explicit and implicit models of topographic microclimate in chalk grassland. *Ecol. Modell.* **2008**, *216*, 47–59. [CrossRef]

83. Tran, A.P.; Rungee, J.; Faybishenko, B.; Dafflon, B.; Hubbard, S.S. Assessment of Spatiotemporal Variability of Evapotranspiration and Its Governing Factors in a Mountainous Watershed. *Water* **2019**, *11*, 243. [CrossRef]
84. Gisolo, D.; Previati, M.; Bevilacqua, I.; Canone, D.; Boetti, M.; Dematteis, N.; Balocco, J.; Ferrari, S.; Gentile, A.; N'Sassila, M.; et al. A calibration free radiation driven model for estimating actual evapotranspiration of mountain grasslands (CLIME-MG). *J. Hydrol.* **2022**, *610*, 127948. [CrossRef]

Disclaimer/Publisher's Note: The statements, opinions and data contained in all publications are solely those of the individual author(s) and contributor(s) and not of MDPI and/or the editor(s). MDPI and/or the editor(s) disclaim responsibility for any injury to people or property resulting from any ideas, methods, instructions or products referred to in the content.

Article

Spatiotemporal Variations in Actual Evapotranspiration Based on LPJ Model and Its Driving Mechanism in the Three Gorges Reservoir Area

Xuelei Zhang [1], Gaopeng Wang [2] and Hejia Wang [3,*]

1. School of Geomatics, Remote Sensing, and Information, Guangdong Polytechnic of Industry and Commerce, Guangzhou 510510, China
2. College of History and Culture, Tianjin Normal University, Tianjin 300387, China
3. State Key Laboratory of Simulation and Regulation of Water Cycle in River Basin, China Institute of Water Resources and Hydropower Research, Beijing 100038, China
* Correspondence: hjwang@iwhr.com

Citation: Zhang, X.; Wang, G.; Wang, H. Spatiotemporal Variations in Actual Evapotranspiration Based on LPJ Model and Its Driving Mechanism in the Three Gorges Reservoir Area. *Water* 2023, *15*, 4105. https://doi.org/10.3390/w15234105

Academic Editors: Sonia Raquel Gámiz-Fortis and Matilde García-Valdecasas Ojeda

Received: 22 September 2023
Revised: 14 November 2023
Accepted: 22 November 2023
Published: 27 November 2023

Copyright: © 2023 by the authors. Licensee MDPI, Basel, Switzerland. This article is an open access article distributed under the terms and conditions of the Creative Commons Attribution (CC BY) license (https://creativecommons.org/licenses/by/4.0/).

Abstract: Under the influence of climate change and human activities, the ecohydrological processes in the Three Gorges Reservoir Area (TGRA) present new evolution characteristics at different temporal and spatial scales. Research on the evolution and driving mechanism of key ecohydrological element in the TGRA under the changing environment has important theoretical and practical values for correctly understanding the ecohydrological situation in the reservoir area and guiding the coordinated development of water and soil resources. In this study, the LPJ (Lund–Potsdam–Jena) model was used to simulate and analyze the spatiotemporal variations in evapotranspiration (AET) from 1981 to 2020. Sen's slope and sensitivity analysis methods were used to quantify individual contributions of climate and human factors to changes in AET in different periods. The results indicate the following: (1) The simulation accuracy of the LPJ model for AET in the TGRA was high, with a certainty coefficient (R^2), Nash efficiency coefficient (NSE), and mean relative error (MRE) of 0.89, 0.76, and 4.32%, respectively. (2) The multiyear average AET was 650.71 mm and increased at a rate of 21.63 mm/10a from 1981 to 2020. The annual distribution of AET showed a unimodal seasonal variation trend. The peak value occurred in July, reaching 113.02 mm, and the valley value occurred in January and December, less than 13 mm. (3) AET increased by 5.60% and 6.28% before and after impoundment, respectively. The contribution rate of human activities increased significantly from −3.75% before impoundment to 26.95% after impoundment, and the contribution ratios of climate change were 89.39% and 73.09%, respectively, during these two periods. From 1981 to 2020, AET increased by 5.28%, in which the contribution ratios of climate and human factors were 89.39% and 10.61%, respectively.

Keywords: actual evapotranspiration (AET); LPJ model; climate change; driving mechanism; the Three Gorges Reservoir Area (TGRA)

1. Introduction

Actual evapotranspiration (AET), which is defined as the synthesis process of evaporation and transpiration, plays a key role in surface energy balance by linking the hydrological cycle, carbon budget, and energy transfer [1–3]. As a crucial component of energy balance and the water–carbon cycle under global climate change [4], AET is widely used in water resource management, irrigation planning, ecological projects, and other practices of agricultural production [5–7]. From a background of continuous global warming and increasing human disturbance, the influence of climatic and human factors on regional hydrological processes continues to deepen [8], and the water cycle process has shown new characteristics. Accurate measurement and estimation of AET, as well as quantitative differentiation of the influence of climate and human factors on the change in AET, are very important for the sustainable relationship of regional water resources and ecological environment protection [9,10].

The acquisition of AET has always been an important and difficult point in ecohydrological research [11–13]. The AET, which is composed of canopy transpiration (EP), vegetation interception (EI), and bare soil evaporation (ES), is not only influenced by climate change but also closely related to vegetation, and it is greatly influenced by the underlying surface and vegetation dynamics [14]. Therefore, it is necessary to consider the growth process of vegetation in AET simulation [7,15]. Dynamic Global Vegetation Models (DGVMs) express land surface biophysics, terrestrial carbon cycle, and global vegetation dynamics through an independent and naturally continuous model framework [16]; these models integrate a wide range of biophysical, vegetation physiological, and ecological processes, and they can simulate land surface physical processes, vegetation canopy physiology, vegetation phenology, vegetation dynamics and competition, carbon and nitrogen cycle, and other processes, which are suitable for research at different temporal and spatial scales [17]. Common DGVMs include CLM (Community Land Model), VIC (Variable Infiltration Capacity), IBIS (Integrated Biosphere Simulator), and LPJ (Lund–Potsdam–Jena) models [18]. CLM, VIC, and IBIS are highly complex and difficult DGVMs, which have high requirements for model users and hardware equipment and are suitable for large-scale and global simulations. The LPJ [19] model was jointly researched and developed by Lund University, Potsdam Climate Research Centre, and Max Planck Institute for Biogeochemistry, Jena. As a moderately complex DGVM, the framework and principles of the early version are derived from the BIOME model [16,17]. The vegetation dynamic process of the model is based on natural vegetation, and the establishment and death of natural vegetation depend on a set of plant functional environmental limiting factors based on the 20-year mobile climate extreme [20]. LPJ explicitly considers key attributes, such as physiological adaptability, phenological characteristics, morphological attributes, resource utilization ability, disturbance response, and photosynthetic efficiency of different plant species, and it determines the distribution and composition of vegetation by population statistics. Subsequently, through continuous development and improvement, fire interference [21] was introduced, and the calculation scheme was improved [22]. Recently, the LPJ model has made great progress in ecosystem research and has been widely used in simulations of vegetation productivity, evapotranspiration, and the hydrological cycle [23–26]. While meeting the needs of ecohydrological simulation on the small and medium scales, it can also ensure high simulation accuracy and computing efficiency, obtaining good applicability in small- and medium-scale research [27,28].

As the world's largest hydroelectric power station, the construction of the Three Gorges Project greatly promoted the economic and social development around the TGRA and played an important role in flood control, shipping, power generation, and other aspects [29,30]. However, as a special human activity, the construction of the project, including dam construction, migration engineering, power generation, and reservoir dispatching, has caused severe disturbance to the underlying surface [31], causing changes in land coverage types, which, in turn, changed the surface albedo [32,33] and affected the energy balance [34]. In addition, the impact of a warming climate on the hydrological cycle processes is increasingly aggravated [35]; water resources, water ecology, and water environment in the reservoir area are also facing severe pressure [36,37]. Under the influence of climate change and human activities, the ecohydrological processes in the TGRA have been marked by the strong interference of "natural and artificial" [38,39]. It is necessary and urgent to study the evolution and driving mechanism of key ecohydrological elements in the TGRA under a changing environment [40].

The Three Gorges Dam site was determined in 1984, and a construction plan was approved in 1992. Following this, the dam entered the construction phase. The water level of the Three Gorges Project reached 135 m in 2003 and 175 m in 2009. After 2010, the Three Gorges Project was officially put into operation, and we entered the Three Gorges era. In the past 40 years, the TGRA has experienced four different periods of development, including the stage before the project (1981–1992), the early stage of the project (1993–2002), the later construction and completion stages (2003–2010), and the formal operational stage (2011–2020). There are evident differences in trends in climate change and the disturbance to the underlying surface in different periods, especially between before (1981–2002) and after (2003–2010) impoundment. The impounding of the reservoir caused the water level to rise, and a large number of farmland and residential settlements on both sides of the river were flooded; the underlying surface was thus strongly disturbed. In addition, the increase in the water surface area produced a certain climate effect, and the climatic factors and ecohydrological factors had significant changes after impounding.

Therefore, this paper takes AET as the key ecohydrological element and attempts to estimate the AET over the TGRA based on the LPJ model with multisource remote sensing data from 1981 to 2020 and to quantify individual contributions of climate and human factors to the variations in AET before and after impoundment. The results of this study can provide a scientific basis and data support for the effective evaluation of ecosystem service functions and provide a decision-making basis for a correct understanding of the ecohydrological condition of the reservoir area and for guiding future spatial coordinated development for the TGRA.

2. Material and Methodology

2.1. Study Area

The TGRA refers to the catchment area between the dam site and the backwater end of the reservoir, which was created after the completion of the Three Gorges Dam and the successful impoundment of the reservoir area. This region is situated in the lower section of the upper reaches of the Yangtze River, within 105°50′–111°40′ E, 28°31′–31°44′ N (Figure 1), covering a total area of 59,326 km², with a main channel length of 660 km. The TGRA mainly includes Chongqing Municipality, Hubei Province, part of Linshui County, Dazhu County and Luxian County in Sichuan Province, and Xishui County and Tongzi County in Guizhou Province. This area spans the south foot of the Daba Mountains to the north and extends to the north edge of the Yunnan-Guizhou Plateau to the south. The TGRA is located at the intersection of the Daba Mountain fold belt, East Sichuan fold belt and Sichuan, Hubei, Hunan, and Guizhou uplift belt, thereby crisscrossing areas containing mountains, hills, basins, and valleys. The region is mainly composed of mountains along the northeast and south margins of the Yangtze River and hills in the Midwest [41]. The elevation of the TGRA ranges from 54 to 3099 m, with an average elevation of 773 m (Figure 1). According to the area proportion of different elevation intervals, the area with elevation between 100–500 m, 500–1000 m, and 1000–2000 m accounted for 37%, 35%, and 26%, respectively. Less than 2% of the area has an elevation of more than 2000 m. The TGRA is subject to a subtropical monsoon climate, with an average annual temperature (T) of 16.84 °C. The rainfall in the reservoir area is abundant, but the temporal distribution is uneven. The annual average precipitation (P) in the TGRA is 1186 mm, dominated by summer P (the total P from June to September accounts for about 54% of the annual P). There are four distinct seasons in the TGRA, characterized by warm winters, early springs, hot summers, and rainy autumns [42].

2.2. LPJ Model

2.2.1. Model Description

Based on physiological, morphological, phenological, and disturbance response attributes and a few bioclimatic limiting factors, the LPJ model defines 10 Plant Functional Types (PFTs) for photosynthesis simulation, including 8 kinds of tree and 2 kinds of herb vegetation. Under the influences of the strong control of East Asian monsoon and the transformation of Qinghai-Tibet Plateau on the westerly flow, a unique natural environment has been formed. Depending on field investigation and reference to relevant information [20,43], Zhao et al. [28] added two PFTs of temperate desert shrub and cold herb and defined their physiological parameters of leaf longevity. Huang [44] defined the distribution ratio of roots between the upper and lower soil layers. Finally, 12 PFTs, including 9 kinds of tree and 3 kinds of herb vegetation, were defined. The parameter values of relevant PFTs and their ecological environmental limiting factors are shown in Table 1.

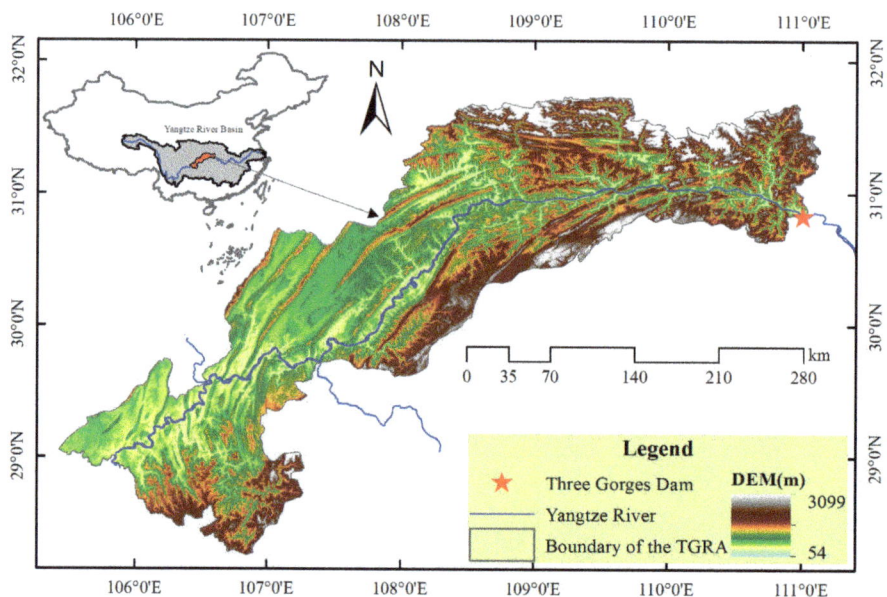

Figure 1. Location of the Three Gorges Reservoir Area (TGRA).

Table 1. Eco-hydrologically relevant PFT parameter values.

PFT (Unit)	T_{c_min}	T_{c_max}	$GDD5_{min}$	T_{w_min}	T_{w_max}	Leaflong	Rootdist
Tropical rainforest	12	-	-	-	-	0.5	0.7/0.3
Tropical broadleaved evergreen tree	12	-	-	-	-	2	0.85/0.15
Temperate needleleaved evergreen tree	−2	22	900	-	-	2	0.6/0.4
Temperate broadleaved evergreen tree	0	14	1500	-	-	1	0.7/0.3
Temperate broadleaved deciduous tree	−17	0	1500	-	-	0.5	0.65/0.35
Northern needleleaved evergreen tree	0	−25	550	23	-	0.5	0.9/0.1
Northern needleleaved deciduous tree	-	−2	350	23	43	2	0.9/0.1
Northern broadleaved deciduous tree	-	−15	350	23	-	0.5	0.9/0.1
Temperate desert scrub	-	−5	350	23	-	1	0.9/0.1
Tropical herb	15	-	-	12	-	1	0.9/0.1
Temperate herb	-	−8	-	-	-	1	0.9/0.1
Cold herb	-	−12	-	-	-	1	0.9/0.1

Note: T_{c_min}, minimum average temperature in the coldest month (°C); T_{c_max}, maximum average temperature in the coldest month (°C); $GDD5_{min}$, minimum annual accumulated temperature above 5 °C required for vegetation settlement and growth (°C); T_{w_min}, minimum average temperature in the hottest month (°C); T_{w_max}, maximum average temperature in the hottest month (°C); leaflong, leaf longevity(a); rootdist, distribution ratio of roots between the upper and lower soil layers.

The simulation of AET by LPJ is mainly divided into three parts, namely, EP, EI, and ES. The water balance simulation in the LPJ model mainly adopts the water tank model. The soil is divided into two layers. The water balance formulas of the upper and lower layers (soil thickness is 0.5 m and 1.0 m, respectively) are as follows:

$$\Delta W_1 = P - ES - EI - \beta_1 \times EP - R_1 - Perc_1 \tag{1}$$

$$\Delta W_2 = Perc_1 - \beta_2 \times EP - R_2 - Perc_2 \tag{2}$$

where ΔW_1 and ΔW_2 are the changes in water content in the upper and lower soil layers (mm/d); β_1 and β_2 are the proportion of water consumed by roots for vegetation transpiration from the upper and lower soil layers; R_1 and R_2 are the surface and underground runoff (mm/d), respectively; $Perc_1$ and $Perc_2$ are the infiltration quantity (mm/d) of the upper and lower soil layers, respectively; P is precipitation (mm/d). The model does not consider the lateral water exchange and the convergence path between the grids.

EI is calculated using the following formula:

$$EI = E_q \times \alpha \times f_{wet} \tag{3}$$

where f_{wet} is the daytime canopy wetness ratio of each vegetation functional type. The remaining canopy $1 - f_{wet}$ is used to calculate vegetation transpiration. E_q and α are the equilibrium evaporation rate and Priestley–Taylor coefficient, respectively. E_q and α, involved in the calculation, are calculated from the Priestley–Taylor formula:

$$E_q = \frac{\Delta}{\Delta + Y}(R_n - G) \tag{4}$$

$$\alpha = \frac{LE}{E_q} = \frac{LE}{[\Delta/(\Delta + Y)](R_n - G)} \tag{5}$$

where R_n is the net radiation (MJ/m^2/d); G is the soil heat flux (MJ/m^2/d); LE is the latent heat flux (MJ/m^2/d); Δ is the slope of the saturated water vapor pressure–temperature relation curve (Claussius–Clapeyron relation—kPa/°C); and Y is the wet and dry table constant, which is equal to 66.5×10^{-6} kPa/°C.

EP is the minimum value of water demand (D) and water supply (S) of vegetation under the condition of adequate water supply, and the formula is as follows:

$$EP = Min[S, D] \times f_v \tag{6}$$

$$S = E_{max} \times W_r \tag{7}$$

$$D = E_q \times \alpha_{max} \times (1 - f_{wet})/(1 + g_m/g_{pot}) \tag{8}$$

where E_{max} is maximum transpiration rate of each vegetation functional type, and the value ranges from 5 to 7 mm/d; W_r is soil moisture content that can be utilized by vegetation roots; α_{max} is the maximum Priestley–Taylor coefficient, which is equal to 1.391 in LPJ model; g_m is the conversion conductance, which is equal to 3.26 mm/s in the LPJ model; g_{pot} is the canopy latent conductance (mm/s); and f_v is the vegetation coverage.

ES occurs only in the surface 20 cm soil layer of bare land $(1 - f_v)$, and the formula is:

$$ES = E_q \times \alpha \times W_{r20} \times (1 - f_v) \tag{9}$$

where W_{r20} is the water content in the surface 20 cm soil layer. All the above factors were calculated based on vegetation functional types.

2.2.2. Model Input

The LPJ model was run for the period 1981–2020, preceded by a 1000-year spin-up period to reach an initial equilibrium with respect to ecosystem and soil structure [24]. In this study, we used climate data from 1981 to 2020 to drive the LPJ model 25 times to reach equilibrium. Taking the equilibrium state as the initial state, the ecohydrological process in the TGRA during the study period was simulated. The simulations were driven by gridded monthly fields (0.1° resolution) of T, P, number of wet days, cloud cover, and soil texture. Non-gridded model inputs include annual CO_2 concentrations. Furthermore, various parameters are assigned to the different PFTs (Table 1). The source, website, spatial and temporal resolution, and time scale of all input datasets are shown in Table 2.

Table 2. Attributes of the input data.

Data Type	Source	Website	Resolution	Time Scale
Monthly temperature (T)	China meteorological forcing dataset [45]	https://data.tpdc.ac.cn/zh-hans/data/8028b944-daaa-4511-8769-965612652c49/ (accessed on 1 January 2023)	0.1°/3 h	January 1981~December 2018
Monthly precipitation (P)	China meteorological forcing dataset [45]	https://data.tpdc.ac.cn/zh-hans/data/8028b944-daaa-4511-8769-965612652c49/ (accessed on 1 January 2023)	0.1°/3 h	January 1981~December 2018
Monthly cloud cover	CRU TS Version 4.07	https://crudata.uea.ac.uk/cru/data/hrg/cru_ts_4.07/cruts.2304141047.v4.07/ (accessed on 1 January 2023)	0.5°	January 1981~December 2020
Monthly wet days	CRU TS Version 4.07	https://crudata.uea.ac.uk/cru/data/hrg/cru_ts_4.07/cruts.2304141047.v4.07/ (accessed on 1 January 2023)	0.5°	January 1981~December 2020
Soil texture	Geographic data platform of Peking university	https://geodata.pku.edu.cn (accessed on 1 January 2023)	$1:10^6$	2009
Annual CO_2	Earth's CO_2	https://www.co2.earth (accessed on 1 January 2023)	Northern Hemisphere/Annually	1981~2020

Monthly T and monthly P are provided by National Tibetan Plateau Data Center. In this study, we extracted the T and P data in the TGRA and synthesized the 3 h data into a monthly scale. Finally, we obtained the monthly average T and P from January 1981 to December 2018. The average error is less than 5% by comparing the synthesized data and the measured data from meteorological stations, indicating that the synthesized data are reliable and can be used for model spin-up and simulation. The P and T data from January 2019 to December 2020 were extended via the interpolation of station data. Monthly cloud cover and monthly wet days were derived from the CRU4.07 data set. We collected monthly cloud cover and monthly wet days in the TGRA from January 1981 to December 2020. The soil data were mainly used to simulate the infiltration in precipitation process, soil moisture content, and the process of absorbing nutrients and minerals in the roots of vegetation, and they objectively reflect the distribution of the surface in the study area. The spatial distribution data of soil type in China adopted in this paper is supported by the Geographic Data Sharing Infrastructure, College of Urban and Environmental Science, Peking University. All these data were resampled to 0.1° for model-driven use. CO_2 concentration data are derived from the Earth's CO_2 Network, which updates CO_2 concentrations on a daily basis and provides annual, monthly, and daily data on Global, Northern, and Southern Hemisphere scales for the past 2000 years. In this paper, the annual CO_2 concentration in the Northern Hemisphere from 1981 to 2020 was captured to drive the model.

2.2.3. Accuracy Validation

MODIS AET products (https://modis.gsfc.nasa.gov; accessed on 1 January 2023) were used to verify the simulation results. MODIS AET is an 8-day composite product with a spatial resolution of 500 m, and the timescale is from 2001 to 2020. The validation of AET simulations was conducted from 2001 to 2020. The quality of AET simulations was determined via the certainty coefficient (R^2), Nash efficiency coefficient (NSE), and mean relative error (MRE). The calculation formula is as follows:

$$R^2 = \frac{\left(\sum_{i=1}^{n}(O_i - O_{avg})(Q_i - Q_{avg})\right)^2}{\sum_{i=1}^{n}(O_i - O_{avg})^2 \sum_{i=1}^{n}(Q_i - Q_{avg})^2} \tag{10}$$

$$NSE = 1 - \frac{\sum_{i=1}^{n}(O_i - Q_i)^2}{\sum_{i=1}^{n}(O_i - O_{avg})^2} \tag{11}$$

$$MRE = \frac{1}{n}\sum_{i=1}^{n}\left|\frac{(Q_i - O_i)}{O_{avg}}\right| \tag{12}$$

where Q_i is the monthly average AET simulated by the LPJ model, O_i is the MODIS AET results for model verification, Q_{avg} is the total monthly average AET in the simulation period (all months), O_{avg} is the total monthly average MODIS AET results for model verification (all months), and n is the total number of months in the simulation period. The closer R^2 and NSE are to 1, and the closer MRE is to 0, the higher the simulation accuracy is.

2.3. Analysis Methods

2.3.1. Change Characteristics Analysis

Sen's slope [46] and the Mann–Kendall (M-K) test method [47–49] are relatively mature statistical methods in the field of geosciences and have been applied to analyze the trends, mutations, and magnitudes of hydrological and meteorological factors in basins. In this study, the analysis of the change characteristics for each element adopted the above two methods.

2.3.2. Driving Mechanism of AET

The sensitivity coefficient (SC) calculation method proposed by Beven [50] and the method by Zhao et al. [51] were used to discuss the driving mechanism of AET and quantify the contribution rate of driving factors to the change in AET; the formula is as follows:

$$Se_{vi} = \lim_{\Delta vi \to 0}\left(\frac{\Delta AET}{AET} \bigg/ \frac{\Delta V_i}{V_i}\right) = \frac{\partial AET}{\partial V_i} \cdot \frac{V_i}{AET} \tag{13}$$

$$G_{vi} = \frac{\Delta V_i}{\overline{V_i}} \cdot Se_{vi} \tag{14}$$

where Se_{vi} is the SC of the ith meteorological factor (V_i). G_{vi} is the contribution rate of the ith meteorological factor (V_i) to the change in AET. The SC of a meteorological element is positive or negative, indicating that AET increases or decreases as the element increases.

3. Results

3.1. Validation of AET Simulation Results

In this study, seasonal statistics and analysis were conducted from March to May for spring, June to August for summer, September to November for autumn, and December to February for winter. From Table 3, in terms of seasonal simulation results, spring has the best effect, followed by autumn and winter, and summer has the biggest error. In spring, R^2, NSE, and MRE were 0.83, 0.72, and 6.39%, respectively, indicating that the LPJ model could accurately simulate the AET characteristics in the TGRA. Both R^2 and NSE in autumn were higher than those in spring, indicating that the description of the change characteristics in AET in autumn was better than that in spring, and MRE was higher than that in spring, indicating that the characterization of AET in autumn was inferior to that in spring. The overall simulation accuracy in winter was slightly worse than that in spring and autumn, and it was the worst in summer, with R^2, NSE, and MRE percentages of 0.76, 0.63, and 17.68%, respectively. For the annual simulation results, R^2 and NSE were 0.89 and 0.76, respectively, indicating a high degree of fitting between simulated and measured AET sequences. The MRE was 4.32%, indicating a small error between the simulated and measured values of AET. From the scatterplot of measured and simulated AET (Figure 2), scatter points are basically distributed around the trend line, and the slope of the trend line is less than 1, indicating that the simulated AET was slightly lower than the measured AET. In general, the LPJ model had good performance in simulating the AET, which can accurately represent the actual trend characteristics of AET in the TGRA.

Table 3. Precision evaluation of seasonal and annual AET simulation results in the TGRA.

Evaluation Index	Spring	Summer	Autumn	Winter	Annual Average
R^2	0.83	0.76	0.86	0.75	0.89
NSE	0.72	0.63	0.74	0.67	0.76
MRE	6.39%	17.68%	7.49%	9.59%	4.32%

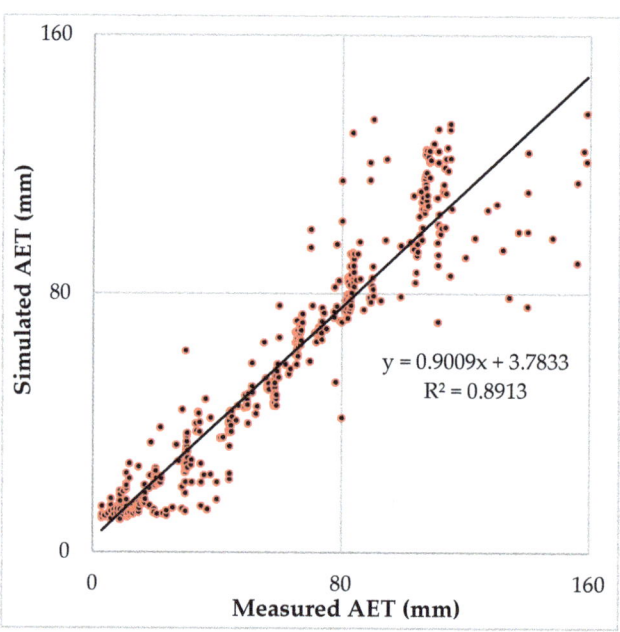

Figure 2. Scatter plot of the simulated and measured AET in the TGRA.

3.2. Spatial and Temporal Characteristics of Variations in AET

From the perspective of the interannual variation trend, the AET showed a fluctuating upward trend (Figure 3a), with a multiyear average value of 651.43 mm. The highest and lowest values were 739.77 mm and 583.04 mm, respectively. After 2002, it was significantly higher than before. The annual distribution of AET showed a unimodal seasonal variation trend (Figure 3b). The peak value occurred in July, reaching 113.02 mm, whereas the valley value occurred in January and December, less than 13 mm. In addition to a slight decreasing trend in May, AET showed an increasing trend in other months, among which Sen's slope statistics in January to April and July to September passed the significance test. AET increased the fastest in July at a speed of 6.64 mm/10a, and the overall change was evident.

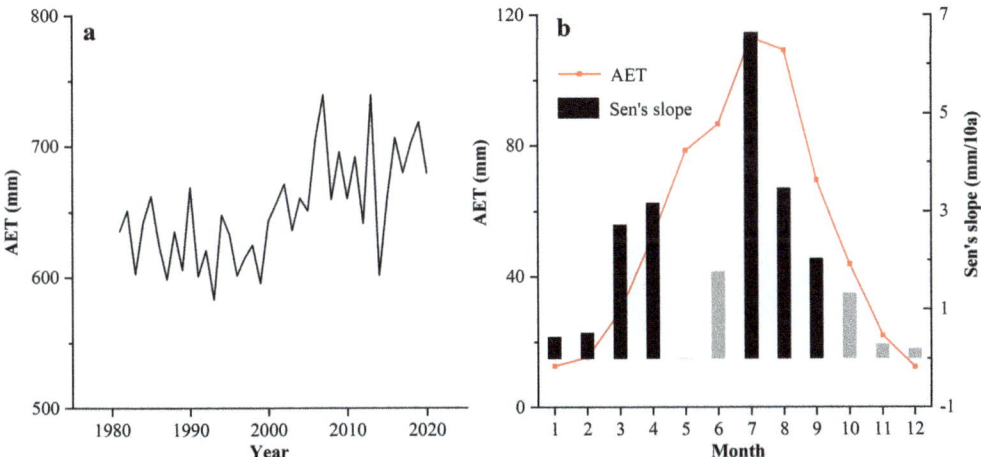

Figure 3. Interannual variation in of the AET (**a**) and monthly trends in AET and Sen's slope tests for the monthly AET in the TGRA (**b**): the black columns represent Sen's slope passed the significance test; the gray columns represent Sen's slope not passed the significance test.

As shown in Figure 4, the spatial distribution of monthly AET showed significant seasonal differences. In spring and summer from March to August, the spatial distribution of AET showed significant characteristics of high in the west and low in the east. The value of monthly AET in summer was higher than other seasons, reflecting the characteristics of increased soil evaporation and vegetation transpiration level under adequate water and heat conditions, which led to good vegetation growth in summer. The spatial distribution of autumn AET from September to November showed similar characteristics of high in the head and the west and lower in the end and the east of the reservoir. In November, the AET clearly decreased, reflecting the characteristics of vegetation fading and evapotranspiration decreasing in late autumn and early winter. The spatial pattern of AET from December to February was similar and reached the lowest level of the whole year, reflecting the characteristics of the low AET level caused by vegetation decay and insufficient hydrothermal conditions in winter.

The multiyear average AET in the TGRA ranged from 492.59 to 719.45 mm (Figure 5a), and the average for the whole area was 645.96 mm. AET exhibited nonsignificant spatial differentiation and showed a decreasing trend from the east and west to the center. As shown in Figure 5b, from 1981 to 2020, the annual change rate of AET in the TGRA ranged from 6.56 to 53.16 mm/10a, and the spatial statistical mean value was 21.63 mm/10a. The changes in AET in the end of tail areas and part of the northern belly areas were significant, and the increased amplitude in other regions was small.

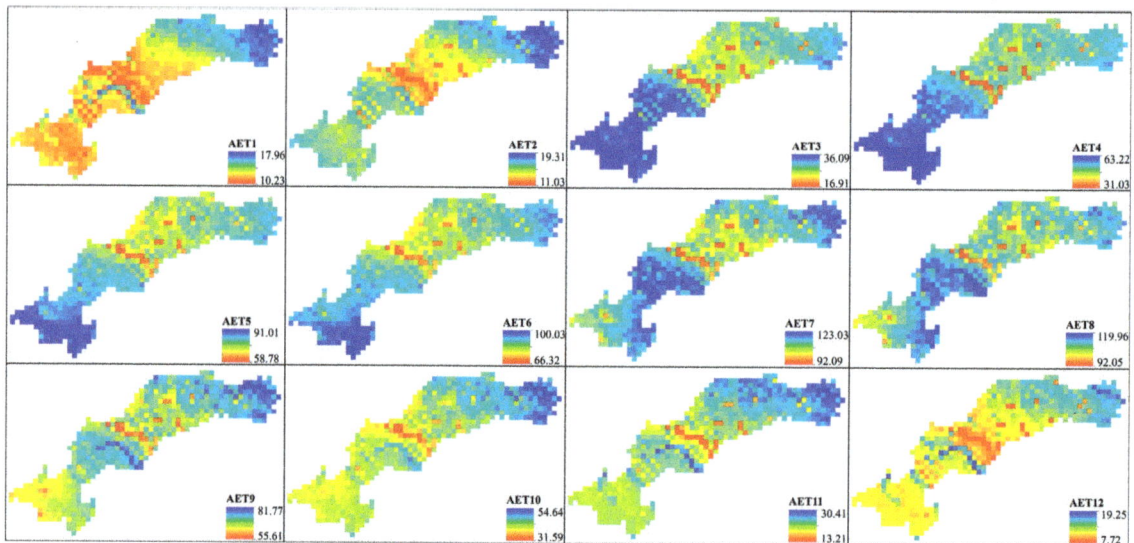

Figure 4. Monthly distribution in AET from 1891 to 2020 in the TGRA.

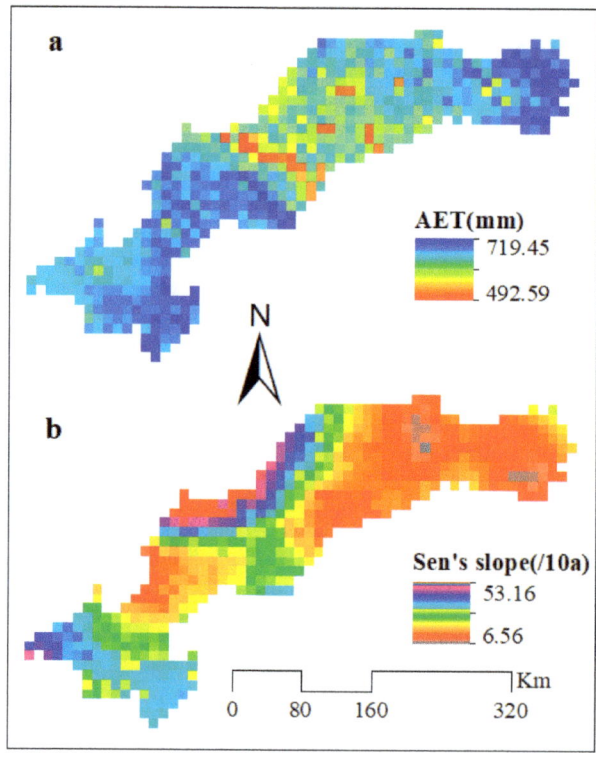

Figure 5. Spatial distribution of the multiyear average of the AET (**a**) and amplitude of the variation in the AET based on Sen's slope (**b**) in the TGRA.

3.3. Analysis of Driving Mechanism for AET in the TGRA

3.3.1. Correlation Analysis

Correlation analysis [52] was conducted between AET and climate factors to identify the key climate factors affecting AET. The results showed a significant linear correlation between AET and P, T, and net radiation (Rs) in the TGRA, with correlation coefficients of 0.74, 0.96, and 0.91, respectively (Table 4). In order to improve the fitting accuracy, the principle of the maximum correlation coefficient was adopted. After many simulations, it was found that the correlation coefficients of exponential regression between AET and P, T, and Rs were relatively higher than linear regression, with correlation coefficients of 0.78, 0.97, and 0.94, respectively (Table 4).

Table 4. Linear and exponential correlation coefficients between AET and P, T, and Rs.

Fitting Relation	P	T	Rs
Linear correlation	0.74	0.96	0.91
Exponential correlation	0.78	0.97	0.94

Figure 6 shows the exponential regression relationships between AET and P, T, and Rs in the TGRA. In comparison, the regression relation between AET and T was the best, followed by Rs and the lowest of P, with an adjusted R^2 of 0.95, 0.88, and 0.61, respectively. It showed that AET in the TGRA was mainly affected by the temperature and radiation term, which is consistent with the conclusion of Yan et al. [53].

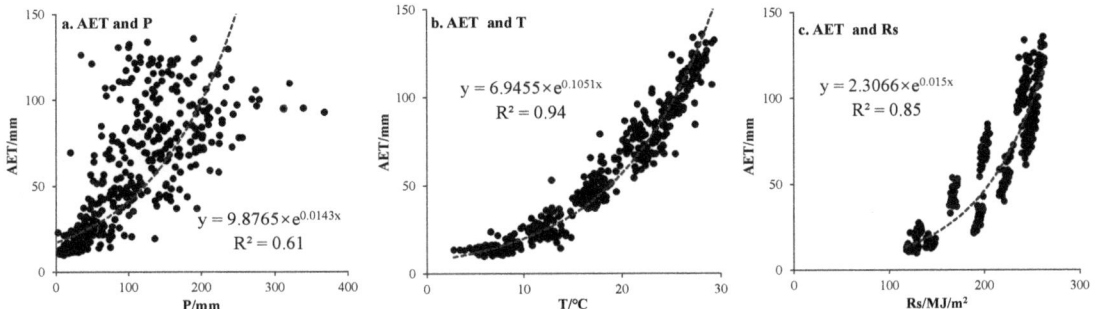

Figure 6. Exponential regression relations and adjusted R^2 of AET and P, T, and Rs in the TGRA ($p \leq 0.01$).

3.3.2. Change Characteristics of Key Climatic Factors

The interannual variation characteristics (Figure 7a,c,e) and M-K test (Figure 7b,d,f) of P, T, and Rs in the TGRA from 1981 to 2020 are shown in Figure 7. Table 5 shows the statistical variation results of each factor based on Sen's slope tests. In the past 40 years, the P has fluctuated significantly (Figure 7a) and increased slightly at a change rate of 11.86 mm/10a (Table 5). There were several abrupt transition points in P, which occurred in 1981, 1983, 2015, and 2018 during the statistical period (Figure 7b). Since 1981, the T has been steadily increasing at a rate of 0.31 °C/10a (Table 5), and an abrupt transition point occurred in 2000 (Figure 7d). From 1981 to 2020, the mean value of Rs was 2297.31 MJ/m^2 (Figure 7e) and decreased at a rate of 2.75 MJ/m^2/10a (Table 5). The abrupt transition points in Rs occurred in 2001, 2003, and 2018 (Figure 7f). In general, there were certain fluctuations in climate change in the TGRA, but the whole trend was relatively stable.

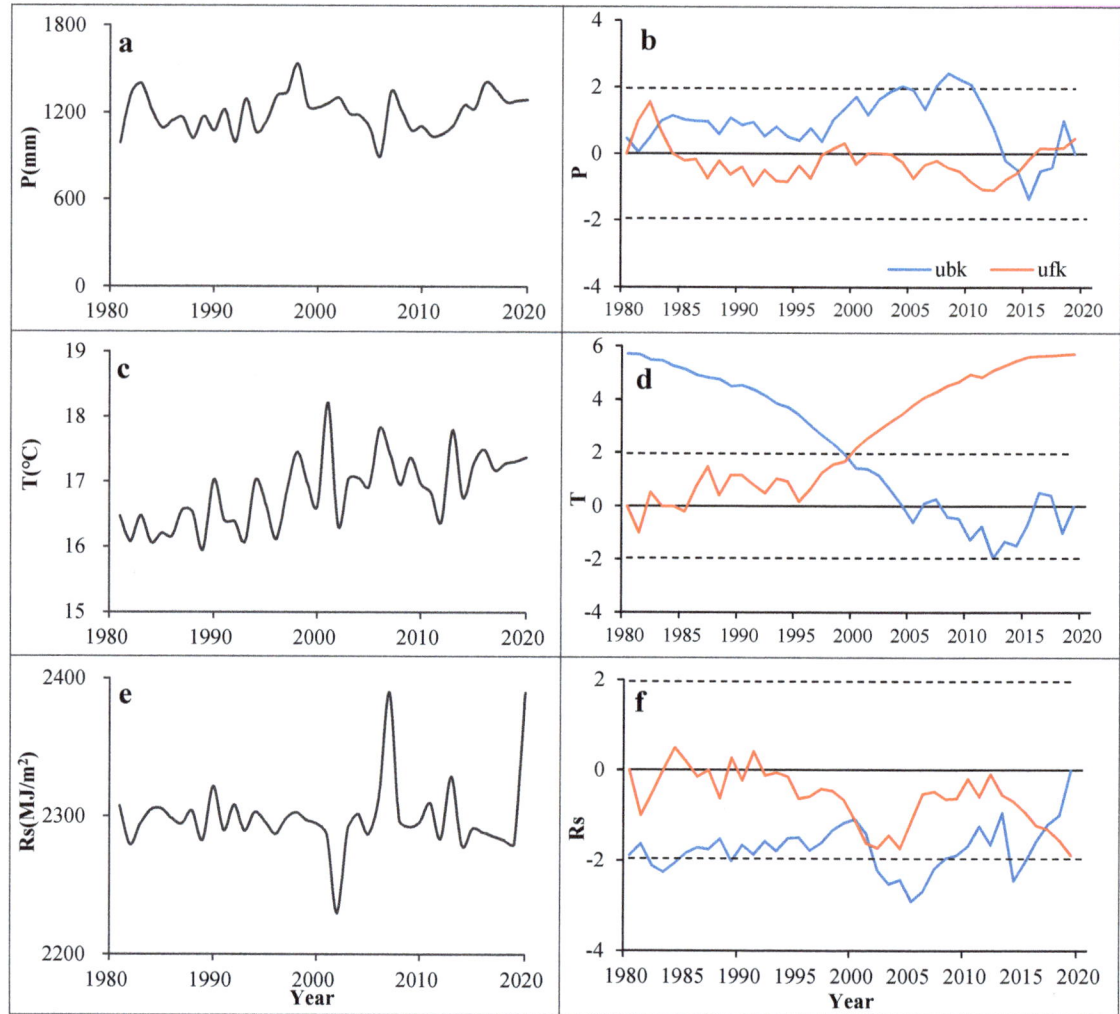

Figure 7. Interannual variations (**a,c,e**) and M-K test (**b,d,f**) of the variation trends in P, T, and Rs during 1981 to 2020 in the TGRA.

Table 5. Sen's slope test for amplitude of the variation in P, T, and Rs in the TGRA (/10a).

	P (mm)	T (°C)	Rs (MJ/m^2)
Sen's slope	11.86 *	0.31 **	−2.75

Note: * The values are significant at $p \leq 0.05$; ** the values are significant at $p \leq 0.01$ (the same below).

3.3.3. Sensitivity Analysis

Combined with the abrupt change time of T (Figure 7d: 2001) and Rs (Figure 7f: 2001 and 2003), which has a large-degree influence on AET, and the change trend in AET (Figure 3a), there is a significant difference between the two periods before and after impoundment for the Three Gorges Project. Therefore, the statistical period (1981–2020) for the driving mechanism of AET was divided into two stages of before (BI: 1981–2002) and after impoundment (AI: 2003–2020).

Table 6 shows the amplitudes of the SCs calculated according to Equation (13) for the P, T, and Rs of the TGRA. The SC of AET to P increased at a rate of 0.07/10a, 0.15/10a, and 0.01/10a during BI, AI, and the whole study period, respectively. The statistical results did not pass the significance test. The SC of AET to T changed by 0.09/10a and −0.03/10a in BI and AI, respectively. The influence of T on AET increased before impoundment and began to weaken after impoundment. During the whole study period, the SC for T showed a weak increasing trend with a change rate of 0.03/10a. The SC of AET to Rs showed a decreasing trend in BI, AI, and the whole study period. From 2003 to 2020, the decrease trend was significant, with a change rate of −0.08/10a, indicating that the influence of Rs on AET clearly weakened during AI.

Table 6. Sen's slope tests for the amplitudes of the SCs for P, T, and Rs in the TGRA (/10a).

Time Interval	P	T	Rs
1981–2002	0.07	0.09 *	−0.03 *
2003–2020	0.15	−0.03	−0.08 *
1981–2020	0.01	0.03 *	−0.10 **

Figure 8 shows the spatial distributions of the annual average SCs of AET to each climate factor in different periods. From 1981 to 2002, the SC for P ranged from 0.75 to 6.21, with a spatial averaged value of 1.07. Except for a few areas in the southern edge of the TGRA reaching a higher value, the SC for P distribution in other areas was even. The averaged SC for T was 1.76, and the spatial distribution was high in the west and low in the east. The averaged SC for Rs was 3.09. The spatial distribution showed a characteristic of 'high in the middle and low at both end'. During 2003–2020, the spatial difference of the SC for P was significant, with maximum and minimum values of 0.51 and 8.45, respectively, and the mean value was 1.12, slightly increased compared with that before 2002. The SC for P was higher in east belly areas and head areas of the reservoir and evenly distributed at a lower level in other areas. The SC for T increased compared with that before 2002, with a mean value of 1.84. The spatial distribution of the SC for Rs was basically the same as that in BI, and the value of SC decreased slightly, with a spatial mean value of 2.84. During the whole study period, the SCs of AET to P, T, and Rs were 1.09, 1.80, and 2.99, respectively. The spatial distributions were generally consistent with that in AI. The SC for Rs was the highest, followed by T, and P was the lowest, which indicates that the change in Rs per unit quantity leads to the largest degree of change in AET, whereas P was the least. It also indicates that the temperature and radiation term was the dominant climatic factor leading to the change in AET in the TGRA.

3.3.4. Contribution of Driving Factors to the Change in AET

The contribution rates of meteorological factors to the change in AET in different statistical periods in the TGRA were calculated according to the sensitivity analysis results and Equation (14), and the results are shown in Table 7. From 1981 to 2002, the changes in P, T, and Rs could cause the change rates of AET to be 5.54%, 0.24%, and 0.03%, respectively. The influences of the three factors on AET were all positive driving, in which the P contributed the most to the change in AET, followed by T, and the contribution of Rs was the smallest. During 2003 to 2020, the contribution of P and T to the change in AET decreased to a certain extent compared with that before 2002, whereas Rs had a certain inhibitory effect on the increase in AET, which was mainly related to the significant decrease trend in Rs during this period. From the perspective of the whole study period, P and T had a great influence on the change in AET. In general, the increases in P and T were the main reasons for the increase in the AET of the TGRA, and the decrease in Rs weakened the increased tendency.

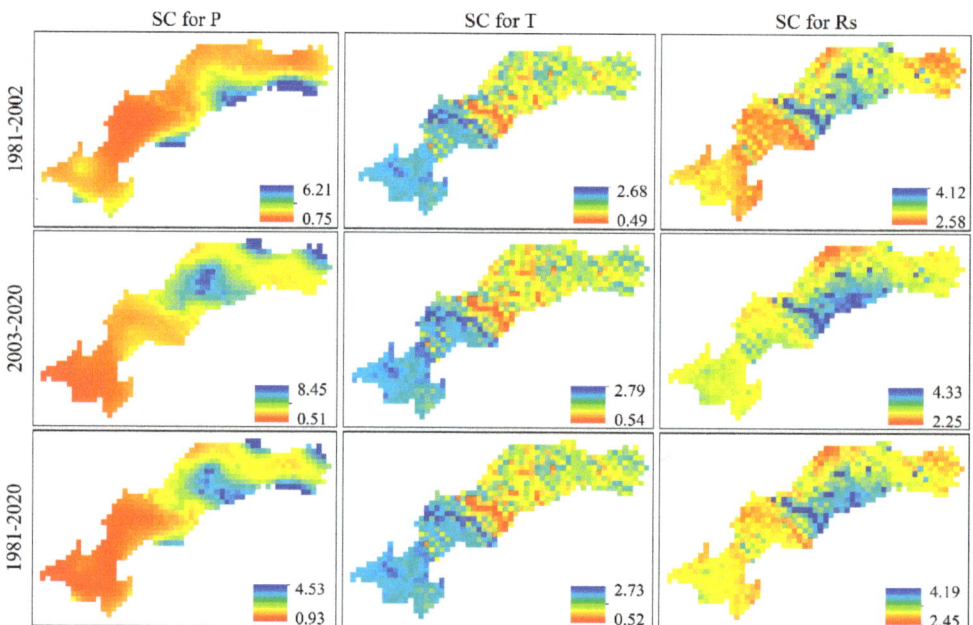

Figure 8. Distributions of the SCs for the P, T, and Rs in different periods in the TGRA.

Table 7. Contribution of P, T, and Rs to AET changes in different statistical periods (%).

Time Interval	P	T	Rs
1981–2002	5.54	0.24	0.03
2003–2020	4.46	0.17	−0.04
1981–2020	4.59	0.19	−0.06

The change rates of AET in BI, AI, and the whole study period were calculated, and the contribution of human activities to the change in AET was obtained by subtracting the change rate caused by climate factors from the total change rate of AET. As shown in Table 8, during 1981–2002, the AET changed by 5.60%, the change rate caused by climate factors was 5.81%, accounting for 103.75%, and the contribution of human activities was −0.21%, accounting for −3.75%, indicating that the effect of human activities on AET during this period was negative driving. In BI, human activities caused a certain degree of disturbance to the underlying surface and vegetation destruction, which inhibited the increase in AET. From 2003 to 2020, the AET changed by 6.28%, in which climate factors and human activities contributed 4.59% and 1.69%, with contribution rates of 73.09% and 26.91%, respectively; the contribution rate of human activities increased significantly. During the whole study period, the change rate of AET was 5.28%, the contributions of climate change and human activities were 4.72% and 0.56%, respectively, and the contribution ratios were 89.39% and 10.61%, respectively, indicating that in the nearly 40 years from BI to AI, human activities contributed 10.61% to the increase in AET, and the climate factor was the dominant factor on the increase in AET.

Table 8. Contribution of climate change and human activities to AET changes in different statistical periods (%).

Time Interval	AET Total Change Rate/%	Climate Change Contribution/%	Rate/%	Human Activity Contribution/%	Rate/%
1981–2002	5.60	5.81	103.75	−0.21	−3.75
2003–2020	6.28	4.59	73.09	1.69	26.91
1981–2020	5.28	4.72	89.39	0.56	10.61

4. Discussion

4.1. Reliability of the Simulated AET

The validation of AET simulated by the LPJ was mainly based on MODIS AET products from 2001 to 2020. In order to further explore the reliability of this study, we compared other scholars' research results about AET in the TGRA and its surrounding areas. The specific comparison results are shown in Table 9. Wang et al. [54] used CLM4.5, which belongs to a kind of DGVM-simulated AET of the TGRA, from 1993 to 2013. The results showed that the multiyear average AET was 606 mm. In our study, the multiyear average AET during the same time scale was 655.11 mm, which is close to the CLM4.5-simulated AET, with a relative error of 7.50%. Cui et al. [55] also used the CLM4.5 to estimate the AET in the TGRA from 1990 to 2015 and obtained a multiyear average AET of 590.75 mm, which had a relative error of 9.17% with this paper. Cao et al. [56] estimated the regional AET in the middle and lower reaches of the Yangtze River by using the principle of water balance and remote-sensed data from 1992 to 2015 and found that the multiyear average AET was 728.70 mm, which was more than our result by 10.57%. Considering that the middle and lower reaches of the Yangtze River contain a part of the TGRA, we think the results are reliable for comparison to a certain extent.

According to the above comparisons, it is concluded that the results of AET in this study match well with previous studies in different research periods and locations around the TGRA. Therefore, we are convinced that the simulation of AET in this study is reliable and credible.

Table 9. The comparison of the LPJ simulated values of AET with previous studies.

Research	Study Area and Period	Methods	Their Values (mm)	This Study's Values (mm)	Relative Error (%)
Wang et al. [54]	TGRA/1993–2013	CLM4.5	606.00	655.11	7.50
Cui et al. [55]	TGRA/1990–2015	CLM4.5	590.75	650.41	9.17
Cao et al. [55]	middle and lower reaches of Yangtze River/1992–2015	water balance and remote sensed data	728.70	651.70	10.57

4.2. Distribution and Variations in AET Components

In the LPJ model, AET consists of EP, EI, and ES. An analysis of the variation characteristics in these three parts is helpful to understand the variation in AET. Figure 9 shows the spatial distribution and variation amplitude of the three parts in the TGRA. The spatial statistical mean values of EP, EI, and ES were 459.33 mm, 78.57 mm, and 108.06 mm, respectively. Among them, EP accounted for the largest proportion of AET, with 71.11%, indicating that the AET was mainly from canopy transpiration. EP ranged from 317.86 to 515.07 mm. Under the background of increasing T and P, which provided a hydrothermal condition, vegetation growth showed a benign trend, and EP increased accordingly. EI accounted for the smallest proportion of AET with 12.16%. The annual averaged EI ranged from 20.59 to 92.21 mm and showed a decreasing trend in most regions, especially in urban areas, where the urban expansion was the main reason for the decrease in EI. ES ranged from 48.51 to 128.36 mm, accounting for 16.72% of AET. Except for a few areas at the end of the reservoir, ES increased in most areas of the TGRA. As for the reservoir impoundment, the water level and water surface area increased; the water supply was sufficient to fully meet the potential evapotranspiration demand. Therefore, ES increased significantly in water bodies and surrounding areas. In general, the AET increased during the study period, and its components presented different spatial and variation characteristics.

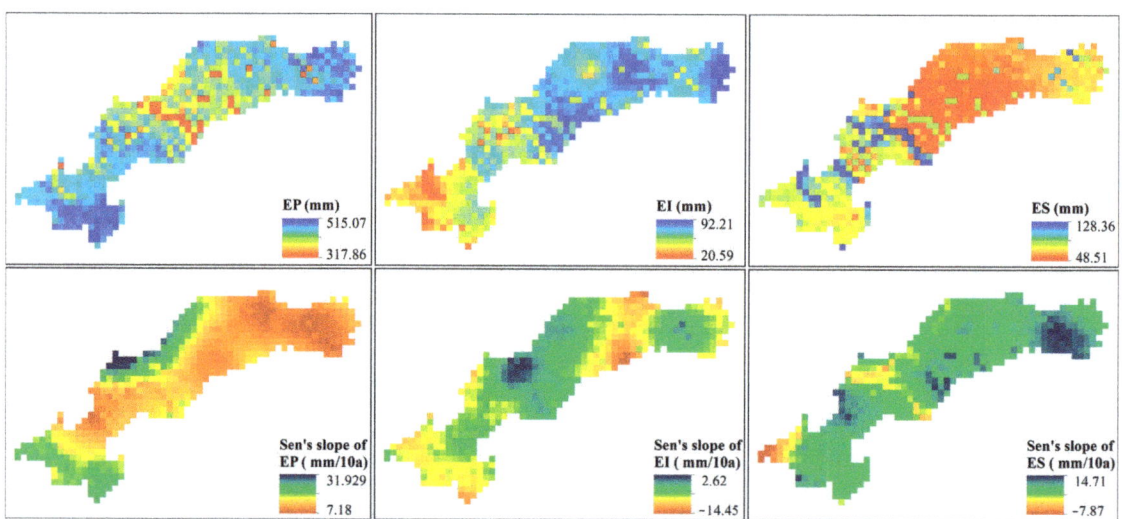

Figure 9. Spatial distribution and amplitude of the variation based on Sen's slope of the EP, EI, and ES in the TGRA.

4.3. Limitations and Future Improvements

Although the LPJ model had a good performance in simulating the AET over the TGRA, it is still necessary to discuss the limitations of the modeling results. Firstly, the spatial resolution of the atmospheric forcing data is 0.1 degrees, which leads to the spatial distribution of AET not being accurately described. Secondly, the parameters in the LPJ model are numerous and difficult to obtain, and the ecological physiological parameters in the model mainly refer to the previous research results; we used just 2009 soil texture data in the LPJ model, which created uncertainty in the simulation of AET. Finally, human activities drive AET, mainly in two ways. The first is to directly affect the dynamic process of the hydrologic cycle by changing the land use types [54], and the second is to make a certain climate effect due to the disturbance to the underlying surface. This climate effect feeds back into climate change, which, in turn, affects the hydrologic cycle [57–59]. The change in the contribution rate of human activities on AET variation indicates that human activities drove AET mainly through the first and second ways before and after impoundment. In fact, the process of producing climate effects by human activities throughout the whole study period and the contribution of climate factors on AET change partially include human effects. Therefore, localization parameters need to be acquired through field detection or experimental observation, and a more accurate input dataset and different ages of soil data for different simulation periods are needed to carry out future research. Based on the higher-resolution simulation results and setting different simulation scenarios, separating the part of the effect produced by human factors from climate change would also be the focus of future research.

5. Conclusions

In this study, we simulated the AET from 1981 to 2020 in the TGRA by using the LPJ model, and we quantified the individual contributions of climate and human factors to changes in AET in different periods using Sen's slope and sensitivity analysis methods. The results of this study demonstrate the following:

(1) The simulation accuracy of the LPJ model for AET in the TGRA was the best in spring and autumn, followed by winter, and the worst in summer. In general, the overall accuracy was high, which can accurately represent the trend characteristics of AET in the TGRA.

(2) During 1981–2020, the AET showed a fluctuating upward trend, with a multiyear average value and increase rate of 645.96 mm and 21.63 mm/10a. The annual distribution of AET showed a unimodal seasonal variation trend, with a peak value in July of 113.02 mm and the valley value in January and December at less than 13 mm. The spatial distribution of AET was even. The changes in AET in the end of tail areas and part of the northern belly areas were significant, whereas the increased amplitude in other regions was small.

(3) Correlation analysis showed that the key climate factors affecting AET changes in the TGRA were mainly P, T, and Rs. The SC of AET to Rs was the highest at 2.99, followed by T of 1.80, and P was the lowest at 1.09, which means that the change in Rs per unit quantity leads to the largest degree of change in AET, whereas P was the least. It also indicates that the temperature and radiation term was the dominant climatic factor, leading to the change in the AET in the TGRA.

(4) From 1981 to 2002, the change rate in the AET was 5.60%; the change rates caused by climate factors and human activities were 5.81% and −0.21%, accounting for 103.75% and −3.75%, respectively. From 2003 to 2020, the AET changed by 6.28%, in which climate and human factors contributed 73.09% and 26.91%, respectively. From the whole study period, the change rate of AET was 5.28%, and the contribution ratios of climate and human factors were 89.39% and 10.61%, respectively. Overall, human activities and climate change, which hold a dominant position, jointly promoted the increase in AET in the TGRA.

Author Contributions: H.W. conceived the idea and helped with language editing. X.Z. performed the analyses and wrote the paper. G.W. helped revise the paper. All authors have read and agreed to the published version of the manuscript.

Funding: This research was funded by the Special Topic on Basic and Applied Research of Guangzhou Science and Technology Bureau (SL2023A04J00927), Special program for high level talents of Guangdong polytechnic of industry and commerce (2022-gc-12), and National Key Research and Development Project of China (Grant 2021YFC3000202).

Data Availability Statement: Data are contained within the article.

Acknowledgments: We thank the anonymous reviewers and the editor for their suggestions, which substantially improved the manuscript. Special thanks need to go to the National Tibetan Plateau Data Center; they provided the China meteorological forcing dataset.

Conflicts of Interest: The authors declare no conflict of interest.

References

1. Wu, B.; Yan, N.; Xiong, J.; Bastiaanssen, W.G.M.; Zhu, W.; Stein, A. Validation of ETWatch using field measurements at diverse landscapes: A case study in Hai Basin of China. *J. Hydrol.* **2012**, *436-437*, 67–80. [CrossRef]
2. Mahmoud, S.H.; Gan, T.Y. Irrigation water management in arid regions of Middle East: Assessing spatio-temporal variation of actual evapotranspiration through remote sensing techniques and meteorological data. *Agric. Water Manag.* **2019**, *212*, 35–47. [CrossRef]
3. Wu, B.; Zhu, W.; Yan, N.; Xing, Q.; Xu, J.; Ma, Z.; Wang, L. Regional Actual Evapotranspiration Estimation with Land and Meteorological Variables Derived from Multi-Source Satellite Data. *Remote Sens.* **2020**, *12*, 332. [CrossRef]
4. Su, W.; Shao, H.; Xian, W.; Xie, Z.; Zhang, C.; Yang, H. Quantification of Spatiotemporal Variability of Evapotranspiration (ET) and the Contribution of Influencing Factors for Different Land Cover Types in the Yunnan Province. *Water* **2023**, *15*, 3309. [CrossRef]
5. Yao, J.; Mao, W.; Yang, Q.; Xu, X.; Liu, Z. Annual actual evapotranspiration in inland river catchments of China based on the Budyko framework. *Stoch. Environ. Res. Risk Assess.* **2016**, *31*, 1409–1421. [CrossRef]
6. Shi, Z.; Xu, L.; Yang, X.; Guo, H.; Dong, L.; Song, A.; Zhang, X.; Shan, N. Trends in reference evapotranspiration and its attribution over the past 50 years in the Loess Plateau, China: Implications for ecological projects and agricultural production. *Stoch. Environ. Res. Risk Assess.* **2017**, *31*, 257–273. [CrossRef]

7. Ji, Y.; Tang, Q.; Yan, L.; Wu, S.; Yan, L.; Tan, D.; Chen, J.; Chen, Q. Spatiotemporal Variations and Influencing Factors of Terrestrial Evapotranspiration and Its Components during Different Impoundment Periods in the Three Gorges Reservoir Area. *Water* **2021**, *13*, 2111. [CrossRef]
8. Jiao, P.; Hu, S. Estimation of Evapotranspiration in the Desert–Oasis Transition Zone Using the Water Balance Method and Groundwater Level Fluctuation Method—Taking the Haloxylon ammodendron Forest at the Edge of the Gurbantunggut Desert as an Example. *Water* **2023**, *15*, 1210. [CrossRef]
9. Wei, L.; Duan, K.; Liu, X.; Lin, Y.; Chen, X.; Wang, X. Assessing human-induced evapotranspiration change based on multi-source data and Bayesian model averaging at the basin scale. *Shuili Xuebao* **2022**, *53*, 433–444. [CrossRef]
10. Duan, K.; Sun, G.; Liu, N. A review of research on watershed water-carbon balance evolution in a changing environment. *Shuili Xuebao* **2021**, *52*, 300–309. [CrossRef]
11. Chen, H.; Zhu, G.; Shang, S.; Qin, W.; Zhang, Y.; Su, Y.; Zhang, K.; Zhu, Y.; Xu, C. Uncertainties in partitioning evapotranspiration by two remote sensing-based models. *J. Hydrol.* **2022**, *604*, 127223. [CrossRef]
12. Mekonnen, K.; Melesse, A.M.; Woldesenbet, T.A. How suitable are satellite rainfall estimates in simulating high flows and actual evapotranspiration in MelkaKunitre catchment, Upper Awash Basin, Ethiopia? *Sci. Total Environ.* **2022**, *806*, 150443. [CrossRef] [PubMed]
13. Zhang, X.; Duan, Y.; Duan, J.; Jian, D.; Ma, Z. A daily drought index based on evapotranspiration and its application in regional drought analyses. *Sci. China Earth Sci.* **2022**, *65*, 317–336. [CrossRef]
14. Zhang, Y.; Yao, L.; Geurink, J.S.; Parajuli, K.; Wang, D. Climatic Control on Mean Annual Groundwater Evapotranspiration in a Three-Stage Precipitation Partitioning Framework. *Water Resour. Res.* **2023**, *59*, e2022WR034167. [CrossRef]
15. Chen, H.; Liu, H.; Chen, X.; Qiao, Y. Analysis on impacts of hydro-climatic changes and human activities on available water changes in Central Asia. *Sci. Total Environ.* **2020**, *737*, 139779. [CrossRef]
16. Sitch, S.; Huntingford, C.; Gedney, N.; Levy, P.E.; Lomas, M.; Piao, S.L.; Betts, R.; Ciais, P.; Cox, P.; Friedlingstein, P.; et al. Evaluation of the terrestrial carbon cycle, future plant geography and climate-carbon cycle feedbacks using five Dynamic Global Vegetation Models (DGVMs). *Glob. Chang. Biol.* **2008**, *14*, 2015–2039. [CrossRef]
17. Che, M.; Chen, B.; Wang, Y.; Guo, X. Review of dynamic global vegetation models (DGVMs). *Chin. J. Appl. Ecol.* **2014**, *25*, 263–271.
18. Wang, G.; Qian, J.; Cheng, G. Current situation and prospet of the ecological hydrology. *Adv. Earth Sci.* **2001**, *16*, 314–323.
19. Smith, B.; Sykes, P.M.T. Representation of vegetation dynamics in the modelling of terrestrial ecosystems: Comparing two contrasting approaches within European climate space. *Glob. Ecol. Biogeogr.* **2001**, *10*, 621–637. [CrossRef]
20. Bonan, G.; Levis, S.; Sitch, S.; Vertenstein, M.; Oleson, K. A dynamic global vegetation model for use with climate models: Concepts and description of simulated vegetation dynamics. *Glob. Chang. Biol.* **2003**, *9*, 1543–1566. [CrossRef]
21. Thonicke, K.; Venevsky, S.; Sitch, S.; Cramer, W. The role of fire disturbance for global vegetation dynamics: Coupling fire into a Dynamic Global Vegetation Model. *Glob. Ecol. Biogeogr.* **2001**, *10*, 661–677. [CrossRef]
22. Venevsky, S.; Maksyutov, S. SEVER: A modification of the LPJ global dynamic vegetation model for daily time step and parallel computation. *Environ. Model. Softw.* **2007**, *22*, 104–109. [CrossRef]
23. Renwick, K.M.; Fellows, A.; Flerchinger, G.N.; Lohse, K.A.; Clark, P.E.; Smith, W.K.; Emmett, K.; Poulter, B. Modeling phenological controls on carbon dynamics in dryland sagebrush ecosystems. *Agric. For. Meteorol.* **2019**, *274*, 85–94. [CrossRef]
24. Sitch, S.; Smith, B.; Prentice, I.; Arneth, A.; Bondeau, A.; Cramer, W.; Kaplan, J.; Levis, S.; Lucht, W.; Sykes, M.; et al. Evaluation of ecosystem dynamics, plant geography and terrestrial carbon cycling in the LPJ dynamic global vegetation model. *Glob. Chang. Biol.* **2003**, *9*, 161–185. [CrossRef]
25. Zhao, M.; Yue, T.; Zhao, N.; Sun, X.; Zhang, X. Combining LPJ-GUESS and HASM to simulate the spatial distribution of forest vegetation carbon stock in China. *J. Geogr. Sci.* **2014**, *24*, 249–268. [CrossRef]
26. Sun, G.; Mu, M. Understanding variations and seasonal characteristics of net primary production under two types of climate change scenarios in China using the LPJ model. *Clim. Chang.* **2013**, *120*, 755–769. [CrossRef]
27. Zhang, F.; Zhang, Z.; Tian, J.; Huang, R.; Kong, R.; Zhu, B.; Zhu, M.; Wang, Y.; Chen, X. Forest NPP simulation in the Yangtze River Basin and its response to climate change. *J. Nanjing For. Univ. (Nat. Sci. Ed.)* **2021**, *45*, 175–181.
28. Zhao, D.; Wu, S.; Yin, Y. Variation trends of natural vegetation net primary productivity in China under climate change scenario. *Chin. J. Appl. Ecol.* **2011**, *22*, 897–904. [CrossRef]
29. Hao, B.; Yang, H.; Ma, M.; Hao, D.; Liu, Y.; Han, X.; Li, S.; Lai, P. Variation in Land Use and Land Surface Parameters in the Three Gorges Reservoir Catchment Based on Google Earth Engine. *Resour. Environ. Yangtze Basin* **2020**, *29*, 1343–1355.
30. Luo, R.; Luo, D.; Xiao, J. A Brief Analysis of the Categories and Characteristics of Reservoir Area Towns' Cityscape Changes along the River in the Post Three Gorges Period: Case Studies of Wanzhou, Wushan and Yunyang. *Ecol. Environ. Landsc.* **2020**, *38*, 5–11+43.
31. Lv, S. Study on Land Use Change and Ecological Service Value in the Three Gorges Reservoir Area. Master's Thesis, Southwest University, Chongqing, China, 2019.
32. Shao, P.; Zeng, X. Progress in the study of the effects of land use and Land cover change on the climate system. *Clim. Environ. Res.* **2012**, *17*, 103–111.
33. Zhai, J.; Liu, R.; Liu, J.; Zhao, G.; Huang, L. Radiative forcing over China due to albedo change caused by land cover change during 1990–2010. *J. Geogr. Sci.* **2013**, *68*, 875–885. [CrossRef]

34. Liu, F.; Tao, F.; Xiao, D.; Zhang, S.; Wang, M.; Zhang, H. Influence of land use change on surface energy balance and climate: Results from SiB2 model simulation. *Prog. Geogr.* **2014**, *33*, 815–824.
35. Ding, X.; Zhou, H.; Wang, Y.; Lei, X. Prediction and evaluation on status of water cycle elements within area of Three Gorges Reservoir. *Water Resour. Hydropower Eng.* **2011**, *42*, 1–5.
36. Yang, H.; Wang, G.; Wang, L.; Zheng, B. Impact of land use changes on water quality in headwaters of the Three Gorges Reservoir. *Environ. Sci. Pollut. Res. Int.* **2016**, *23*, 11448–11460. [CrossRef] [PubMed]
37. Webber, M.; Li, M.T.; Chen, J.; Finlayson, B.; Chen, D.; Chen, Z.Y.; Wang, M.; Barnett, J. Impact of the Three Gorges Dam, the South–North Water Transfer Project and water abstractions on the duration and intensity of salt intrusions in the Yangtze River estuary. *Hydrol. Earth Syst. Sci.* **2015**, *19*, 4411–4425. [CrossRef]
38. Yue, P. Land Use/Cover Change and it's Eco-Environment Effect of Typical Districts in the Three Gorges Reservoir Region (Chongqing). Ph.D. Thesis, Southwest University, Chongqing, China, 2010.
39. Nakayama, T.; Shankman, D. Impact of the Three-Gorges Dam and water transfer project on Changjiang floods. *Glob. Planet. Chang.* **2013**, *100*, 38–50. [CrossRef]
40. Xiao, Q.; Hu, D.; Xiao, Y. Assessing changes in soil conservation ecosystem services and causal factors in the Three Gorges Reservoir region of China. *J. Clean. Prod.* **2017**, *163*, 172–180. [CrossRef]
41. Meng, H.; Zhou, Q.; Li, M.; Chen, P.; Tan, M. Topographic Distribution Characteristics of Vegetation Cover Change in the Three Gorges Reservoir Area Based on MODIS Pixel Scale. *Resour. Environ. Yangtze Basin* **2020**, *29*, 1790–1799.
42. Li, Y.C.; Liu, C.X.; Min, J.; Wang, C.J.; Zhang, H.; Wang, Y. RS/GIS-based integrated evaluation of the ecosystem services of the Three Gorges Reservoir area (Chongqing section). *Acta Ecol. Sin.* **2013**, *33*, 168–178.
43. Bonan, G.; Levis, S.; Kergoat, L.; Oleson, K. Landscapes as patches of plant functional types: An integrating concept for climate and ecosystem models. *Glob. Biogeochem. Cycles* **2002**, *16*, 5-1–5-23. [CrossRef]
44. Huang, W. Impacts of Climate Change on Typical Ecohydrological Variables in the Upper-Middle Reaches of Heihe River Basin. Master's Thesis, China University of Geosciences, Qinhuangdao, China, 2018.
45. Chen, Y.; Yang, K.; Tang, W.; Li, X.; Lu, H.; He, J.; Qin, J. China meteorological forcing dataset (1979–2018). *Natl. Tibet. Plateau Data Cent.* **2015**. [CrossRef]
46. Sen, P. Estimates of the Regression Coefficient Based on Kendall's Tau. *J. Am. Stat. Assoc.* **1968**, *63*, 1379–1389. [CrossRef]
47. Güçlü, Y.S. Improved visualization for trend analysis by comparing with classical Mann-Kendall test and ITA. *J. Hydrol.* **2020**, *584*, 124674. [CrossRef]
48. Hamed, K.H.; Ramachandra Rao, A. A modified Mann-Kendall trend test for autocorrelated data. *J. Hydrol.* **1998**, *204*, 182–196. [CrossRef]
49. Güçlü, Y.S. Multiple Şen-innovative trend analyses and partial Mann-Kendall test. *J. Hydrol.* **2018**, *566*, 685–704. [CrossRef]
50. Beven, K. A Sensitivity Analysis of the Penman-Monteith Actual Evapotranspiration Estimates. *J. Hydrol.* **1979**, *44*, 169–190. [CrossRef]
51. Zhao, J.; Xu, Z.; Zuo, D.; Wang, X. Temporal variations of reference evapotranspiration and its sensitivity to meteorological factors in Heihe River Basin, China. *Water Sci. Eng.* **2015**, *8*, 1–8. [CrossRef]
52. Yang, S.Y.; Meng, D.; Li, X.J.; Wu, X.L. Multi-scale responses of vegetation changes relative to the SPEI meteorological drought index in North China in 2001–2014. *Acta Ecol. Sin.* **2018**, *38*, 1028–1039.
53. Yan, R.; Li, L. Exploring the Influence of Seasonal Cropland Abandonment on Evapotranspiration and Water Resources in the Humid Lowland Region, Southern China. *Water Resour. Res.* **2022**, *58*, e2021WR031888. [CrossRef]
54. Wang, H.; Xiao, W.; Zhao, Y.; Wang, Y.; Hou, B.; Zhou, Y.; Yang, H.; Zhang, X.; Cui, H. The Spatiotemporal Variability of Evapotranspiration and Its Response to Climate Change and Land Use/Land Cover Change in the Three Gorges Reservoir. *Water* **2019**, *11*, 1739. [CrossRef]
55. Cui, H.; Wang, L.; Wang, H.; Xiao, W.; Hou, B.; Gao, B. Temporal and Spatial Changes of Actual Evapotranspiration and Its Relationship with Meteorological Factors in the Three Gorges Reservoir Area. *Res. Soil Water Conserv.* **2021**, *28*, 193–202. [CrossRef]
56. Cao, X.; Xing, W.; Fu, Q.; Yang, L. Evolution characteristics and driving mechanism of actual evapotranspiration in the middle and lower reaches of the Yangtze River. *J. North China Univ. Water Resour. Electr. Power (Nat. Sci. Ed.)* **2023**, 1–11.
57. Song, Z.; Liang, S.; Feng, L.; He, T.; Song, X.-P.; Zhang, L. Temperature changes in Three Gorges Reservoir Area and linkage with Three Gorges Project. *J. Geophys. Res. Atmos.* **2017**, *122*, 4866–4879. [CrossRef]
58. Gao, L.; Chen, H.; Sun, S. Impacts of Three Gorges Project on Land Surface Temperature Based on MODIS Dataset. *Progress. Inquisitionesde Mutat. Clim.* **2014**, *10*, 226–234.
59. Tao, Y.; Wang, Y.; Rhoads, B.; Wang, D.; Ni, L.; Wu, J. Quantifying the impacts of the Three Gorges Reservoir on water temperature in the middle reach of the Yangtze River. *J. Hydrol.* **2020**, *582*, 124476. [CrossRef]

Disclaimer/Publisher's Note: The statements, opinions and data contained in all publications are solely those of the individual author(s) and contributor(s) and not of MDPI and/or the editor(s). MDPI and/or the editor(s) disclaim responsibility for any injury to people or property resulting from any ideas, methods, instructions or products referred to in the content.

Article

Multi-Scale Analysis of Agricultural Drought Propagation on the Iberian Peninsula Using Non-Parametric Indices

Marco Possega [1,*], Matilde García-Valdecasas Ojeda [2,3] and Sonia Raquel Gámiz-Fortis [2,3]

1. Department of Physics and Astronomy, University of Bologna, 40126 Bologna, Italy
2. Department of Applied Physics, Faculty of Sciences, University of Granada, 18071 Granada, Spain; mgvaldecasas@ugr.es (M.G.-V.O.); srgamiz@ugr.es (S.R.G.-F.)
3. Andalusian Inter-University Institute for Earth System Research (IISTA-CEAMA), Avda. Del Mediterráneo s/n, 18006 Granada, Spain
* Correspondence: marco.possega2@unibo.it

Citation: Possega, M.; García-Valdecasas Ojeda, M.; Gámiz-Fortis, S.R. Multi-Scale Analysis of Agricultural Drought Propagation on the Iberian Peninsula Using Non-Parametric Indices. *Water* 2023, 15, 2032. https://doi.org/10.3390/w15112032

Academic Editor: Oz Sahin

Received: 21 April 2023
Revised: 18 May 2023
Accepted: 25 May 2023
Published: 27 May 2023

Copyright: © 2023 by the authors. Licensee MDPI, Basel, Switzerland. This article is an open access article distributed under the terms and conditions of the Creative Commons Attribution (CC BY) license (https://creativecommons.org/licenses/by/4.0/).

Abstract: Understanding how drought propagates from meteorological to agricultural drought requires further research into the combined effects of soil moisture, evapotranspiration, and precipitation, especially through the analysis of long-term data. To this end, the present study examined a multi-year reanalysis dataset (ERA5-Land) that included numerous drought events across the Iberian Peninsula, with a specific emphasis on the 2005 episode. Through this analysis, the mechanisms underlying the transition from meteorological to agricultural drought and its features for the selected region were investigated. To identify drought episodes, various non-parametric standardized drought indices were utilized. For meteorological droughts, the Standardized Precipitation-Evapotranspiration Index (SPEI) was employed, while the Standardized Soil Moisture Index (SSI), Multivariate Standardized Drought Index (MSDI), and Standard Precipitation, Evapotranspiration and Soil Moisture Index (SPESMI) were utilized for agricultural droughts, while their ability to identify relative vegetation stress in areas affected by severe droughts was investigated using the Fraction of Absorbed Photosynthetically Active Radiation (FAPAR) Anomaly provided by the Copernicus European Drought Observatory (EDO). A statistical approach based on run theory was employed to analyze several characteristics of drought propagation, such as response time scale, propagation probability, and lag time at monthly, seasonal, and six-month time scales. The retrieved response time scale was fast, about 1–2 months, and the probability of occurrence increased with the severity of the originating meteorological drought. The duration of agricultural drought was shorter than that of meteorological drought, with a delayed onset but the same term. The results obtained by multi-variate indices showed a more rapid propagation process and a tendency to identify more severe events than uni-variate indices. In general terms, agricultural indices were found to be effective in assessing vegetation stress in the Iberian Peninsula. A newly developed combined agricultural drought index was found to balance the characteristics of the other adopted indices and may be useful for future studies.

Keywords: drought propagation; agricultural drought; meteorological drought; Iberian Peninsula; non-parametric drought index

1. Introduction

Droughts are complex and spatially heterogeneous phenomena, with high variability of conditions between adjacent locations, making it easy to find an area subject to drought while neighbouring regions feature normal or even wet conditions. These spatial characteristics are mainly detectable in climatic transition areas where atmospheric influences are heterogeneous. The Iberian Peninsula (IP) is a notable example of such an area (Figure 1), a Mediterranean region located between temperate and subtropical climates, and is subject to diversified atmospheric patterns that cause a large variability of precipitation [1–3], that has presented recurrent droughts and a significant tendency towards more arid conditions in the last decades [4]. Drought is a multi-scalar phenomenon, as the effects of precipitation deficits occur across different systems and at various time scales. This is described by [5] and is due to the involvement of mechanisms at multiple scales. The drought signal is propagated through a water and energy cycle that involves a multitude of processes. The first transition in the propagation of drought generally occurs from meteorological to agricultural drought, which is driven by the response of soil moisture or crop yield to various meteorological variables such as precipitation and evapotranspiration. The combined effect of water shortage due to a lack of precipitation and the enhanced atmospheric evaporative demand can lead to a significant depletion of soil moisture, which in turn can trigger agricultural drought events.

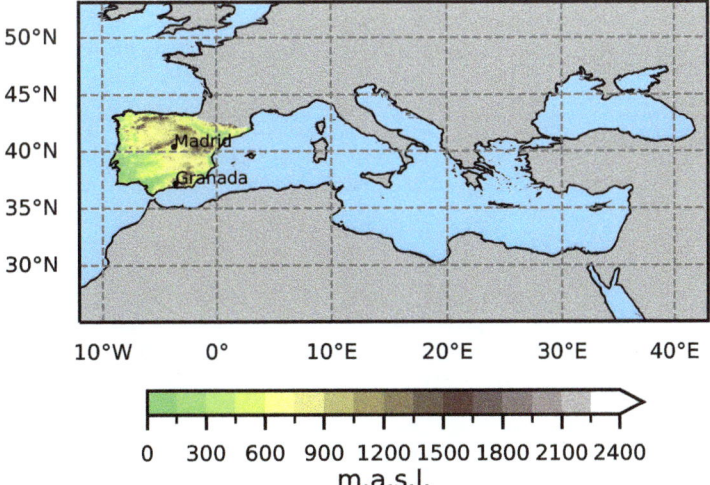

Figure 1. Location of the Iberian Peninsula within the Mediterranean sector. It includes the continental areas of Spain, Portugal, and Andorra.

The propagation from meteorological drought to agricultural drought is an understudied area, requiring further investigation to understand its complex characteristics. According to a recent review by [6], the co-occurrence of multiple driving factors in drought generation is a critical aspect that requires further attention. Currently, most studies only take into account soil moisture [7] or soil water deficit [8] and agricultural reservoir levels [9], and the combined contribution of soil moisture, evapotranspiration, and precipitation to agricultural droughts is not yet fully understood. Accordingly, as suggested in previous studies, the development and application of new multi-variate indices [10] could be helpful in shedding light on the complex relationships between meteorological and agricultural drought. Additionally, for the analysis of propagation from meteorological to agricultural drought, it is appropriate to employ high resolution and long-term data, as highlighted by [11], in particular concerning soil moisture, which can be obtained through three major sources: in-situ observations, remote sensing, and hydrological models. The majority of studies based on in-situ observations involve measuring soil moisture levels at different depths across the globe using various soil moisture networks, but this approach has limitations as the observations are relatively short [12] and unevenly distributed [13], or may be unavailable in some isolated areas. Remote sensing products have been preferred in some studies [14] since they provide better spatial coverage, but they appeared insufficient because they only cover a few centimeters of soil, while soil moisture from hydrological simulations has been found to be possibly affected by discrepancies compared to in-situ data [15]. Therefore, more studies that combine and evaluate different datasets are needed. Note that variations in vegetation health and/or cover may be due not only to rainfall or soil moisture deficits, but also to other stress factors, such as plant diseases. In this sense, indicators of vegetation stress and information on the deficit of precipitation and soil moisture must be considered together.

Given the existing research gaps, this study aimed to contribute to our understanding of agricultural drought over the IP, whose land cover is composed by a large extension of cropland along with other vegetation systems such as tree cover and grassland, specifically in relation to its propagation from meteorological droughts. To achieve this, a long-term dataset containing several drought events was analyzed, providing a comprehensive characterization of both meteorological and agricultural droughts. Various standardized drought indices were used, ranging from uni-variate to multi-variate indices that considered different physical quantities. In addition, a new combined index was proposed to account for the different factors that contribute to drought propagation. Overall, this study significantly advances our knowledge of meteorological and agricultural drought and their propagation process by leveraging a comprehensive set of tools, including meteorological, agricultural, multivariate, and combined drought indices. It offers a comprehensive perspective on the complex dynamics of drought, providing valuable insights for future research and informing effective drought management strategies.

2. Materials and Methods

2.1. Dataset

With the development of data assimilation technology, reanalysis data have become more representative of observed conditions and less limited than in-situ and remote sensing data. Reanalysis data offer global coverage, long time series, no gaps in space and time, and contain subsurface data, making them ideal for assessing agricultural drought. Several reanalysis datasets have been developed, and this study was conducted by employing the state-of-the-art reanalysis dataset for land applications, ERA5-Land [16], provided by the European Centre for Medium-Range Weather Forecasts (ECMWF) and included in the Copernicus Climate Change Service (C3S) of the European Commission. The ERA5-Land dataset was chosen as recommended in [17] due to its demonstrated relatively high accuracy compared to other remote sensing and reanalysis datasets [18] and hydrological models [19]. ERA5-Land offers a detailed record of hourly land surface evolution from several decades ago to the present, providing a vast array of key variables that represent the water and energy cycles. This dataset was chosen due to its superior ability to characterize the water cycle compared to ERA5 [20]. The original spatial resolution of the ERA5-Land dataset is 9 km on a reduced Gaussian grid, but C3S provides re-gridded data on a regular latitude-longitude grid of $0.1° \times 0.1°$, which corresponds to approximately 11 km at mid-latitudes. This study focused on the IP, covering a 72-year period from 1950 to 2021, using monthly-mean averages pre-calculated by C3S since sub-monthly fields were not necessary for our analysis.

To better explain the variables used in this study, we employed three key variables that are essential for computing drought indices:

- Total Precipitation [m]—This variable represents the total amount of rain and snow that has fallen on the Earth's surface between the beginning of the forecast time and the end of the forecast step. The units of precipitation are measured in depth in meters, which represents the extent of water that would be spread uniformly over the grid box;
- Soil Moisture [$m^3 m^{-3}$]—This variable represents the volume of water in different soil layers defined by the ECMWF Integrated Forecasting System. Specifically, for this study, we considered the first two ERA5-Land layers, which range from 0 to 28 cm in depth. Although the depth required for the most adequate representation of soil moisture content for agricultural droughts is still under exploration [21,22], we chose to follow the indication of [23], which suggests removing the lower layers to better

represent the soil moisture conditions due to ancillary sources such as local rainfall or irrigation, so only the first two ERA5-Land layers were used (0–28 cm);
- Potential Evapotranspiration [m]—This variable is usually considered to be the amount of evaporation, under existing atmospheric conditions, from a surface of water having the temperature of the lowest layer of the atmosphere. The ECMWF Integrated Forecasting System computes it for an agricultural surface assuming it is well-irrigated, presuming that it does not significantly impact the atmospheric conditions in the region, such as humidity or cloud formation. This simplification allows for a standardized approach to estimating potential evaporation in agricultural contexts, which can introduce some uncertainties. In this respect, we compared the potential evapotranspiration derived from ERA5-Land with that calculated using the Penman-Monteith [24] equation, which takes into account various meteorological variables and thus eliminates the assumption of zero atmospheric impact, and we found that there were no significant differences in the results between the two methods.

2.2. Drought Indices

To identify meteorological droughts, we used the Standardized Precipitation and Evapotranspiration Index (SPEI) [10], which is based on the water balance of precipitation minus evapotranspiration and was chosen because it has been shown to be suitable for drought detection in Spain [3]. To capture agricultural droughts, we adopted several standardized indices to analyze the different outcomes generated by their distinct formulations. The first agricultural drought index was the Standardized Soil Moisture Index (SSI) [25], which was chosen for its simplicity and well documented capability to detect agricultural drought events [26], besides its reliability at a global scale for studying the propagation from meteorological drought detected by SPEI [13]. In addition, we adopted two indices to evaluate composite drought anomalies (agricultural and meteorological). We computed a multivariate index, the Multivariate Standardized Drought Index (MSDI) [27], which is based on the joint probability of precipitation and soil moisture, considering the effect of different variables in the characterization of agricultural droughts. We also included the Standard Precipitation, Evapotranspiration and Soil Moisture Index (SPESMI), a newly developed index that entirely accounts for the different variables involved in agricultural drought generation, introduced by [28] and formulated depending on both precipitation minus evapotranspiration balance and soil moisture. All indices used in this study were calculated using the non-parametric approach suggested by [25]. This method removes assumptions about the distribution of the variables and avoids the computationally expensive fitting of parametric distributions.

For example, MSDI was computed by treating precipitation and soil moisture at a selected time scale (e.g., 3 months) as random variables X and Y, respectively, and considering their joint distribution,

$$P(X \leq x, Y \leq y) = p. \qquad (1)$$

The empirical joint probability p was estimated with the Gringorten plotting position formula [29] as in [25],

$$P(x_k, y_k) = \frac{m_k - 0.44}{n + 0.12}, \qquad (2)$$

where n was the number of the total input data and m_k was the number of occurrences of the pair (x_i, y_i) with $x_i \leq x_k$ and $y_i \leq y_k$ for $i = 1, 2, \ldots, n$. Similarly, for univariate indices such as SSI the empirical marginal probability was calculated by using the univariate form of the Gringorten plotting position formula,

$$P(x_i) = \frac{i - 0.44}{n + 0.12}, \qquad (3)$$

where n was again the number of total input data and i was the rank of the observed values from the smallest. After obtaining the joint or marginal probability p shown in Equation (1), to compute the drought index, it was only needed to retrieve the inverse of the standard normal distribution function ϕ as in [25], namely,

$$MSDI = \phi^{-1}(P). \tag{4}$$

By applying this methodology, we were able to calculate all the drought indices using the same approach, simply by modifying the variables used in the calculations. For example, in SPESMI, the joint probability p of precipitation minus evapotranspiration and soil moisture was calculated (see Table 1 for all the details).

Table 1. Characteristics of the standardized drought indices constructed with the non-parametric technique. P stands for Precipitation, SM for Soil Moisture, and E for Evapotranspiration.

Drought Index	Structure	Variables	Type of Drought
SPEI	Multivariate	P–E	Meteorological
SSI	Univariate	SM	Agricultural
MSDI	Multivariate	P, SM	Agro-Meteorological
SPESMI	Multivariate	P–E, SM	Agro-Meteorological

This not only made the calculations more straightforward but also ensured that the analysis was consistent across all the different indices. We used time scales of 1-, 3-, and 6-months to capture the immediate to seasonal/semi-annual impacts of precipitation, evapotranspiration, and soil moisture on drought characterization. To classify the drought categories, we followed the system described in [30] for the IP. Extreme drought was defined as indices with values below -2, severe drought as values between -2 and -1.5, and drought as values below -1 but above -1.5. Normal/wet conditions were associated with index values above 0, while dry conditions were indicated by an index value of -0.5. See Table 2 for a summary of the categories.

Table 2. Drought categories for the uni-variate and the multi-variate standardized drought indices.

Drought Index Value	Drought Category	Conditions
Index ≤ -2	-2	Extreme Drought
$-2 <$ Index ≤ -1.5	-1.5	Severe Drought
$-1.5 <$ Index ≤ -1	-1	(Moderate) Drought
$-1 <$ Index ≤ 0	-0.5	Dry
Index > 0	1	Normal/wet

Besides the described uni-variate and multi-variate indices, owing to the complex nature of drought events, another new index was proposed. In order to avoid relying on the information provided by a single index only, which might omit important characteristics of drought phenomena, a Combined Agricultural Drought Index (COMB) was developed adapting the Combined Drought Indicator described by [31]. In detail, COMB was based on the composition of the three agricultural drought indices SSI, MSDI, and SPESMI, and it was structured to favor the predominance of drought conditions over the other possible classes, namely when more than one indicator showed values below -1. To construct COMB, the drought categories reported for individual drought indices were taken into account, following the approach proposed by [32]. The index was not a simple average of the three indices, but rather a value was assigned to it based on the combination of drought categories. The severe drought condition (-1.5) was a refinement with respect to [32], to provide information with higher detail and consistency with [30]. When the three indices belonged to different categories, the arithmetic average was calculated and COMB was assigned to the resulting category. The focus of the combined index is on identifying drought conditions, while normal and wet conditions are only important for detecting the end of drought events. Please refer to Table 3 for details on the COMB classification.

Table 3. Methodology for the calculation of Combined Agricultural Drought Index (COMB).

Drought Indices Values (SSI, MSDI, SPESMI)	COMB	Conditions
2 + indices $\in (-\infty, -2]$	−2	Extreme Drought
2 + indices $\in (-2, -1.5]$	−1.5	Severe Drought
2 + indices $\in (-1.5, -1]$	−1	(Moderate) Drought
2 + indices $\in (-1, 0]$	−0.5	Dry
2 + indices $\in (0, +\infty)$	1	Normal/wet

2.3. Methods of Analysis

In this study, we investigated various aspects of drought phenomena, with a primary focus on the propagation from meteorological to agricultural drought. To this end, we employed different approaches to capture the distinct behaviors of the drought indices used. The first part of our analysis involved characterizing the two types of drought separately, with particular attention paid to agricultural drought events. We began by qualitatively examining the temporal evolution of the different drought indices at the three time scales, spanning the entire time window of the dataset (1950–2021). This involved observing the trends over the years and identifying possible similarities or differences between SPEI and the agricultural indices, as well as among the agricultural indices themselves.

In addition to characterizing meteorological and agricultural drought separately, we compared the agricultural drought indices to observations of vegetation health. To do so, we employed the Fraction of Absorbed Photosynthetically Active Radiation (FAPAR) Anomaly [33], which has been demonstrated to be effective in monitoring and assessing agricultural drought impacts [34]. Specifically, we used the FAPAR Anomaly indicator provided by the Copernicus European Drought Observatory EDO [35], which is computed as deviations from the long-term mean of biophysical FAPAR derived from surface reflectances measured by the MODIS-Terra satellite over a 21-year period (2001–2021). The EDO FAPAR anomalies are available at a spatial resolution of 1 km and are calculated for 10-day intervals. To compare these data to the monthly drought indices obtained from the ERA5-Land dataset, we calculated them on an 11 km grid and computed the mean for every 30-day period. Furthermore, following the recommendation of [36], a re-standardization step was performed on the monthly-averaged FAPAR anomalies obtained. This involved computing a new index, F_s, used to better compare and analyze the data and described by

$$F_s = \frac{F_i - \overline{F}}{\sigma}, \qquad (5)$$

where for each specific year, F_i represents the monthly averaged FAPAR anomaly for month i, while the mean and standard deviation of the monthly averaged FAPAR anomaly across all months i during the entire time period from 2001 to 2021 are represented by $\overline{F}$ and σ, respectively.

The aim of the analysis using FAPAR anomalies was to assess the ability of agricultural indices to identify vegetation stress in areas affected by severe droughts. The analysis was based on verification metrics adapted from [37]. To compare the performance of the drought indices SSI, MSDI, SPESMI, and COMB with FAPAR anomalies, we used the following verification metrics: Probability Of Detection (POD), False Alarm Ratio (FAR), Critical Success Index (CSI), and Effect Of Drought (EOD). The drought categories presented in Tables 2 and 3 were considered, and the metrics were formulated as follows:

$$POD = H/(H + M)$$
$$FAR = F/(H + F)$$
$$CSI = H/(H + M + F)$$
$$EOD = (H + H_N)/(M + F + H + H_N),$$

where H (Hit) denoted the number of grids where the agricultural drought index showed categories −1, −1.5, or −2 and the FAPAR anomaly showed values belonging to the same range of categories; M (Miss) designated the number of grids where the FAPAR anomaly was subjected to categories -1, -1.5, or -2 and the agricultural drought index was subjected to categories higher than -1; F (false alarm) stood for the number of grids where the FAPAR anomaly belonged to categories higher than -1, but the agricultural drought index indicated categories -1, -1.5, or -2; H_N (Hit Null) expressed the amount of grids where the drought indices and FAPAR anomaly revealed categories 0.5 or 1. The total quantity of grids considered was given by the sum of H, M, F, and H_N, and the values of all four verification metrics ranged between 0 and 1, with a perfect fit characterized by $POD = 1$, $FAR = 0$, $CSI = 1$, and $EOD = 1$.

The subsequent step involved examining the relationship between the characteristics of droughts identified using different indices. To extract drought characteristics, we utilized the widely used *run theory* [38] to identify drought events. According to this method, a drought event begins when a drought index falls below a fixed threshold and continues until the index values remain continuously below that threshold (negative run), ending only when the index exceeds the threshold level (positive run). For this study, a threshold value of -1 was adopted for drought indices, as is customary. However, special attention was given to severe and extreme drought events, which are generally the most significant in terms of impacts and consequences.

After identifying the drought events, the next step involved calculating the average values of the agricultural drought indices on each gridpoint over the entire study period for the identified drought events. This was carried out to examine the relationship between the different types of indices and the properties of droughts. Specifically,

- Percentage: spatial fraction of IP affected by severe/extreme droughts;
- Duration: number of months with drought index values below the -1.5 threshold;
- Frequency: number of drought events per year;
- Severity: lowest value of the drought index during the drought periods;
- Intensity: ratio of drought severity to the drought duration;
- Magnitude: sum of absolute values of the drought index during the drought period.

After examining the characteristics of agricultural drought events using various indices, the second part of this study focused on analyzing the propagation of drought from meteorological to agricultural domains. To do this, the response time scale (RT) was calculated, following the approach of [39]. The RT reflects the time it takes for the accumulated deficit in meteorological drought to correspond to agricultural drought, and it is based on a correlation analysis between agricultural and meteorological drought indices at different time scales. Specifically, for each location, the Pearson correlation was calculated between the time series of SSI, MSDI, SPESMI, and COMB at a 1-month time scale and SPEI at time scales of 1-, 2-, 3-, ..., and 48-months. The response time scale for a given agricultural drought index was then determined as the time scale of SPEI with the highest correlation coefficient with the index values. This analysis covered the entire time period of the ERA5-Land dataset.

To analyze specific parameters of drought propagation, this study focused on a significant case study, the 2005 drought event in the IP, which had severe impacts on the entire European continent, reducing cereal yields by 10% [40]. This event is well documented and reported in various databases, including the European Drought Observatory and Emergency Events Database EM-DAT [41], making it a suitable reference for the analysis. The period including the 2005 drought was empirically investigated by examining the sequential spatial extent of drought coverage according to different drought indices at various time scales [42]. This approach allowed for the observation of spatial drought propagation across different systems over the IP, and some locations were selected for further analysis involving the lag time (LT) between the onset of meteorological and agricultural droughts. According to [43], the lag time for two different types of drought event was expressed as

$$LT = T_M - T_A, \qquad (6)$$

where T_M and T_A represented the initial time (in months) of meteorological and agricultural drought, respectively. Therefore, LT was used to characterize drought propagation by measuring the difference in onset timing between the two drought episodes.

3. Results

3.1. Characterization of Agricultural Drought over the IP

The analysis began with a qualitative inspection of the indices computed over the time period from 1950 to 2021. Figure 2 displays the temporal evolution of the meteorological index (SPEI) and agricultural drought (AD) indices (MSDI, SPESMI, SSI, and COMB) at 1-, 3-, and 6-month time scales, which allowed for the distinction of the main features of meteorological and agricultural drought events. The evolution of the indices appeared smoother for larger time scales than for the 1-month time scale, and there was a slight trend towards increased dryness in the last two decades for both SPEI and AD indices, extending the assessment of droughts from meteorological to agricultural droughts. The analysis of Figure 2 revealed that the frequency and severity of meteorological drought episodes have increased since the 1990s, and the recurrence of severe droughts has notably developed in the last ten years. While the similarities of the indices suggested a good correlation between the contributions of precipitation deficit, water balance, and soil moisture, some differences emerged among the AD indices. SSI reached the highest values, while the two multi-variate indices reported the lowest values, with COMB being a compromise between the two types of indices. The multi-variate indices, which include the effect of precipitation, seemed to better reproduce the variations of SPEI, suggesting only a limited impact of soil moisture in their computation.

Figure 3 presents the skill score metric indices for the four AD indices at different time scales (1-, 3-, and 6-month) in comparison to FAPAR anomalies. The metrics include H (hit) and M (miss), which represent the ability of the AD indices to detect drought-affected areas where there were also dry vegetation conditions identified by FAPAR anomalies. A high value of $POD = (H + M)/M > 0.8$ indicates that the AD indices were successful in identifying gridpoints dominated by agricultural droughts characterized by stressed land vegetation, especially for semi-annual droughts (6-month time scale) detected through MSDI and SPESMI. On the other hand, F (false alarm) represents the number of grids where SSI, MSDI, SPESMI, or COMB detected drought conditions but with disagreement compared to the FAPAR anomalies. The non-zero results for $FAR = F/(H + F)$ indicate that there were some areas with dehydrated plants that could not be accurately monitored by AD indices, with almost negligible variations among the time scales and indices. The values of $CSI = H/(H + M + F)$ suggest that the regions classified with a desiccated flora by FAPAR anomalies were not only those where the AD indices identified drought conditions, while the opposite was, almost everywhere, true. The ratio of drought-affected areas detected by AD indices corresponding to high vegetation stress with respect to the total zones of high vegetation stress recognized by FAPAR anomalies was approximately 0.5, with slightly increasing CSI for semi-annual droughts. This suggests that FAPAR could monitor arid areas which could not be successfully captured by SSI, MSDI, SPESMI, or COMB, maybe due to the fact that FAPAR anomalies reveal variations in the vegetation health which can derive not only from rainfall or soil moisture deficits, but also from other stress factors such as plant diseases. Considering H_N (hit null) as the amount of gridpoints that were free of droughts and with healthy foliage, the values of $EOD = (H + H_N)/(M + F + H + H_N)$ lower than 1 implied some areas in mixed conditions, namely affected by drought, which had no relevant impacts on vegetation or unhealthy vegetated regions caused by factors other than droughts. In conclusion, SSI, MSDI, SPESMI, and COMB were efficient in assessing the vegetation stress of IP during drought events, but they were not sufficient to distinguish all the areas identified by FAPAR anomalies, whose stress could be due to different factors other than drought occurrence.

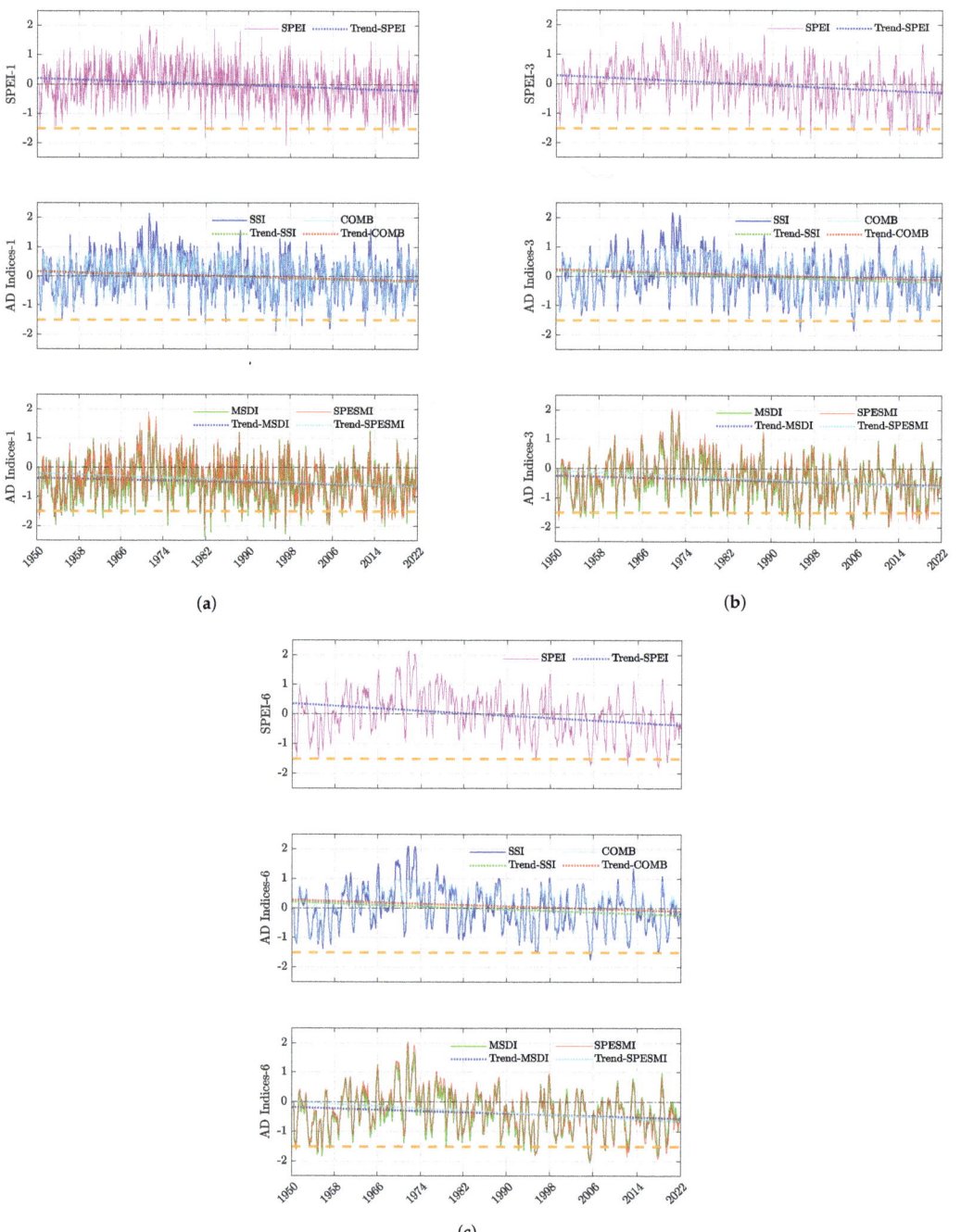

Figure 2. Temporal evolution of Iberian Peninsula averaged SPEI and agricultural drought (AD) indices (SSI, COMB, MSDI and SPESMI) at 1-month (**a**), 3-month (**b**) and 6-month (**c**) time scales for the entire period considered. The linear trend is also represented for each index. The orange dashed line indicates the -1.5 threshold for severe drought events.

To investigate the diverse responses of the AD indices in the IP region during the studied period, various drought characteristics were analyzed. Firstly, the average percentage of the IP region affected by drought events was computed. The results are illustrated in Figure 4, which displays the percentage of the IP region experiencing severe to extreme droughts (i.e., AD indices ≤ -1.5) at 1-, 3-, and 6-month time scales. One key finding was the marked disparity between SSI and other indices, especially the two multi-variate indices. Specifically, while MSDI and SPESMI revealed more than 75% of the IP region suffering from droughts, SSI only covered less than 40%. The integration of precipitation (or water balance) and soil moisture deficits generated drought events in a wider area of the IP region compared to the case of soil moisture deficit alone, with a difference of approximately 40% for each time scale, a logical consequence of the MSDI's capability to detect both meteorological and agricultural events. COMB exhibited a percentage value of around 60%, a trade-off between uni-variate and multi-variate indices. The variations in the values of the same AD index depending on the time scale were small, and the highest percentage was generally observed for 1-month droughts.

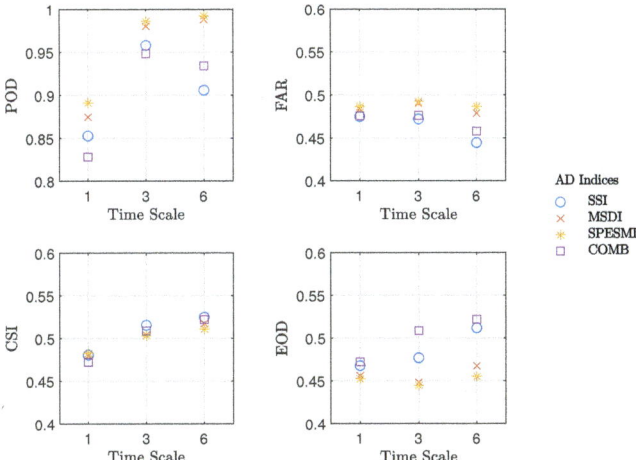

Figure 3. Skill score metrics regarding FAPAR anomalies for AD indices at the 1-, 3-, 6-month time scales.

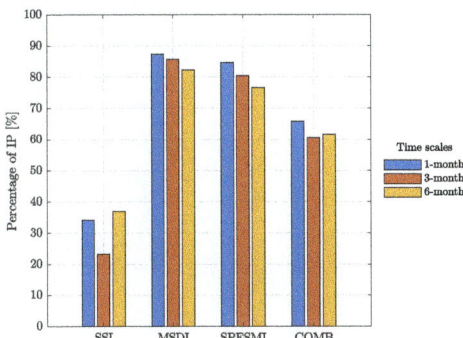

Figure 4. Percentage of the IP affected on average by severe/extreme droughts according to agricultural indices at the 1-, 3-, 6-month time scales.

To gain further insights into the agricultural droughts identified by the AD indices at different time scales, we investigated several other characteristics. One of these is the frequency of severe/extreme droughts per year, which is closely related to their duration, as well as their severity, which is typically used to assess the significance of drought events. Figure 5 displays these two characteristics for each index and time scale. The frequency of severe/extreme drought events per year (Figure 5a,c,e for 1-, 3- and 6-month time scales, respectively) confirmed that multi-variate indices were more sensitive than uni-variate ones in detecting drought events in a larger area. Specifically, SSI identified less than one drought event per year, on average, across all the IP at the 1-month time scale, while SPESMI and MSDI detected multiple drought episodes in several areas, with values ranging from around 1 event/year to 1.2 events/year on average. This is in agreement with the fact that precipitation, evapotranspiration, and soil moisture variables are vital factors to adequately represent agricultural drought conditions [28] and, especially in large areas with different climate characteristics, their combination by using multi-variate indices could better detect the occurrence of drought events. The patterns of COMB showed a balance between SSI and MSDI/SPESMI. The north-west and the Pyrenees were the most affected zones for all indices, indicating that both soil moisture and precipitation/water balance deficits were recurrent. Similar patterns were found at 3- and 6-month time scales, although the frequency values were naturally lower due to the longer duration of the considered events. The average drought severity showed a more complex behavior among the AD indices. SSI exhibited a large area of non-severe droughts (severity smaller than 1.5) at the 1-month time scale (Figure 5b), while COMB, SPESMI, and MSDI showed gradually increasing regions affected by severe drought conditions. However, the distribution of severity values differed among the indices. For instance, a small fraction of the centre-east of the IP was one of the most affected zones according to SSI but was not equivalently accounted for by the other indices, especially by MSDI, which retrieved its lowest severity values in the same area. This suggested a balance between the contribution of water balance variables in multi-variate indices, which could significantly impact the effect of soil moisture. Severity also showed a wider range of values for longer time scales. Bi-annual deficits in water balance and soil moisture resulted in large areas affected by severe droughts, with peaks close to extreme droughts in certain regions. On the other hand, even if the 6-month scale showed severity values similar to shorter time scales on average, severity appeared more pronounced or not depending on the region and the index. For example, in the south of Portugal, COMB-6, MSDI-6, and especially SPESMI described a higher severity than at 1- and 3-month time scales. Other areas, such as the mountain region in the southern part of the IP and the northern peninsular zone, reported the opposite behavior, revealing lower severity values at longer time scales.

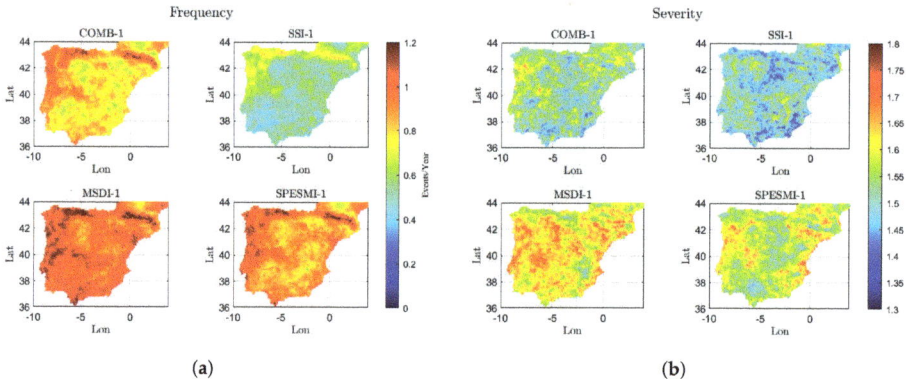

(a) (b)

Figure 5. *Cont.*

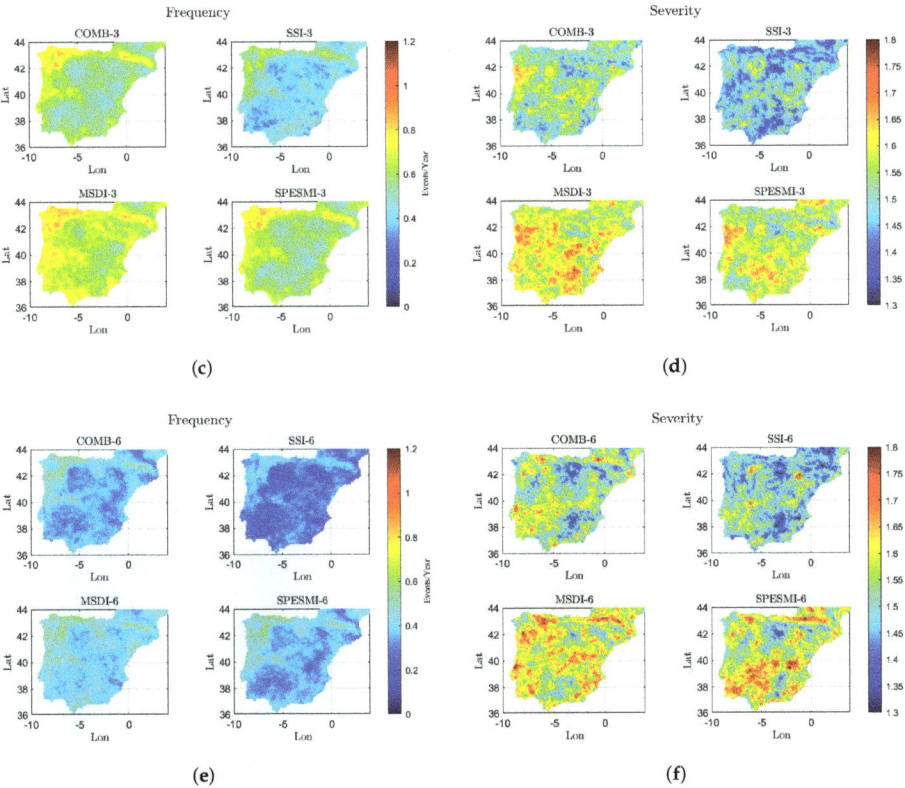

Figure 5. Patterns of the average characteristics of drought events on the Iberian Peninsula according to 1-, 3- and 6-month time scales for the different AD indices: frequency of severe/extreme droughts per year (**a,c,e**), and absolute value of drought severity (**b,d,f**).

3.2. Propagation from Meteorological to Agricultural Drought

This study investigated the response time scale (RT) as a key parameter in drought propagation, which represents the accumulated precipitation deficiency in the antecedent RT months that causes agricultural drought. A shorter RT indicates a faster response to meteorological drought. Figure 6a shows the maximum Pearson correlation between 1-month agricultural drought indices and SPEI computed from 1- to 48-month time scales for the entire 72-year period over the IP. Figure 6b indicates the corresponding RT in months for each gridpoint.

The analysis revealed a high correlation between agricultural and meteorological droughts at small time scales, consistent with previous studies [44]. The RT for SSI was 2 months for most of the IP, except for the Pyrenees where RT was 1 month and 3–4 months in some isolated regions, particularly in the southern coastal areas. MSDI, SPESMI, and COMB had an RT of 1 month, indicating that the contribution of other variables accelerated the response compared to soil moisture alone. In particular, SPESMI, which includes evapotranspiration, presented the highest correlation with meteorological droughts detected by SPEI, which is also based on water balance.

For a detailed analysis of the propagation from meteorological to agricultural drought, we focused on the 2005 drought episode in the IP region. To investigate the spatial evolution of the drought phenomenon, we analyzed the monthly patterns of drought index values during the reference period from October 2004 to December 2005. Specifically, we examined the variations of the AD indices at the 1-month time scale and compared them to the progression of the SPEI at 1-month and 3-month time scales to understand the propagation from the current month and from the seasonal meteorological water balance deficit to the development of agricultural drought. Figure 7 illustrates the temporal evolution of the meteorological drought identified by the SPEI at 1- and 3-month time scales from October 2004 to December 2005 over the entire IP region.

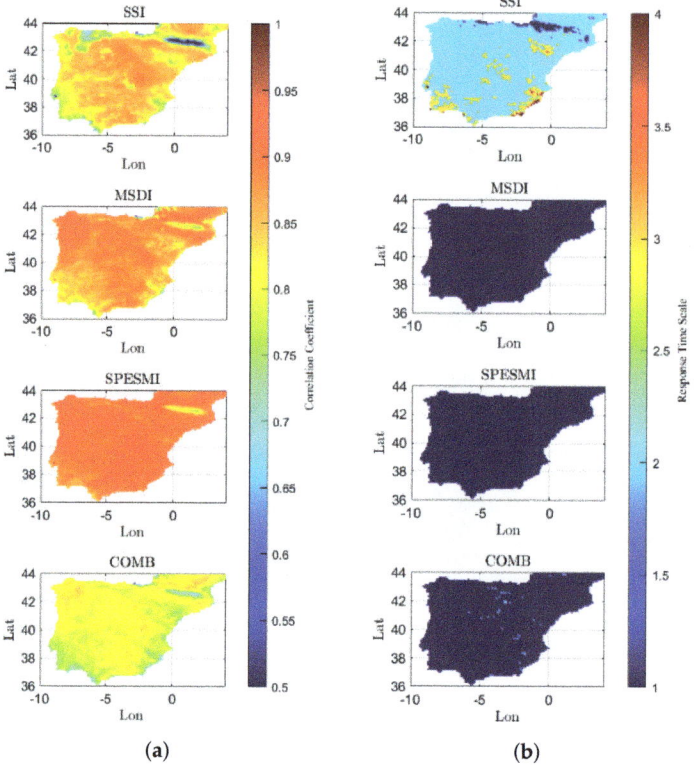

Figure 6. (**a**) Maximum Pearson correlation retrieved between the SPEI at the 1-, 2-,..., 48-month time scales and the 1-month time scale AD indices. (**b**) Corresponding *RT* in months for each gridpoint.

As shown in Figure 7a, the SPEI calculated at monthly intervals indicated a moderate drought condition in December 2004 over the majority of the IP. February 2005 was identified as the driest month, with a significant area experiencing extreme drought conditions. This pattern was followed by two wetter months, after which the severe/extreme drought episode reoccurred in May 2005, initially affecting only the south and eastern coasts of the IP. This condition lasted until the end of summer 2005 (September), with a modified pattern, which was more concentrated in the centre and northern IP. This event had severe impacts on Spain and Portugal, as reported by the European Drought Observatory, with a relaxation during the winter of 2005. The evolution of seasonal meteorological drought, shown in Figure 7b, was similar to the SPEI at monthly intervals, with the initial peak in February 2005. However, the SPEI-3 patterns were more continuous, essentially reporting prolonged drought conditions over the reference period, with a delayed conclusion (extreme drought conditions were observed even in October 2005). The driest region was the southern IP at the beginning, with varying features extending to the central and northern IP.

To examine the progression of agricultural drought during the meteorological drought episodes, the spatial patterns of the four different agricultural drought indices at monthly time scale were analyzed. Figure 8 illustrates the spatial distribution of these indices across the IP during the reference period (October 2004–December 2005). All four AD indices captured the onset of agricultural drought in February 2005. However, there were some differences in their assessment of the drought conditions leading up to that point. While COMB, MSDI, and SPESMI indicated moderately dry conditions over the IP even from December 2004 except for the eastern coastal region, SSI showed normal or wet patterns during that period, suggesting that the incorporation of variables other than soil moisture may have allowed for a more accurate detection of drought impacts. Furthermore, SSI generally indicated less severe drought conditions than the other indices, and its patterns were less uniform and homogeneous compared to MSDI and SPESMI, particularly during the initial month of the drought event.

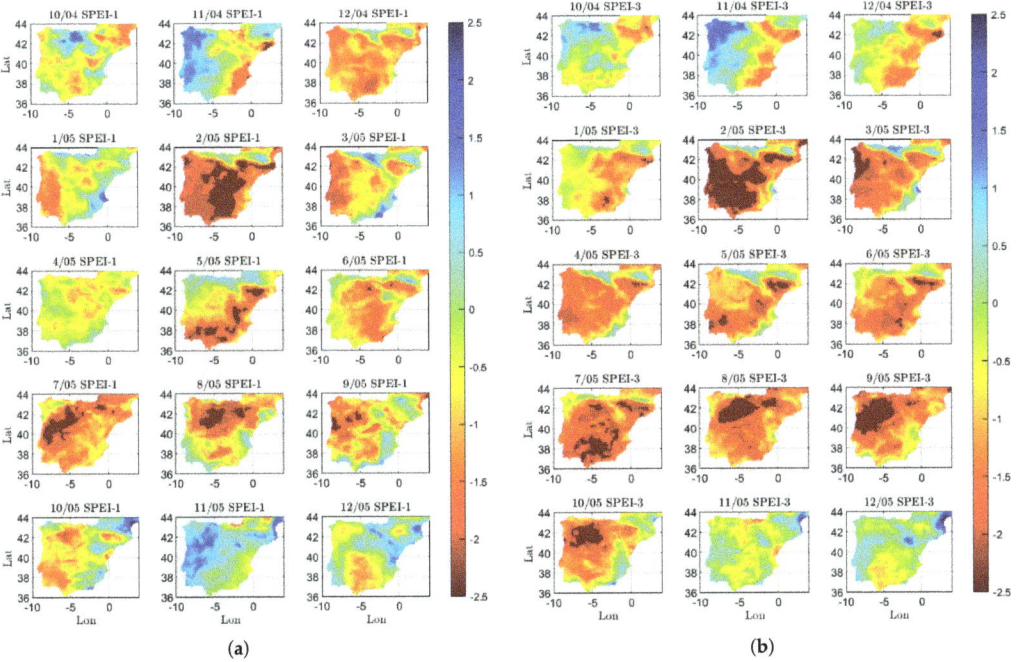

Figure 7. Temporal evolution of meteorological drought pattern over the IP according to SPEI at (**a**) 1- and (**b**) 3-month time scales, from October 2004 to December 2005.

MSDI and SPESMI showed a high degree of similarity in their behaviors, indicating that the inclusion of evapotranspiration had only a marginal impact on the results. COMB exhibited features that were balanced between the other three indices, but its similarity to the two multivariate indices appeared to have a greater influence on its performance than SSI. Overall, the patterns of agricultural drought identified by the AD indices were consistent with those identified by SPEI, particularly SPEI-3, which presented a higher spatial correlation with seasonal meteorological drought than with monthly water balance deficits. However, the severity values of the four AD indices were generally higher than those of SPEI-1 and SPEI-3, indicating that the agricultural drought impacts were more severe and extensive than the meteorological drought.

To understand the propagation of drought from meteorological to agricultural systems, Figures 7 and 8 were useful in providing a qualitative overview of the spatial details of the 2005 drought evolution. However, to derive more quantitative information, a further investigation was conducted. Drawing inspiration from [13], we analyzed the probability of drought propagation from meteorological to agricultural systems under different levels of severity. Specifically, the fraction of the IP experiencing agricultural drought conditioned on the occurrence of meteorological drought was calculated for each month between October 2004 and December 2005. This fraction was defined as the propagation probability (PP) and was computed separately for the four AD indices at the 1-month time scale, distinguishing between three severity thresholds, namely moderate, severe, and extreme drought. To provide a comprehensive understanding of the results, Figure 9 represents the temporal evolution of PP for each AD index. The panels in each row exhibit the PP values of agricultural droughts with gradually increasing severity from left to right, while the different colors refer to the severity threshold of meteorological drought, based on SPEI-1. For instance, the red line in the first panel of row one shows the time evolution of PP for the occurrence of moderate agricultural drought (based on SSI-1) conditioned on the occurrence of severe meteorological drought (based on SPEI-1).

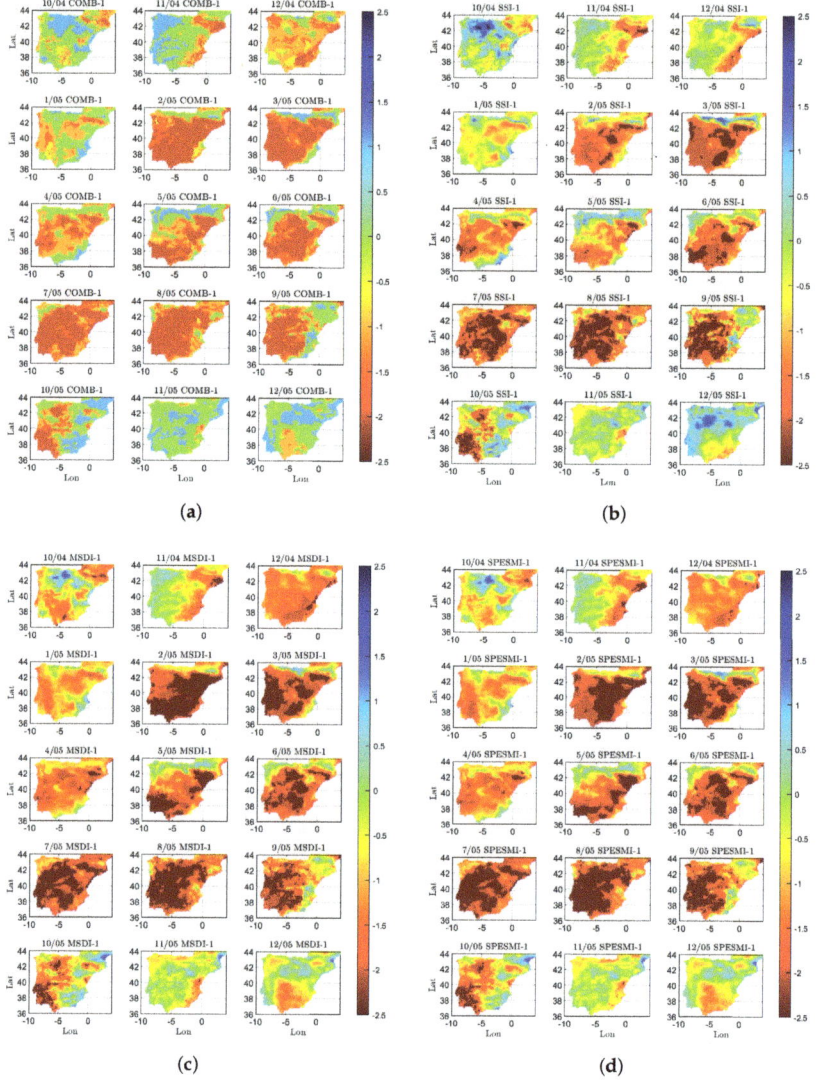

Figure 8. Temporal evolution of drought pattern over the IP according to the four AD indices: (**a**) COMB, (**b**) SSI, (**c**) MSDI, and (**d**) SPESMI at 1-month time scale from October 2004 to December 2005.

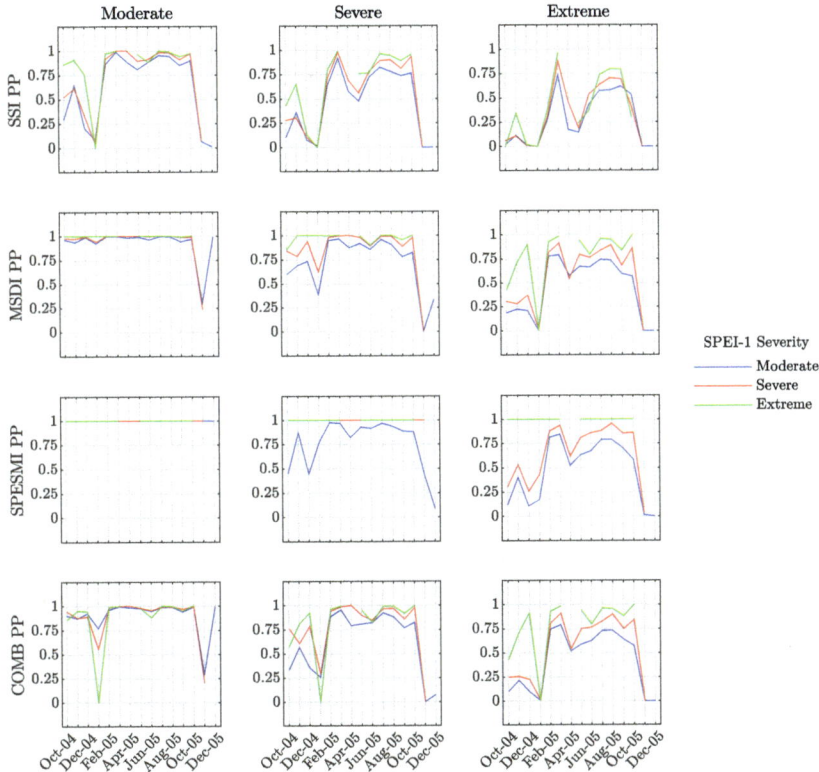

Figure 9. Propagation probability (PP) from meteorological drought detected with SPEI-1 to agricultural drought of different severity levels according to 1-month time scale AD indices. The colors distinguish the severity of generating meteorological drought (blue for SPEI-1 ≤ -1, red for SPEI-1 ≤ -1.5, green for SPEI-1 ≤ -2).

The results of the analysis showed that all four AD indices displayed a significant increase in propagation probability (PP) as the severity levels of meteorological drought increased from moderate to extreme, with PP values approaching 1, indicating that the likelihood of agricultural drought was higher under drier meteorological conditions. For instance, in the first row of Figure 9, SSI indicated that areas affected by extreme meteorological drought from February 2005 were highly susceptible to various levels of agricultural drought, with the highest PP values observed for moderate severity propagation. MSDI and SPESMI consistently showed nearly constant $PP = 1$ values in the left panels, implying that these two multi-variate indices were highly sensitive to moderate agricultural drought propagation and less prone to severe and extreme propagation. On the other hand, SSI generally displayed lower PP values for all three severity levels compared to the multivariate indices, while COMB demonstrated a balance between the two typologies of indices. Except for the deflection observed in March and April 2005, where the meteorological drought did not propagate in all the affected regions, all indices showed overall high PP values during the identified drought event, with $PP > 0.5$ for moderate SPEI-1 severity and $PP > 0.75$ for severe/extreme SPEI-1 severity. Although the PP values of MSDI, SPESMI, and COMB remained above 0.5 during this period, SSI demonstrated a more significant reduction, with $PP < 0.25$.

To investigate the propagation of drought from seasonal meteorological drought, we applied the same procedure as before. Figure 10 presents the behavior of PP for agricultural drought conditioned on SPEI-3 values.

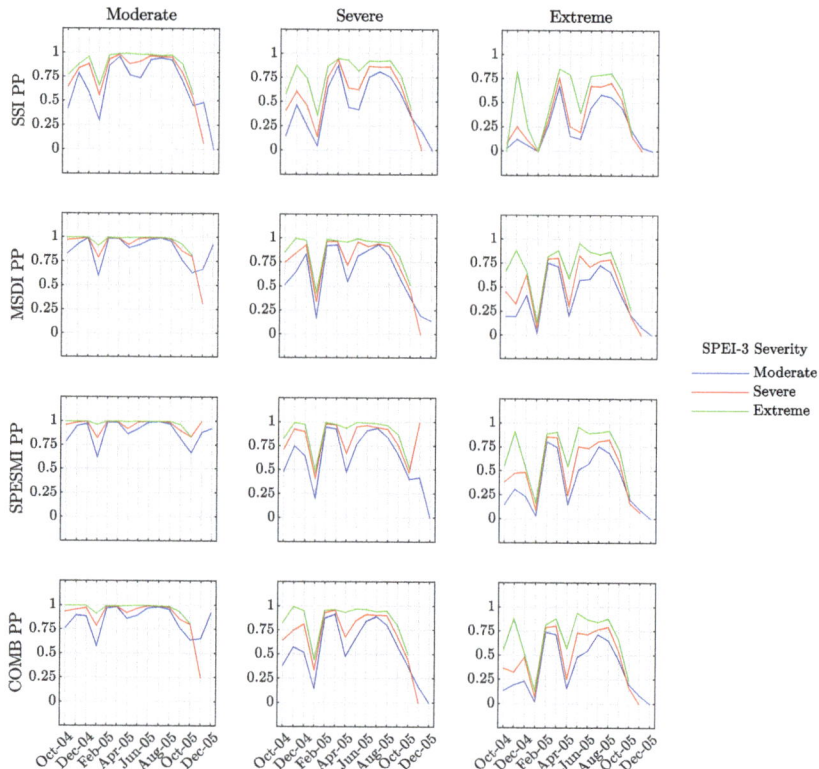

Figure 10. Propagation probability PP from meteorological drought detected with SPEI-3 to agricultural drought of different severity levels according to 1-month time scale AD indices. The colors distinguish the severity of generating meteorological drought (blue for SPEI-3 ≤ -1, red for SPEI-3 ≤ -1.5, green for SPEI-3 ≤ -2).

Similar to the analysis with monthly droughts, we observed a significant increase in PP as the severity levels of meteorological drought enhanced from moderate to extreme. The maximum PP values were found for moderate agricultural droughts, indicating that the propagation to more severe droughts was less likely to occur. Compared to the outcomes reported in Figure 9, the propagation from seasonal meteorological droughts revealed a more homogeneous behavior among the AD indices, with common features in the maxima/minima of PP and the evolution of the PP signal. Similar to the previous case, the lowest value of PP was reached in March–April 2005, suggesting that, during this period, the meteorological drought was present but did not propagate into agricultural drought. Additionally, the panel regarding severe and extreme agricultural drought displayed smaller PP values than the SPEI-1 example, indicating that seasonal drought was, in general, less efficient in propagation compared to drought caused by monthly water balance deficits.

To further investigate the 2005 drought event, two areas of the Iberian Peninsula that showed significant variations during the event were selected: the centre and the south. In order to monitor changes in drought indices and evaluate the lag time (LT), one city was chosen to represent each region. Following the approach of [45], Madrid was chosen as the representative location for the IP centre and Granada was selected for the IP South.

Figure 11 shows the 3-month time scale drought indices for these cities from January 2004 to January 2006. The seasonal drought was considered, as it is characterized by smoother and slower variations than the 1-month drought and provides more significant information for the calculation of LT.

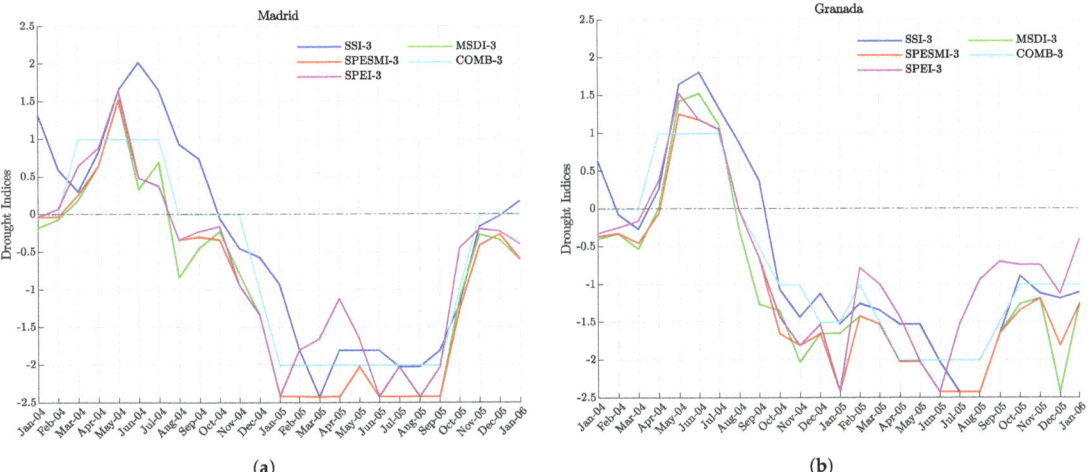

Figure 11. Local temporal evolution of different drought indices at 3-month time scale in Madrid (**a**) and Granada (**b**) from January 2004 to January 2006.

To calculate the LT, we compared the onset times of drought as measured by SPEI and the AD indices. In Madrid, the multivariate indices showed $LT \sim 0$, indicating that they captured agricultural drought onset at the same time as SPEI. However, SSI had a lag time of about 2 months, since agricultural drought in that city began in February 2005, while meteorological drought began in December 2004. Compared to SPEI, the AD indices showed low variability, and changes in meteorological drought conditions did not necessarily result in modifications of agricultural droughts. The propagation of meteorological drought evolution had prolonged and almost constant effects on agricultural drought, especially according to COMB. The situation in Granada was slightly different. Meteorological drought onset preceded that in Madrid by 2 months (October 2004) and was simultaneous for all AD indices, resulting in $LT \sim 0$. However, MSDI showed the onset of agricultural drought even 1 month before SPEI, indicating that the detected event was not solely associated with drought propagation. Although the onset lag time between meteorological and agricultural droughts was approximately zero, we noted that the duration of the two phenomena was not equivalent. While SPEI indicated dry meteorological conditions without the presence of drought in August 2005, SSI and COMB estimated the end of agricultural drought in October 2005, and MSDI and SPESMI required even more time.

4. Discussion

This study examined various types and temporal aspects of drought, with a specific focus on the transition from meteorological to agricultural drought, which marks the initial phase of drought propagation. In the first phase of the study, the aim was to characterize drought events across the IP over various time scales, including monthly, seasonal, and six-month periods, for the entire duration of the study period (1950–2021). The findings indicated a slight trend towards increased aridity during the last two decades. This was supported by both the SPEI, which aligns with the results obtained by [46] using parametric indices to identify meteorological droughts, as well as the AD indices. When comparing the values of the FAPAR anomaly, the AD indices were found to be effective in assessing vegetation stress during drought events in the IP. However, they were not always able to accurately detect all the dehydrated areas identified by FAPAR anomalies. In terms of the average characteristics of agricultural drought events, the SSI captured a noticeably smaller area affected by severe/extreme droughts (40%) compared to other AD indices (75%). The frequency of severe/extreme drought events per year also revealed that multi-variate AD indices were more sensitive than SSI, albeit with lower severity. Additionally, when considering long time scales, the severity of drought events showed a wider range of values than on the monthly time scale, ranging from the most to the least severe values. In the second phase of the study, we analyzed the mechanisms of propagation from meteorological to agricultural drought. The response time scale (RT) for the AD indices was calculated obtaining small values, consistent with findings from other studies [44]. Specifically, the results showed that the SSI had an RT of 2 months in most parts of the IP, while the MSDI, the SPESMI, and the COMB had an RT of 1 month, suggesting that the contribution of water balance accelerated the response compared to soil moisture alone. We analyzed the 2005 drought episode to study the temporal and spatial features of drought propagation in detail. The time evolution of seasonal agricultural drought, as revealed by the SSI, had a 2-month delayed onset compared to other AD indices (February 2005 vs. December 2004), while the end of the drought was almost coincident (October 2005). Furthermore, the severity values obtained by SSI were lower than those obtained by other AD indices. The agricultural drought patterns determined by the AD indices were consistent with those identified by SPEI, particularly showing a higher spatial correlation with seasonal rather than monthly meteorological drought. We evaluated the propagation probability (PP) from monthly meteorological drought and found that PP values were high during the identified drought event period. Monthly agricultural drought was also found to be more likely to occur when meteorological conditions were increasingly dry. However, smaller PP values were obtained from seasonal meteorological to monthly agricultural drought, indicating a reduced propagation efficiency compared to meteorological drought caused by monthly water balance deficits. The computation of lag time (LT) for SSI revealed different outcomes depending on the location, with values ranging from $LT \sim 0$ to $LT \sim 2$ months, while the multi-variate indices consistently showed $LT \sim 0$, regardless of the location. These values close to 0 suggested interest for a future analysis concerning sub-monthly features of LT.

There are currently several issues for analyzing drought propagation, such as the difficulty of comparing studies that use different indices and approaches, and the challenge of isolating single factors for analysis. Future studies on agricultural drought propagation should address these challenges by integrating techniques beyond statistical analysis based on run theory, such as extending the probabilistic approach of [47]. Considering the phenomenon of global warming, it is important to conduct studies that account for non-stationary conditions in future changing environments. In this regard, future work could expand upon the investigation by including the newly developed COMB index and incorporating an ensemble analysis with other meteorological and agricultural drought indices from existing literature. Additionally, future studies could explore the lead time between meteorological drought onset and agricultural drought onset, considering location and crop type. This would contribute to a more comprehensive understanding of agricultural drought propagation and enhance drought monitoring and early warning systems.

5. Conclusions

In conclusion, this study provided valuable information on the propagation of drought phenomena from meteorological to agricultural droughts on the IP. The investigation utilized a long record of data and distinct standardized indices, considering variables beyond just soil moisture such as precipitation and evapotranspiration. Some relevant outcomes were retrieved, in particular:

- Results showed a slight trend towards increased dryness over the last two decades and identified multi-variate AD indices as more effective in identifying severe drought events compared to the uni-variate SSI index.

- The severity values of the four AD indices were generally higher than those of SPEI-1 and SPEI-3, indicating that the agricultural drought impacts were more severe and extensive than the meteorological drought.
- This study introduced a novel combined agricultural drought index that balances the characteristics of other adopted indices and could be a valuable resource for future investigations.
- The response time scale was calculated for the AD indices and small values were obtained, also suggesting that the contribution of water balance accelerated the response compared to the effect of soil moisture alone.
- The analysis of the 2005 episode revealed a 2-month delayed onset compared to other AD indices in the analysis of seasonal agricultural drought, with a higher probability of propagation depending on the severity of the originating meteorological drought.

Author Contributions: Conceptualization, M.P., M.G.-V.O. and S.R.G.-F.; Data curation, M.P.; Formal analysis, M.P.; Funding acquisition, S.R.G.-F.; Investigation, M.P., M.G.-V.O. and S.R.G.-F.; Methodology, M.P. and M.G.-V.O.; Project administration, M.P. and M.G.-V.O.; Resources, M.P.; Software, M.P.; Supervision, M.G.-V.O. and S.R.G.-F.; Validation, M.P.; Visualization, M.P.; Writing—original draft, M.P.; Writing—review and editing, M.P. All authors have read and agreed to the published version of the manuscript.

Funding: This research has been carried out in the framework of the projects P20_00035 funded by FEDER/Junta de Andalucía-Consejería de Transformación Económica, Industria, Conocimiento y Universidades; LifeWatch-2019-10-UGR-01 co-funded by the Ministry of Science and Innovation through the FEDER funds from the Spanish Pluriregional Operational Program 2014–2020 (POPE) LifeWatch-ERIC action line; and PID2021-126401OB-I00, funded by MCIN/AEI/10.13039/501100011033/FEDER Una manera de hacer Europa.

Data Availability Statement: The data presented in this study are available on request from the first author.

Acknowledgments: The authors acknowledge Copernicus for providing open access data. The authors wish to express their appreciation to Silvana Di Sabatino for her instrumental role in initiating and facilitating a mutually beneficial collaboration between the Department of Physics and Astronomy at the University of Bologna and the Department of Applied Physics at the University of Granada.

Conflicts of Interest: The authors declare no conflict of interest.

References

1. Martín-Vide, J.; Barriendos Vallvé, M. The use of rogation ceremony records in climatic reconstruction: A case study from Catalonia (Spain). *Clim. Change* **1995**, *30*, 201–221. [CrossRef]
2. Austin, R.; Cantero-Martınez, C.; Arrúe, J.L.; Playán, E.; Cano-Marcellán, P. Yield-rainfall relationships in cereal cropping systems in the Ebro river valley of Spain. *Eur. J. Agron.* **1998**, *8*, 239–248. [CrossRef]
3. Vicente-Serrano, S.M.; Lopez-Moreno, J.I.; Beguería, S.; Lorenzo-Lacruz, J.; Sánchez-Lorenzo, A.; García-Ruiz, J.; Azorin-Molina, C.; Morán-Tejeda, E.; Revuelto, J.; Trigo, R.; et al. Evidence of increasing drought severity caused by temperature rise in southern Europe. *Environ. Res. Lett.* **2014**, *9*, 044001. [CrossRef]
4. Páscoa, P.; Gouveia, C.M.; Russo, A.; Trigo, R.M. Drought trends in the Iberian Peninsula over the last 112 years. *Adv. Meteorol.* **2017**, *2017*, 4653126. [CrossRef]
5. McKee, T. Drought monitoring with multiple time scales. In Proceedings of the 9th Conference on Applied Climatology, Dallas, TX, USA, 15–20 January 1995.
6. Zhuang, X.; Hao, Z.; Singh, V.; Zhang, Y.; Feng, S.; Xu, Y.; Hao, F. Drought propagation under global warming: Characteristics, approaches, processes, and controlling factors. *Sci. Total Environ.* **2022**, *838*, 156021. [CrossRef]
7. Xu, Y.; Wang, L.; Ross, K.; Liu, C.; Berry, K. Standardized soil moisture index for drought monitoring based on soil moisture active passive observations and 36 years of north American land data assimilation system data: A case study in the southeast United States. *Remote Sens.* **2018**, *10*, 301. [CrossRef]
8. Zhou, K.; Li, J.; Zhang, T.; Kang, A. The use of combined soil moisture data to characterize agricultural drought conditions and the relationship among different drought types in China. *Agric. Water Manag.* **2021**, *243*, 106479. [CrossRef]
9. Bae, H.; Ji, H.; Lim, Y.-J.; Ryu, Y.; Kim, M.-H.; Kim, B.-J. Characteristics of drought propagation in South Korea: Relationship between meteorological, agricultural, and hydrological droughts. *Nat. Hazards* **2019**, *99*, 1–16. [CrossRef]

10. Vicente-Serrano, S.; Beguería, S.; López-Moreno, J. A multiscalar drought index sensitive to global warming: The standardized precipitation evapotranspiration index. *J. Clim.* **2010**, *23*, 1696–1718. [CrossRef]
11. Li, Y.; Huang, S.; Wang, H.; Zheng, X.; Huang, Q.; Deng, M.; Peng, J. High-resolution propagation time from meteorological to agricultural drought at multiple levels and spatiotemporal scales. *Agric. Water Manag.* **2022**, *262*, 107428. [CrossRef]
12. Quiring, S.; Ford, T.; Wang, J.; Khong, A.; Harris, E.; Lindgren, T.; Goldberg, D.; Li, Z. The North American soil moisture database: Development and applications. *Bull. Am. Meteorol. Soc.* **2016**, *97*, 1441–1459. [CrossRef]
13. Zhu, Y.; Liu, Y.; Wang, W.; Singh, V.; Ren, L. A global perspective on the probability of propagation of drought: From meteorological to soil moisture. *J. Hydrol.* **2021**, *603*, 126907. [CrossRef]
14. Tian, Q.; Lu, J.; Chen, X. A novel comprehensive agricultural drought index reflecting time lag of soil moisture to meteorology: A case study in the Yangtze River basin, China. *Catena* **2022**, *209*, 105804. [CrossRef]
15. Xia, Y.; Sheffield, J.; Ek, M.; Dong, J.; Chaney, N.; Wei, H.; Meng, J.; Wood, E. Evaluation of multi-model simulated soil moisture in NLDAS-2. *J. Hydrol.* **2014**, *512*, 107–125. [CrossRef]
16. Muñoz Sabater, J. ERA5-Land Monthly Averaged Data from 1981 to Present, Copernicus Climate Change Service (C3S) Climate Data Store (CDS). 2019. Available online: https://doi.org/10.24381/cds.68d2bb30 (accessed on 4 November 2022). [CrossRef]
17. Zhang, R.; Li, L.; Zhang, Y.; Huang, F.; Li, J.; Liu, W.; Mao, T.; Xiong, Z.; Shangguan, W. Assessment of agricultural drought using soil water deficit index based on ERA5-land soil moisture data in four southern provinces of China. *Agriculture* **2021**, *11*, 411. [CrossRef]
18. Beck, H.E.; Pan, M.; Miralles, D.G.; Reichle, R.H.; Dorigo, W.A.; Hahn, S.; Sheffield, J.; Karthikeyan, L.; Balsamo, G.; Parinussa, R.M.; et al. Evaluation of 18 satellite-and model-based soil moisture products using in situ measurements from 826 sensors. *Hydrol. Earth Syst. Sci.* **2021**, *25*, 17–40. [CrossRef]
19. Almendra-Martín, L.; Martínez-Fernández, J.; González-Zamora, Á.; Benito-Verdugo, P.; Herrero-Jiménez, C.M. Agricultural drought trends on the Iberian Peninsula: An analysis using modeled and reanalysis soil moisture products. *Atmosphere* **2021**, *12*, 236. [CrossRef]
20. Muñoz-Sabater, J.; Dutra, E.; Agustí-Panareda, A.; Albergel, C.; Arduini, G.; Balsamo, G.; Boussetta, S.; Choulga, M.; Harrigan, S.; Hersbach, H.; et al. ERA5-Land: A state-of-the-art global reanalysis dataset for land applications. *Earth Syst. Sci. Data* **2021**, *13*, 4349–4383. [CrossRef]
21. Arora, V.; Boer, G. A representation of variable root distribution in dynamic vegetation models. *Earth Interact.* **2003**, *7*, 1–19. [CrossRef]
22. Qiu, J.; Crow, W.; Nearing, G. The impact of vertical measurement depth on the information content of soil moisture for latent heat flux estimation. *J. Hydrometeorol.* **2016**, *17*, 2419–2430. [CrossRef]
23. Bageshree, K.; Kinouchi, T. A Multivariate Drought Index for Seasonal Agriculture Drought Classification in Semiarid Regions. *Remote Sens.* **2022**, *14*, 3891. [CrossRef]
24. Monteith, J. Evaporation and environment. In *Proceedings of the Symposia of the Society for Experimental Biology*; Cambridge University Press (CUP): Cambridge, UK, 1965; Volume 19, pp. 205–234.
25. Hao, Z.; AghaKouchak, A. A nonparametric multivariate multi-index drought monitoring framework. *J. Hydrometeorol.* **2014**, *15*, 89–101. [CrossRef]
26. Khan, M.I.; Zhu, X.; Arshad, M.; Zaman, M.; Niaz, Y.; Ullah, I.; Anjum, M.N.; Uzair, M. Assessment of spatiotemporal characteristics of agro-meteorological drought events based on comparing Standardized Soil Moisture Index, Standardized Precipitation Index and Multivariate Standardized Drought Index. *J. Water Clim. Change* **2020**, *11*, 1–17. [CrossRef]
27. Hao, Z.; AghaKouchak, A. Multivariate standardized drought index: A parametric multi-index model. *Adv. Water Resour.* **2013**, *57*, 12–18. [CrossRef]
28. Xu, L.; Chen, N.; Yang, C.; Zhang, C.; Yu, H. A parametric multivariate drought index for drought monitoring and assessment under climate change. *Agric. For. Meteorol.* **2021**, *310*, 108657. [CrossRef]
29. Gringorten, I. A plotting rule for extreme probability paper. *J. Geophys. Res.* **1963**, *68*, 813–814. [CrossRef]
30. García-Valdecasas Ojeda, M.; Gámiz-Fortis, S.; Romero-Jiménez, E.; Rosa-Cánovas, J.; Yeste, P.; Castro-Díez, Y.; Esteban-Parra, M. Projected changes in the Iberian Peninsula drought characteristics. *Sci. Total Environ.* **2021**, *757*, 143702. [CrossRef]
31. Spinoni, J.; Vogt, J.; Naumann, G.; Barbosa, P.; Dosio, A. Will drought events become more frequent and severe in Europe? *Int. J. Climatol.* **2018**, *38*, 1718–1736. [CrossRef]
32. Spinoni, J.; Naumann, G.; Vogt, J.; Barbosa, P. *Meteorological Droughts in Europe: Events and Impacts-Past Trends and Future Projections*; Publications Office of the European Union: Brussels, Belgium, 2016.
33. Myneni, R.; Williams, D. On the relationship between FAPAR and NDVI. *Remote Sens. Environ.* **1994**, *49*, 200–211. [CrossRef]
34. Gobron, N.; Pinty, B.; Mélin, F.; Taberner, M.; Verstraete, M.; Belward, A.; Lavergne, T.; Widlowski, J.-L. The state of vegetation in Europe following the 2003 drought. *Int. J. Remote Sens.* **2005**, *26*, 2013–2020. [CrossRef]
35. Spinoni, J.; Muñoz, C.; Masante, D.; McCormick, N.; Vogt, J.; Barbosa, P. *European Drought Observatory*; European Commission: Ispra, Italy, 2018.
36. Peng, J.; Muller, J.-P.; Blessing, S.; Giering, R.; Danne, O.; Gobron, N.; Kharbouche, S.; Ludwig, R.; Müller, B.; Leng, G.; et al. Can we use satellite-based FAPAR to detect drought? *Sensors* **2019**, *19*, 3662. [CrossRef] [PubMed]
37. Zhang, Q.; Li, Q.; Singh, V.; Shi, P.; Huang, Q.; Sun, P. Nonparametric integrated agrometeorological drought monitoring: Model development and application. *J. Geophys. Res. Atmos.* **2018**, *123*, 73–88. [CrossRef]

38. Yevjevich, V. Objective Approach to Definitions and Investigations of Continental Hydrologic Droughts. Ph.D. Thesis, Colorado State University, Fort Collins, CO, USA, 1967.
39. Huang, S.; Huang, Q.; Chang, J.; Leng, G.; Xing, L. The response of agricultural drought to meteorological drought and the influencing factors: A case study in the Wei River Basin, China. *Agric. Water Manag.* **2015**, *159*, 45–54. [CrossRef]
40. Ivanova, M. Assessing the United Nations Environment Programme. *Glob. Commons* **2006**, *2*, 117–158.
41. Sapir, D.; Misson, C. The development of a database on disasters. *Disasters* **1992**, *16*, 74–80. [CrossRef]
42. Tijdeman, E.; Blauhut, V.; Stoelzle, M.; Menzel, L.; Stahl, K. Different drought types and the spatial variability in their hazard, impact, and propagation characteristics. *Nat. Hazards Earth Syst. Sci.* **2022**, *22*, 2099–2116. [CrossRef]
43. Sattar, M.; Lee, J.-Y.; Shin, J.-Y.; Kim, T.-W. Probabilistic characteristics of drought propagation from meteorological to hydrological drought in South Korea. *Water Resour. Manag.* **2019**, *33*, 2439–2452. [CrossRef]
44. Scaini, A.; Sánchez, N.; Vicente-Serrano, S.; Martínez-Fernández, J. SMOS-derived soil moisture anomalies and drought indices: A comparative analysis using in situ measurements. *Hydrol. Process.* **2015**, *29*, 373–383. [CrossRef]
45. García-Herrera, R.; Hernández, E.; Barriopedro, D.; Paredes, D.; Trigo, R.; Trigo, I.; Mendes, M. The outstanding 2004/05 drought in the Iberian Peninsula: Associated atmospheric circulation. *J. Hydrometeorol.* **2007**, *8*, 483–498. [CrossRef]
46. Coll, J.; Aguilar, E.; Ashcroft, L. Drought variability and change across the Iberian Peninsula. *Theor. Appl. Climatol.* **2017**, *130*, 901–916. [CrossRef]
47. Wong, G.; van Lanen, H.; Torfs, P. Probabilistic analysis of hydrological drought characteristics using meteorological drought. *Hydrol. Sci. J.* **2013**, *58*, 253–270. [CrossRef]

Disclaimer/Publisher's Note: The statements, opinions and data contained in all publications are solely those of the individual author(s) and contributor(s) and not of MDPI and/or the editor(s). MDPI and/or the editor(s) disclaim responsibility for any injury to people or property resulting from any ideas, methods, instructions or products referred to in the content.

Article

Water Whiplash in Mediterranean Regions of the World

Citlalli Madrigal [1], Rama Bedri [1], Thomas Piechota [2,*], Wenzhao Li [1,3], Glenn Tootle [4] and Hesham El-Askary [1,3,5]

1. Schmid College of Science and Technology, Chapman University, One University Drive, Orange, CA 92866, USA; cmadrigal@chapman.edu (C.M.); bedri@chapman.edu (R.B.); elaskary@chapman.edu (H.E.-A.)
2. Fowler School of Engineering, Chapman University, One University Drive, Orange, CA 92866, USA
3. Earth Systems Science and Data Solutions Lab, Chapman University, Orange, CA 92866, USA
4. Civil, Construction and Environmental Engineering, University of Alabama, Tuscaloosa, AL 35487, USA; gatootle@eng.ua.edu
5. Department of Environmental Sciences, Faculty of Science, Alexandria University, Moharem Bek, Alexandria 21522, Egypt
* Correspondence: piechota@chapman.edu; Tel.: +1-714-628-2897

Abstract: The presence of weather and water whiplash in Mediterranean regions of the world is analyzed using historical streamflow records from 1926 to 2023, depending on the region. Streamflow from the United States (California), Italy, Australia, Chile, and South Africa is analyzed using publicly available databases. Water whiplash—or the rapid shift of wet and dry periods—are compared. Wet and dry periods are defined based on annual deviations from the historical record average, and whiplash occurs when there is an abrupt change that overcomes an accommodated deficit or surplus. Of all the stations, there are more dry years (56%) than wet years (44%) in these regions, along with similarities in the variances and shifts in extremes (i.e., whiplash). On average, 35% of the years were defined as water whiplash years in all countries, with the highest levels in the US (California), where 42–53% of the years were whiplash years. The influence of the El Niño–Southern Oscillation (ENSO) influences Chile and South Africa strongest during the first quarter of the year. This study found that smaller extreme wet periods and larger and less extreme dry periods are prevalent in Mediterranean regions. This has implications for water management as adaptation to climate change is considered.

Keywords: streamflow; climate; extremes; ENSO; flood; drought; hydrology

Citation: Madrigal, C.; Bedri, R.; Piechota, T.; Li, W.; Tootle, G.; El-Askary, H. Water Whiplash in Mediterranean Regions of the World. *Water* **2024**, *16*, 450. https://doi.org/10.3390/w16030450

Academic Editors: Sonia Raquel Gámiz-Fortis and Matilde García-Valdecasas Ojeda

Received: 22 December 2023
Revised: 22 January 2024
Accepted: 26 January 2024
Published: 30 January 2024

Copyright: © 2024 by the authors. Licensee MDPI, Basel, Switzerland. This article is an open access article distributed under the terms and conditions of the Creative Commons Attribution (CC BY) license (https://creativecommons.org/licenses/by/4.0/).

1. Introduction

Extreme, unpredictable climate events can have adverse impacts on water security, society, and natural ecological systems. Extreme events are predicted to change in certain climate change hotspots of the world, including California in the United States [1,2]. Of interest in the present study is how Mediterranean regions will change, as they are characterized by having mild winters with variable precipitation periods, and hot, dry summers [3,4]. This climate region exists at equal distances from the equator, ranging between 30° North or South and 45° North or South. Furthermore, Mediterranean climate zones fall on the continent's western side. The five Mediterranean regions around the world are in California, the Mediterranean Basin in Europe, Western Australia, Chile, and the Cape of South Africa [3]. Furthermore, these regions are identified in the Koppen–Geiger Climate Classification world map [4], have cool, damp winters and hot dry summers, and are highly desirable due to their ideal weather conditions. These regions host agricultural hotspots, dense populations, and diverse species, making it important to understand their climate patterns.

A study of California precipitation found that dry to wet events are 25% to 100% more frequent in California under future climate projections [2]. A global study of Mediterranean precipitation found a decrease in the frequency of daily precipitation events, combined with increased amounts in rare extreme events, resulting in more year-to-year variability [5]. The Mediterranean basin (identified as the region around the Mediterranean Sea) has been noted as a climate change hotspot, and recent research has shown a strong coupling between temperature and precipitation, with tendencies for warm-dry anomalies in the summer and cold-wet anomalies in the winter [6]. This drying over Mediterranean regions has also been shown in climate change projections (e.g., [7–9]). The impact of hydrological intensification (or whiplash) was investigated for different regions of the world, and it was found that large areas of intensification occurred in areas with large reservoir systems in place, thus allowing for adaption [10].

Previous studies have been limited to extreme precipitation shifts for the regions noted above [1,2,5]. The work presented here augments past studies by focusing on water supply, as represented by streamflow, with an aim to understand weather (water) whiplash phenomena in Mediterranean regions. Weather whiplash is defined as sudden changes in weather conditions from one extreme to another, such as drought to heavy precipitation or flooding [1,5,11]. The study presented here is motivated by California's switch from having the worst drought on record in 2022 to seeing one of the wettest years in 2023, relieving the state of drought. Furthermore, the Emilia Romagna region of Northern Italy has experienced a steady increase in the intensity of rainfall events [12]. The frequency and intensity of destructive, heavy rainfall events are expected to increase in this region [13]. The study here compares all global Mediterranean regions based on the historical streamflow record and whether the whiplash phenomenon is present in other Mediterranean climate zones.

A further aim in this study is to evaluate the impact of the El Niño–Southern Oscillation (ENSO) on streamflow in Mediterranean regions. ENSO has been shown to have global impacts (e.g., [14–19]). These global studies have shown that three of the five Mediterranean regions investigated in this study are drier during El Niño years. The investigation of streamflow in this study will provide a broader perspective on integrated water impacts in these regions. It is important to note that other modes of large-scale climate variability may have a more significant impact on certain Mediterranean regions. For instance, European precipitation has been connected to North Atlantic Oscillation (NAO), the Artic Oscillation (AO), the North Sea Caspian Pattern (NCP), and two indices of Mediterranean Oscillation (MOI2, WeMOI) [20]. A study of North Africa did not find a significant ENSO effect for North Africa (a Mediterranean region) [21].

2. Materials and Methods

2.1. Data

Figure 1 and Table 1 present all the streamflow stations used in this study. This includes four stations in each of the countries of Italy, South Africa, United States (California), Australia, and Chile.

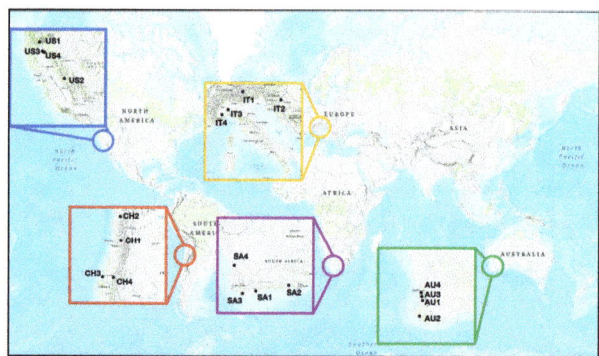

Figure 1. Map of global Mediterranean regions studied and associated streamflow stations. The different colors represent the different regions evaluated in this study.

Table 1. List of streamflow stations with characteristics for all locations shown in Figure 1.

Station Name	ID	Location	Year Measured	Start Year	End Year	Latitude	Longitude	Units
Sacramento Delta at Sacramento River	US1 [a]	California, USA	Water Year	1951	2023	40.94	−122.42	KAF
Happy Isles Bridge near Yosemite at Merced River	US2 [a]	California, USA	Water Year	1951	2023	37.73	−119.56	KAF
Mill Creek Near Los Molinos at Mill River	US3 [a]	California, USA	Water Year	1951	2023	40.05	−122.02	KAF
Deer Creek Near Vina at Deer River	US4 [a]	California, USA	Water Year	1951	2023	40.01	−121.95	KAF
Spondigna at Adige River	IT1 [b]	Italy	Calendar Year	1980	2018	46.63	10.60	mm
Sava Near Catez at Sava River	IT2 [b]	Slovenia	Calendar Year	1926	2020	45.89	15.60	MCM
Montecastello at Fiume Tanaro River	IT3 [c]	Italy	Calendar Year	1936	2008	44.94	8.68	m^3/s
Farigliano at Fiume Tanaro River	IT4 [c]	Italy	Calendar Year	1942	2008	44.51	7.90	m^3/s
Yarragil Brook, Yarragil Formation at Murray River	AU1 [d]	Australia	Calendar Year	1952	2022	−32.80	116.12	ML
Donnelly Near Strickland at Donnelly River	AU2 [d]	Australia	Calendar Year	1952	2022	−34.33	115.77	ML
Big Brook Near O'Neil Rd ay Murray River	AU3 [d]	Australia	Calendar Year	1983	2022	−32.53	116.04	ML
Wungong Brook near Vardi Rd at Swan River	AU4 [d]	Australia	Calendar Year	1981	2022	−32.10	15.98	ML
Chacabuquito at Aconcagua River	CH1 [c]	Chile	Calendar Year	1956	2019	−32.85	−70.51	m^3/s
Algarrobal at Elqui River	CH2 [c]	Chile	Calendar Year	1980	2019	−29.99	−70.58	m^3/s
Desembocadura at Biobio River	CH3 [c]	Chile	Calendar Year	1969	2019	−36.83	−73.07	m^3/s
San Lorezo at Diguillin River	CH4 [c]	Chile	Calendar Year	1960	2019	−36.92	−71.57	m^3/s
Dassjes Klip at Duiwenhoksrivier	SA1 [c]	South Africa	Calendar Year	1968	2022	−34.25	20.99	m^3/s
Grootrivierspoort at Grootrivier	SA2 [c]	South Africa	Calendar Year	1965	2022	−33.71	24.61	m^3/s
Hagedisberg Outspan at Kleinrivier	SA3 [c]	South Africa	Calendar year	1964	2021	−34.40	19.59	m^3/s
Melkboom at Doringrivier	SA4 [c]	South Africa	Calendar Year	1928	2022	−31.86	18.68	m^3/s

Notes: [a] Data obtained from the United States Geological Survey. [b] Data obtained from [18]. [c] Data obtained from Global Runoff Data Center. [d] Data obtained from Hydrologic Reference Stations.

2.1.1. Chile, South Africa and Italy (Global Runoff Data Center)

The streamflow data for Chile (CH1-4), South Africa (SA1-4) and Italy (IT3 and IT4) were obtained from the Global Runoff Data Centre's (GRDC) data portal [22]. This data portal provides historical mean daily and monthly river discharge data for over 10,000 stations globally. From the data portal, four stations from along the Southern and the Western regions of South Africa, four stations throughout Chile, and two stations in Northeastern Italy were selected to conduct this study. The retrieved data were monthly mean discharges (m^3/s), which were used to calculate the annual mean data (m^3/s) for each of the ten stations.

2.1.2. United States (California) (USGS)

For stations US 1 through 4 (Table 1), mean annual streamflow data were collected from the U.S. Geological Survey (USGS) NWISWeb Data retrieval [23] for the water years of 1951–2022, which are measured from October to the following September (defined as a water year). The four stations are located in the north central part of California in the United States. The data retrieved from the USGS are in cubic feet per second (ft^3/s), and were converted into kilo-acre feet (KAF) to conduct the study. The 2023 data were based on a forecast of the water year volume (made on 1 April 2023) provided by the National Oceanic and Atmospheric Administration's (NOAA) California Nevada River Forecast Center [24].

2.1.3. Italy

For station IT1 (Table 1), annual modeled streamflow data were acquired from previous research [25–27]. For station IT2 (Table 1), streamflow data were obtained from the Catez gauge, located on the Sava River in Slovenia near the Croatian border. Although this station is not in Italy, the gauge is a critical measurement of streamflow in the European Mediterranean basin [28]. Annual data are given in MCM (million cubic meter). Past studies (e.g., [28]) have shown that the data are sufficient for studies evaluating climate impacts on water resources of the water basins. In addition, direct observations of rivers are provided for two stations [IT3 and IT4].

2.1.4. Australia (Hydrologic Reference Stations)

For stations in Australia, data were retrieved from the Hydrologic Reference Stations (HRS) catchments by the Australian Government Bureau of Meteorology [29]. For this study, annual streamflow (ML/year) was derived from the HRS, which provided high quality daily streamflow data and corresponding statistics for 467 stations [30]. Four stations were selected in Southwestern Australia, near Perth which is part of the Mediterranean region of Australia.

2.2. Analysis of Whiplash—Wet and Dry Periods

For the study shown here, wet and dry periods were identified in the records based on yearly deviations from the long-term average of the historical record. Wet years are positive departures (y) from the long-term average of the historical record (x), and dry years are negative departures (y) where:

$$y_i = x_i - \overline{x} \qquad (1)$$

A wet (or dry period) is calculated by summing up consecutive dry or wet years where:

$$\sum_{i=1}^{n} y_i = A_j \qquad (2)$$

A dry or wet period continues until the accumulated (A) deficit or surplus condition is switched by a single year (y), with a higher opposite sign of deficit or surplus. This sudden shift in state is referred to as "whiplash" (W), since an accumulated state is abruptly changed in one year where:

$$W \gg \frac{\text{is negative when } y_i > |A_j|}{\text{is positive when } |y_i| > A_j} \quad (3)$$

To compare stations with different magnitudes, an indicator ratio (IR) of accumulated deficit or surplus to the long-term mean is calculated.

$$IR_i = \frac{A_j}{\overline{X}} \quad (4)$$

Indicator values that shift from negative to positive or positive to negative in two consecutive years reflect whiplash years.

2.3. Testing of Differences in Populations

The testing of different populations for this study used the F-test. For instance, tests were conducted between countries to evaluate similarities or differences. The F-test evaluates the variance in two populations. This was carried out for differences (or similarities) between countries. Results are displayed and discussed in Section 3.

The sample populations of wet (surplus) and dry (deficit) periods for each country are also presented as box and whisker plots, where the middle of the box represents the median, the top and bottom represent the 75th and 25th percentiles of the population, and the top and bottom of the whisker represent the 90th and 10th percentiles of the population (see Section 3.1.1). In addition, box plots are used to represent the proportion of years that are whiplash years (see Section 3.1.2).

2.4. Analysis of ENSO Impacts

Streamflow data are tested with four different Oceanic Niño Index (ONI) values, a measure of the El Niño-Southern Oscillation [31]. JFM, AMJ, JAS, and OND are used as representations for the year. Correlation between flow and each ONI index are tested for each station.

3. Results

3.1. Weather Whiplash Results

The analyses to test the similarities of Mediterranean regions/countries to weather whiplash are shown in Tables 2 and 3, Figures 2–5. These results are shown for the historical period of records that vary depending on the region (see Table 1).

Table 2. Frequency of wet and dry periods per station represented by the count of wet and dry years and percentage compared to the total number of years. Values in parentheses are results of analysis using a common period of record (1983–2018). N/A represents analysis that was not available for these stations. * are stations where the common period of record has unequal number of wet and dry years.

Station ID	Total Wet Years	Percentage	Total Dry Years	Percentage
US1 *	36 (13)	49% (36%)	37 (23)	51% (64%)
US2 *	32 (14)	44% (39%)	41 (22)	56% (61%)
US3 *	34 (14)	47% (39%)	39 (22)	53% (61%)
US4 *	35 (14)	48% (39%)	38 (22)	52% (61%)
IT1	19 (18)	49% (50%)	20 (18)	51% (50%)
IT2 *	35 (5)	37% (14%)	60 (31)	63% (86%)
IT3	31(N/A)	43% (N/A)	42 (N/A)	58% (N/A)
IT4	33 (N/A)	50% (N/A)	33 (N/A)	50% (N/A)
AU1 *	20 (3)	29% (8%)	49 (33)	71% (92%)

Table 2. *Cont.*

Station ID	Total Wet Years	Percentage	Total Dry Years	Percentage
AU2 *	31 (10)	45% (28%)	38 (26)	55% (72%)
AU3 *	14 (13)	37% (36%)	24 (23)	63% (64%)
AU4	18 (18)	47% (50%)	20 (18)	53% (50%)
CH1	33 (19)	52% (53%)	31 (17)	48% (47%)
CH2	15 (15)	38% (42%)	25 (21)	63% (58%)
CH3	25 (19)	49% (50%)	26 (18)	51% (50%)
CH4	31 (18)	52% (50%)	29 (18)	48% (50%)
SA1	25 (19)	45% (53%)	30 (17)	55% (47%)
SA2 *	14 (8)	24% (22%)	44 (28)	76% (36%)
SA3 *	31 (23)	53% (64%)	27 (13)	47% (36%)
SA4	38 (17)	40% (47%)	57 (19)	60% (53%)

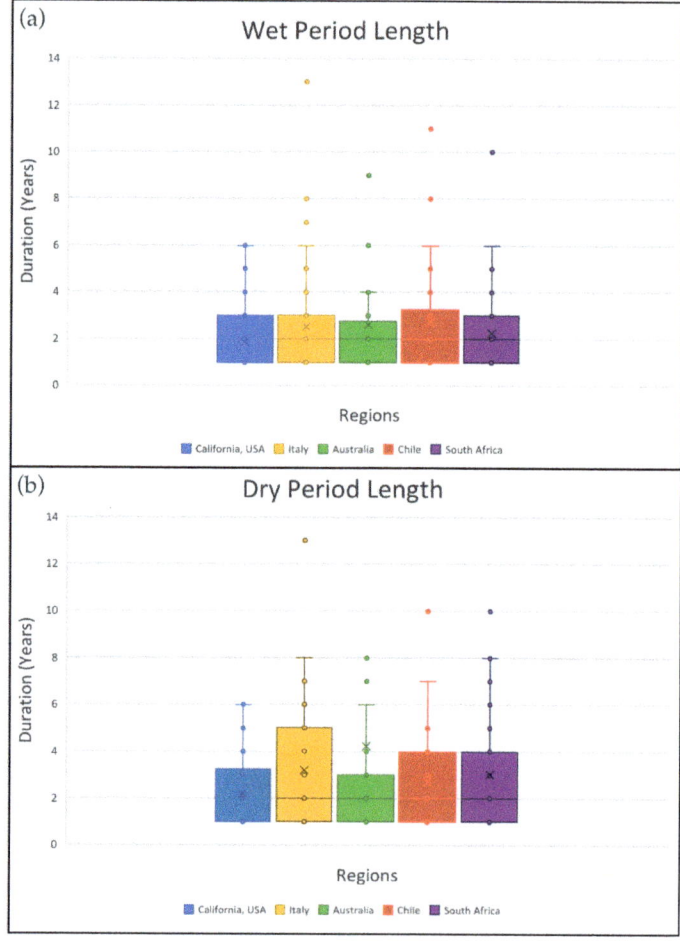

Figure 2. Boxplot of the duration in years of (**a**) wet period length and (**b**) dry period length averaged by region. The X in the boxplot represents the mean.

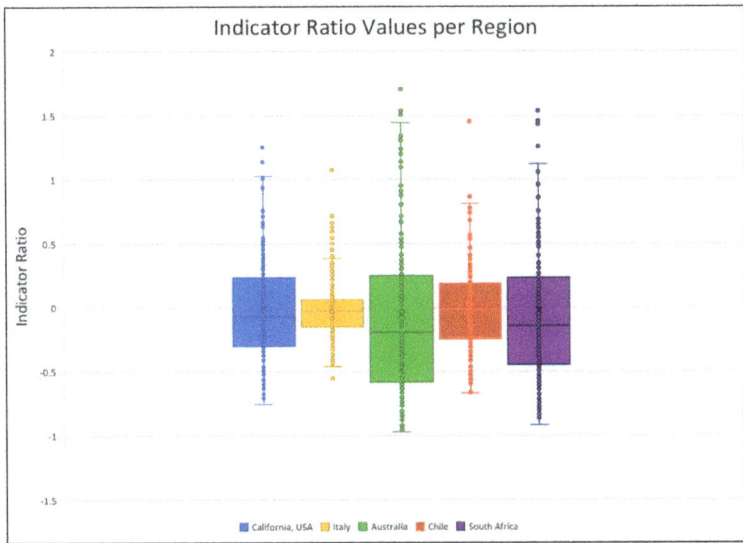

Figure 3. Boxplot of indicator ratio values averaged by region. The ratio is an indicator of the accumulated deficit or surplus to the long-term mean. Indicator values that shift from negative to positive or positive to negative in two consecutive years reflect whiplash years.

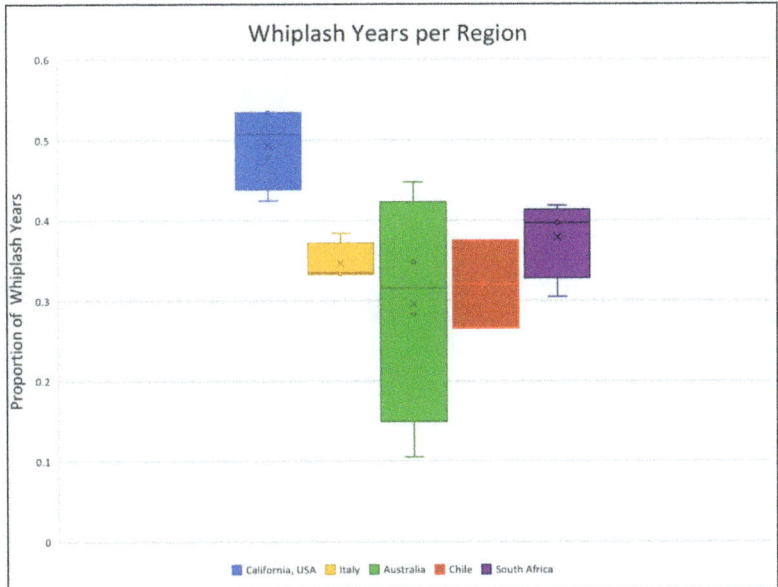

Figure 4. Boxplot of the proportion of years that experienced whiplash averaged by region. For each station, whiplash years are identified, being the years that change from wet to dry or dry to wet. The count of whiplash years is divided by the total years in the record, being the proportion.

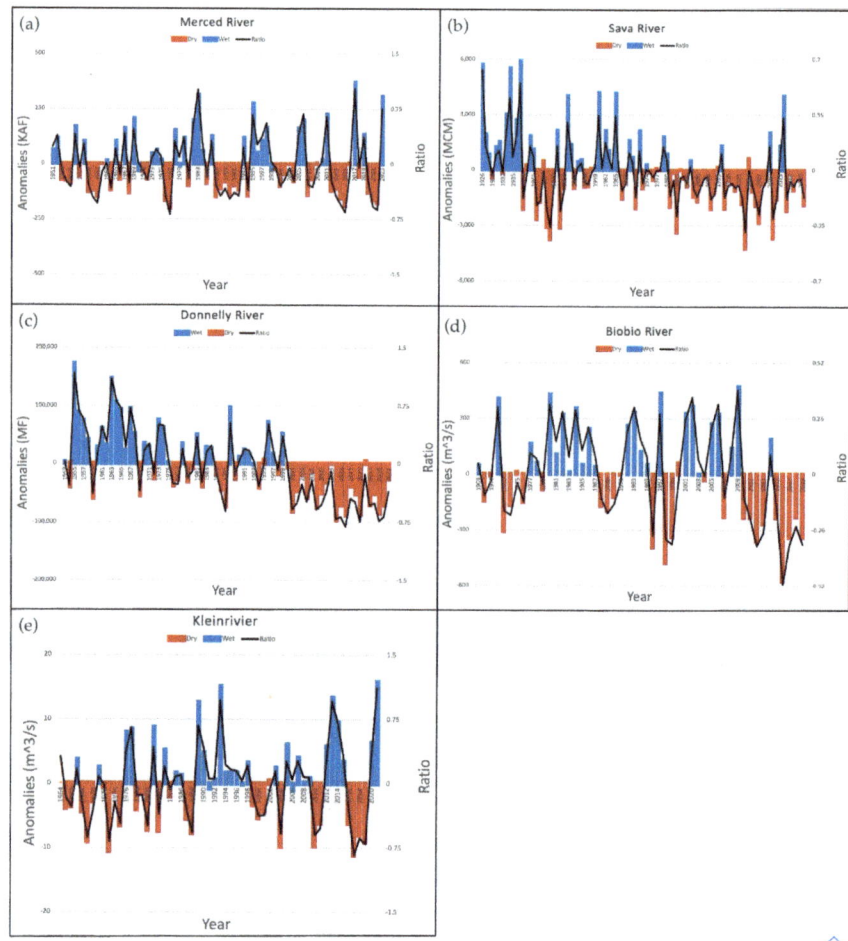

Figure 5. Time series of anomaly (blue and red) and ratio (black) values. Blue areas represent surplus periods and red areas represent deficit periods. One station's plot that is representative of all four stations in the area is chosen per region: (**a**) Merced River in California, USA (US2); (**b**) Sava River in Slovenia (IT2); (**c**) Donnelly River in Australia (AU2); (**d**) Biobio River in Chile (CH3); (**e**) Kleinrivier in South Africa (SA3).

Table 3. F-test results comparing each pair of regions with wet period length, dry period length, and indicator ratio. Statistically significant values (at 1% and 5% level) are indicated as regions that are similar in terms of wet and dry periods.

Region	US-IT	US-AU	US-CH	US-SA	IT-AU	IT-CH	IT-SA	AU-CH	AU-SA	CH-SA
Wet Period Length	<0.01	<0.01	<0.01	<0.05					<0.01	
Dry Period Length	<0.01	<0.01	<0.01	<0.05	<0.01			<0.01	<0.01	
Indicator Ratio	<0.01	<0.01		<0.01	<0.01	<0.01	<0.01	<0.01	<0.05	<0.01

3.1.1. Wet and Dry Periods

The occurrence of wet and dry periods is noted in Table 2 and Figure 2. In Table 2, two analyses are presented using the entire period of record for all the stations, and another analysis using a common period (1983–2018) for 18 of the 20 stations. Using the entire length of record for all stations, the majority of the stations (14 of 20) have an equal likelihood of wet and dry years (i.e., the number of dry and wet years are about the same). A couple stations have a significantly higher percentage of dry years (over 70%), namely Murray River (AU1) in Australia and Grootriver (SA2) in South Africa. Six stations had at least 60% of years drier than average, with a majority (4/6) of those being in Australia and South Africa. It is noteworthy that none of the stations had a larger portion of years that were wetter than drier, representing a general drying of these regions. Finally, the overall average percentage of dry years (56%) is greater than that of wet years (44%). Using a common period in the record, only 8 of the 18 stations had an equal likelihood of wet and dry years, with nine (9) of the stations indicating a larger number of dry years.

Figure 2 presents boxplots for wet and dry period lengths using the entire record. The periods are relatively the same length for wet periods (Figure 2a); however, there is a larger variation for the dry periods (Figure 2b) in each region. F-tests were conducted for both wet and dry periods between each pair of regions. Values less than 0.05 indicate that the probability that two regions are different is low (Table 3). Therefore, highlighted values mean that the regions tested are statistically similar. In both the wet and dry periods, the United States (California) is statistically similar with each region. Furthermore, the F-test analysis shows that Australia is similar to each region for the dry periods only.

3.1.2. Whiplash Results

The rapid change from dry to wet or wet to dry states (whiplash) is evaluated in Figure 3 and Table 3. Figure 3 provides a compilation of all the indicator values expressed as an indicator ratio described in Section 2.2. A positive indicator ratio is a rapid transition from dry to wet state, and a negative indicator ratio is a rapid transition from wet to dry state. There are higher whiplash events from dry to wet, as shown by the larger positive values in Figure 3. Comparing this result to Figure 2, it is concluded that there are shorter intense wet periods and longer and less intense dry periods. Variability is highest in the United States (California), Australia, and South Africa. Results from Table 3 display that most countries have similar ratio values, with exceptions for the United States (California) and Chile.

Figure 4 presents the proportion of whiplash years that occur in each country for all stations. The US (California) locations have the highest proportion of years that are whiplash years—between 42 and 53% of years. Australia has the largest variability, from 10 to 45%. Italy, Chile, and South Africa are similar, and have about 35% of their years as whiplash years.

3.1.3. Regional Analysis

The regional analysis of wet and dry periods and associated whiplash events is shown in Figure 5. Three of the five regions show drying in the latter half of the record (Sava (Figure 5b), Donnelly (Figure 5c), Biobio (Figure 5d)). In California (Merced River, US2); the early part of the record (1951 to 1986) had wet and dry periods that were short and mild. In 1987, the dry period became longer, and wet and dry periods became more intense and more frequent. In Italy (Sava River, IT2), 1926–1980 were predominantly wet years, with a switch in 1980 to dry years. In Australia (Donnelly River, AU2), 1953–2000 were predominantly wet years, with a switch in 2000 to dry years, similar to Italy. In Chile, (Biobio River, CH3) dry periods became longer in recent years after 2000. Finally, in South Africa (Kleinrivier, SA3) the data shows that wet and dry periods became more intense later in the record, similar to California (Merced River).

3.2. ENSO Impacts

Table 4 provides a summary of the impact of ENSO on all the regions and rivers of this study. Table 4 displays correlation coefficients for each test. Values are highlighted, with a significance of $p = 0.10$ and $p = 0.05$. While ENSO has been shown to have global impacts [32], the areas impacted in this study are fairly limited. Most notable were the stations in Chile, with all four stations showing significance at both 95% and 90% confidence levels. South Africa and United States (California) had some significance, with two and one station(s), respectively, displaying statistical significance. Stations in Italy and Australia did not show statistical significance under either level. The selected season does have an impact on whether ENSO influences streamflow. For instance, the ONI values in JFM were significant in five stations, compared to the other seasons having two significant stations each.

Table 4. Correlation coefficients between streamflow and the respective ONI value for each station. Values that are statistically significant at a 10% level ($p < 0.1$) are designated with a * symbol. Values that are significant at the 5% level ($p < 0.05$) are highlighted in grey. Four ONI values were used to represent the year in three-month-long periods (JFM, AMJ, JAS, OND).

Station ID	Flow and JFM	Flow and AMJ	Flow and JAS	Flow and OND
US1	0.20 *	0.17	−0.11	−0.14
US2	0.15	0.19	−0.05	−0.07
US3	0.11	0.14	−0.04	−0.05
US4	0.10	0.15	−0.04	−0.05
IT1	−0.06	0.05	0.05	0.07
IT2	0.03	0.12	0.13	0.17
IT3	−0.12	0.01	0.16	0.21
IT4	−0.04	0.10	0.14	0.15
AU1	−0.05	−0.18	−0.13	−0.18
AU2	−0.06	−0.18	−0.14	−0.18
AU3	−0.12	−0.15	−0.18	−0.16
AU4	0.09	0.03	−0.05	−0.10
CH1	0.26	0.20	0.17	0.07
CH2	0.31	0.14	0.03	−0.10
CH3	−0.13	0.27 *	0.46	0.46
CH4	−0.12	0.20	0.27	0.30
SA1	−0.26 *	−0.19	0.03	0.06
SA2	−0.34	−0.33	−0.14	−0.17
SA3	−0.15	0.00	0.07	0.06
SA4	−0.15	−0.06	0.02	0.00

4. Discussion

In this study, historical streamflow data were used to evaluate the presence of water whiplash in Mediterranean regions of the world. Most regions indicated a general drying in terms of the number of years that were dry as opposed to wet. Additionally, many of the regions had longer dry periods than wet periods in terms of number of years. For instance, up to 53% of the years in the US (California) were defined as whiplash years (Figure 4). Previous work has shown California not drying to the same level as other Mediterranean regions [5]. On average for other regions, approximately 35% of the years were water whiplash (Figure 4).

For all regions, there are shorter intense wet periods and longer and less intense dry periods—switching back and forth between conditions of long, dry periods where the soil is dry, and intense, wet periods where soils saturate and produce excess runoff. This intense flooding and rapid shifting between extremes may have adverse impacts on ecosystems and surrounding communities.

In the face of the abrupt shifts due to weather whiplash, developing nations and poorer regions disproportionately experience the impacts of these events. For instance, regions of Chile and South Africa do not have the same means to adapt as other countries such as the United States, Italy, and Australia. These vulnerable regions are usually not equipped in resources and infrastructure, causing greater inequality due to the compiling of damage from multiple whiplash events [33].

5. Conclusions

The results of this research are consistent with previous work that have identified various Mediterranean regions in the world where weather whiplash in precipitation is persistent in future data. The work presented here focuses on the impacts on water, and has implications for management of resources. Comparative analyses of the durations, conditions, and frequencies of wet and dry periods offer insights into the similarities and disparities among these regions. The results underscore a historical trend wherein wet years were more prevalent until the 1980–2000s, after which dry periods increased in frequency and intensity. This research contributes to the evolving field of hydroclimatology in Mediterranean regions, emphasizing the critical role of understanding streamflow patterns in shaping future water supply dynamics. The implications of this study extend to water resources management, emphasizing the need for proactive measures to address the complex challenges posed by climate-induced hydrological variability in Mediterranean regions.

Author Contributions: Conceptualization, C.M., R.B., T.P., W.L. and H.E.-A.; Data curation, C.M., R.B., T.P., C.M. and G.T.; Formal analysis, C.M., R.B. and W.L.; Funding acquisition, H.E.-A. and T.P.; Investigation, C.M., R.B. and T.P.; Methodology, C.M., R.B., T.P. and W.L.; Project administration, T.P.; Resources, T.P. and H.E.-A.; Supervision, T.P.; Validation, C.M., R.B. and W.L.; Visualization, C.M. and R.B.; Writing—original draft, C.M., R.B. and T.P.; Writing—review and editing, C.M., R.B., T.P., W.L., G.T. and H.E.-A. All authors have read and agreed to the published version of the manuscript.

Funding: This research was funded by Department of Education Grant "Earth Systems Science and Data Solutions Lab Applying Data Science Techniques to Achieve the UN Sustainable Development Goals" (P116Z220190).

Data Availability Statement: Data for this study were accessed at the Global Runoff Data Centre's (GRDC) data portal (Available online: https://portal.grdc.bafg.de/applications/public.html?publicuser=PublicUser, accessed on 10 October 2023), the U.S. Geological Survey (USGS) NWISWeb Data retrieval (Available online: https://waterdata.usgs.gov/nwis/, accessed on 1 February 2023), and the Australian Bureau of Meteorology Hydrologic Reference (HRS) catchments (Available online: http://www.bom.gov.au/water/hrs/, accessed on 25 September 2023). Additional data for stations in Italy were retrieved from previous studies [25,26].

Acknowledgments: We would like to thank Nejc Bezak at University of Ljubljana, and Martin Morlot and Giuseppe Formetta (University of Trento) for providing data for this study.

Conflicts of Interest: The authors declare no conflicts of interest.

References

1. Chen, D.; Norris, J.; Thackeray, C.; Hall, A. Increasing Precipitation Whiplash in Climate Change Hotspots. *Environ. Res. Lett.* **2022**, *17*, 124011. [CrossRef]
2. Swain, D.; Langenbrunner, B.; Neelin, J.; Hall, A. Increasing Precipitation Volatility in Twenty-First-Century California. *Nat. Clim. Chang.* **2018**, *8*, 427–433. [CrossRef]

3. Spano, D.; Snyder, R.L.; Cesaraccio, C. Mediterranean Climates. In *Phenology: An Integrative Environmental Science*; Schwartz, M.D., Ed.; Tasks for Vegetation Science; Springer: Dordrecht, The Netherlands, 2003; pp. 139–156. ISBN 978-94-007-0632-3.
4. Peel, M.; Finlayson, B.; Mcmahon, T. Updated World Map of the Koppen-Geiger Climate Classification. *Hydrol. Earth Syst. Sci. Discuss.* **2007**, *11*, 1633–1644. [CrossRef]
5. Polade, S.D.; Gershunov, A.; Cayan, D.R.; Dettinger, M.D.; Pierce, D.W. Precipitation in a Warming World: Assessing Projected Hydro-Climate Changes in California and Other Mediterranean Climate Regions. *Sci. Rep.* **2017**, *7*, 10783. [CrossRef]
6. De Luca, P.; Messori, G.; Faranda, D.; Ward, P.J.; Coumou, D. Compound Warm–Dry and Cold–Wet Events over the Mediterranean. *Earth Syst. Dyn.* **2020**, *11*, 793–805. [CrossRef]
7. Polade, S.D.; Pierce, D.W.; Cayan, D.R.; Gershunov, A.; Dettinger, M.D. The Key Role of Dry Days in Changing Regional Climate and Precipitation Regimes. *Sci. Rep.* **2014**, *4*, 4364. [CrossRef]
8. Giorgi, F.; Lionello, P. Climate Change Projections for the Mediterranean Region. *Glob. Planet. Chang.* **2008**, *63*, 90–104. [CrossRef]
9. Mariotti, A.; Pan, Y.; Zeng, N.; Alessandri, A. Long-Term Climate Change in the Mediterranean Region in the Midst of Decadal Variability. *Clim. Dyn.* **2015**, *44*, 1437–1456. [CrossRef]
10. Ficklin, D.L.; Null, S.E.; Abatzoglou, J.T.; Novick, K.A.; Myers, D.T. Hydrological Intensification Will Increase the Complexity of Water Resource Management. *Earths Future* **2022**, *10*, e2021EF002487. [CrossRef]
11. Francis, J.A.; Skific, N.; Vavrus, S.J.; Cohen, J. Measuring "Weather Whiplash" Events in North America: A New Large-Scale Regime Approach. *J. Geophys. Res. Atmos.* **2022**, *127*, e2022JD036717. [CrossRef]
12. Tomozeiu, R.; Lazzeri, M.; Cacciamani, C. Precipitation Fluctuations during the Winter Season from 1960 to 1995 over Emilia-Romagna, Italy. *Theor. Appl. Climatol.* **2002**, *72*, 221–229. [CrossRef]
13. Bérubé, S.; Brissette, F.; Arsenault, R. Optimal Hydrological Model Calibration Strategy for Climate Change Impact Studies. *J. Hydrol. Eng.* **2022**, *27*, 04021053. [CrossRef]
14. Rasmusson, E.M.; Arkin, P.A. Chapter 40 Interannual Climate Variability Associated with the El Niño/Southern Oscillation. In *Elsevier Oceanography Series*; Nihoul, J.C.J., Ed.; Coupled Ocean-Atmosphere Models; Elsevier: Amsterdam, The Netherlands, 1985; Volume 40, pp. 697–725.
15. Ropelewski, C.F.; Jones, P.D. An Extension of the Tahiti–Darwin Southern Oscillation Index. *Mon. Weather Rev.* **1987**, *115*, 2161–2165. [CrossRef]
16. Rajagopalan, B.; Cook, E.; Lall, U.; Ray, B.K. Spatiotemporal Variability of ENSO and SST Teleconnections to Summer Drought over the United States during the Twentieth Century. *J. Clim.* **2000**, *13*, 4244–4255. [CrossRef]
17. Chiew, F.H.S.; Piechota, T.C.; Dracup, J.A.; McMahon, T.A. El Nino/Southern Oscillation and Australian Rainfall, Streamflow and Drought: Links and Potential for Forecasting. *J. Hydrol.* **1998**, *204*, 138–149. [CrossRef]
18. Lenssen, N.J.L.; Goddard, L.; Mason, S. Seasonal Forecast Skill of ENSO Teleconnection Maps. *Weather Forecast.* **2020**, *35*, 2387–2406. [CrossRef]
19. Mason, S.; Goddard, L. Probabilistic Precipitation Anomalies Associated with ENSO. *Bull. Am. Meteorol. Soc.* **2001**, *82*, 619–638. [CrossRef]
20. Müller-Plath, G.; Lüdecke, H.-J.; Lüning, S. Long-Distance Air Pressure Differences Correlate with European Rain. *Sci. Rep.* **2022**, *12*, 10191. [CrossRef] [PubMed]
21. Lüdecke, H.-J.; Müller-Plath, G.; Wallace, M.G.; Lüning, S. Decadal and Multidecadal Natural Variability of African Rainfall. *J. Hydrol. Reg. Stud.* **2021**, *34*, 100795. [CrossRef]
22. GRDC Data Portal. Available online: https://portal.grdc.bafg.de/applications/public.html?publicuser=PublicUser (accessed on 10 October 2023).
23. USGS Water Data for the Nation. Available online: https://waterdata.usgs.gov/nwis (accessed on 1 February 2023).
24. National Oceanic and Atmospheric Administration; National Weather Service; California Nevada River Forecast Center. Available online: https://www.cnrfc.noaa.gov/?product=espfcstWY&zoom=7&lat=40.506&lng=-121.896 (accessed on 1 April 2023).
25. Morlot, M.; Rigon, R.; Formetta, G. Hydrological Digital Twin Model of a Large Anthropized Italian Alpine Catchment: The Adige River Basin. *J. Hydrol.* **2023**, *629*, 130587. [CrossRef]
26. Formetta, G.; Antonello, A.; Franceschi, S.; David, O.; Rigon, R. Hydrological Modelling with Components: A GIS-Based Open-Source Framework. *Environ. Model. Softw.* **2014**, *55*, 190–200. [CrossRef]
27. Formetta, G.; Rigon, R.; Chávez, J.L.; David, O. Modeling Shortwave Solar Radiation Using the JGrass-NewAge System. *Geosci. Model Dev.* **2013**, *6*, 915–928. [CrossRef]
28. Tootle, G.; Oubeidillah, A.; Elliott, E.; Formetta, G.; Bezak, N. Streamflow Reconstructions Using Tree-Ring-Based Paleo Proxies for the Sava River Basin (Slovenia). *Hydrology* **2023**, *10*, 138. [CrossRef]
29. Hydrologic Reference Stations. Water Information: Bureau of Meteorology. Available online: http://www.bom.gov.au/water/hrs/ (accessed on 25 September 2023).
30. Amirthanathan, G.E.; Bari, M.A.; Woldemeskel, F.M.; Tuteja, N.K.; Feikema, P.M. Regional Significance of Historical Trends and Step Changes in Australian Streamflow. *Hydrol. Earth Syst. Sci.* **2023**, *27*, 229–254. [CrossRef]
31. Climate Prediction Center Internet Team. NOAA's Climate Prediction Center. Available online: https://origin.cpc.ncep.noaa.gov/products/analysis_monitoring/ensostuff/ONI_v5.php (accessed on 1 October 2023).

32. Ropelewski, C.F.; Halpert, M.S. Global and Regional Scale Precipitation Patterns Associated with the El Niño/Southern Oscillation. *Mon. Weather Rev.* **1987**, *115*, 1606–1626. [CrossRef]
33. Zhang, B.; Wang, S.; Zscheischler, J.; Moradkhani, H. Higher Exposure of Poorer People to Emerging Weather Whiplash in a Warmer World. *Geophys. Res. Lett.* **2023**, *50*, e2023GL105640. [CrossRef]

Disclaimer/Publisher's Note: The statements, opinions and data contained in all publications are solely those of the individual author(s) and contributor(s) and not of MDPI and/or the editor(s). MDPI and/or the editor(s) disclaim responsibility for any injury to people or property resulting from any ideas, methods, instructions or products referred to in the content.

Article

Assessment of the Impacts of Rainfall Characteristics and Land Use Pattern on Runoff Accumulation in the Hulu River Basin, China

Muhammad Imran [1], Jingming Hou [1,*], Tian Wang [1], Donglai Li [1], Xujun Gao [2], Rana Shahzad Noor [3], Jing Jing [1] and Muhammad Ameen [4]

1. State Key Laboratory of Eco-Hydraulics in Northwest Arid Region of China, Xi'an University of Technology, Xi'an 710048, China; imran_badani@hotmail.com (M.I.)
2. Power China Northwest Engineering Corporation Limited, Xi'an 710065, China
3. Faculty of Agricultural Engineering and Technology, PMAS-Arid Agriculture University, Rawalpindi 46000, Pakistan; engr.rsnoor@uaar.edu.pk
4. School of Water Conservancy, Yunnan Agricultural University, Kunming 650201, China
* Correspondence: jingming.hou@xaut.edu.cn

Citation: Imran, M.; Hou, J.; Wang, T.; Li, D.; Gao, X.; Noor, R.S.; Jing, J.; Ameen, M. Assessment of the Impacts of Rainfall Characteristics and Land Use Pattern on Runoff Accumulation in the Hulu River Basin, China. *Water* 2024, 16, 239. https://doi.org/10.3390/w16020239

Academic Editors: Sonia Raquel Gámiz-Fortis and Matilde García-Valdecasas Ojeda

Received: 29 November 2023
Revised: 16 December 2023
Accepted: 17 December 2023
Published: 10 January 2024

Copyright: © 2024 by the authors. Licensee MDPI, Basel, Switzerland. This article is an open access article distributed under the terms and conditions of the Creative Commons Attribution (CC BY) license (https://creativecommons.org/licenses/by/4.0/).

Abstract: Climate change causes the river basin water cycle disorders, and rainfall characteristics frequently result in flood disasters. This study aims to simulate and assess the response behavior of basin floods under the influence of rainfall characteristics and land use changes in the Hulu River basin using a 2D hydrological and hydraulic GAST (GPU Accelerated Surface Water Flow and Transport Model). The peak flow rate and water depth during floods were examined by simulating the evolution process of basin floods and related hydraulic elements under the independent effects of various rainfall characteristics or land use and further simulating the response results of basin floods under the combined effects of rainfall characteristics and land use. The seven scenarios were set to quantify the degree of influence that land use and rainfall characteristics have on the basin flood process based on examining changes in land use and rainfall characteristics in the research area. The results from different rainfall characteristics scenarios depicted that as the rainfall return period is shorter, the peak flow rate is higher, and the peak flow rate is lower as the return period is prolonged. Under different rainfall characteristics, the peak flow rate in scenario R8 is 41.30%, 40.00%, and 34.51% higher than the uniform distribution of rainfall, while water depth is decreased by 0.55%, increased by 4.96% and 2.92% as compared to the uniform distribution of rainfall. While under different land use scenarios, it is observed that the change in land use has increased 2.7% in cultivated land and 1.1% in woodland. In addition, the interactive effect of different rainfall characteristics and land use it can be seen that the scenario with the greatest reduction in flood risk due to rainfall characteristics and land use is RL2-4, representing a 12.55% decrease in peak flow and a 37.69% decrease in peak water depth. In this scenario, the rainfall is heavier in the southeast and northwest regions and lighter in the northeast and southwest regions. The land use type is characterized by reforestation and the return of cultivated land to forests. The changes in rainfall distribution and the increase in grassland contribute to the decrease in flood threat. Future research in the erodible parts of the Hulu River basin, planning for water resources, and soil and water conservation can all benefit from the study's conclusions.

Keywords: rainfall; land use; Hulu River Basin; hydrology; GAST model; flood simulation

1. Introduction

Life requires access to water resources and has enormous socioeconomic significance in various industries, including agriculture, manufacturing, trade, food production, and hydropower generation [1,2]. It is well acknowledged that human activity and climate change significantly impact the hydrological cycle and the spatiotemporal patterns of rainfall [3].

According to [4], streamflow in most rivers has been steadily declining over the past few decades, leading to serious water shortages and ecological and environmental instability. Climate change and human activity typically work together to impact river streamflow changes [5]. One of the most dependable ways to assess changes in water resources is to combine climate prediction data from general circulation models (GCMs) with hydrological models [6]. Many hydrological modeling studies have been conducted since the end of the 1950s to assess the effects of changes in land use on the rainfall–runoff regime and the magnitude and frequency of floods [7]. Study [8] examines how changes in upstream land use affect downstream flood patterns using the HEC-1 model. Study [9] indicates that changes in land use significantly impact the magnitude of flood peaks and runoff volume. The authors of [10] use the hydrological SWAT model to evaluate how changes in land use affect a mesoscale catchment's annual water balance and temporal runoff dynamics. The authors of [11] analyzed the Impact of Land-Use Changes on Flood Exposure. The impact of urban sprawl on eco-environmental quality in China, using the Spatial Durbin model and panel data from 2003 to 2018, highlighting significant regional variations and the predominant influence of land use sprawl [12]. Various scholars have used multiple GCMs to evaluate how climate change affects streamflow, thus confirming its role. For example, in the Kashafrud River Basin (KRB) in Iran [13], streamflow has shown increasing trends under various scenarios. In the Kelantan River Basin, precipitation, temperature and streamflow have been predicted to increase from 2015 to 2044 and from 2045 to 2074 [14]. Currently, the Coupled Model Inter-Comparison Project, which is conducting global climate research, has reached its sixth phase, providing more comprehensive evaluations of future global climate [15]. A combined weighting approach and the super-efficiency Slack-Based Measure (SBM) model to assess urban–rural integration and land-use efficiency in China, revealing a low overall level and a trend towards improved coordination over the decade from 2010 to 2019 [16].

In China, including in the Yellow River Basin [17], Biliu River Basin [18], and Luan River Basin [19], future streamflow is expected to show downward trends at varying degrees. The two main elements influencing flood formation are the climate and the state of the underlying surface. The impact of precipitation and intensity on the production of runoff is a major factor in how climate change will affect the runoff process. Numerous studies have shown that precipitation impacts how runoff changes in a watershed. A rise in the amount of rainfall causes an increase in runoff if the underlying surface remains the same [20,21]. However, the alteration of confluence and runoff generation circumstances brought on by variations within the subsurface conditions are the primary variables influencing the environmental changes that lead to flood formation. Human activity impacts these changes, including clearing forest vegetation, building hydraulic infrastructure on rivers, extensive irrigation and drainage, water and soil conservation measures, modifications to land-use patterns, and urbanization and industrialization [22,23]. The critical issue of optimal city size to enhance Resource and Environment Intensity (REI) in China, analyzing the spatial–temporal impact of city size on REI within the Yangtze River Economic Belt through spatial models and panel data from 2004 to 2019. The literature review likely examines previous findings on urban expansion, resource efficiency, and spatial governance, setting the stage for this study's unique contribution to understanding the nuanced relationship between city size and REI [24].

As a result of agricultural practices, humans have gradually altered the earth's surface in important ways. More than 50% of the earth's surface has recently changed, with estimates placing agriculture on nearly one-third of the planet's surface [25]. The conversion of the naturally occurring agricultural land ratio to other land is still in progress [26]. Land use managers and decision-makers can better understand the connections between human and natural activity by looking at the patterns in change detection, which has drawn the attention of researchers and administrators of land use due to these significant changes [27]. The government competition and behavior within the open economic context of China's Huaihe River ecological economic belt, utilizing data from 2004 to 2016 to explore the determinants of land prices across various uses, revealing population density as a pivotal factor and the differentiated impact of policies due to price distortions in industrial land [28].

The upper Wei River, the main tributary of the Yellow River in China's Loess Plateau, has a significant offshoot known as the Hulu River. SWC measurements are a significant human endeavor that impacts the hydrology of the Yellow River Basin in China. From 1950 to 2009, streamflow change in the watershed decreased [29]. After the 1970s, extensive SWC methods were implemented to minimize runoff and manage soil erosion. Similarly, Wei River Catchment SWC techniques can effectively limit soil and water losses [30,31]. The frequency and destructiveness of floods make them a common occurrence in nature [32], affecting various places differently [33]. Following the 1980s, Yellow River water yield was a discernible decline [34]. The reason for this sudden fall has been a major topic of discussion in academics. According to [35], changes in land use may have a significant impact on how runoff is generated and how much water is stored. The authors of [36] pointed out that the Yellow River basin's extensive vegetation restoration was the primary cause of the decrease in water yield. However, [37] discovered that climate change was responsible for more than 50% of the decline in runoff in the Yellow River's middle reaches. It is still unclear how the significant increase in vegetation coverage brought on by land use changes and different rainfall characteristics will impact the runoff process, even though numerous studies have examined the causes of the water and sediment changes in the middle reaches of the Yellow River from the perspectives of climate and land use. More investigation is required, particularly to ascertain whether it influences both catastrophic rainfall floods and common water floods. Climate and land use can both affect water output in the basin, but in areas where land use has significantly improved, the impact of land use change on water yield is more significant [38]. Studies on surface runoff have been conducted by numerous academics. These studies have examined how variations in surface runoff affect soil erosion and how surface runoff [39,40] soil nutrients, rainfall, land use, and land cover distribution are affected [41,42]. The SCS-CN runoff model is somewhat influenced by rainfall intensity and spatial scale [43], and this model has also been studied for the urban and watershed levels. For example, [44] discovered that in a study of impermeable surfaces and surface runoff in Xuzhou City, the reaction of urban surface runoff is more evident when the intensity of rainfall is low. The authors of [45] assessed the simulation's accuracy using waterlogging locations and discovered that a rainfall intensity of 200 mm/d produced the most accurate surface runoff simulation. The landscape pattern evolution's impact on runoff can be examined by examining the association between landscape index and runoff change. Processing is commonly carried out using ArcGIS, ENVI, and Fragstats software. Alterations in land use and rainfall have been found to have an impact on surface runoff fluctuations in other research [46]. Furthermore, related research has demonstrated that runoff is more impacted by the evolution of landscape patterns than by rainfall [47]. The primary focus of relevant domestic and international research on surface runoff and land use and landscape pattern is on the watershed and urban scale [48]. Surface runoff response becomes increasingly evident when urban impermeable surface grows, and precipitation decreases [44].

Since mid-June 1998, the Yangtze River Basin has been experiencing continuous heavy rains and torrential rains, leading to widespread flooding throughout the basin. The peak flow at the Datong hydrological station reached 82,300 m^3/s, inflicting direct economic losses of CNY 166 billion and affecting 223 million people. In the same year, the Songhua River Basin was also affected by climate change, with increased rainfall leading to an unprecedented flood event in the basin. In 2010, Gansu Province experienced a major flood disaster, resulting in 1434 fatalities; 14.76 million people were affected by the Yangtze River Basin floods in 2016, which also resulted in losses to the economy of CNY 31.14 billion. In 2020, multiple strong rainfall events occurred in southern China, with 198 rivers experiencing floods exceeding warning levels and direct economic losses reaching CNY 86.16 billion. Therefore, the Hulu River Basin is used for this study. Observed rainfall data from 13 September 2019, 0:00 to 14 September 2019, 16:00 is used and the total rainfall duration is 40 h. Land use data with a resolution of 30 m for the years between 1985 and 2020 is being used. This study analyzes change patterns in rainfall characteristics and land use of the Hulu River catchment, China, based on the GAST model (GPU Accelerated Surface Water Flow and Transport Model).

2. Materials and Methods

2.1. Study Area

One of the major tributaries of the Wei River and the Hulu River rises on the southern slopes of Moon Mountain near the boundary between Xiji County and Haiyuan County in the Ningxia Hui Autonomous Region (Figure 1). The watershed is situated between 34°30′ and 36°30′ N and 105°05′ to 106°30′ E, and its height ranges from 1141 to 1908 m. According to [49], it crosses both the Ningxia Hui Autonomous Region and Gansu Province before emptying into the Wei River close to Tianshui City in Gansu Province. The research region has an east-west gradient with a higher north elevation and a lower south elevation. The riverbed resembles a gourd because it is meandering and has a wide range of widths [15]. The Hulu River's main channel is 301 km long, and its entire basin is around 10,700 km^2. With an annual sediment transfer as high as 7270×10^4 tons, the basin is characterized by significant vegetation destruction, loose soil, frequent heavy rainfall, and severe soil erosion. The Hulu River Basin's water resources are dispersed unevenly and show noticeable inter-annual changes. A total of 265.4 million m^3 of runoff are produced annually on average [49].

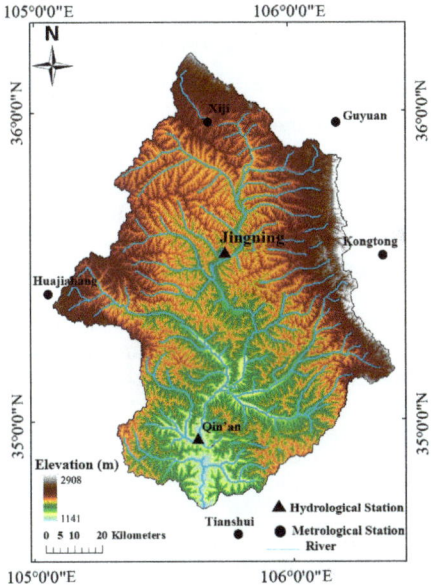

Figure 1. DEM of the study area.

2.2. GAST Model

The GPU Accelerated Surface Water Flow and Transport Model (GAST) simulates basin flood response under various rainfall and land use scenarios using a numerical model that integrates hydrological and hydrodynamic processes. Its great accuracy, quick efficiency, and strong stability make it a good model for complicated networks. Urban waterlogging, sediment transport, and basin flood dynamics are among the processes it can accurately replicate. It can simulate these processes at a fine scale and use GPU-accelerated parallel computing technology, which greatly increases the model's computational performance and offers several benefits [50].

2.2.1. Model Governing Equations

The GAST model utilizes the two-dimensional shallow water equations (SWEs) as the governing equations for surface runoff calculations. The conservation form of the vector equation (neglecting viscous terms of motion, turbulent viscous terms, wind stress, and Coriolis force) is shown in Equations (1)–(3) [51].

$$\frac{\partial q}{\partial t} + \frac{\partial F}{\partial x} + \frac{\partial G}{\partial y} = S \tag{1}$$

$$q = \begin{bmatrix} h \\ q_x \\ q_y \end{bmatrix}, F = \begin{bmatrix} uh \\ uq_x + gh^2/2 \\ uq_y \end{bmatrix}, G = \begin{bmatrix} vh \\ vq_x \\ vq_y + gh^2/2 \end{bmatrix}, \tag{2}$$

$$S = \begin{bmatrix} i \\ -gh\partial z_b/\partial x \\ -gh\partial z_b/\partial y \end{bmatrix} + \begin{bmatrix} 0 \\ -C_f u\sqrt{u^2 + v^2} \\ -C_f v\sqrt{u^2 + v^2} \end{bmatrix} \tag{3}$$

where t is the time, x and y are the Cartesian coordinates, and q is the flow variable vector made up of h, q_x, and q_y, which indicate water depth and unit-width discharges in the x and y directions, respectively; u and v are defined as depth-averaged velocities in the x and y directions, and it is evident that q_x = uh and q_y = vh; f and g are the flux vectors in the x and y directions; S is the source vector that only takes into account the slope source S_b and the friction source S_f; in this case, z_b is the bed elevation and C_f is the bed roughness coefficient that is derived from the Manning coefficient n and h in the form of $gn^2/h^{1/3}$.

2.2.2. Numerical Methods of the Model

The two-dimensional hydrodynamic model GAST employs the Godunov finite volume method to solve the two-dimensional shallow water equations. Riemann's approximate HLLC solver is used to calculate the mass flux and momentum flux at the cell interfaces [52]. The static water reconstruction method is used to handle negative water depths at dry–wet boundaries [53]. The bottom slope flux method is utilized to solve the variation in water depth [54]. An improved splitting-point implicit method is used for the frictional resistance source term to enhance computational stability [55]. A second-order explicit Runge–Kutta method is employed for time integration to ensure second-order accuracy [56]. The MUSCL scheme effectively addresses computational instabilities and non-conservation of mass and momentum caused by non-physical phenomena. Due to the large-scale study area and complex watershed in this study, GPU acceleration and parallel computing techniques are introduced, significantly enhancing the model's computational efficiency to improve its computational speed further.

2.3. Model Setup and Validation

2.3.1. Input Data

The primary input data for the GAST model are the digital elevation model (DEM), land use, soil, and meteorological data. River network features, topographic slope length, and other basin parameters are obtained using a digital elevation model (DEM). In this study, the China Meteorological Data Centre was used to compile rainfall data from the Hulu River Basin's five meteorological stations, Huajialing, Tianshui, Kongtong, Guyuan, and Xiji. Observed rainfall data from 13 September 2019, 0:00 to 14 September 2019, 16:00 are used for model calibration and validation. The rainfall duration is 40 h, as shown in Figure 2. The time interval between data points is 30 min, and the simulation time is set to 40 h. Land use information from various historical eras was acquired from the Resource and Environmental Science and Data Centre of the Chinese Academy of Sciences. Data were collected with a resolution of 30 m for the years 1985–2020. Digital elevation model

(DEM), land use information from various time periods, soil information, and precipitation information are among the fundamental data used in this study. The Geospatial Data Cloud (http://www.gscloud.cn/, accessed on 22 June 2023) provided the DEM data with a grid resolution of 30 m for download.

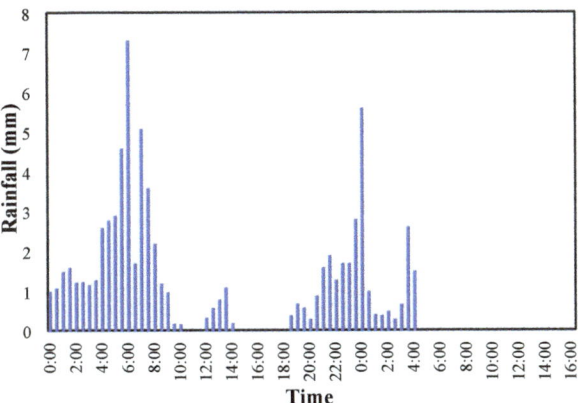

Figure 2. Measured rainfall from 13 September 2019 at 0:00 to 14 September 2019 at 16:00.

The soil information was taken from the global Harmonized World Soil Database (HWSD), which was created by the Food and Agriculture Organization (FAO) and the International Institute for Applied Systems Analysis (IIASA). With a crop to the research area, the soil data are presented at a resolution of 1000 m. An analysis of the Hulu River Basin soil map was performed using the Harmonized World Soil Database (HWSD). The main soil types in the Hulu River Basin are saturated immature soil, unsaturated immature soil, calcareous accumulative black soil, moderately developed highly active leached soil, moderately developed black soil, moderately developed grey soil, moderately developed immature soil, calcareous impacted soil, calcareous immature soil, and sticky chestnut calcareous soil. In the basin, the predominant types of vegetation are cultivated land and grassland, with some woodland, bushes, unused land, water bodies, and impermeable regions also present [57]. The Horton infiltration model is used for the infiltration component of the model. In addition, the infiltration and Manning values for each land use type in the watershed are determined based on various references [58–61].

The commonly used infiltration equation, proposed by Horton in 1939, is as follows:

$$f = f_c + (f_0 - f_c)e^{-kt} \tag{4}$$

In this equation, the infiltration capacity at time t (h) is denoted by f, the starting value of infiltration capacity is represented by f_0, the final or equilibrium infiltration capacity is represented by f_c, and the rate of decline of infiltration capacity (1/h) is determined by an exponent k.

Horton used empirical methods to construct his equation, but it could also be theoretically determined if it were assumed that the decrease in infiltration capacity resulted from an exhaustion process [62].

2.3.2. Model Validation

This study utilizes measured rainfall and streamflow data from the Hulu River Basin to validate the accuracy and reasonability of the constructed GPU-accelerated Surface Water Flow and Associated Transport (GAST) model. The reliability of the model is assessed based on Nash–Sutcliffe Efficiency (NSE) and coefficient of determination (R^2) for the flow at the watershed outlet cross-section. The downstream outlet of the Hulu River Basin in the GAST model is set as an open boundary, while the rest are set as closed boundaries.

The model is validated using rainfall measured data from the Qin'an hydrological station from 13 September 2019 at 0:00 to 14 September 2019 at 16:00, spanning 40 h. The accuracy of the model is evaluated based on the root-mean-square error (RMSE) (Table 1) and relative error between the simulated values and measured values of the flow process at the Qin'an hydrological station.

Table 1. Comparison between measured and simulated values.

Time	Measured Flow/($m^3 \cdot s^{-1}$)	Time	Simulated Flow/($m^3 \cdot s^{-1}$)	Relative Error
13 September 6:03:00	38.9	13 September 6:00:00	39.15	0.64%
13 September 8:00:00	35.5	13 September 8:00:00	38.23	7.68%
13 September 14:39:00	85.8	13 September 14:30:00	78.39	8.64%
14 September 5:54:00	131	14 September 6:00:00	135.68	3.57%
14 September 6:18:00	133	14 September 6:30:00	133.99	0.75%
14 September 8:00:00	92.5	14 September 8:00:00	95.58	3.33%

2.3.3. Root Mean Square Error (RMSE)

$$RMSE = \sqrt{\frac{1}{n}\sum_{i=1}^{n}(Q_{sim,i} - Q_{obs,i})^2} \quad (5)$$

In this equation: $Q_{sim,i}$—simulated value, $Q_{obs,i}$—observed value, and n—number of measurements. When the root-mean-square error (RMSE) is closer to 0, it indicates that the simulated values are in complete agreement with the observed values.

2.3.4. Relative Error

$$\delta = \frac{\triangle}{L} \times 100\% \quad (6)$$

In the equations: δ—relative error, usually expressed as a percentage, $\triangle$—difference between simulated and measured values, and L—measured value.

According to the calculations based on the simulation results, the root-mean-square error (RMSE) between the simulated and observed values is 3.98. The relative errors are 0.64%, 7.68%, 8.64%, 3.57%, 0.75%, and 3.33%, respectively. The RMSE is close to 0, and the relative errors are all less than 10%, indicating that the model has a small error between the simulated and observed values and exhibits high accuracy. Therefore, the model can be used to simulate and analyze the impact of rainfall characteristics and land use changes on flood events in the Hulu River Basin. Additionally, the average absolute value of the relative error is 0.0410, and the average absolute error is 3.1911.

2.4. Simulation of River Flooding under Different Rainfall Characteristics Scenarios

Different rainfall return periods have different impacts on basin floods, and the spatial distribution of rainfall, whether uniform or uneven, also significantly influences basin floods. Therefore, studying the impact of different rainfall spatial distributions on basin floods under different return periods can provide a theoretical basis for disaster prevention, mitigation, and relief work.

Scenario Setup

During the simulation period, the land use data in the Hulu River Basin are assumed to be the current land use while keeping the infiltration and other parameters constant to exclude the influence of other factors on the flood evolution process in the basin. The rainfall is designed using the Chicago rainfall type and the Gansu Tianshui rainfall formula. Short-duration intense rainfall with a rainfall duration of 120 min and return periods of 5 years, 10 years, and 50 years, respectively, is generated with a rain peak coefficient of 0.5. The rainfall formula is as follows:

$$i = \frac{1734.0278 \times (1 + 1.473 \times LgP)}{167 \times (t + 15.3599)^{0.8867}} \tag{7}$$

In equations: i = rainfall intensity (mm/h), P = rainfall return period (years), and t—rainfall duration (minutes).

Two different scenarios are set for the spatial distribution of rainfall under different return periods. Based on the analysis of rainfall distribution and storm intensity distribution, it is known that the rainfall in the Hulu River Basin gradually decreases from southeast to northwest, with higher storm intensity in the southeast and northwest regions and lower intensity in the northeast and southwest regions. Therefore, two different rainfall spatial distribution scenarios are constructed. Additionally, three scenarios are created with the storm center located in the upstream, middle reaches and downstream regions. The specific settings for the rainfall characteristics scenarios are shown in Table 2.

Table 2. Scenario setting of different rainfall characteristics.

Scenario	Scenario Description
Scenario R1	Uniform spatial distribution of 5-year return period rainfall
Scenario R2	Uniform spatial distribution of 10-year return period rainfall
Scenario R3	Uniform spatial distribution of 50-year return period rainfall
Scenario R4	Rainfall gradually decreasing from southeast to northwest
Scenario R5	Relatively high rainfall in southeast and northwest regions, while relatively low rainfall in northeast and southwest regions
Scenario R6	Set upstream rainfall center
Scenario R7	Set middle reaches rainfall center
Scenario R8	Set downstream rainfall center

2.5. Simulation of River Floods Occurance under Different Land Use Scenarios

Different land use types significantly impact the flood evolution process by altering the underlying surface hydrological characteristics, leading to a series of water resource issues such as increased flood disasters, severe soil erosion, and declining groundwater levels. In addition, changes in land use also have certain impacts on the ecological environment and economic conditions, such as implementing policies such as converting cultivated land to forests and grasslands in areas prone to soil erosion. Therefore, studying the impact of land use changes on watershed floods is important for disaster prevention and mitigation and provides technical support for watershed water resource planning. This section uses a two-dimensional hydrodynamic model to simulate and analyze the impact of land use changes in the Hulu River basin on the flood evolution process, peak flow, and peak water depths.

Scenario Setup

Since 1999, multiple water conservation measures have been implemented in the Hulu River basin. Therefore, a scenario of generalized water conservation measures (such as constructing sediment detention dams and terracing projects) is constructed. Seven land use scenarios are established, including the current land use scenario, three historical land use scenarios, two integrated land use scenarios, and the generalized water conservation measures scenario. The response of watershed floods under different land use scenarios is simulated and analyzed. The data for the four integrated land use scenarios are shown in Figure 3, and the specific scenario settings are shown in Table 3. To exclude the influence of rainfall on watershed floods, it is assumed that the rainfall conditions remain unchanged. The rainfall design adopts the Chicago rainfall pattern, and the Gansu Tianshui heavy rain formula (Equation (7)) is used to generate rainfall with a duration of 120 min and a rainfall coefficient of 0.5. The return periods are 5, 10, and 50 years, and the rainfall is assumed to be spatially uniformly distributed.

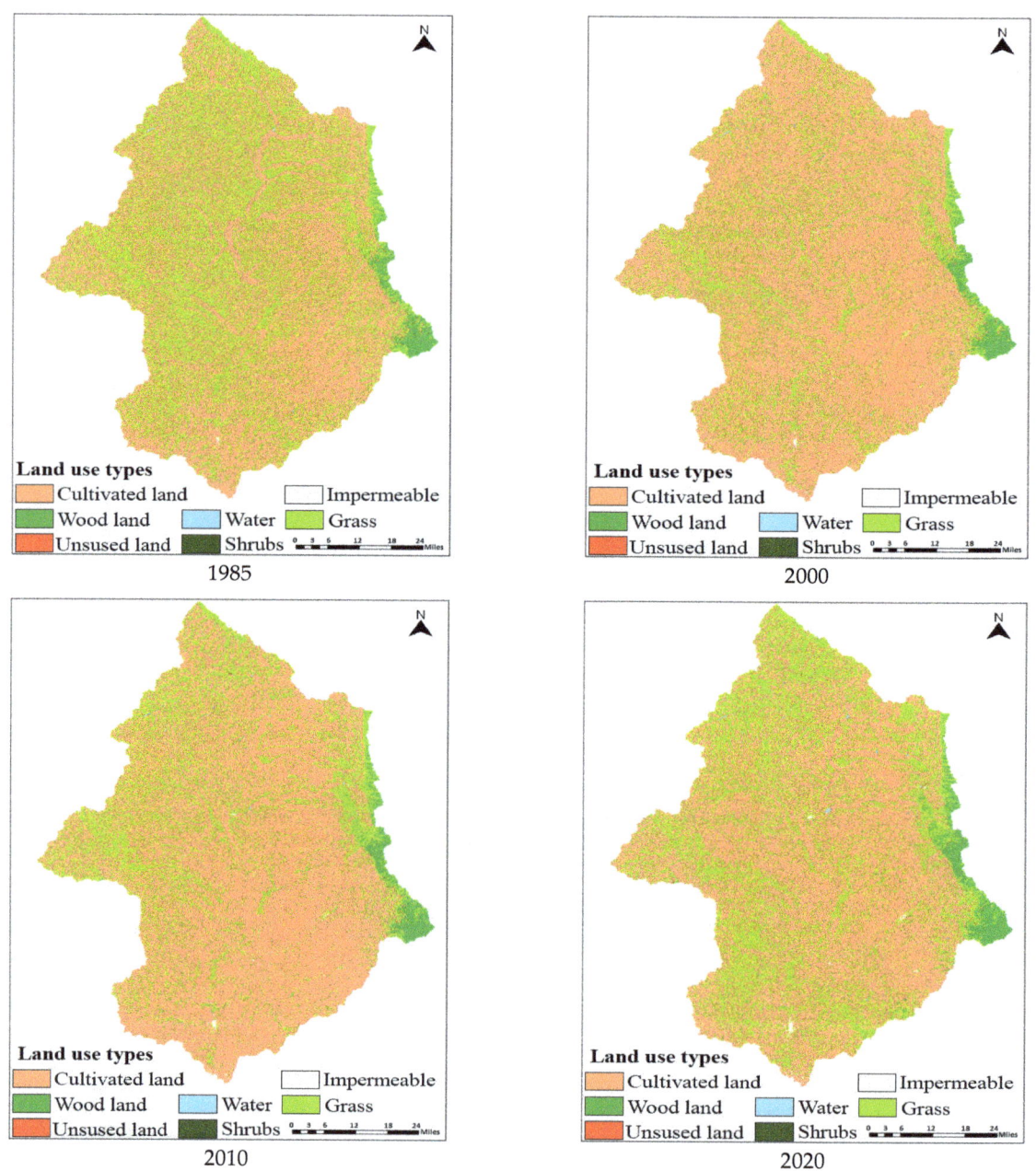

Figure 3. The variation in land use pattern of the study area over time (1985–2020).

Table 3. Set up different land use scenarios.

Scenarios	Scenario Description
Scenario L1	1985 Annual land use data
Scenario L2	2000 Annual land use data
Scenario L3	2010 Annual land use data
Scenario L4	2020 Annual land use data
Scenario L5	2020 Conversion of farmland to forest
Scenario L6	2020 Conversion of farmland to grassland
Scenario L7	Generalize various water conservation measures such as check dams

2.6. Simulation of River Floods under the Combined Influence of Rainfall Characteristics and Land Use

We integrate two influencing factors to simulate the impact of the combined changes in rainfall characteristics and land use on watershed floods. We explore the patterns and characteristics of changes in peak flow and peak water depth.

Scenario Setup

Based on the analysis in the previous section, it was found that the impact of the rainfall center on watershed floods is similar in the upstream and middle reaches of the watershed. Therefore, in this section, besides other scenarios of rainfall characteristics, only the scenario with the rainfall center upstream of the watershed was selected for simulation. The specific scenario settings are shown in Table 4.

Table 4. Scenario setting under the combined action of rainfall characteristics and land use.

Scenario	Scenario Description	
	Rainfall Variation	Land Use Change
Scenario RL1-1	50-year return period rainfall gradually decreases from southeast to northwest	1985
Scenario RL1-2	50-year return period rainfall gradually decreases from southeast to northwest	2000
Scenario RL1-3	50-year return period rainfall gradually decreases from southeast to northwest	2010
Scenario RL1-4	50-year return period rainfall gradually decreases from southeast to northwest	farmland to forest
Scenario RL1-5	50-year return period rainfall gradually decreases from southeast to northwest	farmland to grassland
Scenario RL1-6	50-year return period rainfall gradually decreases from southeast to northwest	soil and water conservation measures
Scenario RL2-1	50-year return period rainfall: The southeast and northwest regions have higher rainfall, while the northeast and southwest regions have smaller rainfall	1985
Scenario RL2-2	50-year return period rainfall: The southeast and northwest regions have higher rainfall, while the northeast and southwest regions have smaller rainfall	2000
Scenario RL2-3	50-year return period rainfall: The southeast and northwest regions have higher rainfall, while the northeast and southwest regions have smaller rainfall	2010
Scenario RL2-4	50-year return period rainfall: The southeast and northwest regions have higher rainfall, while the northeast and southwest regions have smaller rainfall	farmland to forest
Scenario RL2-5	50-year return period rainfall: The southeast and northwest regions have higher rainfall, while the northeast and southwest regions have smaller rainfall	Resume farmland to grassland
Scenario RL2-6	50-year return period rainfall: The southeast and northwest regions have higher rainfall, while the northeast and southwest regions have smaller rainfall	Soil and water conservation measures

Table 4. *Cont.*

Scenario	Scenario Description	
	Rainfall Variation	Land Use Change
Scenario RL3-1	50-year return period rainfall center is upstream	1985
Scenario RL3-2	Set upstream 50-year rainfall center	2000
Scenario RL3-3	Set upstream 50-year rainfall center	2010
Scenario RL3-4	Set upstream 50-year rainfall center	farmland to forest
Scenario RL3-5	Set upstream 50-year rainfall center	farmland to grassland
Scenario RL3-6	Set upstream 50-year rainfall center	soil and water conservation measures
Scenario RL4-1	Set downstream 50-year rainfall center	1985
Scenario RL4-2	Set downstream 50-year rainfall center	2000
Scenario RL4-3	Set downstream 50-year rainfall center	2010
Scenario RL4-4	Set downstream 50-year rainfall center	farmland to forest
Scenario RL4-5	Set downstream 50-year rainfall center	farmland to grassland
Scenario RL4-6	Set downstream 50-year rainfall center	soil and water conservation measures

3. Results

Rainfall and land use are intricately interconnected, influencing each other in several significant ways. The changes that occur in land use, such as urbanization or deforestation, can alter the natural water cycle, impacting both the intensity and rainfall distribution. Conversely, rainfall patterns influence land use decisions, especially in agriculture, where the amount and timing of precipitation often dictate the choice of crops and farming practices. This dynamic relationship highlights the importance of considering both environmental planning and resource management factors.

3.1. Simulation of River Flooding under Different Rainfall Characteristics Scenarios

Rainfall return periods of 5, 10, and 50 years with uniform spatial distribution of rainfall. The flood evolution process under different return periods with a uniform spatial rainfall distribution was simulated. The hydrograph, peak flow, and peak water depth at the outlet section of the basin under different return periods with a uniform spatial distribution of rainfall are shown in Figure 4 and Table 5.

Figure 4 shows that the flow at the outlet section of the basin rapidly increases with time, reaching its peak between 5400 s and 7200 s, and then slowly decreases to a stable level. Both scenario R1 and scenario R2 exhibit a flow value close to the peak flow, which can occur either before or after the peak flow. Scenario R3, on the other hand, does not show a flow value close to the peak flow. Instead, it experiences a sudden drop after the peak flow, followed by a subsequent rise and a slow decline to stability. The peak flow in scenario R1 is 1332.90 m^3/s, occurring at 7200 s, with a peak water depth of 5.42 m at 9000 s. In scenario R2, the peak flow is 1445.33 m^3/s, occurring at 5400 s, with a peak water depth of 5.44 m at 9000 s. Scenario R3 has a peak flow of 2003.04 m^3/s, occurring at 5400 s, with a peak water depth of 8.57 m at 9000 s. It can be observed that both the peak flow and peak water depth increase with the increase in the return period. However, when the return periods are relatively close, such as 5 years and 10 years, the increase in peak flow and peak water depth is not significant. Furthermore, as the return period increases, the timing of the peak flow advances. There will be changes in future precipitation due to global warming and the intensification of the water cycle process [63,64].

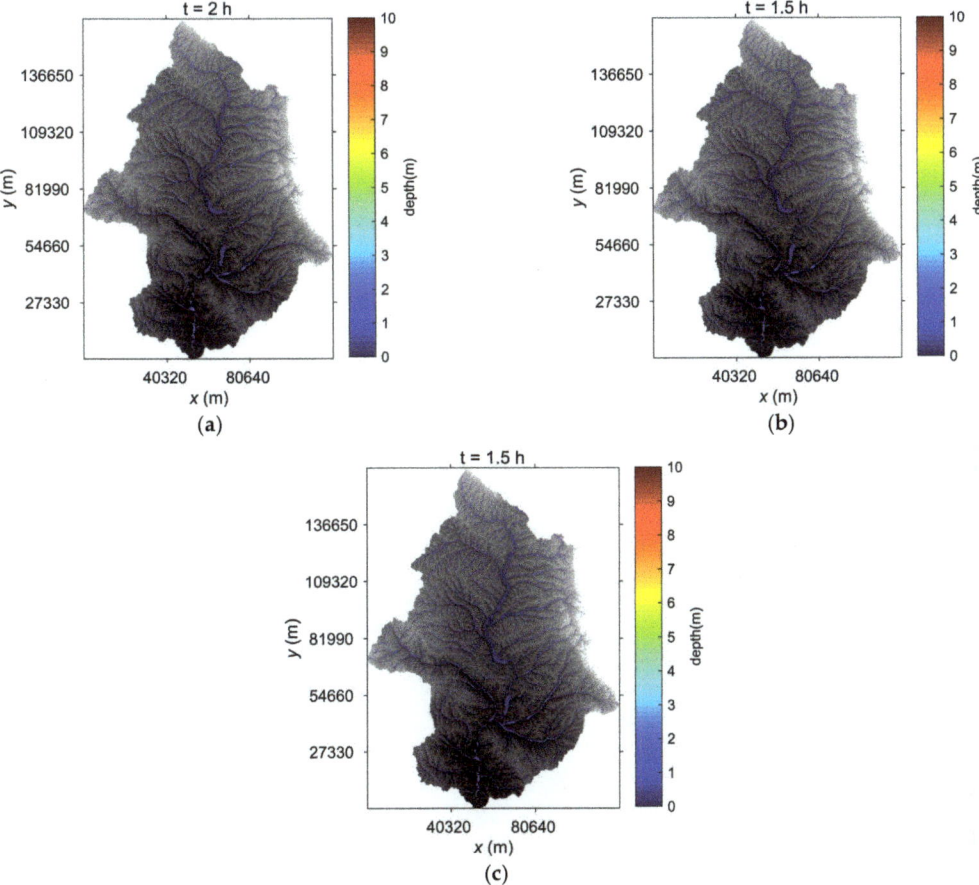

Figure 4. Different rainfall return periods under uniform rainfall distribution. (**a**) Scenario R1; (**b**) Scenario R2; and (**c**) Scenario R3.

Table 5. Peak discharge and peak water depth under different spatial distributions of rainfall.

	Peak Flow Rate/(m^3·s^{-1})	Time of Peak Flow/(s)	Peak Water Depth/(m)	Time of Peak Water Depth Occurrence/(s)
Scenario R1	1332.90	7200	5.42	9000
Scenario R2	1445.33	5400	5.44	9000
Scenario R3	2003.04	5400	8.57	9000

The spatial distribution of rainfall is uneven, with a return period of 5, 10, and 50 years. Simulate the flood evolution process under different return periods of uneven rainfall. The flow hydrograph, peak flow, and peak water depth of the watershed outlet section under different return periods of uneven rainfall are shown in Figure 5 and Table 6. Scenario R4 and Scenario R8 have significant differences in the discharge hydrograph of the watershed outlet section at different return periods, while Scenario R5, Scenario R6, and Scenario R7 have little difference in the discharge hydrograph of the outlet section at different return periods. The peak flow rates of scenarios R4, R5, R6, and R7 with different return periods all appeared at 5400 s and then gradually decreased with time. The peak flow of scenario R8 with different return periods occurs between 5400 s and 7200 s and then slowly decreases to a stable state. The outlet section discharge hydrograph of scenario R6 and scenario R7 is not very different; that is, when the rainfall center is in the upstream and midstream, the impact on the basin flood is relatively close. Scenario R8 is quite different from scenarios R6 and R7, which indicates that when the rainfall center is downstream, the impact on the basin flood is different from that in the middle and upstream.

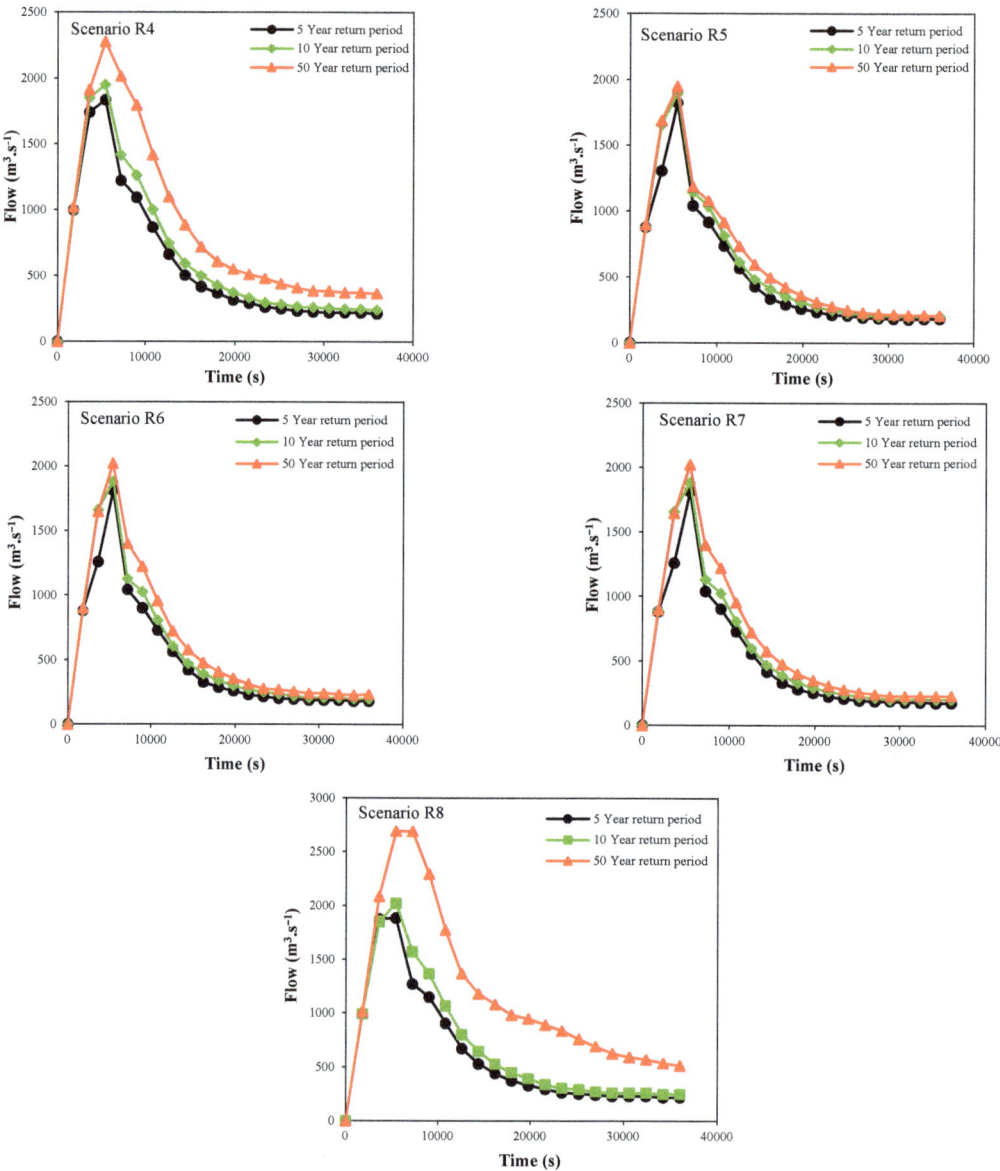

Figure 5. The flow process line of the basin outlet section is under different uneven rainfall distributions.

Table 6. Peak flow and peak water depth under different rainfall scenarios.

Return Period/Scenario	Scenario R4		Scenario R5		Scenario R6		Scenario R7		Scenario R8	
	Flow ($m^3 \cdot s^1$)	Water Depth (m)	Flow ($m^3 \cdot s^1$)	Water Depth (m)	Flow ($m^3 \cdot s^1$)	Water Depth (m)	Flow ($m^3 \cdot s^1$)	Water Depth (m)	Flow ($m^3 \cdot s^1$)	Water Depth (m)
5 years return period	1836.19	5.38	1825.59	5.61	1813.82	5.60	1813.73	5.60	1883.34	5.39
10 years return period	1954.21	5.41	1894.72	5.39	1883.36	5.39	1883.31	5.39	2023.48	5.71
50 years return period	2277.76	7.39	1947.12	5.40	2023.07	5.86	2023.22	5.86	2694.34	8.82

Scenario R4 has a peak flow rate of 1836.19 m³/s, 1954.21 m³/s, and 2277.76 m³/s during the 5, 10, and 50-year return periods, respectively, which is 37.76%, 35.21%, and 13.72% higher than the uniform distribution of rainfall during the same return period. The peak water depths are 5.38 m, 5.41 m, and 7.39 m, respectively, 0.74%, 0.55%, and 13.77% less than the uniform rainfall distribution. Scenario R5 has a peak flow rate of 1825.59 m³/s, 1894.72 m³/s, and 1949.12 m³/s under three different return periods, respectively. Compared with the same return period, the rainfall distribution uniformly increased by 36.96%, 31.09%, and decreased by 2.69%. The peak water depths were 5.61 m, 5.39 m, and 5.40 m, respectively, which increased by 3.51%, decreased by 0.92%, and 36.99% compared to the uniform distribution of rainfall. Scenario R6 has a peak flow rate of 1813.82 m³/s, 1883.36 m³/s, and 2023.07 m³/s during the 5, 10, and 50-year return periods, respectively, which is 36.08%, 30.31%, and 1.00% higher than the uniform distribution of rainfall during the same return period; The peak water depths were 5.60 m, 5.39 m, and 5.86 m, respectively, which increased by 3.32%, decreased by 0.92%, and 31.62% compared to the uniform distribution of rainfall.

Scenario R7 has a peak flow rate of 1813.73 m³/s, 1883.31 m³/s, and 2023.22 m³/s during the 5, 10, and 50-year return periods, respectively, which is 36.07%, 30.30%, and 1.01% higher than the uniform distribution of rainfall during the same return period. The peak water depths were 5.60 m, 5.39 m, and 5.86 m, respectively, which increased by 3.32%, decreased by 0.92%, and 31.62% compared to the uniform distribution of rainfall. Scenario R8 has a peak flow rate of 1883.34 m³/s, 2023.48 m³/s, and 2694.34 m³/s during the 5, 10, and 50-year return periods, respectively, which is 41.30%, 40.00%, and 34.51% higher than the uniform distribution of rainfall during the same return period. The peak water depths were 5.39 m, 5.71 m, and 8.82 m, respectively, which decreased by 0.55%, increased by 4.96%, and 2.92% compared to the uniform distribution of rainfall.

Figure 5 shows that the peak flow values and peak water depths for the 5-year and 10-year return periods are relatively close across different scenarios, while there is a significant difference between the 10-year and 50-year return periods. It is quite different from the peak water depth. It is further explained that when the difference between return periods is larger, the difference between peak flow rate and peak water depth is larger. The peak flow and peak water depth under the above four scenarios with uneven rainfall spatial distribution are quite different from those under uniform rainfall spatial distribution. The largest difference is scenario R8, and the one closest to the uniform rainfall scenario is scenario R7. It can be seen that when the return period is small, the change rate of peak flow is larger, and when the return period is larger, the change rate of peak flow is smaller. It can be seen that when the return period reaches a certain level, such as the 100-year return period, the peak discharge at the outlet section of the basin is relatively close under the scenarios of uneven rainfall distribution and uniform rainfall distribution.

The simulated peak flows of Scenario R4, Scenario R5, Scenario R6, and Scenario R7 under the 5-year return period are relatively close, indicating that when the return period is small, the heavy rain center is in the southeast, southeast, and northwest, the upper reaches of the basin and the middle reaches of the basin. The degree of flood impact in the Hulu River Basin is relatively similar. The peak flows simulated by scenario R8 are significantly different from those of other scenarios in all return periods, indicating that when the center of heavy rain is in the lower reaches of the basin, the impact on floods in the basin is stronger. At this time, special attention should be paid to the safety of the rivers downstream of the basin, and disaster prevention and reduction measures should be taken.

3.2. Simulation of River Flooding under Different Land Use Scenarios

The historical inversion method is used to simulate the four historical land use scenarios, and the changes in the simulation results between two adjacent periods are considered as the degree of land use impact on floods. The peak flow and peak water depth for different return periods under each scenario are shown in Tables 7 and 8, respectively. The flood hydrographs at the watershed outlet for different land use scenarios are shown in Figure 6.

Table 7 shows that compared to Scenario L1, Scenario L2 shows an increase in peak flow of 20.26%, 14.68%, and 17.14% for the 5, 10, and 50-year return periods, respectively. The peak water depth increases by 0.37%, 9.07%, and 3.80%, respectively. Compared to Scenario L2, Scenario L3 shows an increase in peak flow of 4.92%, 4.72%, and 1.86%, and an increase in peak water depth of 1.66%, 6.62%, and 2.63% for the 5, 10, and 50-year return periods, respectively. Compared to Scenario L3, Scenario L4 shows a decrease in peak flow of 25.75%, 26.18%, and 21.39%, and a decrease in peak water depth of 1.45%, 13.38%, and 4.57% for the 5, 10, and 50-year return periods, respectively. It can be observed that there is a significant variation in peak flow between two adjacent land use scenarios, while the variation in peak water depth is less than 15%. This indicates that the changes in land use between two adjacent periods have a greater impact on peak flow and a relatively smaller impact on peak water depth.

Table 7. Historical land use scenarios peak discharge and peak water depth.

Return Period/ Scenario	Scenario L1		Scenario L2		Scenario L3		Scenario L4	
	Flow ($m^3 \cdot s^{-1}$)	Water Depth (m)	Flow ($m^3 \cdot s^{-1}$)	Water Depth (m)	Flow ($m^3 \cdot s^{-1}$)	Water Depth (m)	Flow ($m^3 \cdot s^{-1}$)	Water Depth (m)
5 years return period	1422.8	5.39	1711.02	5.41	1795.2	5.50	1332.9	5.42
10 years return period	1630.3	5.40	1869.7	5.89	1958.04	6.28	1445.33	5.44
50 years return period	2135.57	8.43	2501.61	8.75	2548.2	8.98	2003.04	8.57

In Scenario L1 compared to Scenario L4, the peak flow increased by 6.74%, 12.80%, and 6.62% for different return periods, while the peak water depth decreased by 0.55%, 0.74%, and 1.63% for the 5, 10, and 50-year return periods, respectively. In Scenario L2 compared to Scenario L4, the peak flow increased by 28.37%, 29.36%, and 24.89% for different return periods, and the peak water depth decreased by 0.18% for the 5-year return period but increased by 8.27% and 2.10% for the 10 and 50-year return periods, respectively. In Scenario L3 compared to Scenario L4, the peak flow increased by 34.68%, 35.47%, and 27.22% for the 5, 10, and 50-year return periods, respectively, and the peak water depth increased by 1.48%, 15.44%, and 4.78% for the corresponding return periods. Thus, it can be observed that during the period from 1985 to 2020 [65,66], the changes in land use resulted in an increasing trend followed by a decreasing trend in both peak flow and peak water depth in the watershed.

Table 8. Peak discharge and peak water depth in comprehensive land use scenarios.

Return Period/ Scenario	Scenario L5		Scenario L6		Scenario L7	
	Flow ($m^3 \cdot s^{-1}$)	Water Depth (m)	Flow ($m^3 \cdot s^{-1}$)	Water Depth (m)	Flow ($m^3 \cdot s^{-1}$)	Water Depth (m)
5-year return period	568.63	4.79	720.86	4.93	897.96	5.26
10-year return period	589.79	4.80	755.52	4.97	1035.85	5.30
50-year return period	803.09	4.98	912.13	5.10	1426.29	5.53

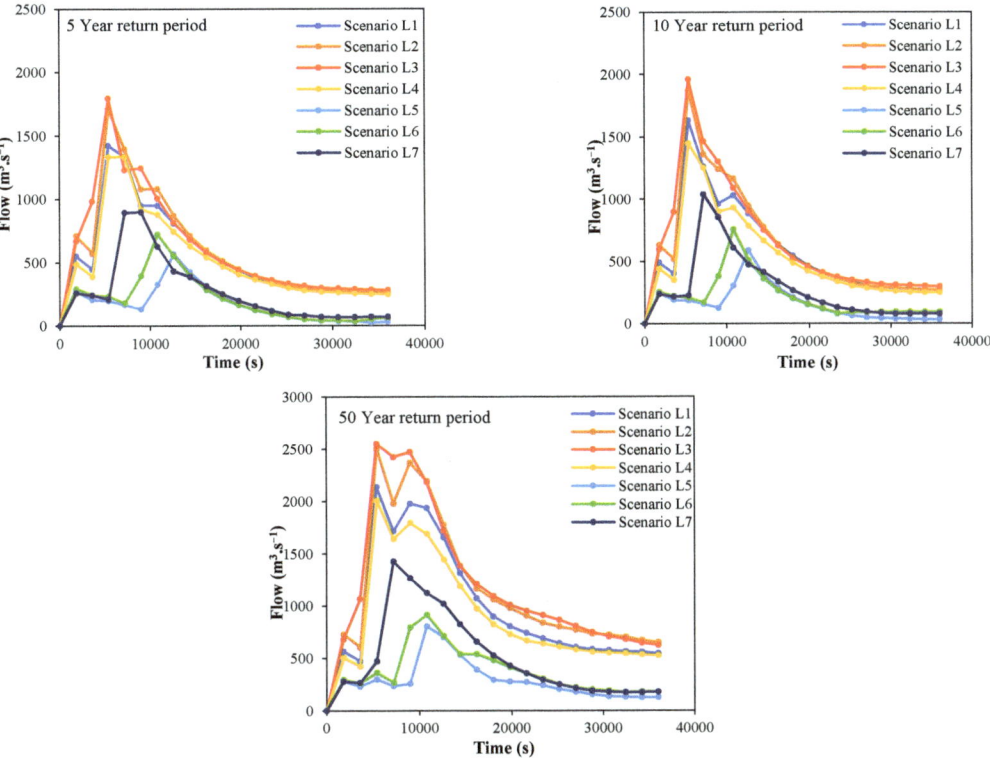

Figure 6. Flow process line of watershed outlet section under different land use scenarios.

The flood evolution process was simulated and analyzed for three land use scenarios, including two integrated land use scenarios and one scenario considering generalized water conservation measures, compared to the current land use scenario (Scenario L4). The simulation results and the changes compared to the current land use scenario are shown in Table 8. From the table, it can be seen that the peak flow differs significantly among the two integrated land use scenarios and the scenario with generalized water conservation measures for different return periods, while the variation in peak water depth is relatively small. Compared to Scenario L4, the peak flow decreased by 57.34%, 45.92%, and 32.63% for the 5-year return period in Scenarios L5, L6, and L7, respectively. Similarly, for the 10-year and 50-year return periods, the peak flow decreased by 59.19%, 47.73%, 28.33%, and 59.91%, 54.46%, and 28.79%, respectively. The corresponding reductions in peak water depth were 11.62%, 9.04%, and 2.95% for the 5-year return period, 11.76%, 8.64%, and 2.57% for the 10-year return period, and 41.89%, 40.49%, and 35.47% for the 50-year return period.

Furthermore, according to Figure 6, the peak flow in Scenarios L5, L6, and L7 occurs later than in Scenario L4, approximately between 3600 s and 7200 s. This indicates that not only do the peak flow and peak water depth decrease significantly for different return periods, but they also occur later. It suggests that the measures of returning farmland to forest and grassland have a pronounced inhibitory effect on both peak flow and peak water depth. The water conservation measures also contribute to a delay in peak flow and peak water depth. This is because the implementation of water conservation measures retains most of the runoff locally, leading to a decreasing trend in peak flow and peak water depth at the outlet section of the watershed.

The main land use types in the Hulu River Basin are cultivated land and grassland, supplemented by woodland, shrubs, water bodies, unused land, and impermeable areas. Cultivated land has the highest proportion in all four periods, exceeding 60% and being the largest land use type. Grassland has the second largest proportion, also exceeding 20%. In these four periods, the proportion of arable land shows a trend of first increasing and then decreasing, accounting for 60.09%, 75.62%, 72.94%, and 62.66% of the total area, respectively. The proportion of grassland shows a trend of decreasing first and then increasing, accounting for 38.02%, 22.30%, 24.78%, and 34.10%, respectively. The proportion of woodland gradually increases, accounting for 1.71%, 1.89%, 2.04%, and 2.81%, respectively. The proportion of shrubs shows a slight increase, accounting for 0.009%, 0.006%, 0.005%, and 0.017% of the total area. The proportion of water bodies and unused land initially decreases and then increases, while the proportion of impermeable areas gradually increases. Between 1985 and 2020, except for a 3.92% decrease in grassland area, the area of cultivated land, woodland, shrubs, water bodies, unused land, and impermeable areas all increased. The increase in proportion is 2.57%, 1.10%, 0.01%, 0.05%, 0.01%, and 0.18%, respectively. In short, the Hulu River Basin has been mainly characterized by cultivated land and grassland, with increasing proportions of woodland and shrubs. Water bodies, unused land, and impermeable areas have also shown changes in their proportions (Table 9).

Table 9. The land use change occurred in the study area during the period 1985–2020.

Type/Area	1985		2000		2010		2020	
	Area/km^2	Proportion/%	Area/km^2	Proportion/%	Area/km^2	Proportion/%	Area/km^2	Proportion%
Cultivated land	6415.3	60.09	8073.5	75.62	7787.25	72.94	6689.71	62.66
Woodland	182.3	1.71	201.3	1.89	217.64	2.04	299.99	2.81
shrubs	0.9	0.009	0.7	0.01	0.55	0.01	1.81	0.02
Grassland	4058.9	38.02	2380.8	22.30	2645.55	24.78	3640.56	34.10
Water	7.9	0.07	5.6	0.050	6.96	0.07	13.04	0.12
Unused land	1.5	0.014	0.5	0.005	0.63	0.01	2.13	0.02
Impermeable	9.04	0.08	13.33	0.13	17.1	0.16	28.44	0.27

3.3. Simulating the Interacting Effects of Rainfall and Land Use Characteristics on River Flooding

By taking the peak flow and peak water depth obtained from the simulation of the 50-year return period uniform rainfall and land use scenario as the baseline and comparing and analyzing the simulation results under different comprehensive scenarios, we can determine the degree of impact of the combined changes in rainfall characteristics and land use on floods. The simulation results for scenarios RL1-1 to RL1-6 are shown in Table 10, and the outflow hydrographs are shown in Figure 7a. The peak flow occurs at 5400 s for all scenarios. In scenario RL1-1, the peak flow is 2313.54 m^3/s, and the peak water depth is 7.47 m, representing a 15.50% increase in peak flow and a 12.84% decrease in peak water depth compared to the baseline period. In scenario RL1-2, the peak flow is 2377.32 m^3/s, representing an 18.69% increase, and the peak water depth is 8.17 m, representing a 4.67% decrease. In scenario RL1-3, the peak flow is 2424.66 m^3/s, and the peak water depth is 8.36 m, representing a 21.05% increase in peak flow and a 2.45% decrease in peak water depth compared to the baseline period. In scenario RL1-4, the peak flow is 1902.56 m^3/s, and the peak water depth is 5.39 m, representing a 5.02% decrease in peak flow and a 37.11% decrease in peak water depth compared to the baseline period. In scenario RL1-5, the peak flow is 2002.31 m^3/s, and the peak water depth is 5.42 m, representing a 0.04% decrease in peak flow and a 36.76% decrease in peak water depth compared to the baseline period. In scenario RL1-6, the peak flow is 2277.76 m^3/s, and the peak water depth is 7.39 m, representing a 13.72% increase in peak flow and a 13.77% decrease in peak water depth compared to the baseline period.

Table 10. Scenarios RL1-1 to RL1-6 peak flow rate and peak water depth.

Scenario	Peak Flow Rate/(m^3·s^{-1})	Peak Water Depth/(m)
Base period	2003.04	8.57
RL1-1	2313.54	7.47
RL1-2	2377.32	8.17
RL1-3	2424.66	8.36
RL1-4	1902.56	5.39
RL1-5	2002.31	5.42
RL1-6	2277.76	7.39

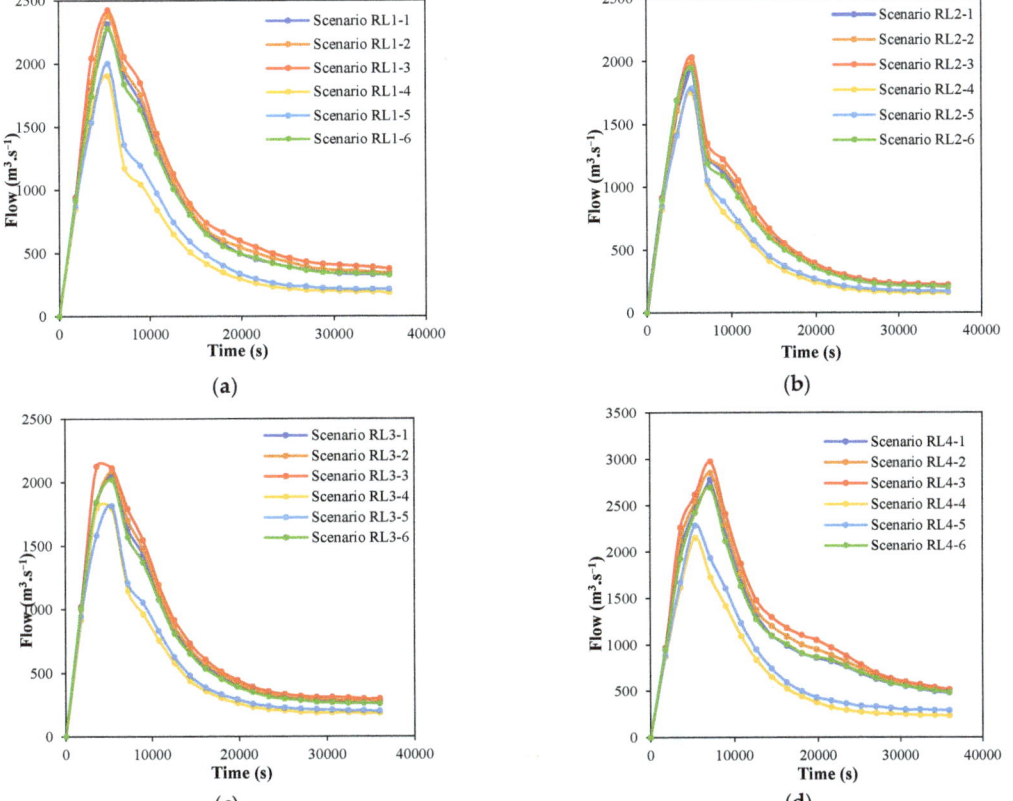

Figure 7. Flow process line of basin outlet section under different comprehensive scenarios. (a) Scenario RL1; (b) Scenario RL2; (c) Scenario RL3; (d) Scenario RL4 figure.

The simulation results for scenarios RL2-1 to RL2-6 are shown in Table 11, and the outflow hydrographs are shown in Figure 7b. The peak flow occurs at 5400 s for all scenarios. In scenario RL2-1, the peak flow is 1989.97 m^3/s, and the peak water depth is 5.42 m, representing a 0.65% decrease in peak flow and a 36.76% decrease in peak water depth compared to the baseline period. In scenario RL2-2, the peak flow is 1990.56 m^3/s, representing a 0.62% decrease, and the peak water depth is 5.41 m, representing a 36.87% decrease. In scenario RL2-3, the peak flow is 2033.76 m^3/s, and the peak water depth is 5.55 m, representing a 1.53% increase in peak flow and a 35.24% decrease in peak water depth compared to the baseline period. In scenario RL2-4, the peak flow is 1751.75 m^3/s, and the peak water depth is 5.34 m, representing a 12.55% decrease in peak flow and a 37.69% decrease in peak water depth compared to the baseline period. In scenario RL2-5, the peak flow is 1786.75 m^3/s, and the peak water depth is 5.34 m, representing a 10.80% decrease in peak flow and a 37.69% decrease in peak water depth. In scenario RL2-6, the peak flow is 1947.12 m^3/s, and the peak water depth is 5.40 m, representing a 2.79% decrease in peak flow and a 36.99% decrease in peak water depth compared to the baseline period.

Table 11. Scenarios RL2-1 to RL2-6 peak flow rate and peak water depth.

Scenario	Peak Flow Rate/(m$^3 \cdot$s^{-1})	Peak Water Depth/(m)
Base period	2003.04	8.57
RL2-1	1989.97	5.42
RL2-2	1990.56	5.41
RL2-3	2033.76	5.55
RL2-4	1751.75	5.34
RL2-5	1786.75	5.34
RL2-6	1947.12	5.40

The simulation results for scenarios RL3-1 to RL3-6 are shown in Table 12, and the outflow hydrographs are shown in Figure 7c. The peak flow occurs between 3600 s and 5400 s for all scenarios. In scenario RL3-1, the peak flow is 2052.46 m^3/s, and the peak water depth is 5.90 m, representing a 2.47% increase in peak flow and a 31.16% decrease in peak water depth compared to the baseline period. In scenario RL3-2, the peak flow is 2088.33 m^3/s, representing a 4.26% increase, and the peak water depth is 6.21 m, representing a 27.54% decrease. In scenario RL3-3, the peak flow is 2127.37 m^3/s, and the peak water depth is 6.49 m, representing a 6.21% increase in peak flow and a 24.27% decrease in peak water depth compared to the baseline period. In scenario RL3-4, the peak flow is 1796.61 m^3/s, and the peak water depth is 5.34 m, representing a 10.31% decrease in peak flow and a 37.69% decrease in peak water depth compared to the baseline period. In scenario RL3-5, the peak flow is 1817.12 m^3/s, and the peak water depth is 5.36 m, representing a 9.28% decrease in peak flow and a 37.46% decrease in peak water depth. In scenario RL3-6, the peak flow is 2023.07 m^3/s, and the peak water depth is 5.71 m, representing a 1.00% increase in peak flow and a 33.37% decrease in peak water depth compared to the baseline period.

Table 12. Scenarios RL3-1 to RL3-6 peak flow rate and peak water depth.

Scenario	Peak Flow Rate/(m$^3 \cdot$s^{-1})	Peak Water Depth/(m)
Base period	2003.04	8.57
RL3-1	2052.46	5.90
RL3-2	2088.33	6.21
RL3-3	2127.37	6.49
RL3-4	1796.61	5.34
RL3-5	1817.12	5.36
RL3-6	2023.07	5.71

The simulation results for scenarios RL4-1 to RL4-6 are shown in Table 13, and the discharge hydrograph at the outlet cross-section is shown in Figure 7d. The peak flow occurs between 5400 s and 7200 s. In scenario RL4-1, the peak flow is 2783.00 m^3/s, and the peak water depth is 8.81 m. Compared to the baseline period, the peak flow has increased by 38.94%, and the peak water depth has increased by 2.80%. In scenario RL4-2, the peak flow is 2853.87 m^3/s, an increase of 42.48% compared to the baseline period, and the peak water depth is 8.82 m, which is an increase of 2.92%. In scenario RL4-3, the peak flow is 2974.82 m^3/s, and the peak water depth is 9.0 m. These values represent an increase of 48.52% and 5.02%, respectively, compared to the baseline period. In scenario RL4-4, the peak flow is 2156.24 m^3/s, and the peak water depth is 5.91 m. The peak flow has increased by 7.65%, while the peak water depth has decreased by 31.04% compared to the baseline period. In scenario RL4-5, the peak flow is 2289.18 m^3/s, and the peak water depth is 6.87 m. The peak flow has increased by 14.29%, while the peak water depth has decreased by 19.84% compared to the baseline period. In scenario RL4-6, the peak flow is 2694.34 m^3/s, and the peak water depth is 8.82 m. The peak flow has increased by 34.51%, and the peak water depth has increased by 2.92% compared to the baseline period.

Table 13. Scenarios RL4-1 to RL4-6 peak traffic and peak water depth.

Scenario	Peak Flow Rate/(m^3·s^{-1})	Peak Water Depth/(m)
Base period	2003.04	8.57
RL4-1	2783.00	8.81
RL4-2	2853.87	8.82
RL4-3	2974.82	9.00
RL4-4	2156.24	5.91
RL4-5	2289.18	6.87
RL4-6	2694.34	8.82

The variations between scenarios RL1-1 to RL1-6, RL2-1 to RL2-6, RL3-1 to RL3-6, and RL4-1 to RL4-6 represent the influence of land use changes on watershed flooding. The range of variation in peak flow is 1.55% to 21.53%, 0.03% to 13.87%, 1.14% to 15.55%, and 2.55% to 27.52%, respectively. The range of variation in peak water depth is 0.56% to 37.11%, 0.00% to 3.78%, 0.37% to 17.72%, and 0.00% to 49.24%, respectively. When the rainfall gradually decreases from southeast to northwest, and the rainfall center is located downstream, the impact of land use changes on watershed flooding is significant, while in the other two land use scenarios, the impact is relatively small.

The variations between scenarios RL1-1, RL2-1, RL3-1, and RL4-1, RL1-2, RL2-2, RL3-2, and RL4-2, RL1-3, RL2-3, RL3-3, and RL4-3, RL1-4, RL2-4, RL3-4, and RL4-4, RL1-5, RL2-5, RL3-5, and RL4-5, RL1-6, RL2-6, RL3-6, and RL4-6 represent the influence of changes in rainfall characteristics on peak flow and peak water depth. The range of variation in peak flow is 3.14% to 39.85%, 4.91% to 43.37%, 4.60% to 46.27%, 2.56% to 23.09%, 1.70% to 28.12%, and 3.90% to 38.38%, respectively. The range of variation in peak water depth is 8.86% to 62.55%, 7.96% to 63.03%, 7.66% to 62.16%, 0.00% to 10.67%, 0.37% to 28.65%, and 5.74% to 63.33%, respectively. When the land use scenario is afforestation and grassland restoration, the impact of changes in rainfall characteristics on watershed flooding is relatively small, while in the other land use scenarios, the impact is more significant. Therefore, except for the land use scenario of afforestation and grassland restoration in the entire region, rainfall characteristics have a greater impact on watershed flooding than land use changes.

In short, it is stated that the most unfavorable scenario for the Hulu River Basin in terms of rainfall characteristics and land use is scenario RL4-3, with a peak flow rate of 2974.82 m^3/s and a peak water depth of 9.00 m. In this scenario, the rainfall distribution is concentrated in the downstream area, and the land use corresponds to the land use in 2010. The basin is threatened by severe floods not only due to changes in rainfall distribution but also because of the reduction in grassland and forest land and the increase in cultivated land.

On the other hand, the scenario with the greatest reduction in flood risk due to rainfall characteristics and land use is RL2-4, with a peak flow rate of 1751.75 m^3/s and a peak water depth of 5.34 m. In this scenario, the rainfall is heavier in the southeast and northwest regions and lighter in the northeast and southwest regions. The land use type is characterized by reforestation and the return of cultivated land to forests. The changes in rainfall distribution and the increase in grassland contribute to the decrease in flood threat.

4. Discussion

The GAST model was found to be suitable for investigating the Impacts Assessment of Rainfall Characteristics and Land use Patterns on Runoff accumulation in the Hulu River

Basin, China. Overall, the GAST model performance classification for the watershed was very good [67]. The GAST model results revealed that rainfall and land use had an effect on the hydrologic process of the Hulu River basin. The observed changes in hydrological processes were attributed to rainfall and land use changes for this study.

4.1. Flood Evolution under Different Rainfall Characteristics Scenarios

With uniform rainfall distribution, an increase in the return period leads to higher peak flow rates and peak water depths, and the occurrence time of peak flow advances. When the difference in return periods is small, such as 5 and 10-year return periods, the increase in peak flow rates and peak water depths is not significant. However, when the rainfall distribution is non-uniform, there is a considerable difference in peak flow rates and peak water depths compared to the scenario with uniform rainfall distribution. As the return period increases, the impact of non-uniform rainfall distribution on peak flow rates decreases. Among the scenarios, scenario R8 has the greatest impact on basin floods. Under a 5-year return period, the peak flow rate increases by 41.30%, and the peak water depth decreases by 0.55%. Under a 10-year return period, the peak flow rate increases by 40.00%, and the peak water depth increases by 4.96%. Under a 50-year return period, the peak flow rate increases by 34.51%, and the peak water depth increases by 2.92%. The scenario that is closest to the uniform rainfall scenario is R7, with a 36.07% increase in peak flow rate and a 3.22% increase in peak water depth under a 5-year return period, a 30.30% increase in peak flow rate and a 0.92% decrease in peak water depth under a10-year return period, and a 1.01% increase in peak flow rate and a 33.37% decrease in peak water depth under 50-year return period.

4.2. Flood Evolution Process under Different Land Use Scenarios

Different land use scenarios result in varying responses in peak flow rates and peak water depths. From 1985 to 2020, the changes in land use led to an initial increase and subsequent decrease in peak flow rates and peak water depths in the Hulu River Basin. Compared to the current land use in 2020, all three historical land use scenarios resulted in increased peak flow rates, while the peak water depths showed variations. Scenario L1 leads to a decrease in peak water depth compared to the current land use scenario. In scenario L2, the peak water depth is smaller under a 5-year return period, while it increases under 10-year and 50-year return periods compared to the current scenario. Scenario L3 results in increased peak water depths compared to the current land use scenario. The integrated land use scenarios and generalized water conservation measures lead to significant differences in peak flow rates compared to the current land use scenario. Across different return periods, there is a significant decrease in peak flow rates, and the occurrence time is delayed. The magnitude of the impacts on peak flow rates and peak water depths is as follows: L5 > L6 > L7.

4.3. Flood Evolution under the Combined Influence of Different Rainfall Characteristics and Land Use

The combined effect of rainfall characteristics and land use has the greatest impact on basin floods in scenario RL4-3, where the rainfall center is located downstream, and the land use corresponds to the 2010 land use type. The resulting peak flow rate and peak water depth are 2974.82 m^3/s and 9.00 m, respectively. The range of the impacts of land use changes on basin floods is between 0.04% and 48.52%, while the range of changes in peak flow rates and peak water depths due to rainfall characteristics is between 0.00% and 63.33%. Under the combined effect of rainfall characteristics and land use changes, the range of changes in peak flow rates and peak water depths is between 0.04% and 48.52%. In general, except for the scenarios with region-wide reforestation and grassland restoration land use, the impact of rainfall characteristic changes on basin floods is greater than the impact of land use changes.

5. Conclusions

In this study, a two-dimensional hydrodynamic GAST model (GPU Accelerated Surface Water Flow and Transport Model) is set up to simulate the flood processes in the Hulu River Basin under different rainfall characteristics and land use scenarios. The impacts of various rainfall and land use scenarios on peak flow rates, peak water depths, and flood propagation processes were analyzed and quantified. In the Hulu River Basin, rainfall distribution and land use significantly affect peak flow and water depth. Uniform rainfall increases peak flow and depth, especially during longer return periods, while uneven rainfall lessens this impact. Concentrated rainfall poses the greatest flood risk in downstream areas, necessitating early emergency response and disaster plans. Land use changes from 1985 to 2020 influenced these patterns, with peak flow initially increasing and then decreasing. Shifting land use towards forestry and grassland and implementing soil and water conservation measures showed a notable decrease in runoff. Future efforts should focus on rainfall prediction and land use policy implementation to mitigate flood risks effectively.

The GAST model developed can predict the river basin's future streamflow conditions. It can guarantee the sustainable use of water resources and provide a theoretical foundation for the planning and management of the total amount of simulated streamflow of water resources in large river basins and regions.

6. Policy Implications

Based on the context provided by the hydrodynamic model GAST and its accuracy assessment through RMSE. It can be used for flood forecasting and management. The model's outputs could inform infrastructure development policies, such as where to build flood defenses or how to design them to withstand predicted flow rates. Policies related to environmental protection might be impacted if the flow simulations suggest changes in water patterns that could affect ecosystems. The research may have implications for urban planning policies. Flood prediction could affect agricultural areas and may lead to changes in agricultural policies, such as crop insurance schemes or water management practices. In summary, the policy implications of a hydrodynamic model simulation study are vast and can influence a wide range of policy areas, from local urban planning to broader environmental and risk management strategies.

Author Contributions: M.I. conceived the conceptualization of the research study, design and development of the experiment, data collection, formal analysis, investigation, methodology, visualization, writing an original draft, reviewed, supervised, and write-up editing. T.W., D.L., X.G., R.S.N., J.J. and M.A. contributed to the review and editing of the draft. J.H. supervised the entire research study and contributed as an internal reviewer for the manuscript. All authors have read and agreed to the published version of the manuscript.

Funding: This research was supported by National Natural Science Foundation of China (52079106, 52009104). Science and Technology Projects of Northwest Engineering Corporation Limited, Power China (XBY-ZDKJ-2022-9).

Data Availability Statement: Data is contained within the article.

Acknowledgments: The authors are thankful and acknowledge the State Key Laboratory of Eco-hydraulics in the Northwest Arid Region of China, Xi'an University of Technology, Xi'an 710048, Shaanxi, China.

Conflicts of Interest: Author Xujun Gao was employed by the company Power China Northwest Engineering Corporation Limited. The remaining authors declare that the research was conducted in the absence of any commercial or financial relationships that could be construed as a potential conflict of interest.

References

1. Wang, X.; Zhang, P.; Liu, L.; Li, D.; Wang, Y. Effects of human activities on hydrological components in the Yiluo River basin in middle Yellow River. *Water* **2019**, *11*, 689. [CrossRef]
2. Zhan, C.; Jiang, S.; Sun, F.; Jia, Y.; Niu, C.; Yue, W. Quantitative contribution of climate change and human activities to runoff changes in the Wei River basin, China. *Hydrol. Earth Syst. Sci.* **2014**, *18*, 3069–3077. [CrossRef]
3. Allen, M.R.; Ingram, W.J. Constraints on future changes in climate and the hydrologic cycle. *Nature* **2002**, *419*, 224–232. [CrossRef] [PubMed]
4. Wang, G.; Xia, J.; Chen, J. Quantification of effects of climate variations and human activities on runoff by a monthly water balance model: A case study of the Chaobai River basin in northern China. *Water Resour. Res.* **2009**, *45*, W00A11. [CrossRef]
5. Wang, S.; Yan, M.; Yan, Y.; Shi, C.; He, L. Contributions of climate change and human activities to the changes in runoff increment in different sections of the Yellow River. *Quat. Int.* **2012**, *282*, 66–77. [CrossRef]
6. Xu, C.-Y. Climate change and hydrologic models: A review of existing gaps and recent research developments. *Water Resour. Manag.* **1999**, *13*, 369–382. [CrossRef]
7. Sharma, C.S.; Behera, M.D.; Mishra, A.; Panda, S.N. Assessing flood induced land-cover changes using remote sensing and fuzzy approach in Eastern Gujarat (India). *Water Resour. Manag.* **2011**, *25*, 3219–3246. [CrossRef]
8. Suwanwerakamtorn, R. Register | Login. *ITC J.* **1994**, *4*, 343–348.
9. Brooks, K.N.; Ffolliott, P.F.; Gregersen, H.M.; Thames, J. *Hydrology and the Management of Watersheds*; Iowa State University Press: Ames, IA, USA, 1991.
10. Fohrer, N.; Haverkamp, S.; Eckhardt, K.; Frede, H.-G. Hydrologic response to land use changes on the catchment scale. *Phys. Chem. Earth Part B Hydrol. Ocean. Atmos.* **2001**, *26*, 577–582. [CrossRef]
11. Liu, J.; Wang, S.-Y.; Li, D.-M. The analysis of the impact of land-use changes on flood exposure of Wuhan in Yangtze River Basin, China. *Water Resour. Manag.* **2014**, *28*, 2507–2522. [CrossRef]
12. Chen, D.; Lu, X.; Hu, W.; Zhang, C.; Lin, Y. How urban sprawl influences eco-environmental quality: Empirical research in China by using the Spatial Durbin model. *Ecol. Indic.* **2021**, *131*, 108113. [CrossRef]
13. Ahmadi, A.; Aghakhani Afshar, A.; Nourani, V.; Pourreza-Bilondi, M.; Besalatpour, A. Assessment of MC&MCMC uncertainty analysis frameworks on SWAT model by focusing on future runoff prediction in a mountainous watershed via CMIP5 models. *J. Water Clim. Chang.* **2020**, *11*, 1811–1828.
14. Tan, M.L.; Yusop, Z.; Chua, V.P.; Chan, N.W. Climate change impacts under CMIP5 RCP scenarios on water resources of the Kelantan River Basin, Malaysia. *Atmos. Res.* **2017**, *189*, 1–10. [CrossRef]
15. O'Neill, B.C.; Tebaldi, C.; Van Vuuren, D.P.; Eyring, V.; Friedlingstein, P.; Hurtt, G.; Knutti, R.; Kriegler, E.; Lamarque, J.-F.; Lowe, J. The scenario model intercomparison project (ScenarioMIP) for CMIP6. *Geosci. Model Dev.* **2016**, *9*, 3461–3482. [CrossRef]
16. Song, M.; Tao, W. Coupling and coordination analysis of China's regional urban-rural integration and land-use efficiency. *Growth Chang.* **2022**, *53*, 1384–1413. [CrossRef]
17. Li, C.; Li, Y.; Wang, P.; Shen, J. Natural runoff prediction of the Yellow River in the future under climate change. In Proceedings of the International Conference on Advanced Education and Management Engineering, Bangkok, Thailand, 29–30 October 2016.
18. Zhu, X.; Zhang, C.; Qi, W.; Cai, W.; Zhao, X.; Wang, X. Multiple climate change scenarios and runoff response in Biliu River. *Water* **2018**, *10*, 126. [CrossRef]
19. Yang, W.; Long, D.; Bai, P. Impacts of future land cover and climate changes on runoff in the mostly afforested river basin in North China. *J. Hydrol.* **2019**, *570*, 201–219. [CrossRef]
20. Asselman, N.E.; Middelkoop, H.; Van Dijk, P.M. The impact of changes in climate and land use on soil erosion, transport and deposition of suspended sediment in the River Rhine. *Hydrol. Process.* **2003**, *17*, 3225–3244. [CrossRef]
21. Ouyang, L.; Liu, S.; Ye, J.; Liu, Z.; Sheng, F.; Wang, R.; Lu, Z. Quantitative assessment of surface runoff and base flow response to multiple factors in Pengchongjian small watershed. *Forests* **2018**, *9*, 553. [CrossRef]
22. Liu, J.; Luo, M.; Liu, T.; Bao, A.; De Maeyer, P.; Feng, X.; Chen, X. Local climate change and the impacts on hydrological processes in an arid alpine catchment in Karakoram. *Water* **2017**, *9*, 344. [CrossRef]
23. Zuo, D.; Xu, Z.; Yao, W.; Jin, S.; Xiao, P.; Ran, D. Assessing the effects of changes in land use and climate on runoff and sediment yields from a watershed in the Loess Plateau of China. *Sci. Total Environ.* **2016**, *544*, 238–250. [CrossRef] [PubMed]
24. Chen, D.; Hu, W.; Li, Y.; Zhang, C.; Lu, X.; Cheng, H. Exploring the temporal and spatial effects of city size on regional economic integration: Evidence from the Yangtze River Economic Belt in China. *Land Use Policy* **2023**, *132*, 106770. [CrossRef]
25. Houghton, R.A. The worldwide extent of land-use change. *BioScience* **1994**, *44*, 305–313. [CrossRef]
26. Hathout, S. The use of GIS for monitoring and predicting urban growth in East and West St Paul, Winnipeg, Manitoba, Canada. *J. Environ. Manag.* **2002**, *66*, 229–238. [CrossRef]
27. Li, F.; Zhang, S.; Yang, J.; Chang, L.; Yang, H.; Bu, K. Effects of land use change on ecosystem services value in West Jilin since the reform and opening of China. *Ecosyst. Serv.* **2018**, *31*, 12–20. [CrossRef]
28. Song, M.; Xie, Q.; Chen, J. Effects of government competition on land prices under opening up conditions: A case study of the Huaihe River ecological economic belt. *Land Use Policy* **2022**, *113*, 105875. [CrossRef]
29. Shi, C.; Zhou, Y.; Fan, X.; Shao, W. A study on the annual runoff change and its relationship with water and soil conservation practices and climate change in the middle Yellow River basin. *Catena* **2013**, *100*, 31–41. [CrossRef]

30. Wang, Y.; Yao, S. Effects of restoration practices on controlling soil and water losses in the Wei River Catchment, China: An estimation based on longitudinal field observations. *For. Policy Econ.* **2019**, *100*, 120–128. [CrossRef]
31. Wang, T.; Li, P.; Li, Z.; Hou, J.; Xiao, L.; Ren, Z.; Xu, G.; Yu, K.; Su, Y. The effects of freeze–thaw process on soil water migration in dam and slope farmland on the Loess Plateau, China. *Sci. Total Environ.* **2019**, *666*, 721–730. [CrossRef]
32. Chen, J.; Li, Q.; Wang, H.; Deng, M. A machine learning ensemble approach based on random forest and radial basis function neural network for risk evaluation of regional flood disaster: A case study of the Yangtze River Delta, China. *Int. J. Environ. Res. Public Health* **2020**, *17*, 49. [CrossRef]
33. Dewan, T.H. Societal impacts and vulnerability to floods in Bangladesh and Nepal. *Weather Clim. Extrem.* **2015**, *7*, 36–42. [CrossRef]
34. Wang, S.; Fu, B.; Piao, S.; Lü, Y.; Ciais, P.; Feng, X.; Wang, Y. Reduced sediment transport in the Yellow River due to anthropogenic changes. *Nat. Geosci.* **2016**, *9*, 38–41. [CrossRef]
35. Yin, J.; He, F.; Xiong, Y.J.; Qiu, G.Y. Effects of land use/land cover and climate changes on surface runoff in a semi-humid and semi-arid transition zone in northwest China. *Hydrol. Earth Syst. Sci.* **2017**, *21*, 183–196. [CrossRef]
36. Feng, X.; Sun, G.; Fu, B.; Su, C.; Liu, Y.; Lamparski, H. Regional effects of vegetation restoration on water yield across the Loess Plateau, China. *Hydrol. Earth Syst. Sci.* **2012**, *16*, 2617–2628. [CrossRef]
37. Zhao, G.; Mu, X.; Tian, P.; Wang, F.; Gao, P. Climate changes and their impacts on water resources in semiarid regions: A case study of the Wei River basin, China. *Hydrol. Process.* **2013**, *27*, 3852–3863. [CrossRef]
38. Hu, C.; Ran, G.; Li, G.; Yu, Y.; Wu, Q.; Yan, D.; Jian, S. The effects of rainfall characteristics and land use and cover change on runoff in the Yellow River basin, China. *J. Hydrol. Hydromech.* **2021**, *69*, 29–40. [CrossRef]
39. Wang, J.; Zhang, R.; Yao, W.; Li, Z. Relationship Between Watershed Landscape Pattern Change and Runoff-Sediment in Wind-Water Erosion Crisscross Region. *J. Landsc. Res.* **2017**, *9*, 53–58.
40. Zhan, F.; Zeng, W.; Li, B.; Li, Z.; Chen, J.; He, Y.; Li, Y. Inhibition of native arbuscular mycorrhizal fungi induced increases in cadmium loss via surface runoff and interflow from farmland. *Int. Soil Water Conserv. Res.* **2023**, *11*, 213–223. [CrossRef]
41. Delgado, M.I.; Carol, E.; Casco, M.A. Land-use changes in the periurban interface: Hydrologic consequences on a flatland-watershed scale. *Sci. Total Environ.* **2020**, *722*, 137836. [CrossRef]
42. Lin, B.; Chen, X.; Yao, H.; Chen, Y.; Liu, M.; Gao, L.; James, A. Analyses of landuse change impacts on catchment runoff using different time indicators based on SWAT model. *Ecol. Indic.* **2015**, *58*, 55–63. [CrossRef]
43. Hernández-Bedolla, J.; García-Romero, L.; Franco-Navarro, C.D.; Sánchez-Quispe, S.T.; Domínguez-Sánchez, C. Extreme Runoff Estimation for Ungauged Watersheds Using a New Multisite Multivariate Stochastic Model MASVC. *Water* **2023**, *15*, 2994. [CrossRef]
44. Liu, R. *Remote Sensing Estimation and Spatiotemporal Evolution Analysis of Urban Surface Runoff in Xuzhou Based on Impervious Surface*; China University of Mining and Technology: Beijing, China, 2022.
45. Fang, G.; Li, H.; Dong, J.; Teng, H.; Pablo, R.D.A.; Zhu, Y. Extraction and Spatiotemporal Evolution Analysis of Impervious Surface and Surface Runoff in Main Urban Region of Hefei City, China. *Sustainability* **2023**, *15*, 10537. [CrossRef]
46. Yonaba, R.; Biaou, A.C.; Koïta, M.; Tazen, F.; Mounirou, L.A.; Zouré, C.O.; Queloz, P.; Karambiri, H.; Yacouba, H. A dynamic land use/land cover input helps in picturing the Sahelian paradox: Assessing variability and attribution of changes in surface runoff in a Sahelian watershed. *Sci. Total Environ.* **2021**, *757*, 143792. [CrossRef] [PubMed]
47. Sheng, F.; Liu, S.-Y.; Zhang, T.; Yu, M.-Q. Runoff effect of precipitation variation and landscape pattern evolution in Lianshui watershed, Jiangxi, China. *Ying Yong Sheng Tai Xue Bao J. Appl. Ecol.* **2023**, *34*, 196–202.
48. Prokešová, R.; Horáčková, Š.; Snopková, Z. Surface runoff response to long-term land use changes: Spatial rearrangement of runoff-generating areas reveals a shift in flash flood drivers. *Sci. Total Environ.* **2022**, *815*, 151591. [CrossRef] [PubMed]
49. Han, H.; Hou, J.; Huang, M.; Li, Z.; Xu, K.; Zhang, D.; Bai, G.; Wang, C. Impact of soil and water conservation measures and precipitation on streamflow in the middle and lower reaches of the Hulu River Basin, China. *Catena* **2020**, *195*, 104792. [CrossRef]
50. Hou, J.; Li, G.; Li, G.; Liang, Q.; Zhi, Z. Research on the application of efficient and high-precision hydrodynamic model in flood evolution. *J. Hydropower* **2018**, *37*, 96–107.
51. Hou, J.; Liang, Q.; Simons, F.; Hinkelmann, R. A stable 2D unstructured shallow flow model for simulations of wetting and drying over rough terrains. *Comput. Fluids* **2013**, *82*, 132–147. [CrossRef]
52. Liu, F.; Hou, J.; Guo, K.; Li, D.; Xu, S.; Zhang, X. Numerical simulation of rainwater and flood processes in watersheds based on full hydrodynamic model. *Hydrodyn. Res. Prog.* **2018**, *33*, 778–785.
53. Sivakumar, P.; Hyams, D.; Taylor, L.K.; Briley, W.R. A primitive-variable Riemann method for solution of the shallow water equations with wetting and drying. *J. Comput. Phys.* **2009**, *228*, 7452–7472. [CrossRef]
54. Hou, J.; Liang, Q.; Simons, F.; Hinkelmann, R. A 2D well-balanced shallow flow model for unstructured grids with novel slope source term treatment. *Adv. Water Resour.* **2013**, *52*, 107–131. [CrossRef]
55. Simons, F.; Busse, T.; Hou, J.; Özgen, I.; Hinkelmann, R. A model for overland flow and associated processes within the Hydroinformatics Modelling System. *J. Hydroinform.* **2014**, *16*, 375–391. [CrossRef]
56. Hubbard, M. Multidimensional slope limiters for MUSCL-type finite volume schemes on unstructured grids. *J. Comput. Phys.* **1999**, *155*, 54–74. [CrossRef]
57. Bai, G.; Hou, J.; Han, H.; Shi, Y.; Guo, K.; Li, B.; Fu, D. Analysis of the Contribution Rates of Factors to Runoff in Hulu River Based on Support vector Machine Regression. *Res. Soil Water Conserv.* **2020**, *27*, 112–117. [CrossRef]

58. Yang, P.; Li, R.; Pan, E.; Wang, Y.; Huang, M.; Zhang, L. Effects of surface roughness and vegetation coverage on the Manning resistance coefficient of slope flow. *Trans. Chin. Soc. Agric. Eng.* **2020**, *36*, 106–114. [CrossRef]
59. Wang, J.; Zhang, K.; Gong, J.; Yang, F.; Dong, X. Laws of water flow resistance on slopes under different coverage conditions. *J. Soil Water Conserv.* **2015**, *29*, 1–6. [CrossRef]
60. Wang, J.; Wu, F.; Meng, Q.; Zhang, Q. Experimental study on soil water infiltration characteristics under different utilization types. *Agric. Res. Arid Areas* **2006**, *24*, 159–162.
61. Gao, P.; Mu, X. Comparative experiment on soil moisture infiltration under different land use patterns in loess hilly areas. *China Soil Water Conserv. Sci.* **2005**, *3*, 27–31.
62. Horton, R.E. An approach toward a physical interpretation of infiltration capacity. *Soil Sci. Soc. Am. Proc.* **1940**, *5*, 24. [CrossRef]
63. Qin, J.; Su, B.; Tao, H.; Wang, Y.; Huang, J.; Jiang, T. Projection of temperature and precipitation under SSPs-RCPs Scenarios over northwest China. *Front. Earth Sci.* **2021**, *15*, 23–37. [CrossRef]
64. Tian, J.; Zhang, Z.; Ahmed, Z.; Zhang, L.; Su, B.; Tao, H.; Jiang, T. Projections of precipitation over China based on CMIP6 models. *Stoch. Environ. Res. Risk Assess.* **2021**, *35*, 831–848. [CrossRef]
65. Li, E.; Mu, X.; Zhao, G.; Gao, P.; Shao, H. Variation of runoff and precipitation in the hekou-longmen region of the yellow river based on elasticity analysis. *Sci. World J.* **2014**, *2014*, 929858. [CrossRef] [PubMed]
66. Ran, D.; Zuo, Z.; Wu, Y.; Li, X.; Li, Z. *Variation of Streamflow and Sediment in Response to Human Activities in the Middle Reaches of the Yellow River*; Science Press: Beijing, China, 2012.
67. Hou, J.; Li, X.; Pan, Z.; Wang, J.; Wang, R. Effect of digital elevation model spatial resolution on depression storage. *Hydrol. Process.* **2021**, *35*, e14381. [CrossRef]

Disclaimer/Publisher's Note: The statements, opinions and data contained in all publications are solely those of the individual author(s) and contributor(s) and not of MDPI and/or the editor(s). MDPI and/or the editor(s) disclaim responsibility for any injury to people or property resulting from any ideas, methods, instructions or products referred to in the content.

Article

Climate Change Effects on Rainfall Intensity–Duration–Frequency (IDF) Curves for the Lake Erie Coast Using Various Climate Models

Samir Mainali and Suresh Sharma *

Civil and Environmental Engineering, Youngstown State University, 1 Tressel Way, Youngstown, OH 44555, USA; smainali01@student.ysu.edu
* Correspondence: ssharma06@ysu.edu; Tel.: +1-(330)-941-1741

Abstract: This study delved into the analysis of hourly observed as well as future precipitation data in the towns of Willoughby and Buffalo on the Lake Erie Coast to examine the variations in IDF relationships over the 21st century. Several regional climate models (RCMs) and general circulation models (GCMs) from the Coupled Model Intercomparison Project (CMIP) Phases 5 and 6 were used. The study evaluated three RCMs with historical and Representative Concentration Pathway (RCP) 8.5 scenarios for each CMIP5 and three GCMs with historical and Shared Socioeconomic Pathways (SSPs) (126, 245, 370, and 585) scenarios for each CMIP6. The results suggested that the town of Willoughby would experience an increase of 9–46%, whereas Buffalo would experience an upsurge of 6–140% in the hourly precipitation intensity under the worst-case scenarios of RCP8.5 for CMIP5 and SSP585 for CMIP6. This increase is expected to occur in both the near (2020–2059) and far future (2060–2099), with a return period as low as 2 years and as high as 100 years when compared to the baseline period (1980–2019). The analysis indicated an increased range of 9–39% in the near future and 20–55% in the far future for Willoughby, while the Buffalo region may experience an increase of 2–95% in the near future and 3–192% in the far future as compared to the baseline period. In contrast to CMIP6 SSP585 models, CMIP5 RCP8.5 models predicted rainfall with an intensity value that is up to 28% higher in the town of Willoughby, while the reverse was true for the Buffalo region. The findings of this study are expected to be helpful for the design of water resource infrastructures.

Keywords: general circulation models (GCM); coupled model inter-comparison project (CMIP); Gumbel extreme value type I distribution; extreme rainfall; rainfall intensity; IDF curves

Citation: Mainali, S.; Sharma, S. Climate Change Effects on Rainfall Intensity–Duration–Frequency (IDF) Curves for the Lake Erie Coast Using Various Climate Models. *Water* 2023, 15, 4063. https://doi.org/10.3390/w15234063

Academic Editors: Sonia Raquel Gámiz-Fortis and Matilde García-Valdecasas Ojeda

Received: 12 October 2023
Revised: 13 November 2023
Accepted: 18 November 2023
Published: 23 November 2023

Copyright: © 2023 by the authors. Licensee MDPI, Basel, Switzerland. This article is an open access article distributed under the terms and conditions of the Creative Commons Attribution (CC BY) license (https://creativecommons.org/licenses/by/4.0/).

1. Introduction

Climate change implies long-term shifts in precipitation and temperature patterns due to anthropogenic influences by generating greenhouse gas emissions such as carbon dioxide and methane gases [1,2]. Future intensification of extreme precipitation events due to greenhouse gas emissions will result in an increase in the frequency and length of rainfall events worldwide [2]. Several studies have reported a significant rise in both total annual precipitation and the frequency of extreme events [3–6]. More specifically, shorter-duration precipitation events are expected to increase significantly across the world [7,8]. For example, the frequency of hourly extreme precipitation events [9] is expected to advance up to 400% in North America [10]. Furthermore, the interaction of higher maximum precipitation rates (15–40% increase) and the expansion of areas affected by heavy rainfall leads to a substantial 80% rise in the overall precipitation volume [10]. Similar trends can also be observed in the United States [11–13]. The Intergovernmental Panel on Climate Change [2] also projects that over the 21st century, heavy precipitation will occur in this area more frequently and with greater intensity.

Future high-intensity rainfalls triggered by climate change will have a more detrimental effect on urban stormwater systems [14,15]. The duration and rainfall intensity are linked to the frequency of the rainfall and such rainfall characteristics can be represented by a curve called the intensity–duration–frequency (IDF) curve. The IDF curve can be mathematically represented in terms of return period, intensity, and rainfall duration The development of the IDF curve was initiated in the nineteenth century and has been widely used across the world.

Since the IDF curves, are frequently utilized to design water infrastructures, it is essential to gain a comprehensive understanding of the alterations in extreme precipitation and subsequently revise the IDF curves in the future [16–18]. The IDF curve has been extensively used across the world for the design of hydraulic structures including urban drainage, culverts, road bridges, and storm sewer systems [19–22].

The pressing need to reexamine the IDF curve arises from potential changes in intense rainfall exacerbated by climate change [23]. Some studies suggest that proactively anticipating design modifications for hydraulic structures would decrease the risk of future issues and uncertainties, resulting in successful and versatile project outcomes [24,25]. Many scientists and professionals have advocated for better knowledge of the possible change in the severity, frequency, and volume of intense rainfall due to climate change [26–30]. This understanding is necessary since the existing drainage systems and hydraulic infrastructures are built to handle historical rainfall time series data on the assumption that past extremes can be used to describe future extremes. This presumption is incorrect given the shifting frequency and amount of intense rainfall triggered by changing climatic variable [31,32]. With these changes, historic IDF curves cannot be used to accurately represent future climatic conditions. Therefore, a changing climate may result in an increase in demand that water management infrastructure built to previous IDF norms may not be able to accommodate [28]. Climate models that integrate greenhouse gas emissions have become increasingly accessible and within reach to foresee future changes in the IDF curve [14,33,34].

To date, the climate models are the primary and most effective tools for past and future climate simulations [35]. However, the prediction of the future climate is location-specific and varies depending on the type of general circulation models (GCMs) and the scenario chosen. For example, according to Coupled Model Intercomparison Project (CMIP) Phase 5 projections, the distribution of temperature and precipitation indices in the northeastern US will undergo significant changes between 2041 and 2070 [36]. Ragno et al. [37] found that densely populated places may experience up to 20% more intense and twice as frequent extreme precipitation events. Cheng and Aghakouchak [38] found that the assumption of extreme precipitation in a stationary climate may lead to an underestimation of extreme precipitation of up to 60%. Coelho et al. [39] conducted a study using CMIP6 projections to assess the impact of changing extreme precipitation on flood engineering designs across the US. By 2100, the northern region is predicted to experience an increase of 10–40% and the southern region, 20–80%. The study showed a meridional dipole-like pattern in the geographical distribution of precipitation changes, with an increase of 10–30% over the US. The results from the CMIP6 models in Tucson, Arizona, show the likely threat of future extreme events being disregarded in stationary-based design frameworks which could pose a significant risk to both safety and the economy by more than 300% [33]

Limited studies have been conducted using predicted precipitation from CMIP6 models in the US, and no future IDF curve has been developed in the Lake Erie Basin using CMIP5 and CMIP6 climate models. As the precipitation pattern of the Lake Erie basin is complex due to lake-enhanced precipitation and rainfall after the snowfall, the future IDF curve due to climate change impacts is crucial in the Lake Erie basin to safely design urban drainage infrastructure and other hydraulic structures. Since climate change effects are region-specific, site-specific evaluations are required to boost local resilience to future extreme precipitation events. As a result, the clear differences in the future IDF curve compared to the existing IDF curve developed based on the historical observed data are needed in order to incorporate such information into urban drainage systems to design climate-resilient infrastructures to mitigate the possible hazardous impact of climate change on infrastructure. Therefore, the objective of this paper is twofold: (i) to derive the future IDF curve for the town of Willoughby (HUC-12) and the Buffalo region using both CMIP5 and CMIP6 models, and (ii) to compare and evaluate the differences in the projected precipitation IDF curves between the two sets of models. The purpose of this paper is to give a thorough understanding of the vulnerabilities associated with future changes in precipitation patterns on the Lake Erie coast.

2. Theoretical Description

2.1. CMIP5 Data Set

Multiple Representative Concentration Pathways (RCPs) experiments have been used with the North American Coordinated Regional Climate Downscaling Experiment (NA-CORDEX) and CMIP5 model data to build various meteorological information at the regional scale [40]. The major benefit of NA-CORDEX is that it uses general circulation models (GCMs) to drive simulations of various regional climate models (RCMs) at higher resolutions (e.g., 50 × 50 km) [41]. Such information is critical for accurately modeling the climate of regions with a complicated topography and small-scale events. The limitations of GCMs, i.e., coarser resolution (100 × 100 km), are often resolved by regional climate model-based projections [42]), further substantiating the assertion that RCMs are frequently used to address the shortcomings of GCMs. Using the western US as an example, [43] demonstrated how the RCM reflects the actual spatial variability in precipitation and snowfall using regional climate simulations at 40 km spatial resolution for the period (2040–2060).

In places with a complicated topography where small-scale phenomena are critical for accurately representing the region's climate, NA-CORDEX's use of GCMs to drive the simulations of several RCMs is a major advantage. The NA-CORDEX has provided simulated precipitation data for two periods, including historical (1980–2005) and future (2006–2099), for CMIP5.

2.2. CMIP6 Data Set

The CMIP6 models provide multi-model climate forecasts based on alternative scenarios that are influenced by a new set of emissions-shared socioeconomic pathways (SSPs) and land use scenarios that are directly related to societal concerns about adaptation, mitigation, or the consequences of climate change [44]. By standardizing socioeconomic and technical assumptions across models, this new paradigm closed crucial gaps in CMIP5's intermediate forcing levels and allowed for a more thorough examination of various pathways. The World Climate Research Program (WRCP) has provided simulated precipitation data for two periods, including historical (1980–2014) and future (2015–2099) for CMIP6.

NA-CORDEX and WCRP both have the goal of improving our understanding of the Earth's climate and its potential future changes [45–48]. While NA-CORDEX focuses on producing high-resolution climate projections specifically for North America, WCRP is broader in its focus, coordinating and conducting research on the fundamental science of the Earth's climate system and its interactions with the environment globally [48,49].

In addition to retaining the CMIP5 emission trajectories RCP2.6, RCP4.5, RCP6.0, and RCP8.5, the CMIP6 data also contain three new emission paths: RCP1.9, RCP3.4, and RCP7.0. As a result, the new scenarios combine SSP1, SSP2, SSP3, SSP4, and SSP5 of five socioeconomic paths with various levels of emissions to form seven future SSP-RCP scenarios, which include SSP1-1.9 (a very low range of scenarios) to SSP5-8.5 (a combination of high societal vulnerability and a high forcing level). The combination of RCPs and shared Socioeconomic Pathways (SSPs) is expected to make future scenarios more realistic.

It is expected that CMIP6 simulations can reproduce historical climate variables, represent smaller biases in sea surface temperature, and be more skillful in capturing the precipitation pattern. The climate model simulations from CMIP6 seem to be more reliable than the earlier CMIP5 in various aspects. Different scientists have reported the limitations of CMIP5, especially in various scenarios and GCM output, due to the large reduction in atmospheric aerosol emissions for RCP scenarios [50]. Since more realistic results can be expected at various locations, especially for extreme precipitation, the application of the latest CMIP6 climate data is more crucial for storm sewer drainage systems. In addition, the multimodal median of CMIP6 (CMIP6-MMM) is expected to perform better than the individual model. Therefore, several models were used for IDF curve development.

2.3. Bias Correction

Before any form of analysis, it is crucial to retrieve the data from climate models such as RCMs and GCMs for a specific location based on latitude and longitude. Since it is not unusual for climate models to produce frequently skewed results, it is necessary to adjust the climate data for bias. This bias correction is essential and recommended in several studies [51–53] to ensure that the bias-corrected data used in hydrological modeling and decision-making processes are accurate and reliable, leading to appropriate results [54–56]. In a study conducted by [57], Standardized Reconstruction (Z) and the Quantile Mapping Method (Q) demonstrated superior simulation skills compared to alternative methods, including Mean Bias-remove (U), Multiplicative Shift (M), Regression (R), and Principal Component Regression (PCR). The Quantile Mapping Method, widely adopted in diverse investigations [58], has emerged as a globally acclaimed choice. Its extensive use is attributed to its proven ability to enhance the precision and consistency of statistical studies, making it the method of choice in this context. Quantile mapping is a technique used to reconcile climate model data with historical observations by transforming the model's data distribution to match the observational data distribution, thereby reducing biases and increasing accuracy in climate predictions [59–65]. The efficiency of this technique has been tested and found to be effective in improving the accuracy for hydrological modeling and decision-making [66–68]. Quantile mapping, which is a well-known approach for bias correction, has been used in generating downscaled GCM data sets for both the United States and global land regions [69]. The approach aims to closely mimic both the statistical distributions of the observed variable and the climatic variable [69,70].

2.4. Intensity Duration Frequency (IDF) Curves

In the 1940s, Gumbel developed the Gumbel distribution, also known as the extreme-value Type I distribution [71]. Since the Gumbel distribution is generally used for the distribution of the maximum of a sample, it is one of the extreme distributions.

The Gumbel theory of distribution is the preferred choice for analyzing intense rainfall events due to its simplicity [72,73] for analyzing extreme events. The Gumbel method has been found to be one of the most credible approaches for hydraulic design, particularly when dealing with high-intensity events due to its focus on extreme occurrences. Several past studies have shown that Gumbel's distribution may reliably anticipate flood magnitudes, enhancing the safety of the design [74–77]. Similarly, ISFRAM (2015) [78] suggests the use of the Gumbel method in practical applications due to its improved accuracy results compared to the Log-Pearson Type III distribution. Nonetheless, the Gumbel distribution was found to be the best fit for the Kelantan River Basin, outperforming the Log-Pearson Type III and normal distributions [79]. It has been observed that the application of Gumbel distribution improves the efficient design and utilization of infrastructure facilities, resulting in improved public safety and cost savings [76].

The following equation [20] calculates the maximum precipitation P_T (in mm) for each duration with a specified return period T (in years).

$$P_T = P_{avg} + KS \qquad (1)$$

where P_{avg} is the average of the maximum precipitation corresponding to a given duration, as stated by:

$$P_{avg} = \frac{1}{n}\sum_{i=0}^{n} P_i \qquad (2)$$

where "P_i" is the specific extreme value of rainfall and "n" is the number of events or years of data available.

K is the Gumbel frequency factor as given by:

$$K = -\frac{\sqrt{6}}{\pi} * \left(0.5772 + \ln\left(\ln\left(\frac{T}{T-1}\right)\right)\right) \quad (3)$$

and S is the standard deviation, which is computed using Equation (4):

$$S = \left[\frac{1}{n-1}\sum_{i=0}^{n}(P_i - P_{avg})^2\right]^{1/2} \quad (4)$$

where S is the standard deviation. The frequency factor (K), when multiplied by the standard deviation, provides the deviation of a specific rainfall event (for a certain period T) from the average. The rainfall intensity (i) in mm/h can then be calculated using this factor and the standard deviation, as follows:

$$I_t = \frac{P_T}{T_d} \quad (5)$$

where T_d is the duration in hours.

While the Gumbel distribution has been popularly used, it has some drawbacks as it is characterized by constant skewness because of its non-tailed distribution. While the modeling using the Gumbel distribution is widely used due to its simplicity, the consideration of independent variables such as the probability of selecting one variable vs selecting another independent variable must be considered carefully [80].

3. Materials and Methods

3.1. Study Area

The Lake Erie region, encompassing the towns of Willoughby in Lake County, Ohio, and Buffalo in Erie County, New York, presents a dynamic climate characterized by distinct seasonal variations and notable precipitation patterns (Figure 1). This geographical area, located in the United States, is situated along the eastern edge of the Great Lakes. The region's climatic conditions and precipitation trends have been the focus of investigation, revealing important insights into changing weather dynamics.

Willoughby is nestled within Lake County, Ohio, and boasts geographical coordinates of 41°38′45″ N latitude and 81°24′35″ W longitude. Covering an area of 26.78 km², with 26.55 km² of land and 0.23 km² of water, the town showcases a blend of natural and aquatic surroundings. The climate of Willoughby exhibits a clear division between its hot, muggy summers and cold, snowy winters. During the warm season, average daily high temperatures soar above 23 °C, peaking at around 28 °C, while lows hover around 20 °C. In contrast, the cold season sees average daily highs of 7 °C with lows plunging to −5 °C, and the high temperatures barely reaching 2 °C. Rainfall is a consistent feature throughout the year, with September holding the record for the wettest month, experiencing an average of 78mm of rain. In contrast, February marks the driest period with an average of 29 mm of rain. This climatic data, meticulously recorded from 2015 to 2023, provides a comprehensive understanding of Willoughby's distinctive weather patterns. The region has witnessed an increase in temperature and rainfall intensity, coupled with a rising trend in extreme weather events, as documented in historical climate data [81].

Similarly, Buffalo, situated in Erie County, New York, is another integral part of the Lake Erie region (Figure 1). The city is positioned at 42°53′11″ N latitude and 78°52′41″ W longitude, encompassing an expansive area of 26.78 km². Buffalo's climate exhibits a distinct contrast between its warm, partly cloudy summers and its freezing, snowy, windy, and mostly cloudy winters. The warm season witnesses average daily high temperatures exceeding 21 °C, peaking at approximately 26 °C, while daily lows stay above 18 °C. In the cold season, average daily high temperatures barely reach 5°C, with lows plummeting to −6 °C and highs only reaching −0 °C. Like Willoughby, Buffalo experiences consistent rainfall throughout the year. September stands as the wettest month, with an average of 72 mm of rain, while February represents the driest month, recording an average of 19 mm of rain. The meticulously recorded climate patterns from 2015 to 2023 provide an in-depth understanding of Buffalo's unique meteorological characteristics (Weather Spark).

Figure 1. The study area showing the two cities, the town of Willoughby and Buffalo, near the Lake Erie Coast.

3.2. Climate Model Data

Past observed precipitation as well as RCM and GCM output data for different models from CMIP5 and CMIP6, respectively, are included in the precipitation data used with historical data and future data under various scenarios. For the town of Willoughby, the historical observations were collected from the Hopkins International Airport station in Cleveland, Ohio, which is 50 km away from the study site.

Similarly, for Buffalo, the observed historical precipitation data were obtained from Buffalo Niagara International Airport. The 1 h precipitation data from the station were utilized to prepare the observed historical data. This station was selected because it provides long records of continuous data sets without any significant interruption.

The historical period from 1980–2019 was considered the baseline period and referred to as Time Span-1 (TS-1), whereas the future period was divided into two time spans 2020–2059 as the near future (TS-2), and 2060–2099 as the far future (TS-3). This was intended because the most recent data were available for the period of 1980–2019 and separating the future period into smaller time frames would allow for a more detailed analysis of potential changes in precipitation patterns with equal time for the near future and distant future, providing a more comprehensive and holistic view of the potential changes in precipitation patterns over time.

For this study, three RCMs with model-generated historical data and RCP8.5 scenarios for each CMIP5 were selected from https://na-cordex.org/, accessed on 1 January 2022. Similarly, for CMIP5, three GCMs with historical and four SSP scenarios, namely, SSP126, SSP245, SSP370, and SSP585, were chosen to examine the potential increase in future precipitation. The projected simulations of precipitation in the future were obtained from three climate models contributing to CMIP6: https://esgf-node.llnl.gov/search/cmip6/, accessed on 1 January 2022. Together, the four CMIP6 scenarios and RCP8.5 from CMIP5 provide historical background and future predictions for the study, with the former serving as a historical baseline for the worst-case climatic scenario and the latter as an attempt to give insights into possible future orientations. The fundamental information for the three selected CMIP5 and CMIP6 models is reported in Table 1.

Table 1. Description of the climate models and climate change scenarios used in the study.

		CMIP5				
Source	Source ID	GCM	Scenario	Grid	Frequency	Resolution
NA-CORDEX	WRF	GFDL-ESM2M	hist, RCP8.5	NAM-22	1 h	$0.44° \times 0.44°$
		HadGEM2-ES				
		MPI-ESM-LR				
		CMIP6				
Source	Source ID		Experiment ID	Variant Label	Frequency	Resolution
WRCP	MIROC6		hist, ssp126, ssp245, ssp370, ssp585	r1i1p1f1	1 h	$1.4° \times 1.4°$
	CNRM-CM6-1-HR			r1i1p1f2		$0.5° \times 0.5°$
	CNRM-ESM2-1			r1i1p1f2		$1.4° \times 1.4°$

The ability of the climate models to contribute hourly data was a primary factor in their selection for this study. In addition, these models have already been widely adopted in the research community, ensuring comparability and consistency with the existing literature and increasing the credibility and reliability of the research. Furthermore, a more noted comprehension of the potential effects of climate change on precipitation patterns was made possible by including both historical and different future scenarios. Such climate scenarios help us understand how precipitation responds to changes in greenhouse gas emissions, which is useful for planning responses to climate change.

3.3. Bias Correction of Raw Data

In this study, the climate data from the climate model were corrected against the observed daily data using the quantile mapping bias-correction approach, also known as probability mapping or distribution mapping.

In this study, the Climate Data Bias Corrector (CDBC) tool developed by Gupta et al. 2019 [82] was used to complete the bias correction. The effectiveness of the tool and its efficacy for bias corrections have been demonstrated in various studies [83–87].

3.4. Development of the IDF Curve

After the raw climate model data were bias-corrected, the next step was to develop an IDF curve using the Gumbel Extreme Distribution method. For this, the raw data were analyzed to determine the maximum precipitation intensity for each year from 1980 to 2099 for different rainfall durations (1 h, 2 h, 6 h, 12 h, and 24 h) at various return periods including 2, 5, 10, 25, 50, and 100 years. For each return period, the intensity of the precipitation for each duration was calculated using the average of the maximum precipitation and the standard deviation corresponding to the time frame. In addition, the Gumbel frequency factor, or K-factor, was used to calculate the probability of the occurrence of an event of a given magnitude.

Finally, the IDF curves were developed by plotting the intensity of precipitation against the duration of the rainfall for each return period using the Multi-Model Ensemble (MME) mean method.

4. Results and Discussion

Since the major objective of this study was to develop IDF curves for both CMIP5 and CMIP6 models and evaluate the differences between them, simulated precipitation data for historical and future periods were used. The data were adjusted to reduce biases using the quantile mapping approach, and the results of the bias correction process are presented in terms of the mean and standard deviation. The comparison of the average and variability (standard deviation) in both the CMIP5 and CMIP6 models, both before and after bias correction for the towns of Willoughby and Buffalo, has been presented in Tables 2 and 3, respectively.

Table 2. Bias in terms of mean and standard deviation (st. dev.) before and after bias correction for CMIP5 and CMIP6 models for the baseline period (TS-1: 1980–2019) for the town of Willoughby.

Statistics	CMIP5 Models						
	Observed	GFDL-ESM2M		HadGEM2-ES		MPI-ESM-LR	
		Before	After	Before	After	Before	After
Average (mm)	2.67	4.04	2.6	3.06	2.68	3.56	2.51
St. Dev. (mm)	6.62	7.24	6.78	7.11	6.58	6.79	6.39
	CMIP6 Models						
Statistics	Observed	GFDL-ESM2M		HadGEM2-ES		MPI-ESM-LR	
		Before	After	Before	After	Before	After
Average (mm)	2.67	3.04	2.69	3.36	2.65	3.30	2.68
St. Dev. (mm)	6.62	6.06	6.77	6.60	6.71	6.13	6.77

Table 3. Bias in terms of mean and standard deviation (st. dev.) before and after bias correction for CMIP5 and CMIP6 models for the baseline period (TS-1: 1980–2019) for Buffalo.

Statistics	CMIP5 Models						
	Observed	GFDL-ESM2M		HadGEM2-ES		MPI-ESM-LR	
		Before	After	Before	After	Before	After
Average (mm)	1.97	3.65	1.93	3.84	2.17	4.01	2.35
St. Dev. (mm)	5.40	7.06	5.66	7.64	6.52	8.05	7.23
	CMIP6 Models						
Statistics	Observed	GFDL-ESM2M		HadGEM2-ES		MPI-ESM-LR	
		Before	After	Before	After	Before	After
Average (mm)	1.97	3.16	1.43	3.27	1.44	3.10	1.44
St. Dev. (mm)	5.40	6.29	4.90	6.11	4.93	6.24	4.87

4.1. CMIP5

A comprehensive analysis of the IDF curves, assembling three CMIP5 models for the RCP8.5 scenario, provides a visual and mathematical representation of the changes in IDF. The IDF curve for the historical baseline period and the near future is presented in Figure 2. The analysis has revealed a considerable rise in rainfall intensity in the near future compared to the historical baseline period, with a projection of 9–39% for various durations and return periods for the town of Willoughby while the results from Buffalo demonstrated an elevation projecting an increase of 4% to 27% across various durations and return periods. It is important to note that the percentage increase was not linear, rather large variations were detected for longer durations and higher return periods. The non-linear nature of the increase in rainfall intensity implies that extreme rainfall events are projected to become even more intense in the near future. The analysis of the trend of precipitation indicated that the increasing pattern observed in the near future could be expected to further increase in the far future, as shown in Figure 3. Precipitation is expected to become more intense and increase by 20–55% compared to the historical baseline period for the town of Willoughby. However, such projections for the Buffalo region were relatively more and indicated a potential surge in precipitation intensity by 38% to 84% relative to the baseline historical period. Instances of extreme rainfall, both in shorter and longer return periods, have surged in both frequency and intensity. This tendency raises concerns about the likelihood of more frequent flash floods and stormwater flooding in the future. To further illustrate this point, Figure 4 presents a graphical comparison of the percentage change in intensity between different time frames. The study revealed that until the final years of the century, hourly precipitation with a 100-year return period would increase by almost 24% and 53% for the town of Willoughby and Buffalo region, respectively. Hourly precipitation intensity could be expected to follow a predictable trend, increasing by 16% in the near future and by a much larger percentage (29%) in the far future for the town of Willoughby while Buffalo exhibited a 17% elevation in the near future and a notably larger increment of 38% in the far future. These divergent tendencies highlight the value of looking across multiple time periods when analyzing climate projections for the future, which provide important clues that help us piece together how precipitation patterns may shift over time. This increasing trend of precipitation in the Lake Erie region that we found in our study is consistent with the findings of previous research [88,89] on the Great Lakes region using CMIP5 models. Notably, the same models were used in the former studies, which suggests the consistency and reliability of our findings.

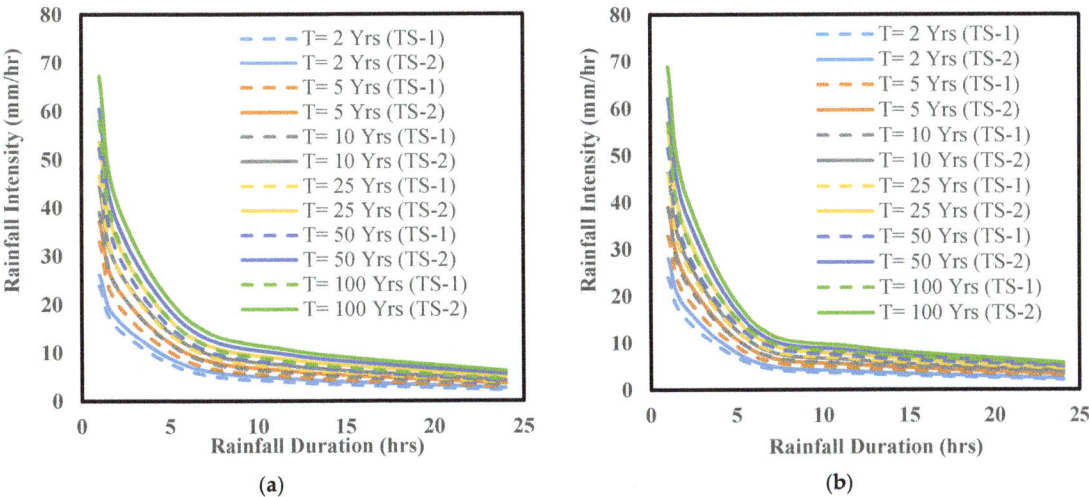

Figure 2. IDF curves for the baseline period from 1980 to 2019 (TS-1) vs. the near future from 2020 to 2059 (TS-2) considering a 2, 5, 10, 25, 50, and 100-year return period ensembling three CMIP5 RCP8.5 models for (**a**) the town of Willoughby (left panel), and (**b**) the city of Buffalo (right panel).

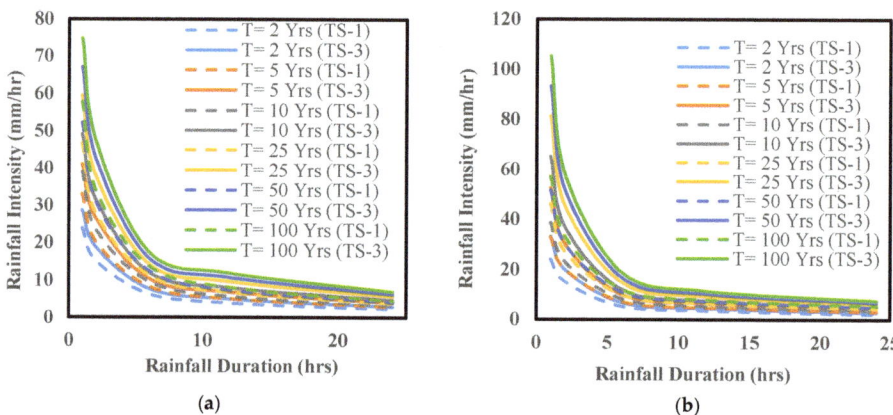

Figure 3. IDF curves for the baseline period from 1980 to 2019 (TS-1) vs. the far-future period from 2060 to 2099 (TS-3) considering a 2, 5, 10, 25, 50, and 100-year return period ensembling three CMIP5 RCP8.5 models for (**a**) the town of Willoughby (left panel), and (**b**) the city of Buffalo (right panel).

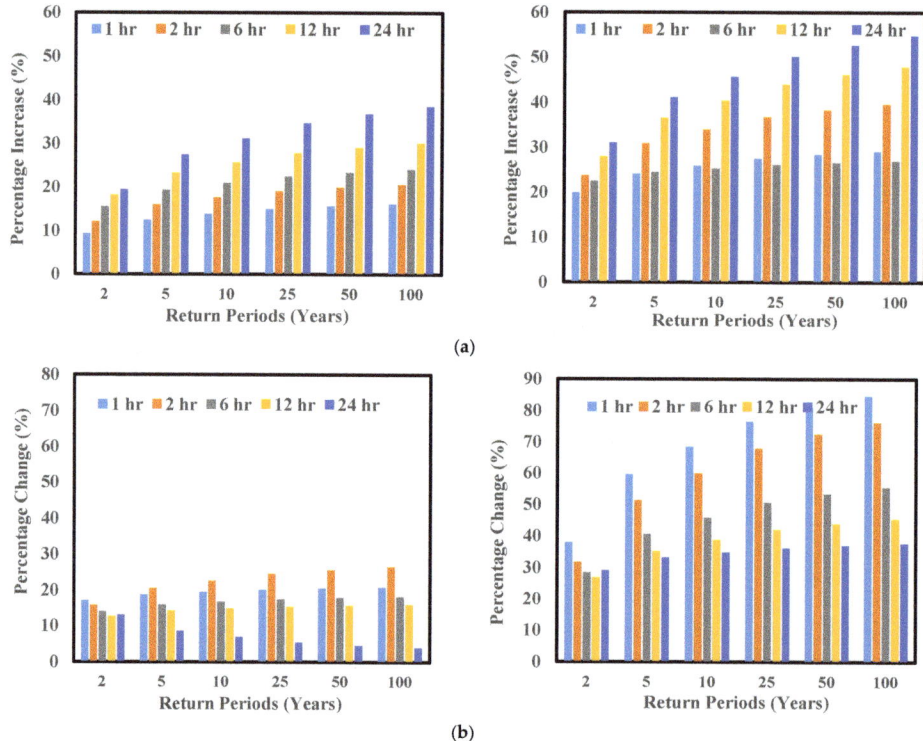

Figure 4. The left and right figures of the upper panel (**a**) represent the town of Willoughby, whereas the left and right figures of the lower panel (**b**) represent the city of Buffalo. The left graph of the upper panel shows the comparison of the percentage change in the rainfall intensity between the baseline period from 1980 to 2019 (TS-1) vs. the near future from 2020 to 2059 (TS-2), whereas the right panel shows the baseline period from 1980 to 2019 (TS-1) vs. the far future period from 2060 to 2099 (TS-3) for the town of Willoughby for different return periods and rainfall durations of CMIP5 RCP8.5. The exact interpretation is true for the city of Buffalo in the lower panel.

4.2. CMIP6

In this study, the most recent climate model, CMIP6, agreed with the earlier versions of the model, i.e., CMIP5, in predicting an increase in precipitation. The findings indicated that even with the lowest SSP scenario (SSP126), there would be an increase in rainfall intensity in the near future, with a range of 3–19% for the town of Willoughby and 4% to 54% for Buffalo (Figure 5). It is interesting to note that the magnitude of the increase in the intensity of rainfall could be expected to vary across different durations and return periods. For the town of Willoughby, the two-year return period for a six-hour rainfall showed the lowest percentage increase in intensity. On the other hand, the return period of 100 years for rainfall lasting 2 h showed the largest percentage increase in intensity. However, for the Buffalo region, the smallest increase could be expected for a 24 h rainfall with a 2-year return period, whereas the largest increment could be expected for a 100-year return period for a 2-h duration rainfall. This trend persists in the far future (Figure 6), with the most pronounced increase as high as 56% anticipated for the 1-h duration of a 100-year return period.

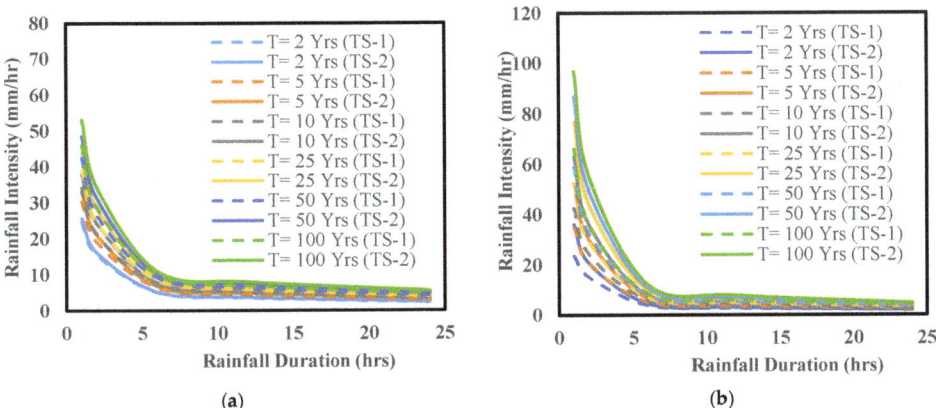

Figure 5. IDF curves for the baseline period from 1980 to 2019 (TS-1) vs. the near future from 2020 to 2059 (TS-2) considering a 2, 5, 10, 25, 50, and 100-year return period ensembling three CMIP6 SSP126 models for (**a**) the town of Willoughby (left panel), (**b**) city of Buffalo (right panel).

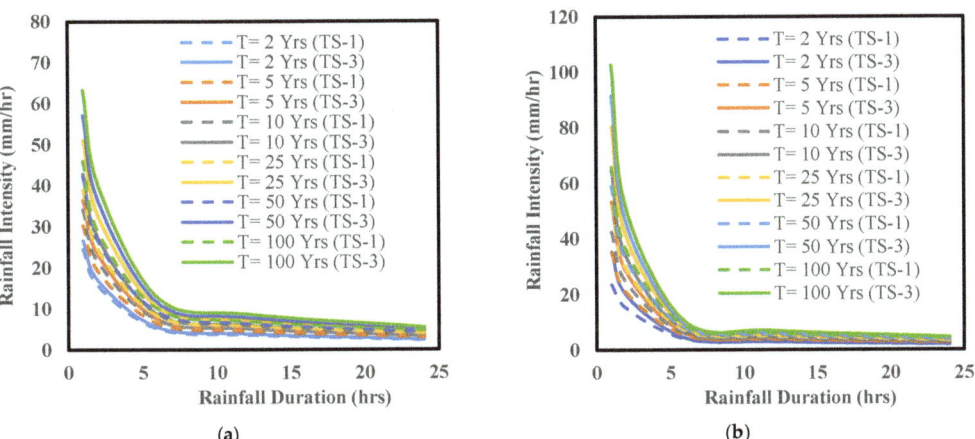

Figure 6. IDF curves for the baseline period from 1980 to 2019 (TS-1) vs. the far future from 2060 to 2099 (TS-3) considering a 2, 5, 10, 25, 50, and 100-year return period ensembling three CMIP6 SSP126 models for (**a**) the town of Willoughby (left panel), and (**b**) city of Buffalo (right panel).

The precipitation intensity for the near future can be expected to rise primarily for shorter durations (Figure 7). Similarly, this trend can be expected for the far future (Figure 8) suggesting a significant future increase in precipitation intensity for both the town of Willoughby and Buffalo, especially for shorter durations. It is interesting to report that a shorter duration of precipitation could be expected significantly in the Buffalo region compared to Willoughby. Comparing the near-future and historical baseline, this study indicated that precipitation intensity might double in the near future and triple in the far future for various durations and return periods. This disparity in trends emphasizes the significance of evaluating various time segments when analyzing future climate projections, enabling a deeper understanding of the evolving patterns of precipitation.

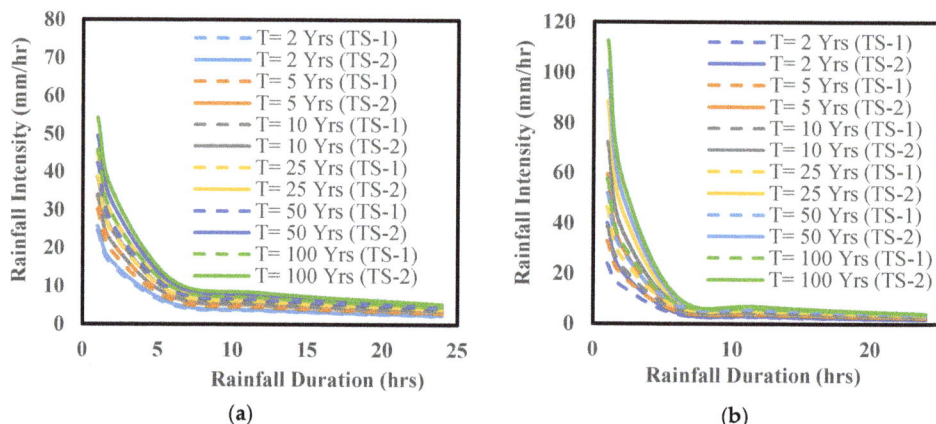

Figure 7. IDF curves for the baseline period from 1980 to 2019 (TS-1) vs. the near future from 2020 to 2059 (TS-2:) considering a 2, 5, 10, 25, 50, and 100-year return period ensembling three CMIP6 SSP245 models for (**a**) the town of Willougby (left panel), (**b**) city of Buffalo (right panel).

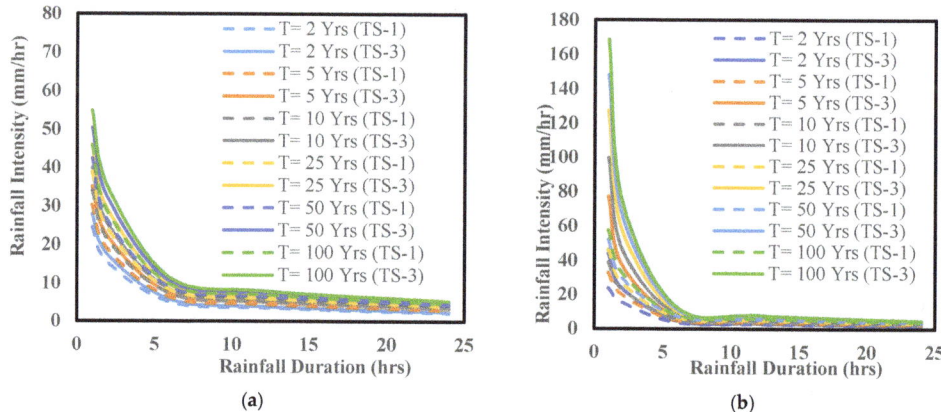

Figure 8. IDF curves for the baseline period (TS-1: 1980–2019) vs. the far future (TS-3: 2060–2099) considering a 2, 5, 10, 25, 50, and 100-year return period ensembling three CMIP6 SSP245 models for (**a**) the town of Willoughby and (**b**) city of Buffalo.

Likewise, the SSP370 scenario predicted intriguing insights about the future of precipitation intensity. In particular, hourly precipitation with a return period of two years is predicted to increase in intensity, with the lowest observed increase of 5% (Figure 9). The most significant increase in intensity, however, is expected for the 2-h duration of precipitation with a 100-year return period, which is projected to increase by 22% for the town of Willoughby. However, for the Buffalo region, it is projected to rise significantly to as high as 55% for a 1-h duration for a 2-year return period in the near future and by 94% for a 100-year return period in the far future. Comparisons of IDF curves for SSP370 near-future and far-future further underscore these trends (Figure 10).

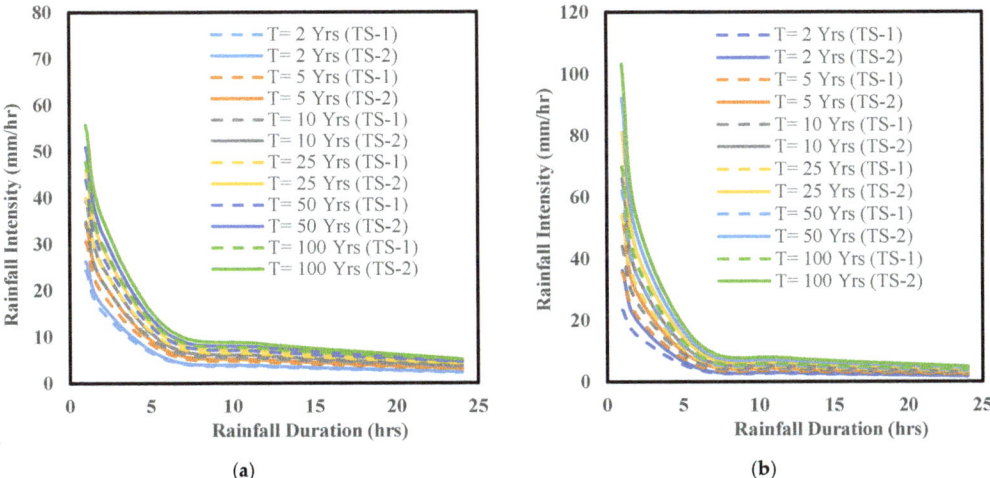

Figure 9. IDF curves for the baseline period (TS-1: 1980–2019) vs. the near future (TS-2: 2020–2059) considering a 2, 5, 10, 25, 50, and 100-year return period ensembling three CMIP6 SSP370 models for (**a**) the town of Willoughby (left panel) and (**b**) city of Buffalo (right panel).

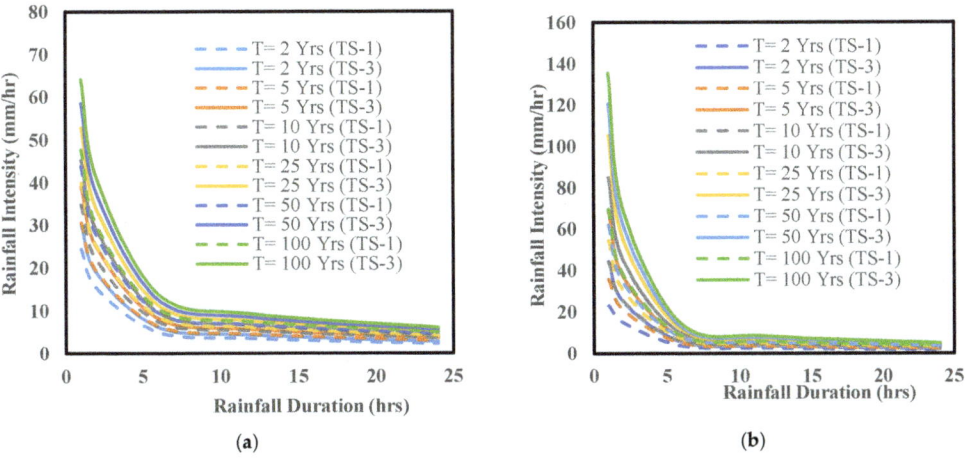

Figure 10. IDF curves for the baseline period (TS-1: 1980–2019) vs. the far future (TS-3: 2060–2099) considering a 2, 5, 10, 25, 50, and 100-year return period ensembling three CMIP6 SSP370 models for (**a**) the town of Willoughby (left panel) and (**b**) city of Buffalo (right panel). In the same manner, the SSP585 scenario under the CMIP6 model demonstrated an increase in precipitation intensity, with a projected range of 6–57% (Figure 11) and 19–140% (Figure 12) for the near-future and far-future, respectively, for various durations and return periods for both the town of Willoughby and the Buffalo region. The results showed that in the most catastrophic scenario (SSP585), hourly precipitation with a 100-year return period would rise by an average of approximately 24% in the future in the town of Willoughby and by around 80% in the Buffalo region (Figure 13).

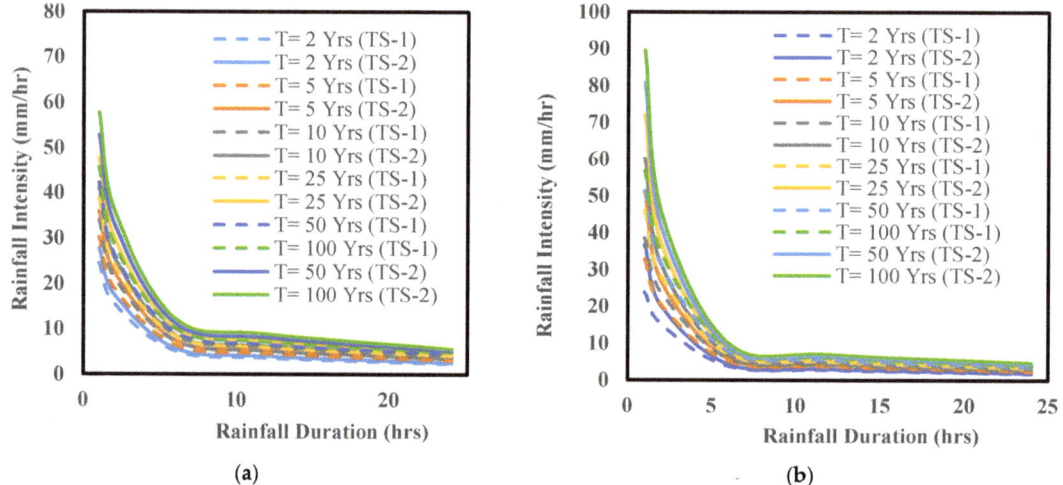

Figure 11. IDF curves for the baseline period from 1980 to 2019 (TS-1) vs. the near future from 2020 to 2059 (TS-2) considering a 2, 5, 10, 25, 50, and 100-year return period ensembling three CMIP6 SSP585 models for (**a**) the town of Willoughby (left panel) and (**b**) city of Buffalo (right panel).

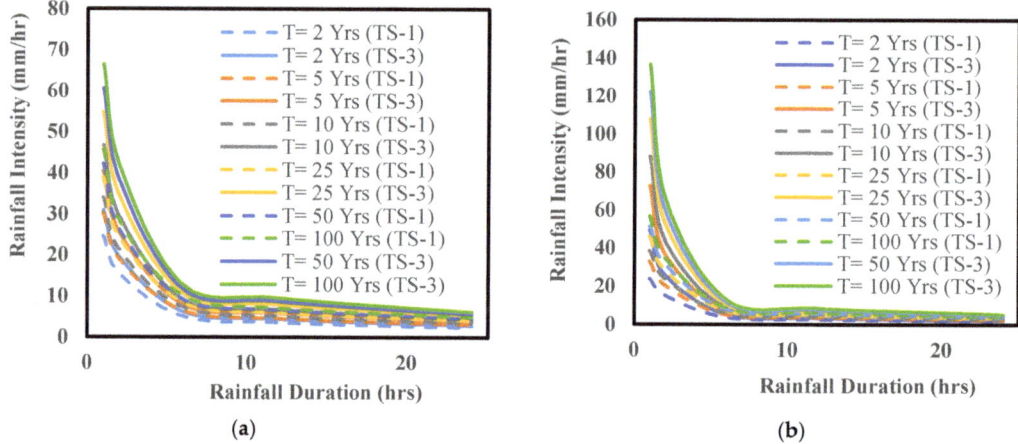

Figure 12. IDF curves for the baseline period from 1980 to 2019 (TS-1) vs. the far future from 2060 to 2099 (TS-3) considering a 2, 5, 10, 25, 50, and 100-year return period ensembling three CMIP6 SSP585 models for (**a**) the town of Willoughby (left panel), (**b**) city of Buffalo (right panel).

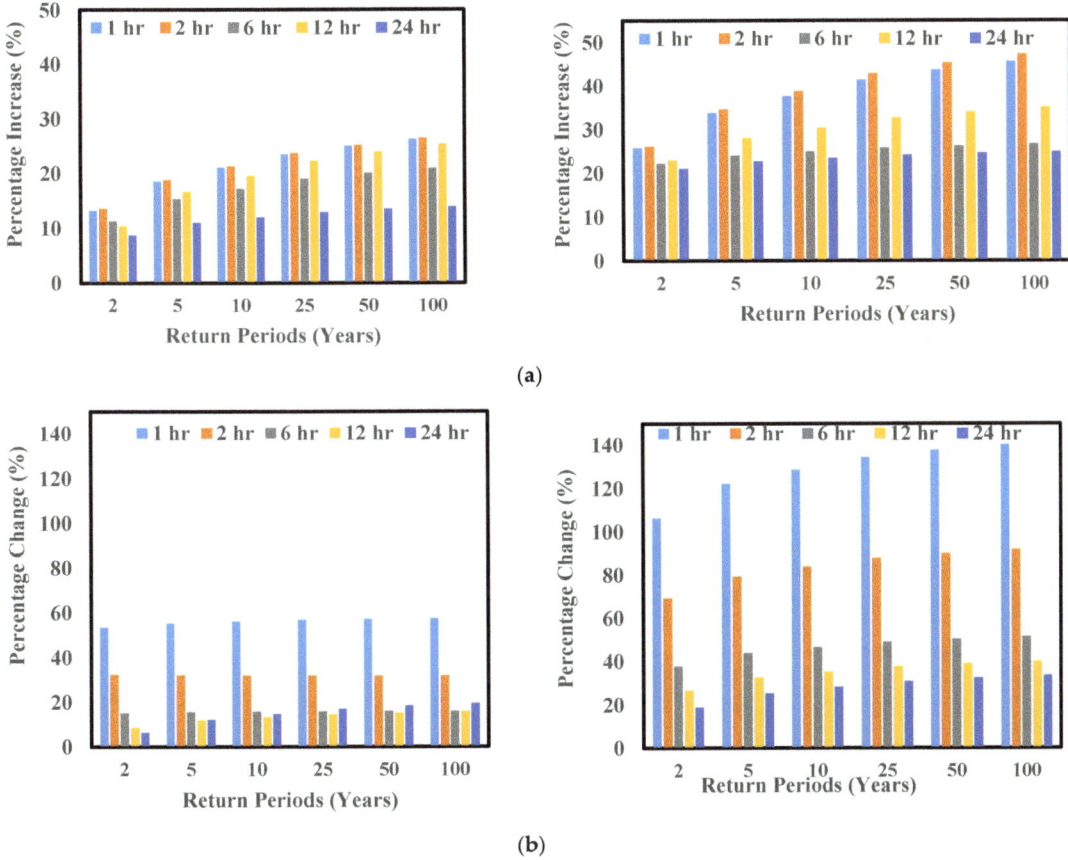

Figure 13. The upper panel (**a**) represents the town of Willoughby and the lower panel (**b**) represents the City of Buffalo. The upper panel shows the graphical comparison showing the rainfall intensity percentage change between the baseline period from 1980 to 2019 (TS-1) vs. the near future from 2020 to 2059 (TS-2), on the left, and the baseline period from 1980 to 2019 (TS-1) vs. the far future from 2080 to 2099 (TS-3), on the right, for different return periods and rainfall duration of CMIP6 SSP585 in the upper panel for the town of Willoughby (**a**). The exact similar comparison is presented in the lower panel (**b**) for the city of Buffalo.

Earlier research in the Great Lakes region [90] found that CMIP6 models' representations of precipitation would vary widely and contrast with those observed in real-world data sets. Nonetheless, the MIROC6 model used in this study agreed with the similar trend in increased precipitation presented by Minallah and Steiner, 2021 [91], indicating the reliability of the findings and validating the predictive ability of the model for future precipitation patterns.

4.3. CMIP5 vs. CMIP6: A Comparison

The comparison of the near future for both CMIP5 RCP8.5 and CMIP6 SSP585 has been presented in Figure 14. The study revealed that the increase in rainfall intensity for various duration hours and return periods for CMIP5 RCP8.5 and CMIP6 SSP585 was projected to be within the range of 9–39% and 20–55% for the near future and the far future, respectively, for the town of Willoughby, whereas much a higher range from 4% to 57% for near future, and 19% to 140% for far future could be expected in the Buffalo region across different durations and return periods. Similarly, Figure 15 shows the plots of the far future for both CMIPs, suggesting a significant increase in precipitation in Lake Erie in the future.

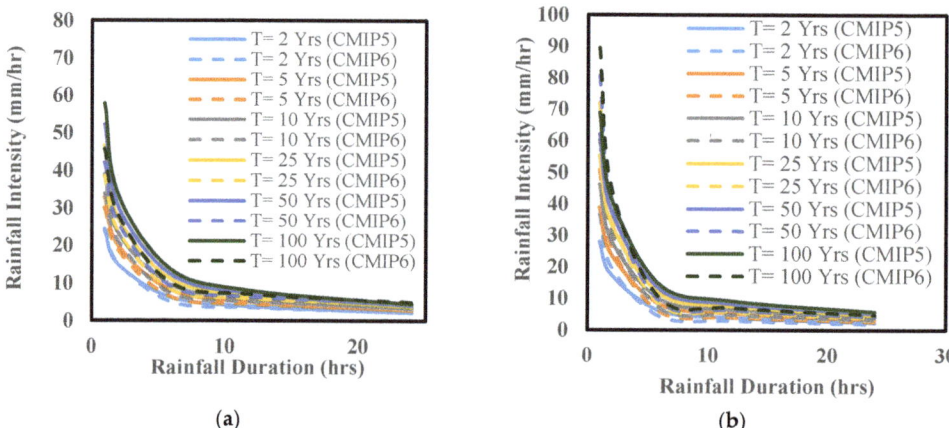

Figure 14. IDF curves for the near future period from 2020 to 2059 (TS-2:) considering a 2, 5, 10, 25, 50, and 100-year return period ensembling three CMIP5 RCP8.5 vs. CMIP6 SSP585 models for (**a**) the town of Willoughby (left panel), and (**b**) city of Buffalo (right panel).

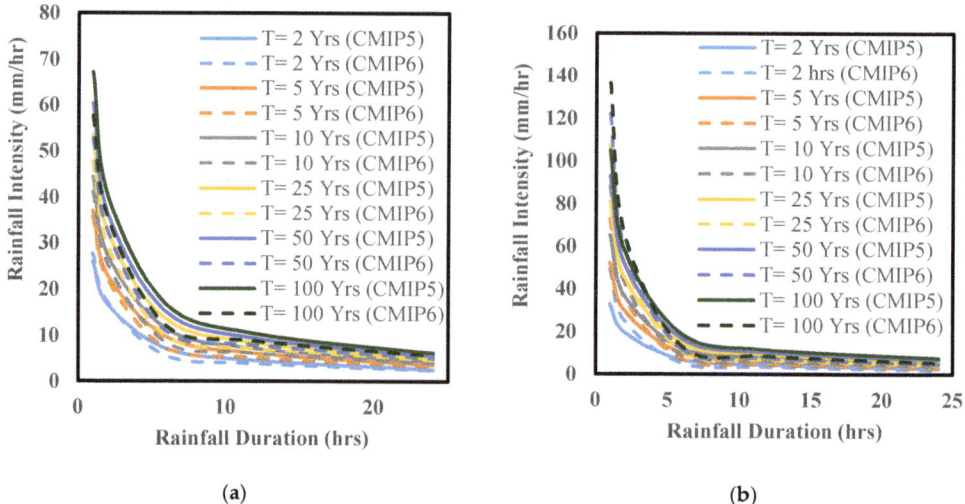

Figure 15. IDF curves for the far future period from 2060 to 2099 (TS-3) considering a 2, 5, 10, 25, 50, and 100-year return period ensembling three CMIP5 RCP8.5 vs. CMIP6 SSP585 models for (**a**) the town of Willoughby (left panel), and (**b**) city of Buffalo (right panel).

The CMIP6 models were assessed under various scenarios, including ssp126, ssp245, ssp370, and ssp585, revealing an increase in precipitation intensity from 2–22% for the near future and 6–40% for the far future across various rainfall durations and return periods for the town of Willoughby, whereas an increase in precipitation intensity from 2% to 95% for the near future and 3% to 192% for the far future was detected in the Buffalo region. Even though both CMIPs indicate an increase in precipitation intensity, the CMIP5 RCP8.5 stands out with a higher rainfall intensity than the CMIP6 SSP585, with an intensity range that exceeds the CMIP6 SSP585 by 28% across varying durations and return periods for the town of Willoughby. Interestingly, in the Buffalo region, the findings highlight the intriguing revelation that CMIP6 projects a more substantial increase in intensity for longer durations and higher return periods, a departure from CMIP5's trend. Notably, CMIP6 SSP scenarios emphasize significant changes, particularly for the future towards the century's end. Furthermore, the contrast between CMIP5 predictions for Willoughby and Buffalo, and CMIP6's higher prediction in Buffalo, underscores the complex regional variability.

During the analysis of meteorological data in this study, it was found that the intensity of precipitation would increase with longer return periods. The hourly precipitation is expected to see an increase in the upper range of extreme values in the future, specifically for the 95th percentile. This means that the most severe precipitation events that happen only 5% of the time are likely to become more intense, with a projected increase in the 95th percentile range of 5% to 24%, and the average hourly rainfall in the near future and far future is expected to increase by 7–28% by both CMIPs, which is a signal that communities need to prepare for the impacts of extreme weather events and invest in measures to build more resilient communities in the face of a changing climate. The results show that extreme weather events will become more intense, requiring sustainable development to mitigate urban flooding.

In Buffalo, the research foresees increased intensity of precipitation for extended return periods. The 95th percentile range, symbolizing severe rainfall events, is expected to triple in intensity by the century's close, reflecting the escalating severity of extreme weather occurrences. There was a discrepancy between the study's findings and the historical data reported by the National Oceanic and Atmospheric Administration (NOAA). One possible explanation for the discrepancies found in the data is that lakes were either simplified or left out entirely from the climate models used to examine potential future climate changes. The credibility of the CMIP5 models' projections was called into question by a previous study by [91], which found that most of them did not accurately capture the impact of the Great Lakes on the regional climate. Inaccurately simulating regional climate patterns requires a thorough understanding of the interaction between lakes, the atmosphere, and the land. This highlights the need for additional research on the accuracy of sub-daily data and casts doubt on the applicability of the models used.

However, it is essential to acknowledge the limitations of this study, such as the fact that it is based on the rainfall estimates of a single location and may not be representative of every location of the Lake Erie basin. Further studies could be accomplished to explore the limitations and make improvements, such as potential uncertainties in the models, data, and bias correction methods. Regardless, the results of this study provide valuable insights for urban planners, engineers, and decision-makers in developing sustainable flood control measures to mitigate the limitations. Additionally, there is a chance that the bias correction methods adopted in this study, data, and models will all have uncertainties that will affect the results. Further studies may explore these limitations and improve upon them.

5. Conclusions

This study aimed to develop and compare the IDF curves for future climate scenarios using data from two different climate phases, CMIP5 and CMIP6, in the Lake Erie basin to evaluate the impact of climate change on rainfall intensity. Since IDF curves are essential tools in designing effective drainage systems for any engineering project, simulated precipitation data from historical and future periods were used to develop the IDF curves and make comparisons. The data were adjusted to reduce biases using the quantile mapping approach, and the bias-corrected climate data were used to develop the IDF curves using the Gumbel Extreme Distribution Type I method.

The results for the town of Willoughby indicated a rise in precipitation intensity in the future, ranging from 9 to 55% across different rainfall durations and return periods for CMIP5 RCP8.5 and CMIP6 SSP585. The analysis of CMIP6 climate scenarios predicted a significant average increase of 27% in the intensity of hourly precipitation for the recurrence interval of 100 years in the future. Specifically, the SSP585 scenario projected an increase of

9–26% in the near future and 21–47% in the far future, while the RCP8.5 scenario predicted an increase of 11% to 24%, respectively. Even under the moderate climate change scenario of SSP126, it can be expected to have an increase (averaging 6%) in hourly precipitation intensity with a 2-year return period.

Similarly, the results from Buffalo indicated a rise in precipitation intensity in the future, ranging from 3 to 140% across different rainfall durations and return periods for CMIP5 RCP8.5 and CMIP6 SSP585. The analysis of CMIP6 climate scenarios predicts a significant average increase of 99% in the intensity of hourly precipitation for the recurrence interval of 100 years in the future. Specifically, the SSP585 scenario projects an increase of 6 to 57% in the near future and 19 to 140% in the far future, while the RCP8.5 scenario predicts increases of 4% to 27% in the near future and 27% to 85% across varying rainfall duration and return periods, respectively. Even under the moderate climate change scenario of SSP126, it can be expected to have an increase (averaging 50%) in hourly precipitation intensity with a 2-year return period.

The reliance on a limited number of models and scenarios may not account for the entire range of uncertainty in future scenarios. By using a variety of models and scenarios, it is possible to ensure a thorough representation of climate projections, which successfully addresses issues with the overestimation or underestimation of climate consequences. This approach reduces biases promoting particular climatic outcomes, improving the study's generalizability across different contexts and periods. In this context, further research is needed to understand the combined effects of these uncertainties with other sources of variability, such as land use change and natural internal weather variability. The large uncertainty is the output of the GCMs, and the RCMs also highlight the need for uncertainty analysis and probability-based IDF curves. Furthermore, the process of bias correction in a climate model is not immune to uncertainties. Climate scientists generally agree that extreme precipitation is intensifying; nevertheless, the phenomenon is complex and depends on a number of elements, including scale dependencies, physical considerations, regional variances, and confidence levels. To get accurate and trustworthy results for climate adaptation and infrastructure development, these aspects must be carefully taken into account during bias-correction processes. Future forecasts of climatic variables may be subjected to uncertainty after being corrected for bias in climate models, even when based on a single reference period. Hence, future climate results may vary depending on the reference period selected. These uncertainties impact research outcomes as they attempt to rectify inaccuracies in the data. The assumptions made during the correction process significantly influence the results and the manner in which data are rectified. Flaws in past data can lead to inaccuracies in future climate forecasts. Uncertainties arise when adjustments are made to data geographically or over different time spans. Different model responses to bias correction can leave behind residual errors. Additionally, in a changing environment, maintaining consistent climatic conditions becomes challenging, complicating future projections. Future research could explore various methods for responding to all these unknowns, such as using the professional analysis of climatologists or utilizing more robust statistical methods or machine learning algorithms. Therefore, in order to improve the current IDF curves in water infrastructure design, it is recommended that many time periods be taken into account in order to accommodate both immediate and long-term demands. In order to achieve dependable IDF curves, it is imperative to emphasize the implementation of strong statistical approaches in the processing of climatic data and bias correction. Furthermore, there is a need for a hybrid approach that makes use of many reference periods due to the complex nature of the interrelationships between climatic variables. To sum up, the study emphasizes the importance of updating the existing IDF curves that guide the design of water management infrastructure to account for the effects of climate change. It makes a substantial contribution to our understanding of how climate change impacts water management by providing information on shifting patterns of rainfall that are essential for developing adaptive infrastructure. The integration of many climate models and scenarios facilitates the development of adaptable infrastructure that can account for a range of potential outcomes. Concentrating on particular regions highlights the significance of customized planning for a range of climate impacts, and addressing uncertainties highlights the necessity of flexible infrastructure to handle a range of future possibilities, guaranteeing long-term climate preparedness.

Author Contributions: S.M. conducted an analysis and prepared the draft. S.S. provided a direction for the research and helped with the analysis and writing the manuscript. All authors have read and agreed to the published version of the manuscript.

Funding: This research was funded by Ohio Sea Grant.

Data Availability Statement: Data can be obtained with a request to the authors.

Acknowledgments: The authors would like to acknowledge Keely Davidson-Bennett from Chagrin River Watershed Partners (CRWP) for sharing her ideas and knowledge about the IDF curve of the region.

Conflicts of Interest: The authors declare no conflict of interest.

References

1. IPCC. *Climate Change 2007: Synthesis Report Summary for Policymakers*; An Assessment of the Intergovernmental Panel on Climate Change; IPCC: Geneva, Switzerland, 2007.
2. IPCC. *Climate Change 2014: Synthesis Report*; Longer Report; IPCC: Geneva, Switzerland, 2014.
3. Allen, M.R.; Ingram, W.J. Insight Review Articles 224. 2002. Available online: www.nature.com/nature (accessed on 1 January 2022).
4. Swain, D.L.; Wing, O.E.J.; Bates, P.D.; Done, J.M.; Johnson, K.A.; Cameron, D.R. Increased Flood Exposure Due to Climate Change and Population Growth in the United States. *Earth's Future* **2020**, *8*, e2020EF001778. [CrossRef]
5. Tabari, H. Climate change impact on flood and extreme precipitation increases with water availability. *Sci. Rep.* **2020**, *10*, 13768. [CrossRef] [PubMed]
6. Trenberth, K.E.; Dai, A.; Rasmussen, R.M.; Parsons, D.B. The Changing character of precipitation. *Bull. Am. Meteorol. Soc.* **2003**, *84*, 1205–1218. [CrossRef]
7. Haerter, J.O.; Berg, P. Unexpected rise in extreme precipitation caused by a shift in rain type? *Nat. Geosci.* **2009**, *2*, 372–373. [CrossRef]
8. Lenderink, G.; Van Meijgaard, E. Increase in hourly precipitation extremes beyond expectations from temperature changes. *Nat. Geosci.* **2008**, *1*, 511–514. [CrossRef]
9. Westra, S.; Fowler, H.J.; Evans, J.P.; Alexander, L.V.; Berg, P.; Johnson, F.; Kendon, E.J.; Lenderink, G.; Roberts, N.M. Future changes to the intensity and frequency of short-duration extreme rainfall. In *Reviews of Geophysics*; Blackwell Publishing Ltd.: Hoboken, NJ, USA, 2014; Volume 52, pp. 522–555. [CrossRef]
10. Prein, A.F.; Rasmussen, R.M.; Ikeda, K.; Liu, C.; Clark, M.P.; Holland, G.J. The future intensification of hourly precipitation extremes. *Nat. Clim. Chang.* **2017**, *7*, 48–52. [CrossRef]
11. Easterling, D.R.; Meehl, G.A.; Parmesan, C.; Changnon, S.A.; Karl, T.R.; Mearns, L.O. Climate Extremes: Observations, Modeling, and Impacts. *Science* **2000**, *289*, 2068–2074. [CrossRef]
12. Groisman, P.Y.; Knight, R.W.; Easterling, D.R.; Karl, T.R.; Hegerl, G.C.; Razuvaev, V.N. Trends in Intense Precipitation in the Climate Record. *J. Clim.* **2005**, *18*, 1326–1350. [CrossRef]
13. Kourtis, I.M.; Tsihrintzis, V.A. Update of intensity-duration-frequency (IDF) curves under climate change: A review. *Water Supply* **2022**, *22*, 4951–4974. [CrossRef]
14. Cook, L.M.; McGinnis, S.; Samaras, C. The effect of modeling choices on updating intensity-duration-frequency curves and stormwater infrastructure designs for climate change. *Clim. Chang.* **2020**, *159*, 289–308. [CrossRef]
15. Thakali, R.; Kalra, A.; Ahmad, S. Understanding the effects of climate change on urban stormwater infrastructures in the Las Vegas Valley. *Hydrology* **2016**, *3*, 34. [CrossRef]
16. Guo, Y.; Asce, M. Updating Rainfall IDF Relationships to Maintain Urban Drainage Design Standards. *J. Hydrol. Eng.* **2006**, *11*, 506–509. [CrossRef]
17. Liu, L. The dynamics of early-stage transmission of COVID-19: A novel quantification of the role of global temperature. *Gondwana Res.* **2023**, *114*, 55–68. [CrossRef] [PubMed]
18. Martel, J.-L.; Brissette, F.P.; Lucas-Picher, P.; Troin, M.; Arsenault, R. Climate Change and Rainfall Intensity–Duration–Frequency Curves: Overview of Science and Guidelines for Adaptation. *J. Hydrol. Eng.* **2021**, *26*, 03121001. [CrossRef]
19. Mohammed, A.; Dan'Azumi, S.; Modibbo, A.A.; Adamu, A.A. Development of Rainfall Intensity Duration Frequency (IDF) Curves for Design of Hydraulic Structures in Kano State, Nigeria. *Platform* **2021**, *5*, 10–22.
20. Elsebaie, I.H. Developing rainfall intensity–duration–frequency relationship for two regions in Saudi Arabia. *J. King Saud Univ.-Eng. Sci.* **2012**, *24*, 131–140. [CrossRef]

21. Kundwa, M.J. Development of Rainfall Intensity Duration Frequency (IDF) Curves for Hydraulic Design Aspect. *J. Ecol. Nat. Resour.* **2019**, *3*, 1–14. [CrossRef]
22. Rashid, M.; Faruque, S.; Rashid, M.M.; Faruque, S.B.; Alam, J.B. Modeling of Short Duration Rainfall Intensity Duration Frequency (SDR-IDF) Equation for Sylhet City in Bangladesh Statistical Downscaling of GCM Outputs to Rainfall View Project Extreme Sea Level Variations along the U.S. Coastlines View Project Modeling of Short Duration Rainfall Intensity Duration Frequency (SDR-IDF) Equation for Sylhet City in Bangladesh. 2. 2012. Available online: http://www.ejournalofscience.org (accessed on 1 January 2022).
23. Singh, R.; Arya, D.S.; Taxak, A.K.; Vojinovic, Z. Potential Impact of Climate Change on Rainfall Intensity-Duration-Frequency Curves in Roorkee, India. *Water Resour. Manag.* **2016**, *30*, 4603–4616. [CrossRef]
24. Prodanovic, P.; Simonovic, S.P. The university of western ontario department of civil and environmental engineering. In *Water Resources Research Report*; John Wiley & Sons: Hoboken, NJ, USA, 2007.
25. Srivastav, R.K.; Schardong, A.; Simonovic, S.P. Equidistance Quantile Matching Method for Updating IDFCurves under Climate Change. *Water Resour. Manag.* **2014**, *28*, 2539–2562. [CrossRef]
26. Hess, J.J.; Malilay, J.N.; Parkinson, A.J. Climate Change. The Importance of Place. *Am. J. Prev. Med.* **2008**, *35*, 468–478. [CrossRef]
27. Hosseinzadehtalaei, P.; Tabari, H.; Willems, P. Climate change impact on short-duration extreme precipitation and intensity–duration–frequency curves over Europe. *J. Hydrol.* **2020**, *590*, 125249. [CrossRef]
28. Peck, A.; Prodanovic, P.; Simonovic, S.P. Rainfall intensity duration frequency curves under climate change: City of London, Ontario, Canada. *Can. Water Resour. J.* **2012**, *37*, 177–189. [CrossRef]
29. Trenberth, K.E. Changes in precipitation with climate change. *Clim. Res.* **2011**, *47*, 123–138. [CrossRef]
30. Rodríguez, R.; Navarro, X.; Casas, M.C.; Ribalaygua, J.; Russo, B.; Pouget, L.; Redaño, A. Influence of climate change on IDF curves for the metropolitan area of Barcelona (Spain). *Int. J. Climatol.* **2014**, *34*, 643–654. [CrossRef]
31. Shrestha, A.; Babel, M.S.; Weesakul, S.; Vojinovic, Z. Developing Intensity-Duration-Frequency (IDF) curves under climate change uncertainty: The case of Bangkok, Thailand. *Water* **2017**, *9*, 145. [CrossRef]
32. Shrestha, S.; Sharma, S. Assessment of climate change impact on high flows in a watershed characterized by flood regulating reservoirs. *Int. J. Agric. Biol. Eng.* **2021**, *14*, 178–191. [CrossRef]
33. Ghasemi Tousi, E.; O'Brien, W.; Doulabian, S.; Shadmehri Toosi, A. Climate changes impact on stormwater infrastructure design in Tucson Arizona. *Sustain. Cities Soc.* **2021**, *72*, 103014. [CrossRef]
34. Lopez-Cantu, T.; Prein, A.F.; Samaras, C. Uncertainties in Future U.S. Extreme Precipitation from Downscaled Climate Projections. *Geophys. Res. Lett.* **2020**, *47*, e2019GL086797. [CrossRef]
35. Chen, H.; Sun, J.; Lin, W.; Xu, H. Comparison of CMIP6 and CMIP5 models in simulating climate extremes. In *Science Bulletin*; Elsevier B.V.: Amsterdam, The Netherlands, 2020; Volume 65, pp. 1415–1418. [CrossRef]
36. Thibeault, J.M.; Seth, A. Changing climate extremes in the Northeast United States: Observations and projections from CMIP5. *Clim. Chang.* **2014**, *127*, 273–287. [CrossRef]
37. Ragno, E.; AghaKouchake, A.; Love, C.A.; Cheng, L.; Vahedifard, F.; Lima, C.H.R. Quantifying Changes in Future Intensity-Duration-Frequency Curves Using Multimodel Ensemble Simulations. *Water Resour. Res.* **2018**, *54*, 1751–1764. [CrossRef]
38. Cheng, L.; Aghakouchak, A. Nonstationary precipitation intensity-duration-frequency curves for infrastructure design in a changing climate. *Sci. Rep.* **2014**, *4*, 7093. [CrossRef] [PubMed]
39. Coelho, G.d.A.; Ferreira, C.M.; Johnston, J.; Kinter, J.L.; Dollan, I.J.; Maggioni, V. Potential Impacts of Future Extreme Precipitation Changes on Flood Engineering Design Across the Contiguous United States. *Water Resour. Res.* **2022**, *58*, e2021WR031432. [CrossRef]
40. Lee, J.W.; Hong, S.Y.; Chang, E.C.; Suh, M.S.; Kang, H.S. Assessment of future climate change over East Asia due to the RCP scenarios downscaled by GRIMs-RMP. *Clim. Dyn.* **2014**, *42*, 733–747. [CrossRef]
41. Rummukainen, M. Added value in regional climate modeling. *Wiley Interdiscip. Rev. Clim. Chang.* **2016**, *7*, 145–159. [CrossRef]
42. Park, J.H.; Oh, S.G.; Suh, M.S. Impacts of boundary conditions on the precipitation simulation of RegCM4 in the CORDEX East Asia domain. *J. Geophys. Res. Atmos.* **2013**, *118*, 1652–1667. [CrossRef]
43. Qian, Y.; Ghan, S.J.; Leung, L.R. Downscaling hydroclimatic changes over the western US based on CAM subgrid scheme and WRF regional climate simulations. *Int. J. Climatol.* **2010**, *30*, 675–693. [CrossRef]
44. O'Neill, B.C.; Tebaldi, C.; Van Vuuren, D.P.; Eyring, V.; Friedlingstein, P.; Hurtt, G.; Knutti, R.; Kriegler, E.; Lamarque, J.F.; Lowe, J.; et al. The Scenario Model Intercomparison Project (ScenarioMIP) for CMIP6. *Geosci. Model Dev.* **2016**, *9*, 3461–3482. [CrossRef]
45. Giorgi, F.; Jones, C.; Asrar, G.R. Addressing climate information needs at the regional level: The CORDEX framework. *WMO Bull.* **2009**, *58*, 175.
46. Gutowski, J.W.; Giorgi, F.; Timbal, B.; Frigon, A.; Jacob, D.; Kang, H.S.; Raghavan, K.; Lee, B.; Lennard, C.; Nikulin, G.; et al. WCRP COordinated Regional Downscaling EXperiment (CORDEX): A diagnostic MIP for CMIP6. *Geosci. Model Dev.* **2016**, *9*, 4087–4095. [CrossRef]
47. McGinnis, S.; Mearns, L. Building a climate service for North America based on the NA-CORDEX data archive. *Clim. Serv.* **2021**, *22*, 100233. [CrossRef]
48. Gutowski, W.J.; Ullrich, P.A.; Hall, A.; Leung, L.R.; O'Brien, T.A.; Patricola, C.M.; Arritt, R.W.; Bukovsky, M.S.; Calvin, K.V.; Feng, Z.; et al. The ongoing need for high-resolution regional climate models: Process understanding and stakeholder information. *Bull. Am. Meteorol. Soc.* **2021**, *101*, E664–E683. [CrossRef]

49. Kirtman, B.; Pirani, A. The State of the Art of Seasonal Prediction: Outcomes and Recommendations from the First World Climate Research Program Workshop on Seasonal Prediction. *Bull. Am. Meteorol. Soc.* **2009**, *90*, 455–458. [CrossRef]
50. Moss, R.H.; Edmonds, J.A.; Hibbard, K.A.; Manning, M.R.; Rose, S.K.; van Vuuren, D.P.; Carter, T.R.; Emori, S.; Kainuma, M.; Kram, T.; et al. The next generation of scenarios for climate change research and assessment. *Nature* **2010**, *463*, 747–756. [CrossRef] [PubMed]
51. Bruyère, C.L.; Done, J.M.; Holland, G.J.; Fredrick, S. Bias corrections of global models for regional climate simulations of high-impact weather. *Clim. Dyn.* **2014**, *43*, 1847–1856. [CrossRef]
52. Donat, M.G.; Lowry, A.L.; Alexander, L.V.; O'Gorman, P.A.; Maher, N. More extreme precipitation in the worldâ€™s dry and wet regions. *Nat. Clim. Chang.* **2016**, *6*, 508–513. [CrossRef]
53. Gao, X.-J.; Wang, M.-L.; Giorgi, F. Climate Change over China in the 21st Century as Simulated by BCC_CSM1.1-RegCM4.0. *Atmos. Ocean. Sci. Lett.* **2013**, *6*, 381–386. [CrossRef]
54. Maraun, D.; Shepherd, T.G.; Widmann, M.; Zappa, G.; Walton, D.; Gutiérrez, J.M.; Hagemann, S.; Richter, I.; Soares, P.M.M.; Hall, A.; et al. Towards process-informed bias correction of climate change simulations. *Nat. Clim. Chang.* **2017**, *7*, 764–773. [CrossRef]
55. Mehrotra, R.; Sharma, A. An improved standardization procedure to remove systematic low frequency variability biases in GCM simulations. *Water Resour. Res.* **2012**, *48*. [CrossRef]
56. Xu, Z.; Yang, Z.L. An improved dynamical downscaling method with GCM bias corrections and its validation with 30 years of climate simulations. *J. Clim.* **2012**, *25*, 6271–6286. [CrossRef]
57. Acharya, N.; Chattopadhyay, S.; Mohanty, U.C.; Dash, S.K.; Sahoo, L.N. On the bias correction of general circulation model output for Indian summer monsoon. *Meteorol. Appl.* **2013**, *20*, 349–356. [CrossRef]
58. Wood, A.W.; Leung, L.R.; Sridhar, V.; Lettenmaier, D.P. Hydrologic Implications of dynamical and statistical approaches to downscaling climate model outputs. *Clim. Chang.* **2004**, *62*, 189–216. [CrossRef]
59. Abatzoglou, J.T.; Brown, T.J. A comparison of statistical downscaling methods suited for wildfire applications. *Int. J. Climatol.* **2012**, *32*, 772–780. [CrossRef]
60. Chen, J.; Brissette, F.P.; Chaumont, D.; Braun, M. Finding appropriate bias correction methods in downscaling precipitation for hydrologic impact studies over North America. *Water Resour. Res.* **2013**, *49*, 4187–4205. [CrossRef]
61. Gudmundsson, L.; Bremnes, J.B.; Haugen, J.E.; Engen-Skaugen, T. Technical Note: Downscaling RCM precipitation to the station scale using statistical transformations – A comparison of methods. *Hydrol. Earth Syst. Sci.* **2012**, *16*, 3383–3390. [CrossRef]
62. Maraun, D. Bias Correcting Climate Change Simulations—A Critical Review. In *Current Climate Change Reports*; Springer: Berlin/Heidelberg, Germany, 2016; Volume 2, pp. 211–220. [CrossRef]
63. Maraun, D.; Wetterhall, F.; Ireson, A.M.; Chandler, R.E.; Kendon, E.J.; Widmann, M.; Brienen, S.; Rust, H.W.; Sauter, T.; Themel, M.; et al. Precipitation downscaling under climate change: Recent developments to bridge the gap between dynamical models and the end user. *Rev. Geophys.* **2010**, *48*, 1–34. [CrossRef]
64. Pierce, D.W.; Cayan, D.R.; Thrasher, B.L. Statistical Downscaling Using Localized Constructed Analogs (LOCA). *J. Hydrometeorol.* **2014**, *15*, 2558–2585. [CrossRef]
65. Tabari, H.; Paz, S.M.; Buekenhout, D.; Willems, P. Comparison of statistical downscaling methods for climate change impact analysis on precipitation-driven drought. *Hydrol. Earth Syst. Sci.* **2021**, *25*, 3493–3517. [CrossRef]
66. Grose, M.R.; Post, D.A.; Ling, F.L.N.; Corney, S.; Bennett, J.C.; Grose, M.R.; Post, D.A.; Ling, F.L.N.; Corney, S.P.; Bindoff, N.L. Performance of Quantile-Quantile Bias-Correction for Use in Hydroclimatological Projections Bioregional Assessment Programme View Project Barwon Water Inflows under Climate Change View Project Performance of Quantile-Quantile Bias-correction for Use in Hydroclimatological Projections. 2011. Available online: http://mssanz.org.au/modsim2011 (accessed on 1 January 2022).
67. Hayhoe, K.; Wake, C.; Anderson, B.; Liang, X.Z.; Maurer, E.; Zhu, J.; Bradbury, J.; Degaetano, A.; Stoner, A.M.; Wuebbles, D. Regional climate change projections for the Northeast USA. *Mitig. Adapt. Strateg. Glob. Chang.* **2008**, *13*, 425–436. [CrossRef]
68. Maurer, E.P.; Duffy, P.B. Uncertainty in projections of streamflow changes due to climate change in California. *Geophys. Res. Lett.* **2005**, *32*, 1–5. [CrossRef]
69. Li, H.; Sheffield, J.; Wood, E.F. Bias correction of monthly precipitation and temperature fields from Intergovernmental Panel on Climate Change AR4 models using equidistant quantile matching. *J. Geophys. Res. Atmos.* **2010**, *115*. [CrossRef]
70. Maurer, E.P.; Pierce, D.W. Bias correction can modify climate model simulated precipitation changes without adverse effect on the ensemble mean. *Hydrol. Earth Syst. Sci.* **2014**, *18*, 915–925. [CrossRef]
71. Obaid, N.; Alghazali, S.; Adnan, D.; Alawadi, H. Fitting Statistical Distributions of Monthly Rainfall for Some Iraqi Stations. *Civ. Environ. Res.* **2014**, *6*, 40–46.
72. AlHassoun, S.A. Developing an empirical formulae to estimate rainfall intensity in Riyadh region. *J. King Saud Univ.-Eng. Sci.* **2011**, *23*, 81–88. [CrossRef]
73. Hailegeorgis, T.T.; Thorolfsson, S.T.; Alfredsen, K. Regional frequency analysis of extreme precipitation with consideration of uncertainties to update IDF curves for the city of Trondheim. *J. Hydrol.* **2013**, *498*, 305–318. [CrossRef]
74. Al Islam, M.; Hasan, H. Generation of IDF equation from catchment delineation using GIS. *Civ. Eng. J.* **2020**, *6*, 540–547. [CrossRef]
75. Mujere, N. Flood Frequency Analysis Using the Gumbel Distribution. *Int. J. Comput. Sci. Eng.* **2011**, *3*, 2774–2778.
76. Solomon, O.; Prince, O. Flood Frequency Analysis of Osse River Using Gumbel's Distribution. *Civ. Environ. Res.* **2013**, *3*, 55–59.
77. Vidal, I. A Bayesian analysis of the Gumbel distribution: An application to extreme rainfall data. *Stoch. Environ. Res. Risk Assess.* **2014**, *28*, 571–582. [CrossRef]

78. Tahir, W.; Bakar, S.H.A.; Wahid, M.A.; Nasir, S.R.M.; Lee, W.K. *ISFRAM 2015*; Springer: Singapore, 2016. [CrossRef]
79. Yong, S.L.S.; Ng, J.L.; Huang, Y.F.; Ang, C.K. Assessment of the best probability distribution method in rainfall frequency analysis for a tropical region. *Malays. J. Civ. Eng.* **2021**, *33*. [CrossRef]
80. Balakrishnan, N.; Cramer, E.; Kundu, D. *Hybrid Censoring Know-How: Designs and Implementations*; Academic Press: Chicago, IL, USA, 2023.
81. Bartels, R.J.; Black, A.W.; Keim, B.D. Trends in precipitation days in the United States. *Int. J. Climatol.* **2020**, *40*, 1038–1048. [CrossRef]
82. Gupta, R.; Bhattarai, R.; Mishra, A. Development of climate data bias corrector (CDBC) tool and its application over the agro-ecological zones of India. *Water* **2019**, *11*, 1102. [CrossRef]
83. Ayugi, B.; Shilenje, Z.W.; Babaousmail, H.; Lim Kam Sian, K.T.C.; Mumo, R.; Dike, V.N.; Iyakaremye, V.; Chehbouni, A.; Ongoma, V. Projected changes in meteorological drought over East Africa inferred from bias-adjusted CMIP6 models. *Nat. Hazards* **2022**, *113*, 1151–1176. [CrossRef] [PubMed]
84. Babaousmail, H.; Ayugi, B.; Rajasekar, A.; Zhu, H.; Oduro, C.; Mumo, R.; Ongoma, V. Projection of Extreme Temperature Events over the Mediterranean and Sahara Using Bias-Corrected CMIP6 Models. *Atmosphere* **2022**, *13*, 741. [CrossRef]
85. Babaousmail, H.; Hou, R.; Ayugi, B.; Sian, K.T.C.L.K.; Ojara, M.; Mumo, R.; Chehbouni, A.; Ongoma, V. Future changes in mean and extreme precipitation over the Mediterranean and Sahara regions using bias-corrected CMIP6 models. *Int. J. Climatol.* **2022**, *42*, 7280–7297. [CrossRef]
86. Lim Kam Sian, K.T.C.; Hagan, D.F.T.; Ayugi, B.O.; Nooni, I.K.; Ullah, W.; Babaousmail, H.; Ongoma, V. Projections of precipitation extremes based on bias-corrected Coupled Model Intercomparison Project phase 6 models ensemble over southern Africa. *Int. J. Climatol.* **2022**, *42*, 8269–8289. [CrossRef]
87. Shrestha, S.; Sharma, S.; Gupta, R.; Bhattarai, R. Impact of global climate change on stream low flows: A case study of the great Miami river Watershed, Ohio. *Int. J. Agric. Biol. Eng.* **2019**, *12*, 84–95. [CrossRef]
88. Xue, P.; Ye, X.; Pal, J.S.; Chu, P.Y.; Kayastha, M.B.; Huang, C. Climate projections over the Great Lakes Region: Using two-way coupling of a regional climate model with a 3-D lake model. *Geosci. Model Dev.* **2022**, *15*, 4425–4446. [CrossRef]
89. Zhang, L.; Zhao, Y.; Hein-Griggs, D.; Janes, T.; Tucker, S.; Ciborowski, J.J.H. Climate change projections of temperature and precipitation for the great lakes basin using the PRECIS regional climate model. *J. Great Lakes Res.* **2020**, *46*, 255–266. [CrossRef]
90. Minallah, S.; Steiner, A.L. Analysis of the Atmospheric Water Cycle for the Laurentian Great Lakes Region Using CMIP6 Models. *J. Clim.* **2021**, *34*, 4693–4710. [CrossRef]
91. Briley, L.J.; Rood, R.B.; Notaro, M. Large lakes in climate models: A Great Lakes case study on the usability of CMIP5. *J. Great Lakes Res.* **2021**, *47*, 405–418. [CrossRef]

Disclaimer/Publisher's Note: The statements, opinions and data contained in all publications are solely those of the individual author(s) and contributor(s) and not of MDPI and/or the editor(s). MDPI and/or the editor(s) disclaim responsibility for any injury to people or property resulting from any ideas, methods, instructions or products referred to in the content.

Article

Advanced Machine Learning Techniques to Improve Hydrological Prediction: A Comparative Analysis of Streamflow Prediction Models

Vijendra Kumar [1,*], **Naresh Kedam** [2], **Kul Vaibhav Sharma** [1], **Darshan J. Mehta** [3,*] **and Tommaso Caloiero** [4,*]

1. Department of Civil Engineering, Dr. Vishwanath Karad MIT World Peace University, Pune 411038, Maharashtra, India; kulvaibhav.sharma@mitwpu.edu.in
2. Department of Thermal Engineering and Thermal Engines, Samara National Research University, Moskovskoye Shosse, 34, Samara 443086, Russia; naresh.kedam@gmail.com
3. Department of Civil Engineering, Dr. S. & S. S. Ghandhy Government Engineering College, Surat 395001, Gujarat, India
4. National Research Council of Italy, Institute for Agricultural and Forest Systems in Mediterranean (CNR-ISAFOM), 87036 Cosenza, Italy
* Correspondence: vijendra.kumar@mitwpu.edu.in (V.K.); ap_darshan_mehta@gtu.edu.in (D.J.M.); tommaso.caloiero@isafom.cnr.it (T.C.); Tel.: +39-0984-841-464 (T.C.)

Citation: Kumar, V.; Kedam, N.; Sharma, K.V.; Mehta, D.J.; Caloiero, T. Advanced Machine Learning Techniques to Improve Hydrological Prediction: A Comparative Analysis of Streamflow Prediction Models. *Water* **2023**, *15*, 2572. https://doi.org/10.3390/w15142572

Academic Editor: Gwo-Fong Lin

Received: 9 June 2023
Revised: 6 July 2023
Accepted: 12 July 2023
Published: 13 July 2023

Copyright: © 2023 by the authors. Licensee MDPI, Basel, Switzerland. This article is an open access article distributed under the terms and conditions of the Creative Commons Attribution (CC BY) license (https://creativecommons.org/licenses/by/4.0/).

Abstract: The management of water resources depends heavily on hydrological prediction, and advances in machine learning (ML) present prospects for improving predictive modelling capabilities. This study investigates the use of a variety of widely used machine learning algorithms, such as CatBoost, ElasticNet, k-Nearest Neighbors (KNN), Lasso, Light Gradient Boosting Machine Regressor (LGBM), Linear Regression (LR), Multilayer Perceptron (MLP), Random Forest (RF), Ridge, Stochastic Gradient Descent (SGD), and the Extreme Gradient Boosting Regression Model (XGBoost), to predict the river inflow of the Garudeshwar watershed, a key element in planning for flood control and water supply. The substantial engineering feature used in the study, which incorporates temporal lag and contextual data based on Indian seasons, leads it distinctiveness. The study concludes that the CatBoost method demonstrated remarkable performance across various metrics, including Mean Absolute Error (MAE), Root Mean Square Error (RMSE), and R-squared (R^2) values, for both training and testing datasets. This was accomplished by an in-depth investigation and model comparison. In contrast to CatBoost, XGBoost and LGBM demonstrated a higher percentage of data points with prediction errors exceeding 35% for moderate inflow numbers above 10,000. CatBoost established itself as a reliable method for hydrological time-series modelling, easily managing both categorical and continuous variables, and thereby greatly enhancing prediction accuracy. The results of this study highlight the value and promise of widely used machine learning algorithms in hydrology and offer valuable insights for academics and industry professionals.

Keywords: hydrological forecasting; machine learning; streamflow prediction; CatBoost; XGBoost; river inflow prediction

1. Introduction

Accurate prediction of daily river inflow is essential for effective water resource management [1]. Inflow predictions play a crucial role in decision-making for water managers and policymakers, influencing water allocation, reservoir operations, flood control measures, and drought mitigation strategies [2]. Accurate predictions enable optimized utilization of water resources by providing insights into availability and distribution. Reservoir operations rely on accurate inflow predictions to make informed decisions on water release and storage, considering downstream demands, flood control, and ecological factors [3,4]. During drought periods, precise inflow predictions help in proactive water supply management by implementing conservation measures, water use constraints, and exploring alternative sources [5]. Accurate inflow predictions support the development of robust drought management plans, ensuring sustainable water provision for communities and ecosystems. The use of accurate inflow predictions aids in mitigating risks, optimizing water storage, and facilitating efficient water resource management practices [6,7].

For estimating streamflow, a variety of techniques have been developed, many of which are physically based models that rely on experimental and statistical analysis [8]. Physically based streamflow forecasting models are based on certain hydrological hypotheses and require a large quantity of hydrological data for calibration [9]. The physical processes involved in the water cycle, such as interactions between rainfall and runoff and river routing, are described by these models. However, the accessibility and dependability of hydrological data could restrict the implementation of these models. Physically based models require accurate hydrological data as inputs, such as rainfall volume, intensity, and dispersion [10]. However, obtaining such data can be difficult, particularly in areas with weak monitoring infrastructure, costly data collection, or convoluted logistics. The calibration and validation processes of these models are hampered by the absence of precise and comprehensive hydrological data, which reduces the forecasting accuracy [11].

The advantage of physically based models is that they faithfully represent the hydrological system and the underlying physical processes. These models reveal information on the mechanics of runoff production and flow dynamics, making them helpful tools for understanding the behavior of watersheds [12]. They are particularly useful when a thorough understanding of the physical processes is necessary, like when analyzing how variations in land use or climatic conditions impact streamflow [13]. However, adopting physically based models has a number of disadvantages. In addition to the already noted data constraints, these models frequently need complicated parameterization, which can be difficult and imprecise. The calibration procedure entails changing model parameters to suit observed data, and the precision of the calibration is strongly influenced by the caliber and representativeness of the available data [14]. Unfortunately, this procedure is costly, involves a lot of work, takes a long time, and requires sample collection. As a result, scientists are becoming more and more interested in enhancing cutting-edge data-driven models for predicting streamflow. These models provide a viable alternative, since they need fewer data and are affordable.

Data-driven models have certain benefits over physically based models. Without using explicit physical equations, these models may discover patterns and connections directly from the available data [15]. Since they can handle a variety of input variables and capture nonlinear interactions, data-driven models are frequently more versatile and flexible [16]. Additionally, they have benefits for streamflow forecasting in data-scarce places, since they can make reasonably accurate forecasts even with limited hydrological data [17]. Data-driven models do, however, have certain drawbacks. They lack the ability to represent the underlying physical processes explicitly, which may limit their interpretability and generalizability in certain cases [18]. Data-driven models are also sensitive to the quality and representativeness of the training data. Biases or outliers in the data can significantly affect the model's performance, and it may be challenging to identify and address these issues without a good understanding of the underlying hydrological processes [19,20].

Streamflow predictions may be divided into short-term and long-term predictions, depending on the time period [21]. For flood control systems, hourly and daily forecasting, often known as short-term or real-time forecasting, is very valuable [22]. In the case of a flood, these projections allow for prompt action and decision making. Authorities can decide on evacuation, emergency response, and resource allocation in accordance with projections that are provided on an hourly or daily basis [23]. Real-time predictions assist in keeping an eye on flood-prone areas and sending out early warnings, therefore reducing the loss of life and property [24]. Long-term forecasting, however, covers the weekly, monthly, and yearly timescales [25]. It helps in managing irrigation systems, operating reservoirs, and producing electricity [26]. These projections are essential for controlling irrigation systems, maximizing the use of water for agriculture, and preserving ecological harmony. Furthermore, precise long-term projections aid in the planning of hydropower generation, permitting the best use of water resources for the development of renewable energy [27]. Streamflow forecasting has significantly advanced with the introduction of data-driven models. These models evaluate historical streamflow data and uncover patterns and correlations using computational methods like machine learning (ML) and artificial intelligence (AI) [28].

The potential for improving the precision and dependability of daily river inflow projections is enormous. With the aid of these methods, it is possible to evaluate sizable amounts of historical data, spot trends, and build intricate connections between meteorological factors, hydrological parameters, and river inflows [29]. ML models may learn and generalize from the patterns by being trained on previous data, which enables these models to produce precise forecasts for upcoming inflow circumstances [30]. The management of water resources will directly benefit from increasing the daily river inflow projections' accuracy with ML. The ability to make educated decisions that assure the best possible use of water resources, reduce the effects of floods and droughts, and promote sustainable development is a key capability of water managers and policymakers. By utilizing ML approaches, it can improve the accuracy of inflow predictions and contribute to better and more efficient methods of managing water resources, which will eventually be advantageous to society, the environment, and the economy [31]. Artificial neural networks (ANNs), support vector machines (SVMs), Random Forests (RFs), gradient boosting machines (GBMs), deep learning (DL) [32], long short-term memory (LSTM) [33], Gaussian processes (GPs), and physics-informed ML [34,35] are a few ML techniques utilized in streamflow forecasting. To accurately anticipate streamflow, these techniques take into account temporal dependencies, manage nonlinear patterns, and capture complicated linkages. They provide a variety of methods for better water resource management and impact reduction from floods.

1.1. Literature Review

1.1.1. Traditional Methods for River Inflow Prediction

For predicting river inflows, traditional methods have been applied in the area of hydrology. Statistical or empirical models based on historical data and certain hydrological factors are frequently used in these strategies [36]. Even while these conventional approaches have proved useful for understanding river inflow patterns and guiding water resource management decisions, they may have shortcomings in terms of capturing complicated non-linear interactions and managing huge datasets with a variety of influencing elements [37]. The autoregressive integrated moving average (ARIMA) model is a typical classical approach [38]. The temporal patterns and trends in data on river inflows may be captured using ARIMA models, which are often used in time series analysis [39]. They take into account the moving average (MA) component for accounting for the impact of prior prediction errors, the integrated (I) component for addressing non-stationary factors, and the auto-regressive (AR) component for modeling the dependency on previous inflow values. For predicting river inflow, physical based models like the Soil and Water Assessment Tool (SWAT) are frequently used in hydrology [40]. These models use elements including rainfall, land cover, soil properties, and terrain to mimic the hydrological processes, based on physical principles [41]. SWAT and similar models estimate river inflows by using mathematical equations to simulate the movement of water through the terrain.

Traditional approaches may have problems capturing non-linear relationships and managing large, complex datasets, even though they have been effective for hydrological forecasting. Since they typically rely on assumptions and simplifications of the underlying mechanics, their accuracy may occasionally be constrained [42]. Additionally, traditional methods with high labor and computational costs are less suitable for real-time forecasting applications. To manage these restrictions, researchers have adopted ML techniques, which provide more adaptability and flexibility in collecting complex patterns and processing enormous datasets. By automatically discovering patterns and correlations from data, ML techniques like ANN, SVM, and RF have shown promise in enhancing the accuracy and resilience of river inflow estimates.

1.1.2. Machine Learning Approaches for River Inflow Prediction

In recent years, there has been a lot of interest in the ability of ML algorithms to manage enormous datasets and capture intricate relationships in hydrological systems. These methods provide a data-driven approach to hydrological modeling, allowing for the creation of prediction models that are more precise [43,44]. Different ML techniques, including ANN, SVM, and decision trees, have been used in the context of river flow prediction to improve forecasting abilities [45,46]. Popular ML models for hydrological modeling include ANNs. ANNs are capable of capturing non-linear correlations between the goal variable of the river flow and the input variables of precipitation, temperature, and soil moisture [47]. They can generalize from prior data patterns to produce forecasts for upcoming timespans. Another ML method for predicting river flow is SVM. Finding the ideal decision boundary that divides several classes or forecasts river flow values based on input data is the goal of SVM algorithms. SVM models are efficient at capturing complicated correlations in hydrological processes and can handle high-dimensional data [48–50].

River flow prediction has also used decision trees and their ensemble approaches, including Random Forests (RFs). These algorithms create decision trees based on past data and employ them to anticipate future events. In order to increase forecast resilience and accuracy, RF merges numerous decision trees. It has been applied to streamflow forecasting to better capture interactions between different hydrological factors [51,52]. In streamflow forecasting, gradient boosting machines (GBMs) like the extreme gradient boosting regression model (XGBoost) [53] and LGBM [54] have grown in popularity. They focus on samples with large prediction errors and repeatedly incorporate weak models to produce a strong predictive model. GBMs are renowned for their capacity to handle missing data and complicated connections.

A special kind of recurrent neural network (RNN) called long short-term memory (LSTM) is made for sequential data. For short-term forecasting applications in particular, LSTMs have proved effective in capturing temporal relationships in streamflow data and producing precise forecasts [55,56]. Probabilistic models known as Gaussian processes (GPs) are capable of capturing errors in forecasts of streamflow. They have been applied to streamflow forecasting to offer not just point predictions but also prediction intervals that show the forecasts' level of uncertainty [57]. Hybrid models mix several machine learning (ML) methods or incorporate ML with physical models [58]. For instance, data assimilation methods may be applied to merge physically based models with ML methods to increase prediction accuracy or incorporate actual streamflow data into ML models. To enhance model performance, [59] created hybrid particle swarm optimization (PSO) and the group method of data handling for short-term prediction of daily streamflow, [60] developed ML-based grey wolf optimization for the short-term prediction of streamflows, [61] used hybrid LSTM-PSO for the streamflow forecast, [62] combined different ML methods for daily streamflow simulation, and [63] used an LSTM-based DL model for streamflow forecasting using Kalman filtering.

For predicting river flow, ML techniques provide a number of benefits. They have the ability to manage non-linear relationships and adjust to shifting hydrological circumstances. A more thorough investigation of the hydrological processes is possible because of ML models' ability to handle huge datasets with many impacting elements. Additionally, ML methods may combine several data sources, such as meteorological data, remote sensing data, and historical streamflow records, to increase forecast accuracy. But it is crucial to remember that ML models have their limits as well. For efficient model building, they need a large volume of high-quality training data. To make sure the models reflect pertinent hydrological processes, care must be taken in the selection of acceptable input variables and feature engineering. Additionally, if the training dataset is too short or the model complexity is not adequately managed, ML models may experience overfitting. A variety of machine learning methods, such as CatBoost, ElasticNet, k-Nearest Neighbors (KNN), Lasso, light gradient-boosting machine regressor (LGBM), Linear Regression (LR), multilayer perceptron (MLP), Random Forest (RF), Ridge, stochastic gradient descent (SGD), and the extreme gradient-boosting regression model (XGBoost), have been used to create models for predicting river inflow in the article. The most efficient method for forecasting river inflow has been determined after the compared results of their investigations into the efficacy of each methodology.

This research makes several contributions that highlight its novelty:

a. Comparative Evaluation: the study provides a comprehensive comparative evaluation of multiple machine learning models for predicting river inflow. While previous studies have explored individual models, this research systematically compares the performance of CatBoost, ElasticNet, KNN, Lasso, LGBM, Linear Regression, MLP, Random Forest, Ridge, SGD, and XGBoost. Such a comprehensive comparative analysis is novel in the context of river inflow prediction.

b. Time Series Analysis: the study specifically focuses on time series analysis for river inflow prediction. Time series data present unique challenges, due to temporal dependencies. By applying different machine learning techniques to this specific domain, the research contributes to the advancement of time series prediction methodologies in the context of water resource management.

c. Application to River Inflow Prediction: while machine learning models have been applied in various domains, their application to river inflow prediction is of significant importance for water resource management. Predicting river inflow accurately is crucial for making informed decisions regarding water allocation, flood management, and hydropower generation.

d. Performance Evaluation on Multiple Datasets: the study evaluates the performance of the models on multiple datasets, including training, validation, and testing data. This comprehensive evaluation provides a robust assessment of the models' performance and their ability to generalize to unseen data, contributing to the understanding of their efficacy in real-world scenarios.

1.2. Objectives of the Study

The primary objective is to develop models for predicting river inflow using the different machine learning methods mentioned, including CatBoost, ElasticNet, k-Nearest Neighbors (KNN), Lasso, light gradient-boosting machine regressor (LGBM), Linear Regression (LR), multilayer perceptron (MLP), Random Forest (RF), Ridge, stochastic gradient descent (SGD), and the extreme gradient-boosting regression model (XGBoost). The models attempt to forecast river inflow based on relevant input characteristics.

2. Methodology and Methods

The steps involved in developing and analyzing a machine learning (ML) model for predicting daily river inflow are outlined. Several important parts of the procedure are included. First, data from credible sources are used to compile historical data on daily river inflow. To guarantee data quality, the obtained data go through preprocessing, which includes cleaning and addressing missing values. Then, using feature engineering approaches, pertinent characteristics are extracted, including seasonal and temporal trends. A piece of the dataset is used to construct and train the models, while a different subset is used to validate their performance and evaluate their correctness. Common evaluation metrics, such as mean squared error (MSE), mean absolute error (MAE), root mean squared error (RMSE), root mean square percentage error (RMSPE) and R-squared (R^2), are used to quantify the model's performance.

To learn more about the model's predictive skills and the importance of various characteristics in predicting river input, the generated data are carefully studied. The model's implications for managing water resources are examined, along with suggestions for more study and possible practical application. By following this methodology, the study aims to contribute to the development of a robust and accurate model for daily river inflow prediction, which can provide valuable insights for effective water resource management and decision-making processes. Figure 1 shows the flowchart of the methodology of the study.

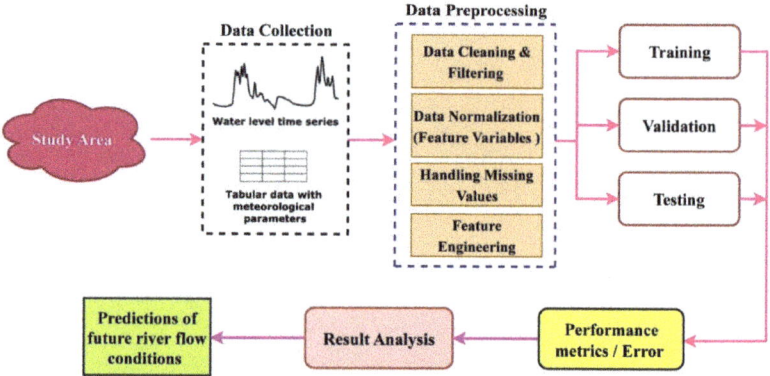

Figure 1. Shows the flowchart of the methodology.

2.1. CatBoostRegressor Algorithm

CatBoostRegressor is an ML technique that predicts continuous values using gradient-boosted decision trees. It is a relatively new algorithm [64]. CatBoostRegressor is known for its efficiency, precision, and capacity for handling categorical characteristics. In order for the CatBoostRegressor algorithm to function, a set of weak decision trees must first be built. A powerful model is then built by combining these trees. Gradient boosting is the method used to join the trees. Gradient boosting works by adding additional trees to the model that fix the mistakes created by the earlier trees. To predict continuous values, CatBoostRegressor applies the following formula, as shown in Equation (1):

$$y = f(x) = \sum_{i=1}^{n} \alpha i\, hi(x) \qquad (1)$$

where the output function $f(x)$ is a linear combination of the basis functions $hi(x)$, and coefficients αi define the weight of each basis function in the linear combination; y is the predicted value, x is the input features.

The gradient descent method is used to calculate the model coefficients. The loss function must be minimized in the CatBoost. The difference between the values that were predicted and the actual values is measured by the loss function. A number of regression problems may be solved with the potent ML method CatBoost. It works especially effectively for issues involving categorical characteristics.

2.2. k-Nearest Neighbors

The KNN algorithm is a non-parametric regression method used for predicting the target variable based on the average of the target values of its k nearest neighbors [65]. Here are the key steps:

1. Prepare the training data with input features and target values.
2. Determine the value of k, the number of nearest neighbors to consider.
3. Calculate the distance between the new data point and the training data points.
4. Select the k nearest neighbors, based on the distances.
5. Calculate the target values' average among the k closest neighbors. Use the average value as the new data point's estimated goal value.

In Equation (2), the target variable prediction formula is shown, where ($\hat{y}$) is the predicted target value, k is the number of nearest neighbors, and $\sum y_i$ is the sum of the target values of the k nearest neighbors.

$$\hat{y} = \frac{1}{k}\sum y_i \tag{2}$$

The k-Neighbors Regressor technique is useful for detecting local patterns, managing non-linear connections, and making the fewest assumptions possible regarding the distribution of the data. However, it can be computationally demanding, sensitive to the selection of k and distance metric, and may call for feature scaling or regularization methods.

2.3. Light Gradient-Boosting Machine Regressor (LGBM)

The effectiveness and adaptability of the LGBM gradient-boosting method are well recognized. It provides a number of features and enhancements to optimize the performance of gradient boosting on big datasets [66]. In the data preparation stage of the method, the training data are divided into input characteristics and target values for regression. Target values and metric characteristics are recommended. The learning rate, number of trees, maximum depth, and feature fraction are then initialized. The LGBM model's behavior is governed by these variables, which can be changed to enhance performance. Making a series of decision trees is part of the model creation and training process. A gradient-based optimization approach that minimizes the loss function is used to construct each tree. The ensemble of trees is iteratively expanded, and the predictions of the model are modified in accordance with the gradients of the loss function. After the model has been trained, additional data points may be predicted by using it. The LGBM method uses a weighted sum to aggregate the forecasts from each tree in the ensemble. During the training phase, the weights are chosen depending on the gradients of the loss function. In LGBM, the target variable may be predicted using the following formula:

$$\hat{y} = \sum \alpha i \, hi(x) \tag{3}$$

where αi indicates the weight given to the ith tree, $\hat{y}$ predicts the target value, and $hi(x)$ the prediction of the ith tree for the input characteristics x. The LGBM can capture complex non-linear correlations between characteristics and the target variable, is quite effective, and can handle enormous datasets. The loss function is optimized via gradient-based optimization, which creates an ensemble of trees that collectively provide precise predictions.

2.4. Linear Regression (LR)

LR method that deals with a set of records having X and Y values. These values are utilized to learn a function that can predict Y for an unknown X. In regression, the aim is to find the value of Y, given that XY is continuous. Here, Y is referred to as the criterion variable, and X is called the predictor variable. Different types of functions or models can be employed for regression, wherein a linear function is the simplest one [67]. In this case, X can be a single or multiple features that represent the problem.

$$Y = C_1 + C_2 \times X \tag{4}$$

where, X = input training data, Y = predicted value of Y for a given X, C_1 = intercept, and C_2 = coefficient of X. Once the optimal values of C_1 and C_2 are determined, the best fit line can be obtained.

2.5. Multilayer Perceptron

The Multilayer Perceptron (MLP) is a sort of artificial neural network that is made up of several layers of linked nodes, or neurons [68]. Since it is a feed-forward neural network, data goes from the input layer to the hidden layers and finally to the output layer. Each neuron in the MLP conducts a weighted sum of its inputs, applies an activation function to the sum, and then transmits the outcome to the neurons in the next layer. The following is a description of the MLP:

(a) Assign random weights to the connections between the neurons as part of the initialization process.
(b) The input layer: Take in input data and send them to the top-most hidden layer.
(c) Hidden layers: Each hidden layer neuron computes the weighted sum of its inputs using the current weights and then applies an activation function (such as a sigmoid) to the sum.
(d) Output layer: The neurons in the output layer compute the same activation function and weighted sum as the neurons in the hidden layers.
(e) The MLP's final output is derived from the neurons in the output layer.

During the training phase, the MLP's weights are modified using optimization methods like gradient descent. A loss function that calculates the difference between the output that was expected and the output that was actually produced must be minimized. In order to produce predictions or categorize data based on fresh input, the MLP must first understand the underlying patterns and relationships in the data.

2.6. Random Forest

Random Forest (RF) is a highly accurate and versatile regression model widely used in ML. It belongs to the ensemble learning category, where multiple decision trees are built during the training phase. Each tree predicts the mean value of the target variable [69]. The steps involved in the Random Forest algorithm are as follows:

1. Random Subset Selection: a random subset of data points is chosen from the training set. This subset typically contains a fraction of the total data points, denoted by 'p'.
2. Construction of a Decision Tree: using the subset of data points that was chosen, a decision tree is built. This procedure is repeated using various subsets of the data for a total of 'N' trees.
3. Prediction Aggregation: each of the 'N' decision trees predicts the value of the target variable for a new data point. The outcomes of all the predictions from the trees are averaged to provide the final forecast.

When using environmental input factors to forecast rainfall data, Random Forest is highly effective. The technique uses the combined predictive capability of the trees to decide the resultant class by creating a large number of decision trees during training. It is known for its effectiveness in handling large datasets and can produce reliable results even when dealing with missing data.

2.7. Lasso

Lasso, also known as $L1$ regularization, is a linear regression model that adds a penalty term based on the $L1$ norm of the coefficients [70]. It is used to encourage sparsity in the coefficient values, effectively performing feature selection by driving some coefficients to exactly zero. The formula for Lasso regression can be represented as follows:

$$y = \beta_0 + \beta_1 x_1 + \beta_2 x_2 + \ldots + \beta_p x_p \quad (5)$$

In addition to the mean squared error (MSE) factor, the objective function of Lasso regression also contains a regularization term:

$$Lasso\ Objective\ Function = \text{MSE} + \alpha \times L1\ Norm \quad (6)$$

where y stands for the dependent variable, and the independent variables (input characteristics) are represented by $x_1, x_2, \ldots,$ and x_p. The independent variables' coefficients (parameters) are $\beta_0, \beta_1, \beta_2, \ldots, \beta_p$. The $L1$ regularization's strength is determined by the regularization parameter, which is α. It chooses the appropriate ratio between punishing the size of the coefficients ($L1$ norm) and fitting the training data (MSE term).

The objective function's $L1$ norm term is calculated as the sum of the absolute values of the coefficients.

$$L1\ Norm = |\beta_1| + |\beta_2| + \ldots + |\beta_p| \tag{7}$$

Lasso regression searches for the best values of the coefficients to minimize the MSE term while maintaining the $L1$ norm term as minimal as possible by minimizing the goal function. Thus, certain coefficients may be reduced to absolute zero, thus removing the related characteristics from the model. Because of this characteristic, Lasso regression may be used to handle high-dimensional datasets and feature selection.

2.8. Ridge

Ridge regression is an ML method frequently applied to regression analysis in the context of supervised learning. Regression analysis frequently uses Ridge regression, commonly referred to as Tikhonov regularization, to address the multicollinearity and overfitting issues [71]. It is an extension of ordinary least squares (OLS) regression that modifies the loss function by including a punishment component. The Ridge regression formula is as follows:

$$minimize = ||Y - X\beta||^2 + \lambda||\beta||^2 \tag{8}$$

Here, the target variable is denoted by Y, the predictor variables are denoted by X, the coefficients are denoted by β, the regularization parameter is denoted by λ controlling how much shrinkage is done to the coefficients, and the Euclidean norm is denoted by $||\beta||$. Ridge regression seeks to reduce the sum of squared discrepancies between predicted and observed values $(Y - X)$, while also penalizing the size of the coefficients $(||\beta||^2)$.

2.9. ElasticNet

ElasticNet is a linear regression model that combines the $L1$ (Lasso) and $L2$ (Ridge) regularization techniques [72]. It is designed to overcome some limitations of each individual method by introducing a penalty term that includes both $L1$ and $L2$ norms.

The formula for ElasticNet regression can be represented as follows:

$$y = \beta_0 + \beta_1 x_1 + \beta_2 x_2 + \ldots + \beta_p x_p \tag{9}$$

The objective function of ElasticNet includes two regularization terms, one for $L1$ regularization and another for $L2$ regularization, along with the mean squared error (MSE) term:

$$ElasticNet\ Objective\ Function = \text{MSE} + \alpha * [\lambda_1 * L1\ Norm + \lambda_2 * L2\ Norm] \tag{10}$$

where y represents the dependent variable (the target variable we want to predict). x_1, $x_2, \ldots, x_p$ represent the independent variables (input features). $\beta_0, \beta_1, \beta_2, \ldots, \beta_p$ are the coefficients (parameters) of the independent variables. α is the mixing parameter that controls the balance between $L1$ and $L2$ regularization. It is between 0 and 1. Ridge regression is represented by a value of $\alpha = 0$, Lasso regression is represented by a value of $\alpha = 1$, and values in between represent a mixture of both. The regularization parameters λ_1 and λ_2 regulate the potency of $L1$ regularization and $L2$ regularization, respectively.

2.10. Stochastic Gradient Descent (SGD) Regressor

For regression challenges, ML algorithms like the Stochastic Gradient Descent (SGD) Regressor are utilized. It is a modification of the common Gradient Descent technique and is especially helpful in cases involving online and massively multi-user learning [73]. A randomly chosen subset of training data (mini-batches) is used to iteratively update the model's parameters via the SGD Regressor. It is computationally effective and appropriate for big datasets, since it calculates the gradients of the loss function with respect to the model's parameters using just the samples in the mini-batch. The SGD Regressor's update formula for the model's parameters is the same as the normal SGD's:

$$\theta\_new = \theta\_old - \alpha * \nabla J(\theta\_old; x_i, y_i) \qquad (11)$$

Here, the parameters of the model are represented by their current values (θ_old), their updated values (θ_new), the learning rate (α), the gradient of the loss function J with respect to the parameters evaluated at the current parameter values ($J(\theta\_old; x_i, y_i)$), and one training example (x_i, y_i). To achieve optimal convergence and performance, it is crucial to carefully choose the learning rate and mini-batch size. Additionally, the performance and stability of the algorithm may be enhanced by using strategies like learning rate schedules, momentum, and regularization. The SGD Regressor works well when faced with massive data volumes, high-dimensional feature spaces, and a steady stream of new data.

2.11. Extreme Gradient-Boosting Regression Model (XGBoost)

XGBoost is a regression model, a potent ensemble learning technique which uses gradient boosting and decision trees to make precise predictions. The XGBoost approach delivers a variety of performance-improving improvements while sharing a similar structure with other gradient-boosting regressors [74]. The XGBoost algorithm is described in the sections below:

1. Choosing the XGBoost model's parameters, such as the learning rate, the number of trees, the maximum depth, and the feature fraction, is the step-one process. These variables can be altered to improve performance and regulate how the model behaves.
2. Create the model and train it: the XGBoost model is produced by the construction of several decision trees. A gradient-based optimization technique that minimizes the loss function is used to build each tree. The ensemble of trees is continuously expanded throughout the training phase, and predictions are updated in line with gradients in the loss function.
3. After model training, the model may be used to make predictions about fresh data points. The XGBoost method incorporates the predictions from each tree in the ensemble to obtain the final regression prediction. The particular method for combining the predictions is determined by the loss function that is used.

3. Model Training and Validation

Model training and validation are crucial steps in the machine learning process. In these stages, a dataset is modelled for training, and the model's effectiveness is assessed on a separate dataset for validation. The goal is to develop a model that accurately predicts the future and generalizes well to new inputs. The model training and validation procedure is summarized as follows:

1. Data Split: a training set, a validation set, and a test set are each provided as separate datasets. The model is trained using the training set. The validation set is used to fine-tune the model and assess model performance throughout training, whereas the test set is used to measure the trained model's final performance on unseen data.
2. Model Selection: select the most effective model architecture or machine learning technique for the particular job. The kind of data, the task (classification, regression, etc.), and the resources available are all factors in the model selection process.
3. Model Training: develop the selected model using the training dataset. During the training phase, the model parameters are frequently repeatedly improved in order to minimize a chosen loss or error function. In order to do this, training data are fed into the model, predictions are generated and compared to actual values, and model parameters are updated, depending on computed errors. This procedure continues until a convergence requirement is satisfied, after a certain number of epochs.
4. Model Evaluation: using the validation dataset, evaluate how well the trained model performed. The validation data is used to generate predictions, which are then

compared to the actual results. There are several assessment measures employed, including mean squared error (MSE), mean absolute error (MAE), root mean square error (RMSE), root mean square percent error (RMSPE), and R-squared (R^2) [75].

$$\text{MSE} = (1/n) * \sum [(y_i - \hat{y}_i)^2] \quad (12)$$

$$\text{MAE} = (1/n) * \sum |y_i - \hat{y}_i| \quad (13)$$

$$\text{RMSE} = \sqrt{(\text{MSE})} = \sqrt{[(1/n) * \sum [(y_i - \hat{y}_i)^2]]} \quad (14)$$

$$\text{RMSPE} = \sqrt{[(1/n) * \sum [((y_i - \hat{y}_i)/y_i)^2]]} \quad (15)$$

$$R^2 = 1 - (\sum [(y_i - \hat{y}_i)^2] / \sum [(y_i - \bar{y})^2]) \quad (16)$$

where the overall number of data points is n. The dependent variable's actual (observed) value for the ith data point is represented by y_i. The predicted value of the dependent variable for the ith data point is represented by $\hat{y}_i$. Σ stands for the total sum, or the sum of the squared differences for each data point. The dependent variable's mean is represented by the symbol $\bar{y}$.

5. Iterative Refinement: to enhance performance, modify the model architecture or data preparation stages based on the evaluation findings. Until a suitable performance is attained, this iterative procedure is continued.
6. Final Assessment: after the model has been adjusted, its performance is evaluated using the test dataset, which simulates unseen data. This offers a neutral assessment of how well the model performs in realistic situations.

To guarantee accurate and trustworthy model training and assessment, it is crucial to remember that correct data preparation, including managing missing values, feature scaling, and controlling class imbalance, should be carried out during the training and validation process. These processes may be efficiently used to train, validate, and assess machine learning models, in order to create reliable and accurate prediction models.

4. Study Area, Data Collection and Preprocessing

4.1. Study Area

One of the largest rivers in central India, the Narmada River, passes through the states of Gujarat, Maharashtra, and Madhya Pradesh. The significance of it for ecology, history, and culture is widely known. Hindus adore the river's waters and a variety of flora and animals call it home. In the Narmada River basin, the Garudeshwar gauging station is an important study location. The gauging station serves as a monitoring station for identifying and analyzing the river's different hydrological properties. It is located close to the Gujarat town of Garudeshwar. The primary duty of the Garudeshwar gauging station is to gauge and track the water levels and flow rates of the Narmada River. The gauging station is equipped with instruments that gather data on a variety of elements, such as water level, discharge, and velocity. The research region around a gauging station is frequently defined by the gauging station's measurement range of impact. This might alter, based on the objectives of the research specifically or the requirements of the water management authority. The research region may extend both upstream and downstream of the gauging station in order to completely comprehend the hydrological characteristics and dynamics of the river. Researchers, hydrologists, and managers of water resources routinely evaluate water availability, look into flood patterns, and make informed judgments regarding water distribution and management using the data collected from the gauging station and the study region. An overview of watershed areas and their placement on a map of India is shown in Figure 2.

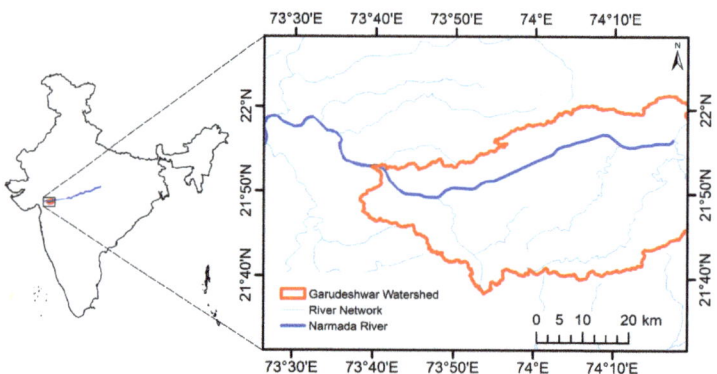

Figure 2. Shows the Garudeshwar watershed area.

4.2. Data Collection

Daily river inflow measurements in cubic meters per second were gathered from a river gauge station and utilized as the dataset for this investigation. The data, which span the years 1980 to 2019, were gathered from India's Water Resources Information System (WRIS) for the time series analysis. A thorough record of the river's inflow across time is provided by the dataset, allowing for examination of flow fluctuations and trends. Table 1 shows the descriptive statistics of the data.

Table 1. Descriptive statistics of data.

	Flow
Mean	784.8985221
Standard Error	18.28637548
Median	184.0000428
Mode	23.19005239
Standard Deviation	2210.307722
Sample Variance	4,885,460.225
Kurtosis	128.7110287
Skewness	8.786730848
Range	60,640.72647
Minimum	1.270052203
Maximum	60,641.99652

4.3. Techniques for Preprocessing Data

Several preprocessing procedures can be used for the dataset from the Garudeshwar gauging station in order to guarantee the correctness and dependability of the data. To resolve errors, outliers, and missing numbers, the data must first be cleaned. This procedure comprises validation, cross-checking with trustworthy sources, and using statistical techniques and subject-matter expertise to spot and fix flaws and inconsistencies. Depending on their relevance, outliers can either be corrected or removed. The dataset's integrity can be preserved by imputing missing values using techniques like mean imputation or interpolation. To improve the models' ability to anticipate outcomes, feature engineering approaches can be used. This entails generating fresh features from preexisting variables. In the context of predicting river inflow, temporal characteristics can be derived from the date variable to identify trends in the data. Lagged features, which represent past inflow values, will also be generated to capture the influence of historical data on future predictions. The first seven days of 1980 (from 1 January to 7 January) are not taken into account to create lagged characteristics, so data here is available from 8 January 1980 to 31 December 2019. Also, no outliers and all peak data points have been taken into account, since there is no elimination of any data points.

An augmented Dickey–Fuller (ADF) statistic is used to check the stationarity or non-stationarity of the data. The ADF statistic is a test statistic used in time series analysis to determine the presence of a unit root in the data. The unit root refers to the presence of a stochastic trend that can cause non-stationarity in the series. If the series is found to be stationary, it implies that there is no significant linear trend present. In the given scenario, the ADF statistic has a value of −13.045793. This indicates a highly negative value, suggesting strong evidence against the presence of a unit root in the data. The p-value associated with the ADF statistic is reported as zero, which further supports the rejection of the null hypothesis of a unit root. To assess the significance level of the ADF statistic, critical values are considered. The critical values at 1%, 5%, and 10% significance levels are −3.431, −2.862, and −2.567, respectively. Since the ADF statistic value of −13.045793 is much lower (in absolute terms) than these critical values, it can conclude that the data is statistically significant and the result of the ADF statistic is shown in Figure 3. Therefore, based on the ADF statistic and its associated p-value, we can infer that the data under consideration are stationary. Stationary data implies that the statistical properties of the series, such as mean, variance, and autocorrelation, remain constant over time. This is an important characteristic for many time series analysis techniques and modeling approaches. It is significant to note that, depending on the location and features of the area under examination, the stationarity of river flow series might change. River flow series do occasionally display stationary qualities, despite the fact that seasonal patterns, trends, and other variables frequently cause river flow series to behave in a non-stationary manner. The particular location under consideration in this study may have unique characteristics that contribute to the observed stationarity. The stationarity of river flow series can be influenced by elements including the hydrological parameters of the river basin, climatic circumstances, land use patterns, and water management techniques. Furthermore, it is worth mentioning that even if the river flow series is stationary, it does not imply that the series is entirely predictable or that it lacks variability. The presence of other forms of variability, such as short-term fluctuations or irregular patterns, can still exist within a stationary series.

```
ADF Statistic: -13.045793
p-value: 0.000000
Critical Values:
        1%: -3.431
        5%: -2.862
        10%: -2.567
Data is stationary
```

Figure 3. Shows the result of the ADF statistic.

The original time series, trend, seasonality, and residual time series are displayed in Figure 4. With regard to the combined influences of trend, seasonality, and random fluctuations, the original data offer a thorough assessment of the real observations. The long-term, regular movement or direction of the river flow is represented by the trend flow component. It shows if the flow is increasing or decreasing over time. It can observe the general behavior of the river flow and spot any enduring alterations by focusing on the trend. In this instance, the trend flow indicates a declining pattern in the data of the river flow. This information is helpful in determining the general trend and making future plans for the management of water resources. Seasonality describes recurring, predictable fluctuations that take place at predetermined times. Seasonality in the context of river flow refers to regular patterns or fluctuations that take place over the course of a year. By examining the seasonality component, it locates any recurring patterns in the river flow data. In this case, the seasonality component varies by up to 4000 m^3/s, demonstrating that the river flow displays significant patterns and changes throughout the year. Understanding seasonality can aid in forecasting future flow patterns and preparing for the demands placed on water resources throughout particular seasons. The residuals are the variations between the values that were seen and those that were anticipated by the trend and seasonality components. They stand for the arbitrary and unpredictable variations in river flow that neither trends nor seasonality can account for. Any remaining anomalies or out-of-the-ordinary events in the data can be understood by analyzing the residuals. The residuals allow us to determine the trend and seasonality components' goodness of fit as well as any other variables affecting the river flow.

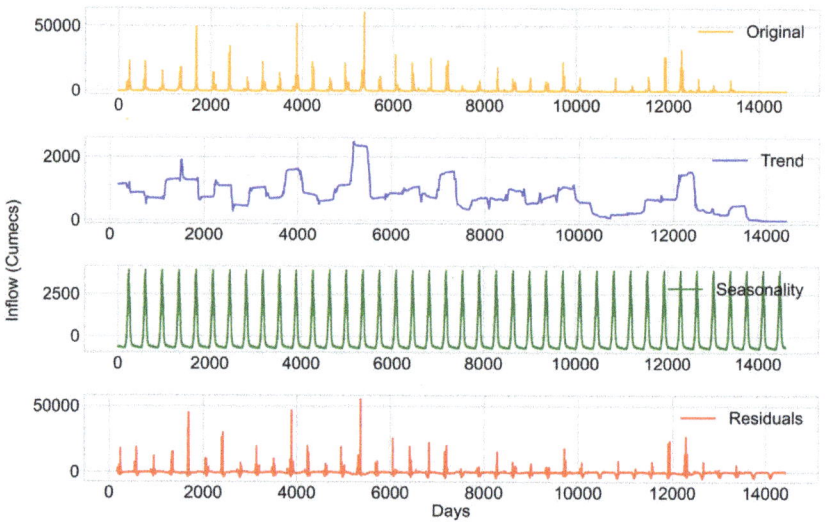

Figure 4. Original time series, trend, seasonality and residuals.

4.3.1. Creating Lagged Features

When working with time series data, the idea of "lagged features" is very pertinent. A value from a previous time period is a lagged characteristic from a time series. A lagged characteristic may be the river input from today, yesterday, or even a week ago if we are forecasting river inflow for tomorrow. These are known, correspondingly, as lag-1, lag-2, and lag-7 characteristics. Lagged features can be used to capture the temporal relationships present in the data. In other words, they offer a method of providing the model with information about previous values, which may be useful for forecasting future values. The lag order, which refers to the number of lagged data to include, is often established empirically, frequently by employing methods like autocorrelation plots or depending on domain knowledge. For this study, lagged features are implemented according to domain knowledge; daily data of a week are taken to predict next-day data.

4.3.2. Date Feature Engineering

The development of date features was a crucial preprocessing step in this work. In order to do this, more pertinent information must be extracted from the timestamp data. The study's date characteristics included the weekday, the month, the Indian month, and the Indian season. These elements were included because they may have a large impact on river input. For instance, because of weather patterns, some months or seasons may see higher or lesser influx. Depending on the timestamp's data format, different processes can be used to create various properties. Before these properties can be retrieved, the timestamp may need to be transformed from a string format into a datetime object. Once the features are finished, they may be used as any other model input.

4.3.3. One-Hot Encoding

One-hot encoding is the last preprocessing step. Categorical variables are handled using this technique. The categorical data must be translated into a format that can be used by these methods, since many machine learning algorithms cannot deal directly with categorical data. One-hot encoding is a typical method. Each distinct category of a categorical variable is represented as a binary vector in one-hot encoding. One-hot encoding would produce seven new features (one for each day of the week) if, for instance, the feature "day of the week" had seven categories (Monday, Tuesday, ... , Sunday). If Monday were the day of the week, "Monday" would have a value of 1, while all other days would have a value of 0. If the day was Tuesday, the "Tuesday" feature would be set to 1, and all other day features would be set to 0, and so on. One-hot encoding completely eliminates any ordinal link between categories (i.e., it prevents the model from assuming that "Monday" is less than "Tuesday" just because we encode Monday as 1 and Tuesday as 2). This is advantageous when there is no ordinal link between the categories, as there is when talking about the days of the week, months, or seasons.

5. Model Preparation

In this investigation, the data were divided into training, validation, and test sets using a time series split. The temporal order of the observations is crucial in time series data; therefore, this approach of data splitting is very appropriate. The data are separated into time periods in a time series split. The earliest observations make up the training set, the sequence observations make up the validation set, and the latest observations make up the test set. This makes sure that each piece of data accurately depicts the chronological order of the actual occurrences. It is crucial to keep in mind that time series splits preserve the temporal dependencies and autocorrelation inherent in time series data, unlike random splits, which forbid the inclusion of any future data in the training set. On the basis of the patterns found in the historical data, the models were trained on the training set to predict the target variable. The models were then tested on the validation set, which contained data that were not utilized during training but temporally followed the training period. This stage allowed us to retain the data's chronological integrity while monitoring the models' performance on previously unknown data and making any required adjustments. The test set, which represented the most current data in the series, was used to evaluate the models. This provided a fair assessment of the models' performance on brand-new, previously unobserved data, and an estimate of how well the models would perform when making predictions about upcoming real-world data. To retain the temporal structure of the data while assessing the predictive performance of our models by using a time series split, guaranteed that the models had the capacity to provide accurate future projections.

6. Results and Discussion

The prediction models in this research were meticulously evaluated, offering insightful information. Several machine learning models, including CatBoost, ElasticNet, KNN, Lasso, LGBM, Linear Regression, MLP, Random Forest, Ridge, SGD, and XGBoost, were assessed for their ability to predict river inflow. A range of error metrics and R-squared values were used to evaluate their performance.

6.1. Performance Metrics of Training Data

The performance indicators for several models based on training data are shown in Table 2. Each model is assessed using the metrics of MAE, MSE, RMSE, RMSPE, and R^2. These metrics evaluate each model's performance on the training data. Higher R^2 values indicate a better fit of the model to the data, while lower MAE, MSE, RMSE, and RMSPE values denote superior performance. A comparison of the models in Table 2 reveals that CatBoost, XGBoost, and RF demonstrate improved prediction accuracy and model fit on the training data, due to their lower MAE, MSE, RMSE, RMSPE values and high R^2. ElasticNet, KNN, Lasso, LR, MLP, Ridge, and SGD perform less effectively on the training data, having lower R^2 and higher MAE, MSE, RMSE, RMSPE values. LGBM also performs well, exhibiting relatively low values across all the criteria. Models with the lowest errors (MAE, MSE, RMSE, RMSPE), highest R^2, and best performance on the training data are CatBoost, XGBoost, and RF. These models fit the training data well, and have excellent predictive capabilities. It is crucial to note that a model's performance on training data might not necessarily generalize to new data. Therefore, further assessment of the models' overall performance using validation and test data is necessary to select the most suitable model for prediction tasks.

Table 2. Performance metrics for various models on the training data.

Sr No.	Model	MAE_Train	MSE_Train	RMSE_Train	RMSPE_Train	R^2_Train
1	CatBoost	124.89	131,672.45	362.87	150.28	0.98
2	ElasticNet	414.90	2,304,350.42	1518.01	853.11	0.61
3	KNN	320.95	1,773,732.98	1331.82	310.48	0.70
4	Lasso	327.18	1,923,781.45	1387.00	568.25	0.67
5	LGBM	215.89	863,329.16	929.16	256.82	0.85
6	LR	434.94	1,979,323.29	1406.88	1005.55	0.67
7	MLP	298.63	1,599,712.13	1264.80	276.29	0.73
8	RF	117.58	332,086.13	576.27	295.72	0.94
9	Ridge	330.27	1,923,316.06	1386.84	584.78	0.68
10	SGD	366.52	1,973,385.04	1404.77	980.74	0.67
11	XGBoost	**75.04**	**38,693.90**	**196.71**	**142.99**	**0.99**

Bold value shows the better solution.

6.2. Performance Metrics of Validation Data

The performance characteristics of several models on the validation data are displayed in Table 3. For each model, the metrics are MAE, MSE, RMSE, RMSPE, and R^2. After reviewing the performance of the models using validation data, the following conclusions can be drawn: LGBM, Lasso, MLP, and Ridge perform better on the validation data as a result of having comparatively lower values for MAE, MSE, RMSE, RMSPE, and higher R^2. CatBoost, ElasticNet, LR, RF, SGD, and XGBoost also exhibit acceptable performance, with moderate metric values. KNN performs poorly on the validation data, with higher values for MAE, MSE, RMSE, RMSPE, and lower R^2. LGBM, Lasso, MLP, and Ridge outperform the other models on the validation data. Their continuously decreased errors (MAE, MSE, RMSE, and RMSPE) and improved R^2 on the validation set indicate increased model fit and prediction accuracy. However, it is crucial to consider the possibility that model performance on the validation data may not generalize to new data. Therefore, additional testing on other datasets, such as a different test set, is required.

Table 3. Performance metrics for various models on the validation data.

Sr No.	Model	MAE_Val	MSE_Val	RMSE_Val	RMSPE_Val	R^2_Val
1	CatBoost	261.90	1,430,686.30	1196.11	346.56	0.65
2	ElasticNet	385.08	1,555,769.49	1247.30	778.53	0.61
3	KNN	329.22	1,960,894.83	1400.32	446.31	0.51
4	Lasso	293.32	1,156,911.27	1075.60	538.62	0.71
5	LGBM	**243.10**	1,181,938.31	1087.17	**287.91**	**0.71**
6	LR	393.23	1,194,250.83	1092.82	992.99	0.70
7	MLP	249.45	**1,069,732.66**	**1034.28**	307.27	0.73
8	RF	259.75	1,386,585.60	1177.53	368.38	0.66
9	Ridge	296.56	1,157,972.15	1076.09	579.68	0.71
10	SGD	345.98	1,183,130.23	1087.72	908.38	0.71
11	XGBoost	264.54	1,349,874.60	1161.84	419.95	0.67

Bold value shows the better solution.

6.3. Performance Metrics of Testing Data

The performance metrics of several models on the testing data are shown in Table 4. For each model, the metrics are MAE, MSE, RMSE, RMSPE, and R^2. The following findings may be drawn from examining how well the models performed on the testing data: with lower MAE, MSE, RMSE, and RMSPE values and greater R^2, LGBM, CatBoost, and MLP demonstrate improved performance on the test data. In addition to ElasticNet, Lasso, RF, Ridge, XGBoost, and others exhibit acceptable performance, with modest values for the metrics. The MAE, MSE, RMSE, RMSPE, and lower R^2 values for KNN, LR, and SGD are comparatively greater, indicating poor performance on the testing data. LGBM, CatBoost, and MLP perform better on the testing data when compared to the other models. They routinely achieve reduced errors (MAE, MSE, RMSE, RMSPE), greater R^2, and better model fit on the testing set, all of which indicate enhanced prediction accuracy.

Table 4. Performance metrics for various models on the testing data.

Sr No.	Model	MAE_Test	MSE_Test	RMSE_Test	RMSPE_Test	R^2_Test
1	CatBoost	108.24	135,853.97	368.58	**327.13**	0.66
2	ElasticNet	267.84	195,282.23	441.91	1308.04	0.52
3	KNN	163.42	257,940.28	507.88	1067.24	0.36
4	Lasso	183.20	141,977.14	376.80	959.14	0.65
5	LGBM	**105.68**	**115,456.65**	**339.79**	332.76	**0.71**
6	LR	292.27	209,780.42	458.02	1424.00	0.48
7	MLP	131.03	123,120.76	350.89	466.30	0.69
8	RF	123.84	152,710.94	390.78	831.76	0.62
9	Ridge	187.82	146,634.81	382.93	996.15	0.64
10	SGD	252.24	195,665.92	442.34	1451.56	0.51
11	XGBoost	129.03	171,242.26	413.81	1102.39	0.58

Bold value shows the better solution.

6.4. Comparison of the Models

A comparison of the performance metrics across the three datasets (training, validation, and testing) was conducted to identify the best-performing model. The performance measures from each of the Tables 2–4 were observed.

a. Training Data: XGBoost has the highest R^2 and the lowest MAE, MSE, RMSE, and RMSPE values, indicating the best performance on the training data. The time series prediction for XGBoost is shown in Figure 5, where predicted streamflow inflows are depicted alongside the actual data. The fundamental patterns and fluctuations in streamflow across the dataset are largely captured by the XGBoost model, as can be seen in this figure.

b. Validation Data: the LGBM model has the highest R^2 and the lowest MAE, MSE, RMSE, and RMSPE values, demonstrating the best performance on the validation data. The time series prediction for LGBM against the actual data is shown in Figure 6.

c. Testing Data: LGBM has the highest R^2 and the lowest MAE, MSE, and RMSE values, showing the best performance on the testing data.

The study's findings provide strong evidence regarding the performance of different models on various datasets, with noticeable differences potentially attributable to overfitting or underfitting. In particular, the results suggest that XGBoost may have overfit the training dataset, resulting in less impressive performance on the test dataset, despite its excellent performance on the training data. Conversely, LGBM performed better on both the validation and testing datasets, suggesting its ability to generalize well to unseen data, although it showed poorer performance on the training set.

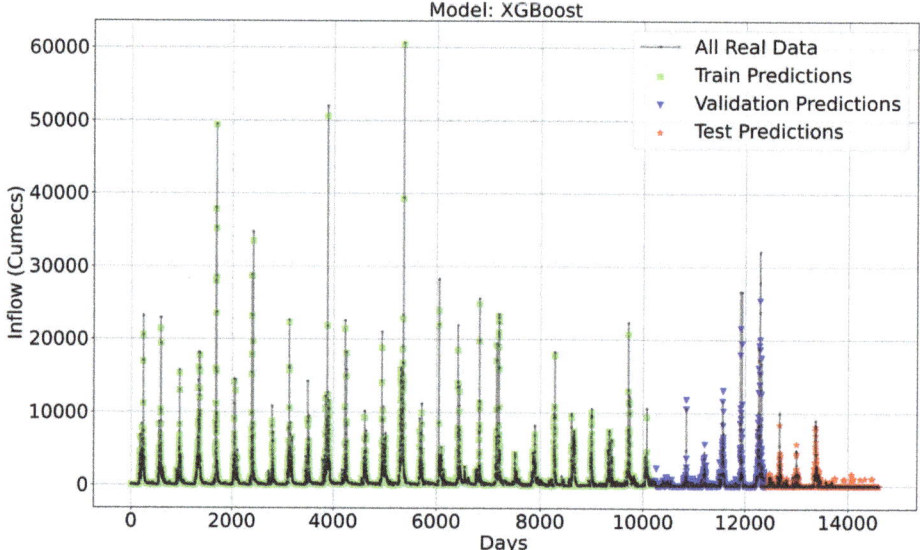

Figure 5. Time-series prediction for the XGBoost.

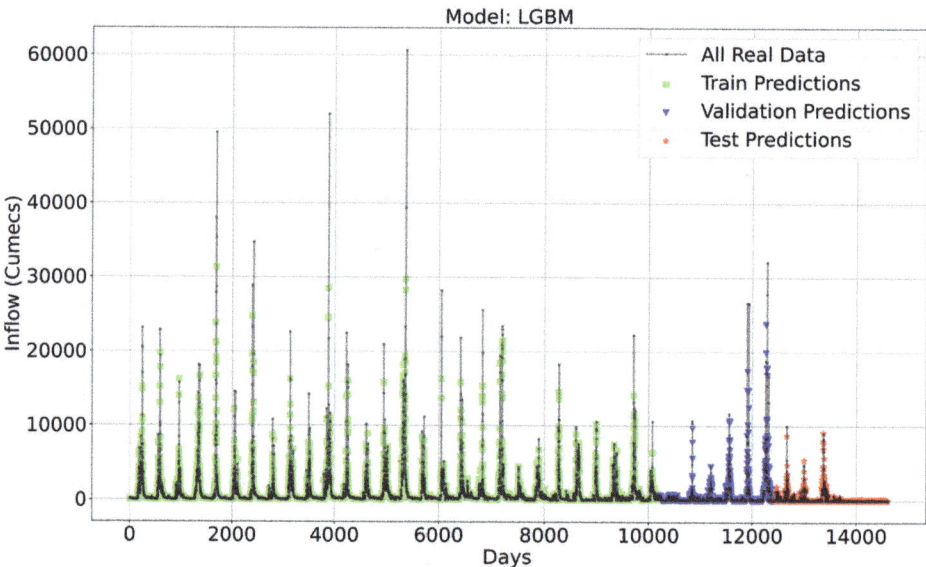

Figure 6. Time series prediction for the LGBM.

Among all, the CatBoost model demonstrated reliable generalization ability, showcased by its robust performance on the training and testing datasets. This suggests that CatBoost is capable of producing accurate predictions even for novel and untested data, as illustrated in Figure 7. However, based on these results, it remains challenging to definitively determine which model performed best in this study.

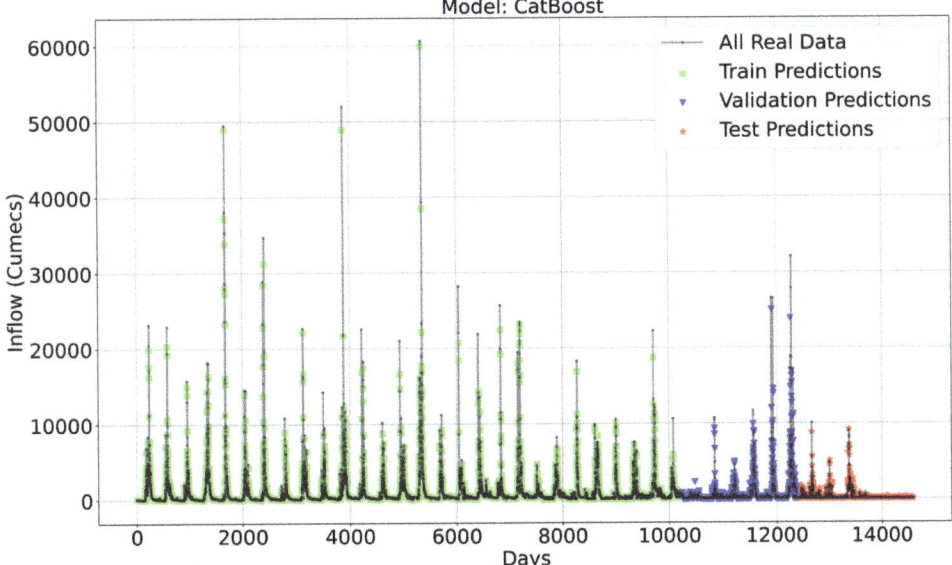

Figure 7. Time series prediction for the CatBoost.

For clearer understanding, scatter plots (shown in Figures 8–10) were generated to illustrate the correlation between the predicted and actual streamflow inflow for XGBoost, LGBM, and CatBoost. An examination of these figures reveals that most data points indicate an error of less than 10% for larger inflow values and less than 20% for moderate inflow levels. In contrast, both XGBoost and LGBM show a higher percentage of data points with errors exceeding 35% for moderate inflow levels above 10,000. Similarly, for CatBoost, inflow levels below 6000 exhibit a larger error rate, of about 35%. It is crucial to note that these lower inflow levels were not the primary focus of this investigation.

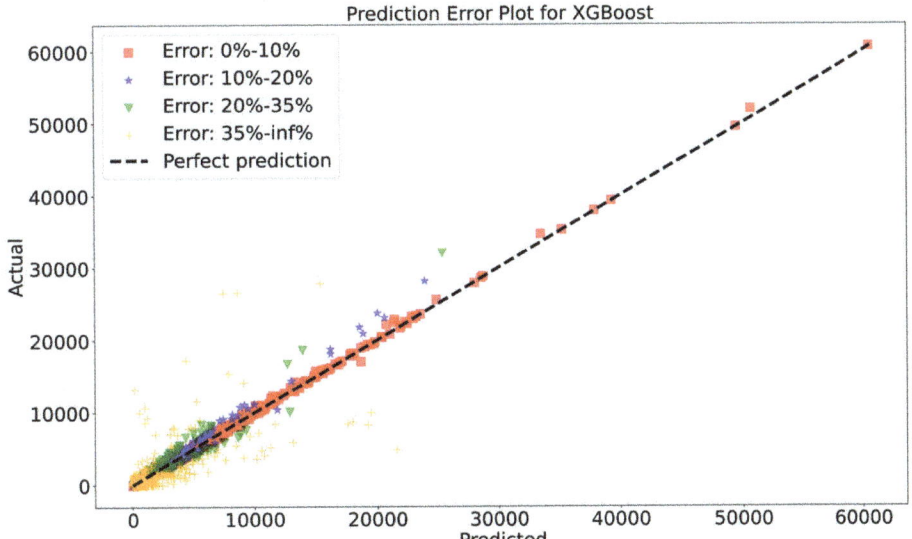

Figure 8. Scatter plots of streamflow prediction for the XGBoost.

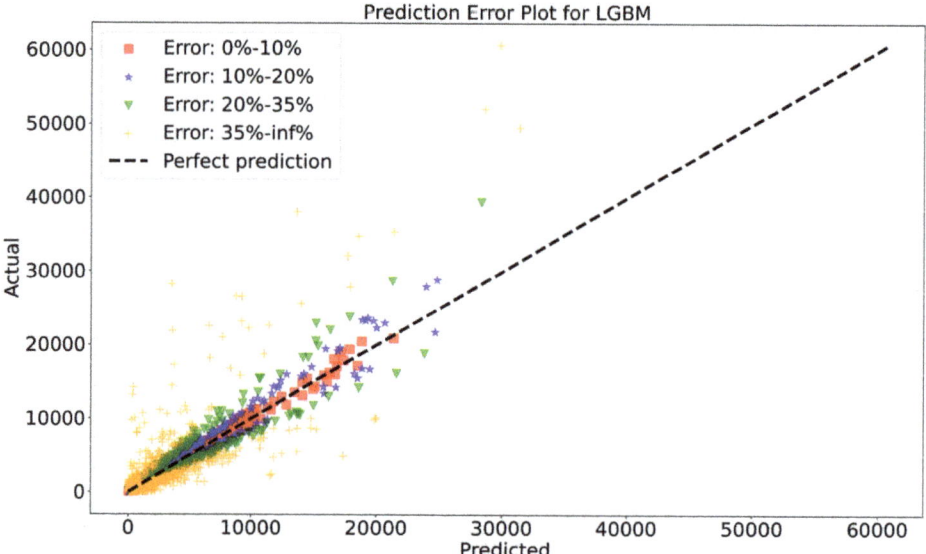

Figure 9. Scatter plots of streamflow prediction for the LGBM.

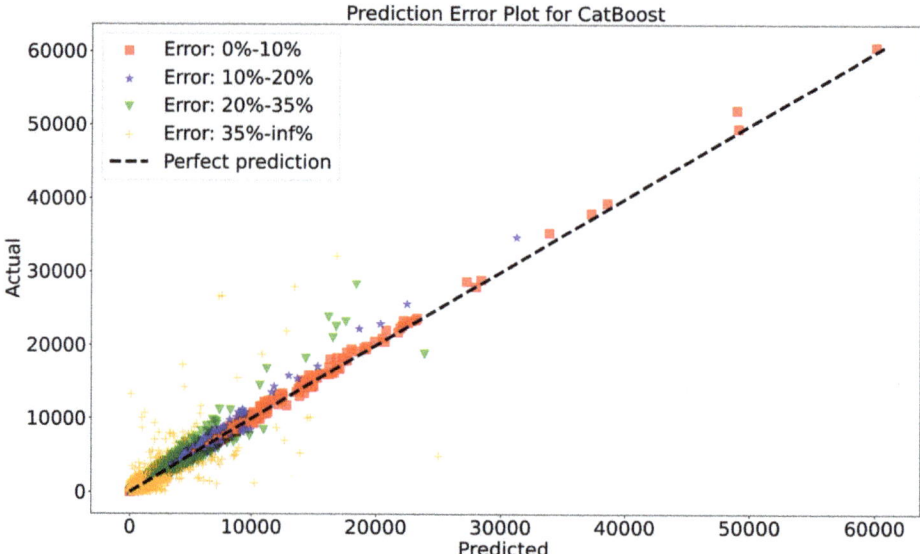

Figure 10. Scatter plots of streamflow prediction for the CatBoost.

Additionally, as demonstrated in Figure 8, XGBoost not only exhibits evidence of overfitting to the training data but also generates inaccurate predictions for higher inflow values in the test data. This raises questions about the accuracy of XGBoost's predictive capabilities under certain circumstances. However, as illustrated in Figure 9, LGBM struggles to accurately predict key factors related to higher inflow levels.

Taking all these factors into account, it can be confidently stated that the CatBoost model outperforms both XGBoost and LGBM in terms of robustness and reliability for inflow predictions. CatBoost is a particularly suitable choice for applications requiring accurate prediction of inflow quantities under specific circumstances. In summary, CatBoost emerges as the most reliable model and a viable option for predicting inflow.

6.5. Limitations of the Study

While the study has provided a comprehensive analysis of various machine learning models for river inflow prediction and identified the most reliable model, it is indeed essential to address the limitations of the study.

(a) One limitation of our research is the reliance on a specific dataset from the Garudeshwar gauging station. The generalizability of the findings may be limited to this particular location, and may not directly apply to other river systems. Future studies should consider incorporating data from multiple gauging stations or rivers to validate the performance of the models across different regions.

(b) Another limitation is the time frame of the dataset used in the study, which spans from 1980 to 2019. Although this provides a substantial historical perspective, it may not capture recent changes or evolving patterns in river inflow. Incorporating more up-to-date data would enhance the accuracy and relevance of the predictions.

(c) Additionally, the study focused primarily on machine learning models and did not consider other factors that could influence river inflow, such as climate change, land use changes, or anthropogenic activities. Incorporating these factors into the modeling process may provide a more comprehensive understanding of the dynamics of river inflow.

(d) Lastly, the performance of the models may be influenced by the quality and completeness of the data. Data quality issues, such as measurement errors, could impact the accuracy of the predictions. It is crucial for future research to address data preprocessing and quality control techniques to mitigate such limitations.

7. Conclusions

To effectively manage water resources, this study compared the efficacy of several machine learning models for predicting river inflow. Models including CatBoost, ElasticNet, KNN, Lasso, LGBM, LR, MLP, RF, Ridge, SGD, and XGBoost were all investigated. CatBoost consistently outperformed other models across all three datasets, displaying remarkable performance across various metrics. It achieved impressive R^2 values on both the training and validation data, demonstrating a strong fit to the data and accurately capturing the variation in the target variable. Additionally, it performed well on the testing data, with relatively low MAE and RMSE values. LGBM also performed well across all three datasets, achieving competitive results for MAE, MSE, RMSE, and R^2 on both the testing and validation data, and demonstrated reasonable MAE and RMSE on the testing data. LGBM, renowned for its effective gradient-boosting implementation and its ability to handle large datasets and capture intricate correlations, showcased these strengths in this study. Results from XGBoost were encouraging, especially when applied to the training and validation data. It achieved the lowest MAE, MSE, RMSE, and RMSPE values on the training set, demonstrating an excellent fit. It also displayed reasonably low MAE and RMSE on the validation data, indicating strong generalization. However, it performed somewhat worse than CatBoost and LGBM in terms of R^2 scores on the testing data. Based on careful investigation and comparison from error plots, CatBoost was determined to have the best performance among the models. CatBoost performed optimally on the test data, demonstrating its ability to make accurate predictions on new, unseen data. Future studies should explore ensemble approaches, which combine the strengths of multiple models to enhance prediction accuracy. Incorporating domain knowledge and additional pertinent factors may also improve the performance of the models. To maintain the efficacy of these models in hydrological forecasting, continuous updating of the models with fresh data will be necessary.

Author Contributions: Conceptualization, V.K.; Software, V.K.; Validation, D.J.M.; Formal analysis, K.V.S. and D.J.M.; Investigation, N.K.; Writing—original draft, T.C.; Writing—review & editing, T.C. All authors have read and agreed to the published version of the manuscript.

Funding: This research received no external funding.

Data Availability Statement: The data presented in this study are available on request from the corresponding authors.

Conflicts of Interest: The authors declare no conflict of interest.

References

1. El-Shafie, A.; Taha, M.R.; Noureldin, A. A neuro-fuzzy model for inflow forecasting of the Nile river at Aswan high dam. *Water Resour. Manag.* **2007**, *21*, 533–556. [CrossRef]
2. Stakhiv, E.; Stewart, B. Needs for Climate Information in Support of Decision-Making in the Water Sector. *Procedia Environ. Sci.* **2010**, *1*, 102–119. [CrossRef]
3. Kumar, V.; Yadav, S.M. Multi-objective reservoir operation of the Ukai reservoir system using an improved Jaya algorithm. *Water Supply* **2022**, *22*, 2287–2310. [CrossRef]
4. Chabokpour, J.; Chaplot, B.; Dasineh, M.; Ghaderi, A.; Azamathulla, H.M. Functioning of the multilinear lag-cascade flood routing model as a means of transporting pollutants in the river. *Water Supply* **2020**, *20*, 2845–2857. [CrossRef]
5. Venkataraman, K.; Tummuri, S.; Medina, A.; Perry, J. 21st century drought outlook for major climate divisions of Texas based on CMIP5 multimodel ensemble: Implications for water resource management. *J. Hydrol.* **2016**, *534*, 300–316. [CrossRef]
6. Hanak, E.; Lund, J.R. Adapting California's water management to climate change. *Clim. Chang.* **2012**, *111*, 17–44. [CrossRef]
7. Sharma, K.V.; Kumar, V.; Singh, K.; Mehta, D.J. LANDSAT 8 LST Pan sharpening using novel principal component based downscaling model. *Remote Sens. Appl. Soc. Environ.* **2023**, *30*, 100963. [CrossRef]
8. Cho, K.; Kim, Y. Improving streamflow prediction in the WRF-Hydro model with LSTM networks. *J. Hydrol.* **2022**, *605*, 127297. [CrossRef]
9. Nearing, G.S.; Kratzert, F.; Sampson, A.K.; Pelissier, C.S.; Klotz, D.; Frame, J.M.; Prieto, C.; Gupta, H.V. What Role Does Hydrological Science Play in the Age of Machine Learning? *Water Resour. Res.* **2021**, *57*, e2020WR028091. [CrossRef]
10. Liang, J.; Li, W.; Bradford, S.; Šimůnek, J. Physics-Informed Data-Driven Models to Predict Surface Runoff Water Quantity and Quality in Agricultural Fields. *Water* **2019**, *11*, 200. [CrossRef]
11. Dinic, F.; Singh, K.; Dong, T.; Rezazadeh, M.; Wang, Z.; Khosrozadeh, A.; Yuan, T.; Voznyy, O. Applied Machine Learning for Developing Next-Generation Functional Materials. *Adv. Funct. Mater.* **2021**, *31*, 2104195. [CrossRef]
12. Clark, M.P.; Fan, Y.; Lawrence, D.M.; Adam, J.C.; Bolster, D.; Gochis, D.J.; Hooper, R.P.; Kumar, M.; Leung, L.R.; Mackay, D.S.; et al. Improving the representation of hydrologic processes in Earth System Models. *Water Resour. Res.* **2015**, *51*, 5929–5956. [CrossRef]
13. Legesse, D.; Vallet-Coulomb, C.; Gasse, F. Hydrological response of a catchment to climate and land use changes in Tropical Africa: Case study South Central Ethiopia. *J. Hydrol.* **2003**, *275*, 67–85. [CrossRef]
14. Yang, S.; Wan, M.P.; Chen, W.; Ng, B.F.; Dubey, S. Model predictive control with adaptive machine-learning-based model for building energy efficiency and comfort optimization. *Appl. Energy* **2020**, *271*, 115147. [CrossRef]
15. Wang, Z.; Yang, W.; Liu, Q.; Zhao, Y.; Liu, P.; Wu, D.; Banu, M.; Chen, L. Data-driven modeling of process, structure and property in additive manufacturing: A review and future directions. *J. Manuf. Process.* **2022**, *77*, 13–31. [CrossRef]
16. Hernández-Rojas, L.F.; Abrego-Perez, A.L.; Lozano Martínez, F.E.; Valencia-Arboleda, C.F.; Diaz-Jimenez, M.C.; Pacheco-Carvajal, N.; Garcia-Cardenas, J.J. The Role of Data-Driven Methodologies in Weather Index Insurance. *Appl. Sci.* **2023**, *13*, 4785. [CrossRef]
17. Feng, D.; Lawson, K.; Shen, C. Mitigating Prediction Error of Deep Learning Streamflow Models in Large Data-Sparse Regions With Ensemble Modeling and Soft Data. *Geophys. Res. Lett.* **2021**, *48*, e2021GL092999. [CrossRef]
18. San, O.; Rasheed, A.; Kvamsdal, T. Hybrid analysis and modeling, eclecticism, and multifidelity computing toward digital twin revolution. *GAMM-Mitt.* **2021**, *44*, e202100007. [CrossRef]
19. Aliashrafi, A.; Zhang, Y.; Groenewegen, H.; Peleato, N.M. A review of data-driven modelling in drinking water treatment. *Rev. Environ. Sci. Bio/Technol.* **2021**, *20*, 985–1009. [CrossRef]
20. Singh, K.; Singh, B.; Sihag, P.; Kumar, V.; Sharma, K.V. Development and application of modeling techniques to estimate the unsaturated hydraulic conductivity. *Model. Earth Syst. Environ.* **2023**. [CrossRef]
21. Yang, D.; Chen, K.; Yang, M.; Zhao, X. Urban rail transit passenger flow forecast based on LSTM with enhanced long-term features. *IET Intell. Transp. Syst.* **2019**, *13*, 1475–1482. [CrossRef]
22. Nagar, U.P.; Patel, H.M. Development of Short-Term Reservoir Level Forecasting Models: A Case Study of Ajwa-Pratappura Reservoir System of Vishwamitri River Basin of Central Gujarat. In *Hydrology and Hydrologic Modelling—HYDRO 2021*; Timbadiya, P.V., Patel, P.L., Singh, V.P., Sharma, P.J., Eds.; Springer: Singapore, 2023; pp. 261–269. [CrossRef]
23. Mehta, D.J.; Eslamian, S.; Prajapati, K. Flood modelling for a data-scare semi-arid region using 1-D hydrodynamic model: A case study of Navsari Region. *Model. Earth Syst. Environ.* **2022**, *8*, 2675–2685. [CrossRef]
24. Gangani, P.; Mangukiya, N.K.; Mehta, D.J.; Muttil, N.; Rathnayake, U. Evaluating the Efficacy of Different DEMs for Application in Flood Frequency and Risk Mapping of the Indian Coastal River Basin. *Climate* **2023**, *11*, 114. [CrossRef]

25. Omukuti, J.; Wanzala, M.A.; Ngaina, J.; Ganola, P. Develop medium- to long-term climate information services to enhance comprehensive climate risk management in Africa. *Clim. Resil. Sustain.* **2023**, *2*, e247. [CrossRef]
26. Kumar, V.; Yadav, S.M. A state-of-the-Art review of heuristic and metaheuristic optimization techniques for the management of water resources. *Water Supply* **2022**, *22*, 3702–3728. [CrossRef]
27. Rivera-González, L.; Bolonio, D.; Mazadiego, L.F.; Valencia-Chapi, R. Long-Term Electricity Supply and Demand Forecast (2018–2040): A LEAP Model Application towards a Sustainable Power Generation System in Ecuador. *Sustainability* **2019**, *11*, 5316. [CrossRef]
28. Singh, D.; Vardhan, M.; Sahu, R.; Chatterjee, D.; Chauhan, P.; Liu, S. Machine-learning- and deep-learning-based streamflow prediction in a hilly catchment for future scenarios using CMIP6 GCM data. *Hydrol. Earth Syst. Sci.* **2023**, *27*, 1047–1075. [CrossRef]
29. Mohammadi, B. A review on the applications of machine learning for runoff modeling. *Sustain. Water Resour. Manag.* **2021**, *7*, 98. [CrossRef]
30. Ibrahim, K.S.M.H.; Huang, Y.F.; Ahmed, A.N.; Koo, C.H.; El-Shafie, A. Forecasting multi-step-ahead reservoir monthly and daily inflow using machine learning models based on different scenarios. *Appl. Intell.* **2023**, *53*, 10893–10916. [CrossRef]
31. Rajesh, M.; Anishka, S.; Viksit, P.S.; Arohi, S.; Rehana, S. Improving Short-range Reservoir Inflow Forecasts with Machine Learning Model Combination. *Water Resour. Manag.* **2023**, *37*, 75–90. [CrossRef]
32. Cai, H.; Liu, S.; Shi, H.; Zhou, Z.; Jiang, S.; Babovic, V. Toward improved lumped groundwater level predictions at catchment scale: Mutual integration of water balance mechanism and deep learning method. *J. Hydrol.* **2022**, *613*, 128495. [CrossRef]
33. Jiang, S.; Zheng, Y.; Wang, C.; Babovic, V. Uncovering Flooding Mechanisms Across the Contiguous United States Through Interpretive Deep Learning on Representative Catchments. *Water Resour. Res.* **2022**, *58*, e2021WR030185. [CrossRef]
34. Herath, H.M.V.V.; Chadalawada, J.; Babovic, V. Hydrologically informed machine learning for rainfall–runoff modelling: Towards distributed modelling. *Hydrol. Earth Syst. Sci.* **2021**, *25*, 4373–4401. [CrossRef]
35. Chadalawada, J.; Herath, H.M.V.V.; Babovic, V. Hydrologically Informed Machine Learning for Rainfall-Runoff Modeling: A Genetic Programming-Based Toolkit for Automatic Model Induction. *Water Resour. Res.* **2020**, *56*, e2019WR026933. [CrossRef]
36. Lima, C.H.R.; Lall, U. Spatial scaling in a changing climate: A hierarchical bayesian model for non-stationary multi-site annual maximum and monthly streamflow. *J. Hydrol.* **2010**, *383*, 307–318. [CrossRef]
37. Turner, S.W.D.; Marlow, D.; Ekström, M.; Rhodes, B.G.; Kularathna, U.; Jeffrey, P.J. Linking climate projections to performance: A yield-based decision scaling assessment of a large urban water resources system. *Water Resour. Res.* **2014**, *50*, 3553–3567. [CrossRef]
38. Ab Razak, N.H.; Aris, A.Z.; Ramli, M.F.; Looi, L.J.; Juahir, H. Temporal flood incidence forecasting for Segamat River (Malaysia) using autoregressive integrated moving average modelling. *J. Flood Risk Manag.* **2018**, *11*, S794–S804. [CrossRef]
39. Banihabib, M.E.; Bandari, R.; Valipour, M. Improving Daily Peak Flow Forecasts Using Hybrid Fourier-Series Autoregressive Integrated Moving Average and Recurrent Artificial Neural Network Models. *AI* **2020**, *1*, 263–275. [CrossRef]
40. Demirel, M.C.; Venancio, A.; Kahya, E. Flow forecast by SWAT model and ANN in Pracana basin, Portugal. *Adv. Eng. Softw.* **2009**, *40*, 467–473. [CrossRef]
41. Chen, J.; Wu, Y. Advancing representation of hydrologic processes in the Soil and Water Assessment Tool (SWAT) through integration of the TOPographic MODEL (TOPMODEL) features. *J. Hydrol.* **2012**, *420–421*, 319–328. [CrossRef]
42. Yaseen, Z.M.; El-shafie, A.; Jaafar, O.; Afan, H.A.; Sayl, K.N. Artificial intelligence based models for stream-flow forecasting: 2000–2015. *J. Hydrol.* **2015**, *530*, 829–844. [CrossRef]
43. Dong, N.; Guan, W.; Cao, J.; Zou, Y.; Yang, M.; Wei, J.; Chen, L.; Wang, H. A hybrid hydrologic modelling framework with data-driven and conceptual reservoir operation schemes for reservoir impact assessment and predictions. *J. Hydrol.* **2023**, *619*, 129246. [CrossRef]
44. Kumar, V.; Sharma, K.V.; Caloiero, T.; Mehta, D.J.; Singh, K. Comprehensive Overview of Flood Modeling Approaches: A Review of Recent Advances. *Hydrology* **2023**, *10*, 141. [CrossRef]
45. Ikram, R.M.A.; Ewees, A.A.; Parmar, K.S.; Yaseen, Z.M.; Shahid, S.; Kisi, O. The viability of extended marine predators algorithm-based artificial neural networks for streamflow prediction. *Appl. Soft Comput.* **2022**, *131*, 109739. [CrossRef]
46. Ni, L.; Wang, D.; Wu, J.; Wang, Y.; Tao, Y.; Zhang, J.; Liu, J. Streamflow forecasting using extreme gradient boosting model coupled with Gaussian mixture model. *J. Hydrol.* **2020**, *586*, 124901. [CrossRef]
47. Meresa, H. Modelling of river flow in ungauged catchment using remote sensing data: Application of the empirical (SCS-CN), Artificial Neural Network (ANN) and Hydrological Model (HEC-HMS). *Model. Earth Syst. Environ.* **2019**, *5*, 257–273. [CrossRef]
48. Adnan, R.M.; Kisi, O.; Mostafa, R.R.; Ahmed, A.N.; El-Shafie, A. The potential of a novel support vector machine trained with modified mayfly optimization algorithm for streamflow prediction. *Hydrol. Sci. J.* **2022**, *67*, 161–174. [CrossRef]
49. Meng, E.; Huang, S.; Huang, Q.; Fang, W.; Wu, L.; Wang, L. A robust method for non-stationary streamflow prediction based on improved EMD-SVM model. *J. Hydrol.* **2019**, *568*, 462–478. [CrossRef]
50. Noori, R.; Karbassi, A.R.; Moghaddamnia, A.; Han, D.; Zokaei-Ashtiani, M.H.; Farokhnia, A.; Gousheh, M.G. Assessment of input variables determination on the SVM model performance using PCA, Gamma test, and forward selection techniques for monthly stream flow prediction. *J. Hydrol.* **2011**, *401*, 177–189. [CrossRef]
51. Tyralis, H.; Papacharalampous, G.; Langousis, A. A Brief Review of Random Forests for Water Scientists and Practitioners and Their Recent History in Water Resources. *Water* **2019**, *11*, 910. [CrossRef]

52. Tyralis, H.; Papacharalampous, G.; Langousis, A. Super ensemble learning for daily streamflow forecasting: Large-scale demonstration and comparison with multiple machine learning algorithms. *Neural Comput. Appl.* **2021**, *33*, 3053–3068. [CrossRef]
53. Song, Z.; Xia, J.; Wang, G.; She, D.; Hu, C.; Hong, S. Regionalization of hydrological model parameters using gradient boosting machine. *Hydrol. Earth Syst. Sci.* **2022**, *26*, 505–524. [CrossRef]
54. Akbarian, M.; Saghafian, B.; Golian, S. Monthly streamflow forecasting by machine learning methods using dynamic weather prediction model outputs over Iran. *J. Hydrol.* **2023**, *620*, 129480. [CrossRef]
55. Luo, P.; Luo, M.; Li, F.; Qi, X.; Huo, A.; Wang, Z.; He, B.; Takara, K.; Nover, D.; Wang, Y. Urban flood numerical simulation: Research, methods and future perspectives. *Environ. Model. Softw.* **2022**, *156*, 105478. [CrossRef]
56. Kumar, V.; Azamathulla, H.M.; Sharma, K.V.; Mehta, D.J.; Maharaj, K.T. The State of the Art in Deep Learning Applications, Challenges, and Future Prospects: A Comprehensive Review of Flood Forecasting and Management. *Sustainability* **2023**, *15*, 10543. [CrossRef]
57. Niu, W.; Feng, Z. Evaluating the performances of several artificial intelligence methods in forecasting daily streamflow time series for sustainable water resources management. *Sustain. Cities Soc.* **2021**, *64*, 102562. [CrossRef]
58. Bhasme, P.; Vagadiya, J.; Bhatia, U. Enhancing predictive skills in physically-consistent way: Physics Informed Machine Learning for Hydrological Processes. *arXiv* **2021**, arXiv:2104.11009. [CrossRef]
59. Souza, D.P.M.; Martinho, A.D.; Rocha, C.C.; da S. Christo, E.; Goliatt, L. Hybrid particle swarm optimization and group method of data handling for short-term prediction of natural daily streamflows. *Model. Earth Syst. Environ.* **2022**, *8*, 5743–5759. [CrossRef]
60. Martinho, A.D.; Saporetti, C.M.; Goliatt, L. Approaches for the short-term prediction of natural daily streamflows using hybrid machine learning enhanced with grey wolf optimization. *Hydrol. Sci. J.* **2023**, *68*, 16–33. [CrossRef]
61. Haznedar, B.; Kilinc, H.C.; Ozkan, F.; Yurtsever, A. Streamflow forecasting using a hybrid LSTM-PSO approach: The case of Seyhan Basin. *Nat. Hazards* **2023**, *117*, 681–701. [CrossRef]
62. Hao, R.; Bai, Z. Comparative Study for Daily Streamflow Simulation with Different Machine Learning Methods. *Water* **2023**, *15*, 1179. [CrossRef]
63. Bakhshi Ostadkalayeh, F.; Moradi, S.; Asadi, A.; Moghaddam Nia, A.; Taheri, S. Performance Improvement of LSTM-based Deep Learning Model for Streamflow Forecasting Using Kalman Filtering. *Water Resour. Manag.* **2023**, *37*, 3111–3127. [CrossRef]
64. Prokhorenkova, L.; Gusev, G.; Vorobev, A.; Dorogush, A.V.; Gulin, A. Catboost: Unbiased boosting with categorical features. In Proceedings of the 32nd International Conference on Neural Information Processing Systems, Montréal, QC, Canada, 2–8 December 2018; Volume 31, pp. 6638–6648.
65. Kramer, O. K-Nearest Neighbors. In *Dimensionality Reduction with Unsupervised Nearest Neighbors*; Springer: Berlin/Heidelberg, Germany, 2013; Volume 51, pp. 13–23. [CrossRef]
66. Fan, J.; Ma, X.; Wu, L.; Zhang, F.; Yu, X.; Zeng, W. Light Gradient Boosting Machine: An efficient soft computing model for estimating daily reference evapotranspiration with local and external meteorological data. *Agric. Water Manag.* **2019**, *225*, 105758. [CrossRef]
67. Su, X.; Yan, X.; Tsai, C.-L. Linear regression. *Wiley Interdiscip. Rev. Comput. Stat.* **2012**, *4*, 275–294. [CrossRef]
68. Gardner, M.; Dorling, S. Artificial neural networks (the multilayer perceptron)—A review of applications in the atmospheric sciences. *Atmos. Environ.* **1998**, *32*, 2627–2636. [CrossRef]
69. Biau, G.; Scornet, E. A random forest guided tour. *TEST* **2016**, *25*, 197–227. [CrossRef]
70. Luo, X.; Chang, X.; Ban, X. Regression and classification using extreme learning machine based on L1-norm and L2-norm. *Neurocomputing* **2016**, *174*, 179–186. [CrossRef]
71. McDonald, G.C. Ridge regression. *Wiley Interdiscip. Rev. Comput. Stat.* **2009**, *1*, 93–100. [CrossRef]
72. Ryali, S.; Chen, T.; Supekar, K.; Menon, V. Estimation of functional connectivity in fMRI data using stability selection-based sparse partial correlation with elastic net penalty. *Neuroimage* **2012**, *59*, 3852–3861. [CrossRef]
73. Song, S.; Chaudhuri, K.; Sarwate, A.D. Stochastic gradient descent with differentially private updates. In Proceedings of the 2013 IEEE Global Conference on Signal and Information Processing, Austin, TX, USA, 3–5 December 2013; pp. 245–248. [CrossRef]
74. Sheridan, R.P.; Wang, W.M.; Liaw, A.; Ma, J.; Gifford, E.M. Extreme Gradient Boosting as a Method for Quantitative Structure–Activity Relationships. *J. Chem. Inf. Model.* **2016**, *56*, 2353–2360. [CrossRef]
75. Chadalawada, J.; Babovic, V. Review and comparison of performance indices for automatic model induction. *J. Hydroinform.* **2019**, *21*, 13–31. [CrossRef]

Disclaimer/Publisher's Note: The statements, opinions and data contained in all publications are solely those of the individual author(s) and contributor(s) and not of MDPI and/or the editor(s). MDPI and/or the editor(s) disclaim responsibility for any injury to people or property resulting from any ideas, methods, instructions or products referred to in the content.

Article

A Model for Assessing the Importance of Runoff Forecasts in Periodic Climate on Hydropower Production

Shuang Hao [1,*], Anders Wörman [1], Joakim Riml [1] and Andrea Bottacin-Busolin [2]

1. Department of Sustainable Development Environmental Science and Engineering, Royal Institute of Technology (KTH), 100 44 Stockholm, Sweden; worman@kth.se (A.W.); riml@kth.se (J.R.)
2. Department of Industrial Engineering, University of Padua, Via Venezia 1, 35121 Padova, Italy; andrea.bottacinbusolin@unipd.it
* Correspondence: shuangha@kth.se

Abstract: Hydropower is the largest source of renewable energy in the world and currently dominates flexible electricity production capacity. However, climate variations remain major challenges for efficient production planning, especially the annual forecasting of periodically variable inflows and their effects on electricity generation. This study presents a model that assesses the impact of forecast quality on the efficiency of hydropower operations. The model uses ensemble forecasting and stepwise linear optimisation combined with receding horizon control to simulate runoff and the operation of a cascading hydropower system. In the first application, the model framework is applied to the Dalälven River basin in Sweden. The efficiency of hydropower operations is found to depend significantly on the linkage between the representative biannual hydrologic regime and the regime actually realised in a future scenario. The forecasting error decreases when considering periodic hydroclimate fluctuations, such as the dry–wet year variability evident in the runoff in the Dalälven River, which ultimately increases production efficiency by approximately 2% (at its largest), as is shown in scenarios 1 and 2. The corresponding potential hydropower production is found to vary by 80 GWh/year. The reduction in forecasting error when considering biennial periodicity corresponds to a production efficiency improvement of about 0.33% (or 13.2 GWh/year).

Keywords: ensemble forecasting; biennial periodic climate; hydropower optimisation; hydropower management; production efficiency; forecasting error

Citation: Hao, S.; Wörman, A.; Riml, J.; Bottacin-Busolin, A. A Model for Assessing the Importance of Runoff Forecasts in Periodic Climate on Hydropower Production. *Water* **2023**, *15*, 1559. https://doi.org/10.3390/w15081559

Academic Editors: Sonia Raquel Gámiz-Fortis and Matilde García-Valdecasas Ojeda

Received: 16 March 2023
Revised: 9 April 2023
Accepted: 13 April 2023
Published: 16 April 2023

Copyright: © 2023 by the authors. Licensee MDPI, Basel, Switzerland. This article is an open access article distributed under the terms and conditions of the Creative Commons Attribution (CC BY) license (https://creativecommons.org/licenses/by/4.0/).

1. Introduction

Hydropower is the largest renewable source of electricity by power capacity and the second largest by annual energy production. Hence, it has great potential for remediating the transition towards a future renewable electric production system, particularly due to the regulatory role played by the energy storage capacity of hydropower reservoirs. However, the planning and management of hydropower regulations remain complex, as these depend not only on matching the electricity demand to the availability of other renewables but also on the management of significant climate variations. Climate variation affects precipitation and temperature patterns and thus stream flows, which, in turn, reduces the reliability of hydropower planning and generation [1–5]. Studies have shown that climate change will alter the temporal and spatial distribution of water resources worldwide [6–8], and there are strong indications of significant periodicity in historical hydrologic records. The authors of [9–11] showed that hydropower availability in Scandinavia varies over a spectrum of periods, with robust periodicity identified at approximately 0.5, 2, and 8–11 year intervals, respectively. The authors of [12] found that 18 selected drainage basins worldwide have periodic water availability variations, with periodic waves ranging from 2.1 to 2.5 years. In [13], river discharge time series of Colombian streamflow were examined, and Fourier analysis showed a spectral peak at a periodicity of 2.17 years. Furthermore, statistically significant periodicity was found in a time series in the 5–7 and 2–3 year bands when performing a spectral analysis of precipitation in the US [14]. The authors of [15] used the maximum entropy method and Fourier spectral analysis to show temperature and precipitation periodicities of approximately 2–3 years for Siberia and East Asia. Consequently, there are strong indications that biennial periodicity exists among short- and long-term periodicities in many basins worldwide [9–15]. The two year period indicates a predominant pattern of sequential dry and wet years, but this pattern is not necessarily recognised in long-term forecasting and hydropower production management. Thus, an important question is how the forecasting of hydroclimatic variations with biennial periodicity impacts hydropower production operational planning. The dry and wet periods will have basic control over the availability of water, hence the potential electricity production during such periods; however, forecasting these conditions can have an important secondary effect on production efficiency, which is the topic of the present study. For example, statistically better knowledge of whether the coming year will be relatively wet or dry will likely lead to better decisions in operational planning and less water spillage. To investigate the relative importance of forecast skills [16] for the planning of hydropower operations, a model framework that can simulate and assess this operational process is needed, including hydrologic forecasting, decision optimisation, and the estimation of production efficiency in an independent future hydrologic scenario.

Previous research has developed model frameworks for studying the impacts of climate variations on hydropower; however, it has not sufficiently acknowledged a model framework that can assess the importance of forecasting periodic hydroclimatic fluctuations for hydropower planning and generation. The authors of [7] developed a global hydrological–electricity modelling framework that focused on the physical impacts of water constraints on current power plant capacities. General circulation models (GCMs) and the variable infiltration capacity model were implemented to generate water availability [17–19]. However, none of these studies used stochastic forecasting, such as historic ensemble forecasting, nor separated the forecast from the applied future scenario. The basic ideas behind stochastic forecasting are that nature is difficult to physically predict because of both aleatory and epistemic uncertainties, but historically unbiased samples are also likely to apply well (as forecasts) to the future [20,21]. These samples provide data points, such as a set of experimental outcomes that satisfy a range of statistical measures appearing during the sample period; hence, they can represent various properties of climate periodicity, that are also likely to be representative of near-future scenarios [22]. The authors of [23,24] applied ensemble forecasting in their research but focused on hydrologic predictions rather than its implications for hydropower production. Thus, a model framework for assessing and analysing the importance of forecasting hydroclimatic periodicities for the efficiency of hydropower planning and generation is generally missing.

In this study, we attempt to bridge the aforementioned knowledge gap by developing a model framework that can assess (simulate) the importance of forecasting periodic hydroclimate fluctuations for the efficiency of hydropower planning and generation. The innovations of this study comprise the following aspects: (a) the model framework facilitates the investigation of the impact of forecasting periodic climate on hydropower operations; (b) we introduce an ensemble forecasting method based on the classification of dry and wet 12 month periods in historical records; and (c) the assessment model is implemented in a MATLAB environment, including stochastic forecasting, and applied to a cascade hydropower system. The forecast and management are focused on the long-term (seasonal) time horizon, thus neglecting some of the complexities of production management that occur on a time scale of up to a few days or slightly longer. This study examines the Dalälven River basin, which exhibits a typical biennial fluctuation in water availability (as well as other periodicities). This means that the study investigates how the dry and wet years in the river basin affect the management efficiency of hydropower. The main contributions are: (a) a model framework that differentiates stochastically between runoff forecasts and simulated real runoff scenarios, enabling the analysis of the importance of the forecast approach for managing hydropower production in light of the uncertainties in climate variability; (b) an ensemble forecasting method that recognises the hydroclimatic biennial periodicity present in the Dalälven River basin; and (c) an application of the simulation framework to a cascade hydropower system with typical biennial climatic fluctuations. Hydropower production optimisation is based on linear programming combined with a receding horizon approach, which converts the nonlinearity in the hydropower production problem to a stepwise linear problem via a system update step. Section 2 provides details on the model's development.

2. Model Framework
2.1. Assessment Model

The overall purpose of the proposed model is to simulate the management efficiency of hydropower generation resulting from the uncertainty of water availability forecasts that reflect the long-term periodicity observed in hydroclimatic time series. The model addresses the fundamental question of the extent to which improvements in forecasting ability lead to better hydropower operational planning and higher hydropower generation. Therefore, this approach distinguishes forecasted runoff availability from real runoff availability and quantifies the efficiency of operational planning as a function of forecast error and forecast biennial scenarios. In this context, runoff availability is defined as all water added to the system, for example, by precipitation or snowmelt, i.e., water generated through runoff processes, while there is also water available for production because of its storage within the river basin in streams, lakes, and reservoirs. In wet years, water availability is particularly high, leading to high hydropower production, but the model should evaluate the importance of forecast quality for production efficiency by optimising reservoir operations and maximising turbine discharge. Consequently, the simulation process consists of two main submodules: (1) the optimisation of hydropower planning and generation under a several month future forecast horizon, and (2) the updating of model variables (i.e., actual runoff and reservoir level) after a much shorter updating time step while considering the real periodic runoff scenario (Figure 1). In this study, both the generation of the forecasted runoff time series and that of the real runoff time series were based on the sampling of historical data (simulated runoff time series from the Sweden-Hydrological Predictions for the Environment (S-HYPE) model, provided by the Swedish Meteorological and Hydrological Institute). The principles of the ensemble forecasting approach are described in Section 2.1. Furthermore, the optimisation of production operations considers the forecasted future runoff for J future states along a time horizon T_H, the water conservation of the river basin, and power production (called the optimisation model in Figure 1). The optimal production decisions are applied to the duration of an updating period, t_u, based on several stochastic runoff forecasts covering a future horizon, T_H, where q, s, and h represent the turbine discharge, spillage, and water head, respectively. To provide a decision process that is statistically representative, the optimisation is carried out N times for each stochastic runoff forecast. Subsequently, the average values of these N optimal decisions of production and spillage discharges are applied as decided values for an updating time step, t_u. The river basin water availability status is then updated in the system updating module, including updating the real runoff and water head (step 3 in Figure 1). The updated water head is used to set the initial conditions of the reservoir levels for the upcoming optimisation horizon based on the immediate past. The updating aims to ensure that the effects of forecasting errors do not accumulate over time and thus represent the actual operational planning process over a more extended period.

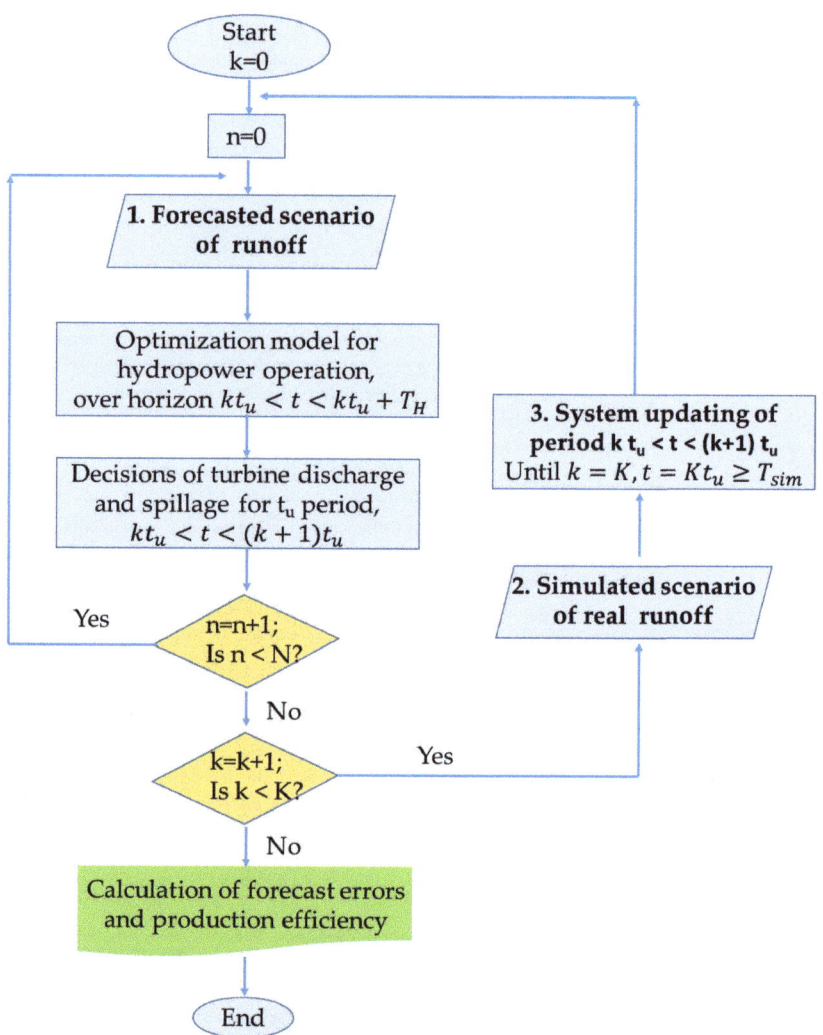

Figure 1. Flowchart of the model framework.

To simplify the optimisation problem and speed up the calculation, we converted the nonlinear statement of power production (see Section 2.2) into a stepwise linear statement by assuming that the water head in the reservoirs are constant during each updating time step, which is acceptable when t_u is sufficiently short. The updating of the water head in the reservoirs after one updating time step t_u ensures that an essential long-term component of nonlinearity is accounted for—the dependence of power production on reservoir levels and the fall height. Linear optimisation is significantly less computationally demanding than nonlinear optimisation. Using a receding horizon approach, after the simulation of each updating time step k, the simulation moves on to the next updating period along the timeline until K updating steps have been executed in total, which forms the entire simulation period, $T_{sim} = K \times t_u$. Consequently, the optimisation problem is solved using linear optimisation programming combined with receding horizon control. After finishing one realisation of the entire simulation period, T_{sim}, the forecasting error and production efficiency are estimated for the entire assessment simulation. The definitions of forecasting error and production efficiency are given in Section 2.3.

2.2. Ensemble Forecasting with Biennial Periodicity

Ensemble forecasting uses a set of possible runoff time series selected from historical data and applies these as runoff forecasts for the future. The forecasted runoff abstracted from such an ensemble time series aims to represent hydroclimatic fluctuations and is therefore divided into periodic segments and classified in terms of particularly wet or dry periods. Dry and wet years generally follow biennial periodicity in Scandinavia [10], but the spectrum of periodicities of hydrologic processes makes it difficult to match the forecast class to the current climate type. However, many studies have shown important cross-correlations with related teleconnections expressed by different climate indices [10,12,15,25,26] that can, in principle, be used as predictors for the current hydrologic regime. The authors of [12] discussed the links between El Niño–La Niña events and biannual discharge fluctuations at the basin scale, as well as global climate indices such as the Southern Oscillation Index, the North Atlantic Oscillation Index, and the Pacific Decadal Oscillation Index. In particular, the authors of [11] used coherence spectra to define the strength of correlation over a range of periods between the two latter climate indices and the overall energy level of the runoff system. Hence, by including the biennial periodicity of the forecasts with the corresponding present climate type, the suggested model investigates how improvements in the forecasted runoff with biennial periodicity can enhance the management of the cascade hydropower system. The goal here is to ensemble runoff data representing the previously identified hydroclimatic fluctuations in runoff, especially the biennial (two year) periodicity, and incorporate such information in a production management system linking forecast class to the current hydrologic regime.

Previous studies using spectral analysis have indicated a biennial periodicity in runoff [10,11], which suggests, for example, that there could be dry and wet years. Such biennial periodicity should be possible to identify directly in runoff time series by assessing the daily average using a different start month and then classifying segments of runoff time series into the categories of odd and even years. One question is the degree to which the start month of the yearly segment division affects the clarity and strength of the dry and wet year periodicities. We investigated the significance of the start month by successively shifting it by one month and using one year-long time-series segments to calculate 12 month discharge statistics. The results show that hydropower production management in December is more sensitive to recognising biennial periodicity. Detailed results can be found in Section 4.2.

Consequently, we suggest that the classification of ensemble members should be conducted in two primary yearly runoff time series from either wet or dry ensemble members that can represent biennial runoff periodicity. As the time horizon of the forecast, T_H, is shorter than the classified segments, there is a possibility of varying the start month. The time series for each subwatershed was abstracted from the 1961–2011 time series, and yearly time series were kept as a statistical repository for stochastic sampling. Each sampled yearly time series starts in December. The random samples of the forecasts and real runoff scenarios were taken from this statistical repository (see step 1 in Figure 1).

2.3. Optimisation Model for Cascade Hydropower Stations

A model based on the stepwise linear optimisation approach for the operation of cascade hydropower stations in a river network was developed in MATLAB 2018b. The model simulates the planning of hydropower generation in a cascade of reservoirs with the aim of maximising electricity production, thus minimising water spillage and maintaining the highest possible water head in the reservoirs. The objective function F represents the energy production plus the water energy in reservoirs for future production. The objective function is stated as energy maximisation based on both produced and stored quantities without considering economic value:

$$\begin{aligned}
F_{n,k} &= \sum_{j=1}^{J}\sum_{i=1}^{NP} P_{i,j,n,k}\Delta t + \sum_{i=1}^{M} E_{s,i,J,n,k} - \sum_{i=1}^{M} E_{s,i,1,n,k} \\
&= \sum_{j=1}^{J}\sum_{i=1}^{NP} \rho g \eta_i (\widetilde{h_{i,k}} - h_i) q_{i,j,n,k} \Delta t \\
&\quad + \sum_{i=1}^{M} \rho g A_i (\widetilde{h_{i,k}} - h_i)\left(hd_{i,J,n,k} - h_i\right) \\
&\quad - \sum_{i=1}^{M} \rho g A_i (\widetilde{h_{i,k}} - h_i)\left(hd_{i,1,n,k} - h_i\right),
\end{aligned} \quad (1)$$

where the dependent variables of the optimisation are hydraulic head h(m) and turbine discharge q (m^3/s). P (W) is the power from hydropower stations; E_s (J) and A (m^2) are the energy stored and the surface water area of each reservoir, respectively; ρ (kg/m^3) is the density of water; g (m/s^2) is the acceleration because of gravity; and η is the generation efficiency of a hydropower plant, which is assumed to be constant over time. The hydraulic head of the stations, $\tilde{h}$ (m), is assumed to be constant in the optimisation problem but is updated after each decision in the updating time step t_u. Furthermore, $\underline{h}$ (m) is the minimum water level in each reservoir, the so-called dead water level, which cannot be used for regulation purposes. In the calculation of the energy of stored water, the routine subtracts the minimum water level from the actual water level and considers the remaining water non-usable. The potential production of water stored in a reservoir depends on the downstream fall height hd (m), including the waterfall height at all downstream stations to the sea. The indices j, n, and k are used to represent different indexes of time steps and forecasts in the programming; j is the index of the numerical time step; J is the total number of j, i.e., $J = T_H/\Delta t$; Δt(days) is the time length of one numerical time step, which is used to represent the flow dynamics; n is the index of the stochastic forecasts that are N in total; and i is the index of each of the multiple hydropower stations. M is the total number of stations, including reservoirs and hydropower plants, and NP is the total number of hydropower stations.

The objective function in Equation (1) contains two parts: the production of energy at each individual station and the change in the stored water energy in relation to the period from time step 1 to the end of time step J. The optimisation constraints include the conservation of water that dynamically flows in the river basin between hydropower stations and the limitations (bounds) of the variables. Flow dynamics reflect the flow between hydropower stations through spills and turbine discharges, runoff from connected watersheds, and changes in water storage in reservoirs. The water travel time, i.e., the lag time between reservoirs, is neglected in this model. The stations are connected by a river network, and water discharged from upstream stations travels to the nearest downstream station; i.e., this can be used to produce electricity at several stations along the network of hydropower stations. Here, the following constraints are recognised:

Maximise F;
Subject to:

$$V_{i,j,n,k} = V_{i,j-1,n,k} + \left(q_{r,i,j,n,k} + q_{up,i,j,n,k} + s_{up,i,j,n,k} - q_{i,j,n,k} - s_{i,j,n,k}\right)\Delta t, \quad (2)$$

$$V_{i,j,n,k} = A_i h_{i,j,n,k}, \quad (3)$$

$$\underline{h_i} \leq h_{i,j,n,k} \leq \overline{h_i}, \quad (4)$$

$$0 \leq q_{i,j,n,k} \leq \overline{q_i}, \quad (5)$$

$$s_i \leq s_{i,j,n,k} \leq \overline{s_i}, \quad (6)$$

where V (m^3) is the water storage in the multireservoir system, the inflow q_r (m^3/s) is the runoff water from the subcatchment that connects directly as the inflow to one reservoir, and q_{up} (m^3/s) and s_{up} (m^3/s) are turbine discharges and water spillages, respectively, that come from the nearest upstream connected stations, which means that water released at one station is retained at the next reservoir regardless of the transport time along the river. The water spillage discharge is s (m^3/s). Water conservation (Equation (2)) indicates that the inflow to a specific reservoir comes from natural runoff as a result of precipitation and snowmelt and outflow from connected upstream multireservoirs. Note that the runoff q_r is implemented both in forecasting step 1 as part of the above optimisation problem and in updating step 2 (Figure 1), in which we use the notations q_{rf} for the forecasted runoff and q_{ra} for the actual runoff applied in the management scenario. In the simulation routine (step 1 in Figure 1), the runoff discharge in Equation (2) is q_{rf}, while in the updating routine (step 2 in Figure 1), the runoff discharge is q_{ra}.

Equations (4)–(6) describe the physical limitations of the operation of a hydropower system. One limitation is that the reservoir water head must not exceed the maximum reservoir level for safety reasons or drop below the dead water level for environmental reasons. Hence, Equation (4) describes the lower and upper boundaries of the water head, denoted $\underline{h}$ (m) and $\overline{h}$ (m), respectively. In addition, hydroturbines have an upper limit on spinning, which implies that a maximum discharge is allowed, denoted as $\overline{q}$ (m^3/s). To sustain the downstream water demand and meet environmental requirements to some extent, a lower limitation of spillage discharge $\underline{s}$ (m^3/s) and an upper limitation $\overline{s}$ (m^3/s) are given.

The mathematical optimisation problem above, in Equations (1)–(6), is a nonlinear optimisation problem because the objective function in Equation (1) is an equation with the product of the variables h and q, which are coupled through the water volume conservation equation (Equation (2)). However, the variation in the water head in a reservoir over a short period is generally sufficiently small to justify the reservoir water head being set as a constant. Hence, to enhance computational efficiency and reduce the complexity of this problem, Equation (1) is linearised by keeping the reservoir levels constant during every updating time step t_u. The model subsequently updates the reservoir head $\tilde{h}$ (m) after every updating time step t_u, thus recognising nonlinearity as an explicit numerical approximation. Updating after each time step t_u keeps the operational decisions of production and spillage discharges combined with the water runoff input of the real scenario.

2.4. Performance Indicators

This section presents the definition of performance indicators expressing the error of forecasting and the efficiency of hydropower production in comparison to the maximum potential. A proposed criterion to evaluate the accuracy of forecasted runoff is the mean absolute scaled error (MASE), first proposed by Hyndman and Koehler (2006) [27]. It never gives undefined or infinite values and is suitable for intermittent demand series, such as when there are periods of zero data in a forecast [28]. The MASE (unitless) defined for the simulation period T_{sim} can be expressed as:

$$\text{Error} = \frac{1}{K} * \sum_{k=1}^{K} \left(\frac{\frac{1}{N*M \cdot J} \sum_{n=1}^{N} \sum_{i=1}^{M} \sum_{j=1}^{J} \left| q_{ra,i,j} - q_{rf,i,j,n,k} \right|}{\frac{1}{M \cdot (J-1)} \sum_{i=1}^{M} \sum_{j=2}^{J} \left| q_{ra,i,j} - q_{ra,i,(j-1)} \right|} \right). \quad (7)$$

Furthermore, the efficiency of production management in the entire watershed depends on the decisions regarding turbine and spillage discharges resulting from the application of forecasted runoff in the optimisation procedure for the entire T_{sim} period. Therefore, to estimate the dependency between the forecasting error and production efficiency, we introduced a production efficiency factor (ηd), which represents the energy production efficiency for all hydropower stations in the watershed. The potential production efficiency factor is formulated as potential production divided by the difference in potential runoff energy and potential storage energy:

$$\eta_d = \text{mean}\left(\frac{Ep_d}{Er_d - \Delta Es_d}\right) = \frac{1}{M * T_{sim}/\Delta t} * \frac{\rho g \Delta t \sum_{i=1}^{M} \sum_{j=1}^{T_{sim}/\Delta t} \widetilde{hd_{i,j}}\widetilde{q_{i,j}}}{\rho g \Delta t * \sum_{i=1}^{M} hd_{max,i} \sum_{j=1}^{T_{sim}/\Delta t} \left(\widetilde{q_{i,j}} + \widetilde{s_{i,j}}\right)}, \quad (8)$$

where Ep_d (J) is the simulated downstream production of energy for all stations for the entire simulation period, Er_d (J) is the downstream potential runoff energy, Es_d (J) is the downstream potential storage energy, hd is the downstream height, which is the water level from the station to the sea, and hd_{max} (m) is the maximum of the downstream height if there are no regulations of the reservoir levels. $\widetilde{hd_{i,j}}$ and $\widetilde{q_{i,j}}$, $\widetilde{s_{i,j}}$ are the decisions from the simulation model based on the simulation period of T_{sim}. The potential energy production in the above expression consists of both potential downstream production from runoff and energy stored in reservoirs. The potential downstream production indicates the estimated production generated at each station and the potential production from stations along the water path towards the sea. Energy stored in the water reservoir can be seen as the initial value of the scenario and can be used in production, which is the reason for the definition given by Equation (8). This study elucidates the relationship between forecasting error and production efficiency.

3. Case Study

The case study involves 36 hydropower plants and 13 reservoirs in the Dalälven River basin, located in central Sweden, stretching from the Scandinavian mountains in the west to the effluence at the Baltic Sea (see Figure 2). The hydropower system in Dalälven produces 4 TWh of energy with a capacity of 970 MW. Lake Siljan and Trängsletsjön are the main reservoirs, and the Trängslet dam that connects with Trängsletsjön is the highest earth-filled dam in Sweden [29].

Figure 2. The Dalälven River basin with hydropower stations marked with red dots.

The historical runoff data in the Dalälven River basin were derived from model simulations using the S-HYPE model, which is the continuously developed version of the HYPE model. S-HYPE is a semi-distributed catchment model that simulates water flow and substances from precipitation through different storage compartments and fluxes to the sea [30]. The historical data in this study start on 1 January 1961 and end on 31 December

2011, spanning a total of 51 years distributed among 64 subwatersheds located in the Dalälven River basin.

4. Results

4.1. Assessment of the Effects of Forecast Error on Production Efficiency in the Dalälven River Basin

The model framework developed here can be used to assess the importance of runoff forecasting for hydropower production (Figure 1), especially when separating the model procedure into three parts: (a) ensemble forecasting recognising historic runoff statistics, (b) optimisation of production management and independent system updating, and (c) assessment of importance by calculating performance indicators. As a demonstration of this model framework, we applied it to 36 cascade hydropower plants and 13 reservoirs in the Dalälven River basin, as described in Section 2.2. In this analysis, forecasts were randomly drawn from historic records classified with a strict biennial period, hence reflecting a wet and dry year classification (Figure 3); each classification had 25 years of data. To represent the uncertainty of the forecasts, the simulated real runoff was drawn independently from one of the two statistical classes. As a test of the assessment model framework, we designed three scenarios for this example application. In scenario 1, both the forecasted and real runoff were both from classification 1 (wet years); scenario 1 is a wet-to-wet scenario. In scenario 2, the forecasted runoff was from classification 2 (dry years), and the real runoff was from classification 1; scenario 2 is a dry-to-wet scenario. In scenario 3, which is a control scenario, the forecasted runoff was from both dry and wet years using the entire data record, and the real runoff was from classification 1; scenario 3 is a neutral-to-wet scenario. Consequently, the optimisation time horizon was 90 days, and the receding horizon control was 90 days, which meant that we needed samples of discrete time series covering 180 days. The parameters applied in this example application are listed in Table 1.

Table 1. List of parameters.

Parameter	Definition	Parameter Value in the Example Application
T_H	Time horizon of optimisation: the duration of the forecasted time series placed into one optimisation procedure.	T_H = 90 (days)
T_{sim}	Period of simulation: the maximum shift in time of the horizon in the receding horizon approach; $T_{sim} = K \times t_u$.	T_{sim} = 90 (days)
t_u	Updating period: the time during which the decided turbine discharges are applied, whereafter the reservoir levels are updated and new decisions are taken; $t_u = T_{sim}/K$.	t_u = 2 (days)
Δt	Numerical time step used to represent the watershed dynamics and to move between the states used in the optimization.	Δt = 0.5 (days)
j	$j = 1 : J$. Index for the numerical time step for water dynamics; $J = T_H/\Delta t$.	J = 180
i	$i = 1 : M$ Index for the reservoirs.	M = 49
n	$n = 1 : N$. Index for the repetition number of one updating period simulation with different stochastic runoff forecasts, which was used to make the average decision.	N = 10
k	$k = 1 : K$. Index for the simulation time step in order to progress over the simulation period T_{sim}. The number of updating time steps is $K = T_{sim}/t_u$.	K = 45

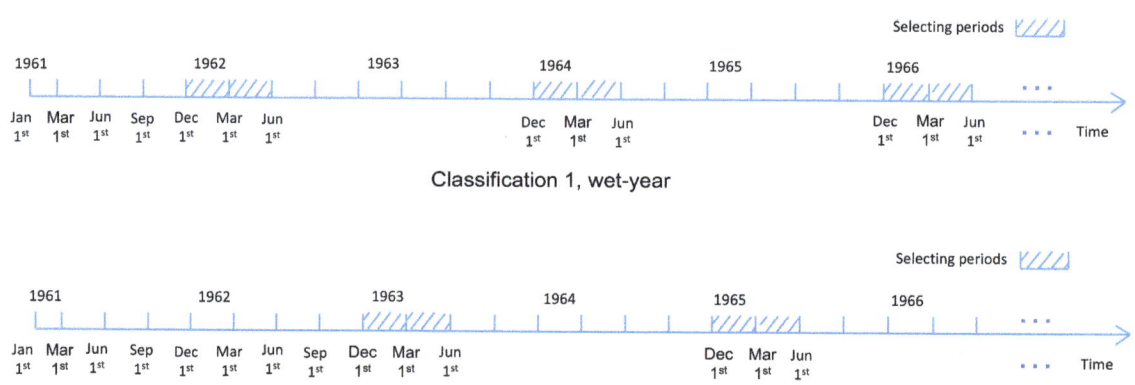

Figure 3. Ensemble classifications for runoff forecasting in wet years (odd years) and dry years (even years).

In part b, the hydropower operational model calculates the optimal decisions regarding turbine discharge and spillage discharge based on the forecasted runoff for a three month horizon (T_H). However, the optimal decisions were applied only under the updating period t_u, which was taken as two days in this application. Within each updating period t_u, the optimisation was repeated N times with forecasts randomly drawn from the same biennial ensemble classification according to the defined scenario, and the mean value of the N optimal decisions in the t_u period provided the final decisions on hydropower operation. After each t_u period calculation, the reservoir levels were updated using a real runoff scenario, as shown in Figure 1. Only one time series of real runoff was used in this assessment model. After updating, the simulation proceeded to simulate the management of the next updating period, where the procedure comprises T_{sim}/t_u steps, which cover three months in this application.

Part c is the assessment of the importance and calculation of the two performance indicators, which were estimated for the entire simulation period T_{sim} using Equations (7) and (8). The forecast error used in this study adopts the mean absolute scaled error to examine forecast accuracy. Based on the simulation structure of the assessment model, the error calculation contains M stations, $T_H/\Delta t$ time steps along the time horizon, N repetitions in each update period, and K progression steps to cover the entire simulation period. Hence, the forecast error is a mean value over all K progression time steps and M stations. One realisation was completed using a K progression step simulation covering the entire T_{sim} period. In addition, 100 Monte Carlo runs of the realisation of the entire assessment of T_{sim} were applied to reduce uncertainty.

As can be seen in Figure 4, there is a tendency for the production efficiency factor ηd to decrease with an increase in the forecast error. The production efficiency factor represents the potential hydropower production in the entire watershed compared to the theoretical maximum production during the same period. ηd was found to vary for the hydropower system in the Dalälven River basin, ranging from 78.5% to 80.5% in scenarios 1 and 2, depending on the three month forecast error. This range of variation depends on the forecast error and could appear to be small but can, in principle, represent a substantial economic value for the management of the hydropower production system. For example, if the periodic nature of water runoff in forecasting can be improved by the entire forecasting error range (2.65–2.35 = 0.4) considering the best and the worst possible cases of scenarios 1 and 2, the production efficiency varies by 2%. It can be expected that the yearly energy production of the Dalälven River basin would be enhanced by 80 GWh/year (4 TWh/year × 2% = 80 GWh/year), based on the current production of approximately 4 TWh/year. The reduction in forecasting error from scenario 2 (with forecasts representing a dry year under wet year conditions) to scenario 1 (with forecasts representing a wet year under wet year conditions) corresponds to a production efficiency improvement of about 0.33% (or 13.2 GWh/year), which is obtained by calculating the difference between the mean production efficiency factors of scenarios 1 and 2.

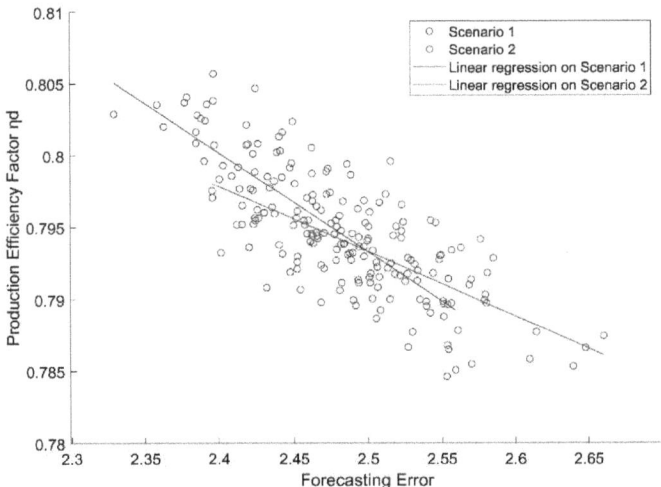

Figure 4. Comparison of scenario 1 (forecasts of wet years) and scenario 2 (forecasts of dry years), with both scenarios using simulated real runoff from a wet year.

The blue and red lines in Figure 4 present the linear regressions of scenarios 1 and 2 that are bounded by the scenario intervals. The regression lines indicate that scenario 1 has a higher production efficiency and lower forecast error than scenario 2. Production is more efficient for scenario 1 than scenario 2 when the forecasting error is smaller than 2.5, but the opposite prevails when the forecasting error is larger than 2.5. The range of the forecast error is relatively similar in scenario 1, but the ranges of production efficiency factors are larger in scenario 1 than in scenario 2. Figure 5 shows all three scenarios and their forecasting error boundaries. The error range of scenario 3 is obviously larger than those of scenarios 1 and 2, and it roughly matches the non-overlapping parts of scenarios 1 and 2. As a control scenario, scenario 3 does not have any periodic treatment on forecasting, so it can provide a neutral solution that should comprise the solutions from scenarios 1 and 2. Figure 5 shows this.

4.2. Start-Month Impact on the Biennial Periodicity

Figure 6 shows the daily mean runoff (m^3/s) of 64 subcatchments from the Dalälven River basin. Runoff data were deduced for every set of odd (blue dots) and even (red dots) years using different start months from January to December in the time series that covers daily data from 1961 to 2011. The green line is the average daily runoff of the 64 subcatchments over 51 years. The graph clearly shows that the biennial classification results in differences in the daily mean runoff from the Dalälven River basin, but the pattern depends on the start month of the segments. Note that the odd–even classification used in this figure denotes the year of the start month, but the yearly data selected for a start month later than January cover both odd and even years. The two curves, odd year (blue line) and even year (black line), in Figure 6 highlight an interchange of the implication of odd and even for the dry–wet year characteristic. At the crossing of the curves in Figure 6, both odd and even years are statistically equally wet, whereas the difference becomes larger when the curves are farther apart. The vertical bars show a 95% confidence interval using a t-distribution and indicate somewhat weak significance in the differences between the curves, which can be explained by the prevalence of other periodic climatic phenomena and some degree of randomness. Nevertheless, the results show the systematic biennial hydrological periodicity of the annual mean runoff in the Dalälven River basin, which varies with the start month of the selection and is most vital with a December start month. This circumstance implies that the long-term forecasts used in hydropower production management in December are more sensitive to recognising biennial periodicity compared with planning conducted in a summer month. Compared with the mean runoff of all years (green line), odd years present a higher mean daily runoff than the average, while even years show a relatively low mean daily runoff regardless of the start month.

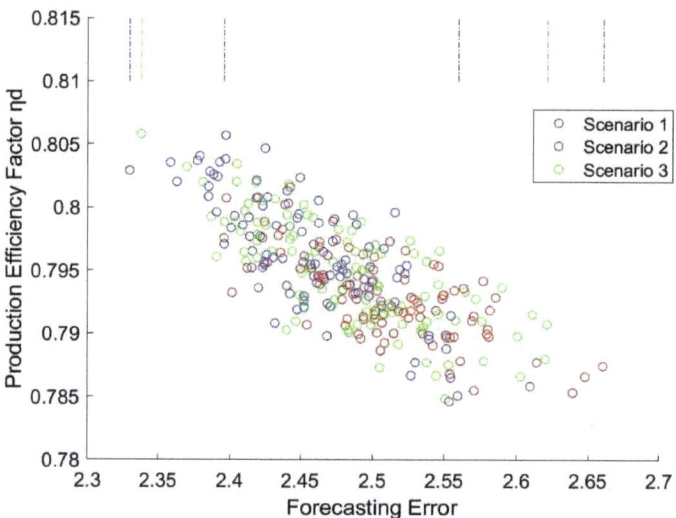

Figure 5. Scenarios 1, 2, and 3 and their forecasting error boundaries (dashed lines).

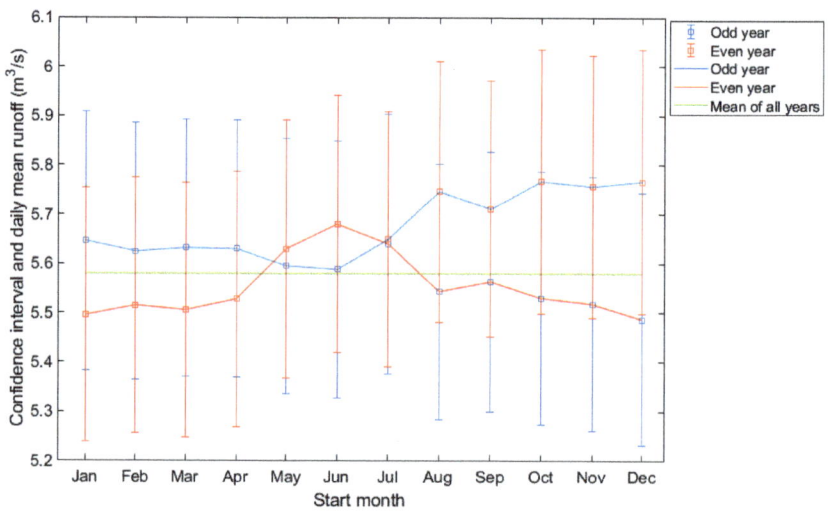

Figure 6. Averaged daily runoff (m^3/s) over 1961–2011 for the 64 subcatchments of the Dalälven River basin in units of m^3/s using different start months. Each start month provides up to 25 yearly samples in odd or even years, and the bars indicate the 95% confidence interval of the mean value estimate (using a t-distribution).

5. Conclusions and Discussions

This study developed a model framework that assesses the importance of forecasting periodic hydroclimate runoff fluctuations for hydropower planning and generation. Based on 51 years of historical records, the ensemble classification of runoff was shown to possess biennial periodicity, with a strength that was dependent on the start month of the year. This emphasises the importance of focusing on the annual seasonal variation in sub-arctic regions in addition to examining long-term climatic modes of variability. While the model approach is not limited by the selected forecasting method, the historical data sampling representing the complex statistical nature of runoff allows for an evaluation of the effect of periodic errors in runoff forecasts on production efficiency. Forecasting, in combination with the stepwise linear simulation framework, is essential for understanding the implications of periodicity in hydroclimatic variations for the proper operational planning of reservoir storage in cascade hydropower systems. It was found that the forecast errors for the Dalälven River were associated with three month future horizons based on historical data in the range of 3–4% when comparing scenarios 1 and 2, and recognising the biennial periodicity in Dalälven can enhance production efficiency by 1–2%. This may seem small, but it represents substantial economic value in terms of annual production compared to what could be expected from, for example, refurbishing the rock tunnel systems and penstocks of all 36 hydropower stations in the Dalälven River basin, which could also lead to efficiency improvements in the order of a few percent. The main result diagrams (Figure 4) show production efficiency versus forecast errors for two scenarios combining wet year forecasts with wet year scenarios (scenario 1) and dry year forecasts with wet year scenarios (scenario 2). The results indicate that the ensemble classification from which the forecasted runoff is selected plays an important role in enhancing production efficiency. Ignoring the biennial periodicity or the failure to associate future hydrology with a dry or wet year may cause an increase in the forecasting error, thereby decreasing the production efficiency. As a control scenario, scenario 3 verified the results of scenarios 1 and 2 by matching the non-overlapping parts of scenarios 1 and 2 (Figure 5).

The hydropower potential depends on the availability of water in a given stream, and the runoff pattern thus governs the energy generation capacity. This study shows that categorising historical runoff time series from the Dalälven River basin into dry and wet years is statistically possible if the yearly separation is conducted in the appropriate month, generally in mid-winter, but this pattern was found to be much less pronounced when using a mid-summer month. The difference in the annual mean discharge over the 51 year period was in the order of 5% if the separation of wet and dry years started in December. This difference would also be a measure of the possible relative forecast error that can arise in the mean runoff value if the biennial periodicity in water availability as a result of climate fluctuations is neglected.

While this study did not cover techniques for forecasting near-future climatic regimes, it is well known that both GCMs combined with hydrologic downscaling, as well as the statistical assessment of suitable climate indices, are powerful tools to predict current and near-future climate regimes and expected precipitation patterns. Such tools can provide an indication of whether a future year will be dry or wet, but they have lower accuracy when representing short-term fluctuations of importance for the regulatory behaviour in hydropower operations, which is why historical ensemble forecasts can provide essential statistical information in the management process. This assessment model framework can be used to assess the impact of biennial periodicity and can be applied to various periodicities by selecting different periodic ensembles. It could be future research. Some assumptions, like the constant water head and stable reservoir area in the optimisation model, can be improved in future work.

Assessing how biennial hydroclimate fluctuations can be accounted for in the management of hydropower generation and operational planning is essential. The model-based methodology developed in this study can be used to assess the impacts of such fluctuations on hydropower generation and enhance the management of hydropower operations in order to assist the hydropower operator with better planning of water and energy resources.

Author Contributions: Conceptualization, S.H. and A.W.; methodology, S.H. and A.W; software, S.H.; validation, S.H., A.W. and J.R.; formal analysis, S.H. and A.W.; investigation, S.H.; resources, A.W. and J.R.; data curation, S.H.; writing—original draft preparation, S.H.; writing—review and editing, S.H., A.W., J.R. and A.B.-B.; visualization, S.H. and J.R.; project administration, A.W. All authors have read and agreed to the published version of the manuscript.

Funding: This research was funded by the China Scholarship Council, no. 201700260188 and the Swedish Energy Agency (Energimyndigheten), no. 019-005552.

Data Availability Statement: The data, code, and models that support the conclusions of this research are available from the corresponding author upon request.

Acknowledgments: The authors thank the China Scholarship Council and the Swedish Energy Agency (Energimyndigheten) for their financial support. They are also grateful to the Swedish Meteorological and Hydrological Institute and Lund University for collecting the data and providing comments. Moreover, the authors acknowledge the valuable comments of Mikael Amelin. The computations involving high-performance computers were enabled by resources provided by the National Academic Infrastructure for Supercomputing in Sweden (NAISS) and the Swedish National Infrastructure for Computing (SNIC) at https://www.snic.se/ and were partially funded by the Swedish Research Council through grant agreements no. 2022-06725 and no. 2018-05973. We are grateful for these resources.

Conflicts of Interest: The authors declare no conflict of interest.

References

1. Chang, J.; Wang, X.; Li, Y.; Wang, Y.; Zhang, H. Hydropower plant operation rules optimization response to climate change. *Energy* **2018**, *160*, 886–897. [CrossRef]
2. Haguma, D.; Leconte, R.; Côté, P.; Krau, S.; Brissette, F. Optimal Hydropower Generation Under Climate Change Conditions for a Northern Water Resources System. *Water Resour. Manag.* **2014**, *28*, 4631–4644. [CrossRef]
3. Tarroja, B.; AghaKouchak, A.; Samuelsen, S. Quantifying climate change impacts on hydropower generation and implications on electric grid greenhouse gas emissions and operation. *Energy* **2016**, *111*, 295–305. [CrossRef]
4. de Queiroz, A.R.; Faria, V.A.D.; Lima, L.M.M.; Lima, J.W.M. Hydropower revenues under the threat of climate change in Brazil. *Renew. Energy* **2019**, *133*, 873–882. [CrossRef]
5. Anghileri, D.; Botter, M.; Castelletti, A.; Weigt, H.; Burlando, P. A Comparative Assessment of the Impact of Climate Change and Energy Policies on Alpine Hydropower. *Water Resour. Res.* **2018**, *54*, 9144–9161. [CrossRef]
6. Forrest, K.; Tarroja, B.; Chiang, F.; AghaKouchak, A.; Samuelsen, S. Assessing climate change impacts on California hydropower generation and ancillary services provision. *Clim. Change* **2018**, *151*, 395–412. [CrossRef]
7. van Vliet, M.T.H.; Wiberg, D.; Leduc, S.; Riahi, K. Power-generation system vulnerability and adaptation to changes in climate and water resources. *Nat. Clim. Chang.* **2016**, *6*, 375–380. [CrossRef]
8. IPCC. *Climate Change 2013—The Physical Science Basis*; Cambridge University Press: Cambridge, UK, 2014. [CrossRef]
9. Foster, K. *Hydrological Seasonal Forecasting*; Lund University: Lund, Sweden, 2019.
10. Uvo, C.B.; Foster, K.; Olsson, J. The spatio-temporal influence of atmospheric teleconnection patterns on hydrology in Sweden. *J. Hydrol. Reg. Stud.* **2021**, *34*, 100782. [CrossRef]
11. Wörman, A.; Lindström, G.; Riml, J. The power of runoff. *J. Hydrol.* **2017**, *548*, 784–793. [CrossRef]
12. Schmidt, R.; Petrovic, S.; Güntner, A.; Barthelmes, F.; Wünsch, J.; Kusche, J. Periodic components of water storage changes from GRACE and global hydrology models. *J. Geophys. Res. Solid Earth* **2008**, *113*, 1–14. [CrossRef]
13. Poveda, G.; Mesa, O.J. Feedbacks between Hydrological Processes in Tropical South America and Large-Scale Ocean–Atmospheric Phenomena. *J. Clim.* **1997**, *10*, 2690–2702. [CrossRef]
14. Rajagopalan, B.; Lall, U. Interannual variability in western US precipitation. *J. Hydrol.* **1998**, *210*, 51–67. [CrossRef]
15. Krokhin, V.; Luxemburg, W.M.J. Temperatures and precipitation totals over the Russian Far East and Eastern Siberia: Long-term variability and its links to teleconnection indices. *Hydrol. Earth Syst. Sci.* **2007**, *11*, 1831–1841. [CrossRef]
16. Lopez, M.G.; Crochemore, L.; Pechlivanidis, I.G. Benchmarking an operational hydrological model for providing seasonal forecasts in Sweden. *Hydrol. Earth Syst. Sci.* **2021**, *25*, 1189–1209. [CrossRef]

17. Voisin, N.; Dyreson, A.; Fu, T.; O'Connell, M.; Turner, S.W.; Zhou, T.; Macknick, J. Impact of climate change on water availability and its propagation through the Western U.S. power grid. *Appl. Energy* **2020**, *276*, 115467. [CrossRef]
18. Chowdhury, A.F.M.K.; Dang, T.D.; Nguyen, H.T.T.; Koh, R.; Galelli, S. The Greater Mekong's Climate-Water-Energy Nexus: How ENSO-Triggered Regional Droughts Affect Power Supply and CO_2 Emissions. *Earth's Futur.* **2021**, *9*, 1–19. [CrossRef]
19. Siala, K.; Chowdhury, A.K.; Dang, T.D.; Galelli, S. Solar energy and regional coordination as a feasible alternative to large hydropower in Southeast Asia. *Nat. Commun.* **2021**, *12*, 4159. [CrossRef]
20. Marco, J.B.; Harboe, R.; Salas, J.D. *Stochastic Hydrology and its Use in Water Resources Systems Simulation and Optimization*; Springer: Dordrecht, The Netherlands, 1993. [CrossRef]
21. Feiring, B.R.; Sastri, T.; Sim, L.S.M. A stochastic programming model for water resource planning. *Math. Comput. Model.* **1998**, *27*, 1–7. [CrossRef]
22. Yevjevich, V. Stochastic models in hydrology. *Stoch. Hydrol. Hydraul.* **1987**, *1*, 17–36. [CrossRef]
23. Tiberi-Wadier, A.L.; Goutal, N.; Ricci, S.; Sergent, P.; Taillardat, M.; Bouttier, F.; Monteil, C. Strategies for hydrologic ensemble generation and calibration: On the merits of using model-based predictors. *J. Hydrol.* **2021**, *599*, 126233. [CrossRef]
24. Wu, L.; Seo, D.-J.; Demargne, J.; Brown, J.D.; Cong, S.; Schaake, J. Generation of ensemble precipitation forecast from single-valued quantitative precipitation forecast for hydrologic ensemble prediction. *J. Hydrol.* **2011**, *399*, 281–298. [CrossRef]
25. Yuan, F.; Berndtsson, R.; Zhang, L.; Uvo, C.B.; Hao, Z.; Wang, X.; Yasuda, H. Hydro Climatic Trend and Periodicity for the Source Region of the Yellow River. *J. Hydrol. Eng.* **2015**, *20*, 05015003. [CrossRef]
26. Engström, J.; Uvo, C.B. Effect of Northern Hemisphere Teleconnections on the Hydropower Production in Southern Sweden. *J. Water Resour. Plan. Manag.* **2016**, *142*, 5015008. [CrossRef]
27. Hyndman, R.J.; Koehler, A.B. Another look at measures of forecast accuracy. *Int. J. Forecast.* **2006**, *22*, 679–688. [CrossRef]
28. Prestwich, S.; Rossi, R.; Tarim, S.A.; Hnich, B. Mean-based error measures for intermittent demand forecasting. *Int. J. Prod. Res.* **2014**, *52*, 6782–6791. [CrossRef]
29. Kokko, V.; Hjerthén, P.; Ingfält, H.; Löwen, K.E.; Sjögren, A. Development of Dalälven hydro power scheme in Sweden. *Houille Blanche* **2015**, *6368*, 5–14. [CrossRef]
30. Lindström, G.; Pers, C.; Rosberg, J.; Strömqvist, J.; Arheimer, B. Development and testing of the HYPE (Hydrological Predictions for the Environment) water quality model for different spatial scales. *Hydrol. Res.* **2010**, *41*, 295–319. [CrossRef]

Disclaimer/Publisher's Note: The statements, opinions and data contained in all publications are solely those of the individual author(s) and contributor(s) and not of MDPI and/or the editor(s). MDPI and/or the editor(s) disclaim responsibility for any injury to people or property resulting from any ideas, methods, instructions or products referred to in the content.

Article

Increasing Trends in Discharge Maxima of a Mediterranean River during Early Autumn

George Varlas [1,*], Christina Papadaki [1], Konstantinos Stefanidis [1], Angeliki Mentzafou [1], Ilias Pechlivanidis [2], Anastasios Papadopoulos [1] and Elias Dimitriou [1]

1. Hellenic Centre for Marine Research, Institute of Marine Biological Resources and Inland Waters, 46.7 km of Athens-Sounio Ave., 19013 Anavyssos, Greece
2. Hydrology R&D, Swedish Meteorological and Hydrological Institute, SE-60176 Norrköping, Sweden
* Correspondence: gvarlas@hcmr.gr; Tel.: +30-22910-76399

Abstract: Climate change has influenced the discharge regime of rivers during the past decades. This study aims to reveal climate-induced interannual trends of average annual discharge and discharge maxima in a Mediterranean river from 1981 to 2017. To this aim, the Pinios river basin was selected as the study area because it is one of the most productive agricultural areas of Greece. Due to a lack of sufficient measurements, simulated daily discharges for three upstream sub-basins were used. The discharge trend analysis was based on a multi-faceted approach using Mann-Kendall tests, Quantile-Kendall plots, and generalized additive models (GAMs) for fitting non-linear interannual trends. The methodological approach proposed can be applied anywhere to investigate climate change effects. The results indicated that the average annual discharge in the three upstream sub-basins decreased in the 1980s, reaching a minimum in the early 1990s, and then increased from the middle 1990s to 2017, reaching approximately the discharge levels of the early 1980s. A more in-depth analysis unraveled that the discharge maxima in September were characterized by statistically significant increasing interannual trends for two of the three sub-basins. These two sub-basins are anthropogenically low affected, thus highlighting the clear impact of climate change that may have critical socioeconomic implications in the Pinios basin.

Keywords: climate change; hydrological extremes; GAM; non-linear trends; Quantile-Kendall; Pinios river

Citation: Varlas, G.; Papadaki, C.; Stefanidis, K.; Mentzafou, A.; Pechlivanidis, I.; Papadopoulos, A.; Dimitriou, E. Increasing Trends in Discharge Maxima of a Mediterranean River during Early Autumn. *Water* **2023**, *15*, 1022. https://doi.org/10.3390/w15061022

Academic Editors: Sonia Raquel Gámiz-Fortis and Matilde García-Valdecasas Ojeda

Received: 6 February 2023
Revised: 1 March 2023
Accepted: 6 March 2023
Published: 8 March 2023

Copyright: © 2023 by the authors. Licensee MDPI, Basel, Switzerland. This article is an open access article distributed under the terms and conditions of the Creative Commons Attribution (CC BY) license (https://creativecommons.org/licenses/by/4.0/).

1. Introduction

Anthropogenic climate change has influenced mean and extreme river flow [1], while it has increased flood risk [2], thus having severe implications worldwide. The Mediterranean region has been characterized as a climate change "hot spot" [3] because its climate is especially responsive to global changes. Climate change has strong effects on temperature, precipitation, and other parameters which determine the local climate. The climatic effects are very intense and complex in the Mediterranean countries such as Greece, characterized by large variabilities in the landscape, orography, and land use [4–6]. According to the Sixth Assessment Report of the Intergovernmental Panel on Climate Change (IPCC) published in 2021 [7], the average precipitation over land on a global scale has likely increased since 1950. However, the Mediterranean region seems to suffer from an increase in droughts. A previous study corroborates this finding as it concluded that the annual discharge volume of rivers in the Mediterranean region had presented declining trends for a period from 1950 to 2013 [8]. Moreover, another study showed declining trends of flooding in the Mediterranean region from 1960 to 2010 [9]. Furthermore, climate change influences the alternation of seasons by increasing the disparity between wet and dry ones [10–12]. This is another factor that may impact water availability and river discharge. It is noteworthy that several studies have shown that the water resources will probably continue the decreasing trend in the future, causing amplification of droughts and aridity [13–16].

In Greece, the most recent significant dry period characterized by decreased precipitation and, thus, increased drought in several areas lasted from the late 1980s to the early 1990s [4,17–19]. The drought during this period influenced not only Greece but also other Mediterranean countries [20–22], having severe implications for the whole Mediterranean region with an economic cost that was estimated to exceed 2.1 billion Euros [23]. Pinios river, which is one of the most productive agricultural areas of Greece and located in central-northern Greece, also suffered from this severe drought which lasted from 1988 to 1993 in this area. It is important to note that especially the drought during 1989–1990 was characterized by an average return period of 88 years for the Pinios river basin [24]. Despite the fact that studies focusing on various Greek rivers [25] and Pinios river [26] have reported climatic variabilities of discharge and drought periods, they did not show significant linear trends during the past decades, at least until 2002 [26] and 2010 [25].

Except for the increasing drought, one of the most significant issues regarding climate change is the change in precipitation extremes [7,27], that under favorable conditions, may trigger catastrophic flash floods in Mediterranean countries [28–32]. There are studies that report the intensification of short-term precipitation extremes induced by anthropogenic climate impacts [33]. Greece has been affected by several extreme precipitation events and floods during the last decades, affecting Thessaly and Pinios river, among other regions, especially during autumn [34–36]. It is noteworthy that the number of flood occurrences in Pinios river presented a rising trend during the period 1990–2010 [36]. Precipitation extremes during autumn in Greece have sometimes been related to Mediterranean tropical-like cyclones ("medicanes"), although they are generally rare phenomena, with approximately three medicanes in the whole Mediterranean per 2 years [37]. For example, Numa medicane in November 2017 had indirect effects on the flash flood of Mandra town, causing 24 fatalities [29,30,32]. Zorbas medicane in September 2018 brought torrential rainfall, causing three deaths and severe damage in several Greek areas [38]. Ianos medicane in September 2020 was the most intense medicane ever recorded in the Mediterranean, causing four fatalities and extensive damage in many regions, including Karditsa city and Mouzaki town located in the southwestern parts of the Pinios river basin [35]. An overall assessment of the flash flood in Karditsa city showed that the exceedance probability of the Ianos-induced flood ranged from 1:400 years in the low-lying catchments to 1:1000 years in the upstream mountainous catchments.

Therefore, there is the assumption that discharge maxima in Pinios river have increased during the last decades, especially in autumn. These maxima sometimes imply severe hydrometeorological phenomena and floods associated with high socioeconomic impacts. Multi-year analyses, also including a period after 2010 that is critical in terms of climate change effects, are needed to reveal significant trends in discharge maxima if they exist. However, Pinios river has not been covered by sufficient monitoring during the last decades, thus making long-term studies difficult and uncertain. The Institute of Marine Biological Resources and Inland Waters (IMBRIW) of the Hellenic Centre for Marine Research (HCMR) has developed a monitoring network in the river basin during the last few years, but this cannot facilitate long-term studies yet. The lack of sufficient measurements poses the necessity of alternative research efforts exploiting other sources of data. In this context, modeled discharge data from one of the most reliable and famous hydrological models, namely E-HYPE, were used in this study due to the lack of continuous, long-term, and reliable discharge measurements during the last decades.

In this framework, the scope of this study is to address two scientific questions, using the Pinios river basin as the study area and employing the E-HYPE discharges. The first scientific question is how climate change has influenced monthly discharge maxima over the years? The second one is what is the amplitude of interannual variabilities of average annual discharge? The trend analysis was performed using Mann-Kendall tests, Quantile-Kendall plots, and generalized additive models (GAMs) for fitting non-linear interannual trends. It is important to note here that the methodological approach adopted in this study may have a wider application in other areas and scales. Moreover, this basin is influenced by both climate variabilities and multiple anthropogenic pressures. Hence, we considered two upstream and partially mountainous sub-basins of Pinios to exclude the potential effects of direct anthropogenic pressures and one similar sub-basin that, however, is partially covered by some agricultural areas to investigate possible trend sensitivity by comparing with the two other sub-basins.

The article is structured as follows: Section 2 presents the study area and data used and also described the methodology used to analyze the data. Section 3 presents the results from addressing the scientific questions and objectives. Section 4 presents the discussion, and finally, Section 5 concludes the work.

2. Materials and Methods

2.1. Study Area

Pinios river basin is located in the administrative region of Thessaly in central-northern Greece. The most populated cities in the area are Larissa, Karditsa, Trikala, and Tirnavos. Pinios was selected because it is one of the most productive areas of Greece, mainly regarding the agricultural sector. The total area of the basin is about 11,000 km^2 and is mainly occupied by agricultural land (51% of the total basin area) and secondarily by vegetated areas (45%) [39]. It is noteworthy that the agricultural sector of the Pinios river basin represents 14% of the gross value added of the agricultural industry of Greece [40]. Pinios basin can be subdivided into two separate endorheic hydrographic networks, in the western (Karditsa plain) and the eastern (Larisa plain) basins, respectively, separated by an internal low-lying hill area [41,42]. The hydrographic network of the Pinios basin is complex and can be characterized as dendritic [32]. The climate of the Pinios basin in the western part is typical Mediterranean, with cold winters and moderate precipitation rate, followed by relatively hot and dry summers, while the eastern and northern parts of the basin have a cold semi-arid climate, with warm to hot, dry summers and cold winters [43]. The average annual precipitation at the Pinios basin is about 700 mm, with large spatial variabilities [4], ranging from 450 mm in the central area to 1850 mm in the western-mountainous part of the basin [44].

Pinios is affected by both interannual climate variabilities and multiple anthropogenic pressures [45], including irrigation processes, bridges, technical works, industries, buildings, pollution, etc. Therefore, in the context of this study, we selected three partially mountainous sub-basins upstream of the river to study the net effects of climate change on discharge for the 37-year period. In this way, we reduced the direct anthropogenic effects in our analysis, which are intense mainly in the lowland areas of the basin and could have variable effects on our results for each year of the 37-year period, thus hiding climatic effects. Two of the investigated sub-basins are located in the western part of the Pinios basin and one in the eastern one. More specifically, the 9729462 (Pinios upstream) sub-basin is located at the north-western part of Karditsa plain, the 9728383 (Titarisios tributary) sub-basin is located at the northern part of the Larisa plain, and the 9728538 (Mega Rema tributary) sub-basin is located at the southwestern part of Karditsa plain (Figure 1). The names of the sub-basins (i.e., 9729462, 9728383, and 9728538) used in this study were based on the encoding used in the European Hydrological Predictions for the Environment (E-HYPE) model data. E-HYPE data were used in this study and are described in the next subsection. In particular, the names 9729462, 9728383, and 9728538 refer to the outflow sub-basins as defined in the E-HYPE model data, but we considered these names for the entire sub-basins under study. For the convenience of readers, the 9729462, 9728383, and 9728538 sub-basins are referred to in the text, also using the code names PINUP, TITAR, and MREMA, respectively. The main characteristics of the three sub-basins, including area, surface waterbody, altitude [46], and land cover [39], are shown in Table 1. It is worthwhile noting that the MREMA sub-basin was selected to slightly differ from the other two sub-basins having similarities with low-lying areas of Pinios to explore probable implying effects induced by its lower altitude and increased agriculture (Table 1).

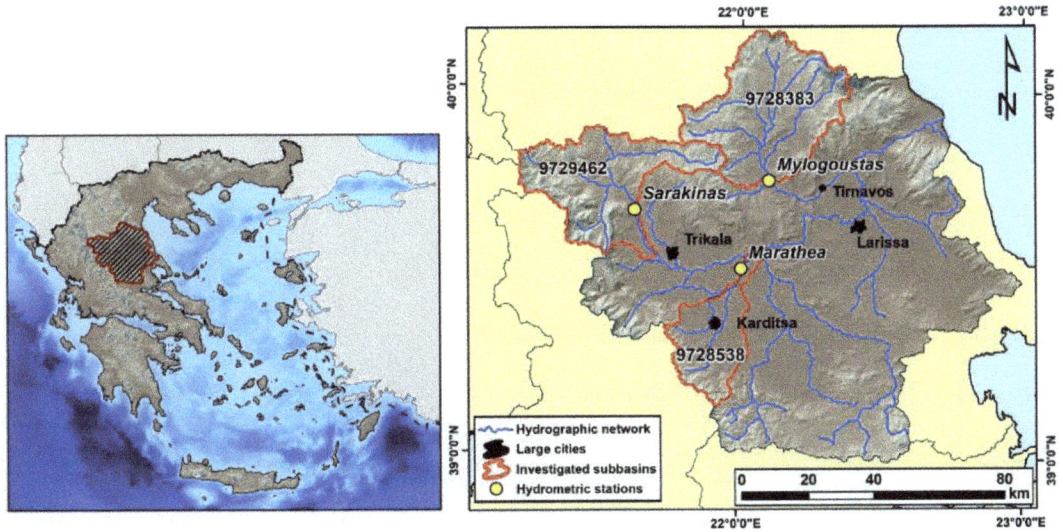

Figure 1. Orientation map of the study area of Pinios river basin and the 9729462 (PINUP), 9728383 (TITAR), and 9728538 (MREMA) sub-basins studied. The locations of the hydrometric stations used for evaluation are also illustrated.

Table 1. Main characteristics of the sub-basins studied.

ID	Area (km^2)	Surface Waterbody	Altitude (m)			Land Cover			
			Minimum	Maximum	Average	Artificial Surfaces	Agricultural Areas	Forest and Seminatural Areas	Other
9729462 (PINUP)	1205.9	Pinios P12	105	2167	775	1%	27%	71%	1%
9728383 (TITAR)	1439.0	Titarisios P2	137	2804	703	1%	40%	60%	0%
9728538 (MREMA)	586.8	Mega Rema 1	80	1484	279	4%	67%	29%	0%

2.2. European Hydrological Predictions for the Environment (E-HYPE) Model Data

For the purpose of the present study, we employed time series of daily discharge values for a 37-year period from 1 January 1981 to 31 December 2017 for the three selected sub-basins. The study area is characterized by a lack of a dense, continuous, long-term, and consistent network of hydrological stations, and thus, the use of long-term modeled discharge time series is considered suitable for the climatic investigations of this study. In this context, the discharge data were obtained from the results of the E-HYPE pan-European hydrological model [47,48] for the outflow of the sub-basins (i.e., 9729462, 9728383, and 9728538). E-HYPE is a model application of the HYPE (i.e., Hydrological Predictions for the Environment) model for the entire European continent, whereby hydrological flows and nutrient processes are calculated daily for each class within a sub-basin level [49]. HYPE model is a semi-distributed catchment model, which simulates the flow of water and substances beginning from precipitation through various storage compartments and fluxes to the sea, i.e., snow accumulation and melting, evapotranspiration, soil moisture, streamflow generation, and routing through rivers and lakes [47]. We note here that the groundwater accounts for contribution from the upper soil layers (1.5 m) and not deep aquifers. HYPE has been used in several applications related to climate change and hydrological extremes in various areas worldwide, presenting very good performance [50–53]. For this reason, the results of the HYPE model application focusing on Europe (i.e., E-HYPE) were selected to be employed in the present study. Moreover, E-HYPE uses the HydroGFD meteorological forcing dataset [54], while also irrigation [55,56] and crop water demand [57] are taken into consideration. Information regarding land use characteristics was retrieved from CORINE Land Cover 2000 [58].

2.3. Model Evaluation Methodology

Prior to the local model evaluation, we note that hydrological modeling based on the E-HYPE setup has the potential to encompass all European river basins, considering cross-regional and international boundaries, and to represent a number of different hydroclimatic conditions. Although the model was spatiotemporally calibrated and evaluated, and parameter identification has considered both in-situ data and earth observations, it is apparent that various river systems are still ungauged [59]. However, the usability of multi-basin modeling lies in the hypothesis that a good performance in space (different stations) and time (different periods) relates to the model's potential to predict the hydrological response at interior ungauged basins [60,61]. This is linked to the parameter regionalization to ungauged regions, which is acceptable if the model performs adequately in the gauged locations over the entire model domain [62].

The overall performance of E-HYPE is reasonably acceptable. For the 115 (538) discharge stations used in model calibration (validation), the median Nash-Sutcliffe Efficiency (NSE; Equation A5) is 0.54 (0.53), and the relative volume error is −1.6% (−1.3%). This indicates that the E-HYPE performance is consistent, and model outputs can be explored (yet with caution) even in ungauged regions. More details can be found in [48].

In Greece, most of the measured discharge data are not released to the public domain at high resolution due to a lack of coordinated water resources management, confidentiality, and/or business cases, which consequently limits the potential to assess the hydrological model performance at the local scale. The E-HYPE hydrological model monthly results of the older version 2.1 have been evaluated only against discharge measurements of large rivers in Greece [25]. Therefore, it was considered useful to evaluate the model's results against available monthly averaged discharge measurements in the three sub-basins before further analysis. The evaluation mainly aimed to show a comparison between the model and measurements in terms of monthly hydrological variabilities and not to demonstrate short-term (e.g., daily) discharge comparison because such detailed measurements were not available. The behavior and the performance of the model were examined with efficiency criteria, which are defined as a quantitative measure of performance, goodness of fit, or likelihood [63,64].

The criteria used to investigate the model reliability were the following: Pearson's Correlation Coefficient (R), Percent Bias (PBIAS), Root Mean Square Error (RMSE), Ratio of the Root Mean Square Error to the Standard Deviation of observed data (RSR) and NSE [65]. The equations of performance criteria can be found in Appendix A (Equations (A1)–(A5)). R = 1 or −1 means the existence of a perfect positive or negative linear correlation. PBIAS (%) shows if the simulated data are larger or smaller than the observations with a perfect value equal to 0, while positive/negative values indicate model underestimation/overestimation. RMSE is higher than 0, and low RMSE values indicate good model performance. RSR is the ratio between the RMSE and the standard deviation of the observations, thus including the benefits of error index statistics and a scaling/normalization factor. RSR ranges from 0 (perfect value), which implies zero RMSE or residual variation, to a high positive value. NSE values range between $-\infty$ and 1, and a value of NSE equal to 1 is considered the perfect value. It should be noted that the RMSE and NSE are sensitive to extreme values (outliers) and timing errors in the predictions [63]. Based on the well-established model evaluation criteria proposed, model simulation can be considered to be satisfactory if NSE > 0.50, RSR < 0.70, and PBIAS ± 25% for discharge [65].

The main characteristics of the monitoring stations used for the evaluation of the simulated discharge are presented in Table 2. All data series used for the evaluation were retrieved from the Ministry of Environment and Energy of Greece, while the quality control, as well as the data screening and processing operations, have been conducted under past projects (Table 2). The data have been statistically elaborated and are the most long-term and reliable measurements available in this region. The observed data were based on discharge measurements taken by humans, and thus, they imply increased uncertainties, especially in medium-high flows. The measured datasets used for the model-measurement comparisons consist of monthly measurements in locations near the outflow of the sub-basins under study (Figure 1).

Table 2. The main characteristics of the monitoring stations of the Ministry of Environment and Energy of Greece used for the evaluation.

	Sub-Basins		Evaluation Stations					
ID	Area (km^2)	Surface Waterbody	Name	Latitude (°)	Longitude (°)	Upstream Area (km^2)	Period	Reference
9729462 (PINUP)	1205.9	Pinios P12	Sarakinas	39.6690	21.6330	1058.5	1981–2001	[66]
9728383 (TITAR)	1439.0	Titarisios P2	Mylogoustas	39.7544	22.0987	1416.7	1981–1993	[67]
9728538 (MREMA)	586.8	Mega Rema 1	Marathea	39.5136	22.005	571.9	1981–1990	[68]

2.4. Analysis of Non-Linear Trends Using Generalized Additive Models

GAMs were employed to fit the interannual trends of discharges of the three studied sub-basins. GAMs have been used widely for time series analysis of environmental data because they can model non-linear trends and deal with the irregular spacing of samples in time [69–71]. The components of a time series are represented as smooth functions, which are non-linear representations of the covariates, composed by the sum of K simpler basis functions [69]. A general form of a generalized additive model is:

$$g(Y) = \beta + f_1(x_1) + f_2(x_2) + \ldots + f_n(x_n) \tag{1}$$

where Y is the expected response value, β is the model intercept, and $f_1, f_2,$ and f_n are smooth functions of the predictors x_1, x_2, x_n [72].

Here we used as a response value the monthly discharge, and as predictors, we defined the "trend" (time step of the series) and the intra-annual variation (named here "month"). For the "trend" smooth term, we used the cubic regression smoothing spline with k = 3, while for "month", we used the cyclic cubic spline with k = 12. We also created separate models for each month using as a predictor the time step ("trend"). Finally, for each model, we plotted the first derivative over time to visually assess how the rate of change of the discharges changes across time and to identify whether and when the rate of change shifts from negative to positive values and vice versa.

2.5. Methods for Discharge Maxima Analysis

It is important to study interannual changes in the maximum discharges that are usually related to climate change effects while also implying higher vulnerability for more frequent and intense floods under favorable conditions. The use of average annual values in temporal analyses of discharge usually hides the source and amplitude of variabilities. In this context, we explored interannual changes in daily discharge statistics of each sub-basin for every month of the year, considering the 37-year period from 1981 to 2017. Two hydrological parameters (indices) were calculated for every month, which is the maximum daily (1-day) discharge and the maximum 7-day average discharge of each month [73,74]. These indices were selected to describe aspects of the high flows. The use of such indices is important to provide a clear picture of interannual changes in high flow patterns and how these changes may impact various sectors. For instance, changes in high flows may have implications on the flood risk assessment that is critical for society, the economy, and ecosystems. Regarding the maximum 1-day discharges, simply the highest daily average discharges for every month were calculated. The maximum 7-day average discharge values were computed by examining each consecutive day of each month and calculating the maximum of the average values for the last 7 days up to and including each given day. Additionally, special focus was given to a selected continuous period from July to October, when the lowest flows usually occur in our study area. Except for the abovementioned hydrometeorological indices, the maximum 30-day average discharge values [73,74] were also calculated for this 4-month period following a similar methodology as in 7-day ones but considering the previous 30 days up to and including each given day. The R-package Exploration and Graphics for RivEr Trends (EGRET) was used for the calculation and analysis of long-term changes in discharge [75,76].

Afterward, to further investigate the strength of the statistical evidence of the high discharges, we used the Mann-Kendall trend test [77,78]. The strength of the evidence is characterized by the likelihood that the direction of the estimated trend is correct, computed from the Mann-Kendall test p-values as $[1 - (p/2)]$. Moreover, we used Quantile-Kendall plots [79] to investigate and describe discharge trends over the period of 37 years at the three sub-basins. The Quantile-Kendall plots were created using daily discharge records and were used to evaluate discharge trends across the range of discharge values in certain months (September and October) of the year. These plots are designed to give an overall impression of the nature of the discharge (streamflow) trend over some period/month of record over the entire flow duration curve. Here we describe the analysis in months. For each year, the daily discharge values of the month we are looking at are sorted from smallest to largest. These values are assigned a rank (k) where k = 1, 2, ... 30 (or 31, depending on the month), with 1 being the smallest and 30 (or 31) being the largest discharge value. For the full-time series of 37 years, all rank 1 discharges are evaluated as a time series of 37 years in length, and the results of that analysis are summarized by a slope and by a two-sided significance level (or p-value). The graphic shows the trend slope for each of the ranks, and color coding is used to indicate the likelihood that the estimated trend direction (upwards or downwards) is correct. The results are arrayed on the plot with low discharges to the left, median discharge in the middle, and high discharges to the right.

3. Results
3.1. Evaluation of E-HYPE Data Using Measurements

Table 3 presents the statistical characteristics and the efficiency criteria considered for the evaluation of monthly discharges simulated by the hydrological model E-HYPE using measured discharges. Based on the results, in all cases, the correlation coefficient R can be characterized as high, based on previously proposed criteria for correlation interpretation [80], indicating sufficient performance for the model. The NSE was in all cases positive, and in sub-basins PINUP and MREMA, near 0.50, which is considered indicative of satisfactory model performance. Additionally, RSR was lower than 0.70 in all cases, and PBIAS was 22% in the case of the PINUP sub-basin. Overall, the model performance at PINUP and MREMA sub-basins can be considered satisfactory (Table 3, Figure A1). The performance in sub-basin TITAR is acceptable, although it is characterized by low NSE and quite a negative PBIAS that could be partially attributed to the increased altitude gradient of this sub-basin (Table 1).

Table 3. Statistical characteristics and efficiency criteria for the evaluation of monthly discharges simulated by the hydrological model E-HYPE using measured discharges.

Sub-Basin		9729462 (PINUP)	9728383 (TITAR)	9728538 (MREMA)
Station		Sarakina	Mylogoustas	Marathea
Number of measurements	N	249	153	101
Pearson's correlation coefficient	R	0.723	0.718	0.723
Root Mean Square Error	RMSE	12.26	7.78	3.93
Nash-Sutcliffe Efficiency	NSE	0.49	0.11	0.49
Percent bias	PBIAS	22%	−49%	5%
Ratio of the root mean square error to the standard deviation	RSR	0.04	0.05	0.04

3.2. Interannual Distribution of Average Annual and Seasonal Discharge

After the statistical evaluation of the model data, we constructed time series of average annual and seasonal discharge from 1981 to 2017 for the three sub-basins (PINUP, TITAR, and MREMA) (Figure 2). Then, we estimated annual and seasonal maximum, minimum, and average discharge values in the three sub-basins for the 37-year period from 1981 to 2017 (Figure 2).

In more detail, Figure 2a–c presents time series of average annual discharge for the three sub-basins. To facilitate the interpretation of temporal variabilities of discharge, the respective time series of total annual precipitation is also illustrated. The results of this first analysis show that the average annual discharge in the three sub-basins is characterized by high temporal variabilities over the years. This indicates that the upstream Pinios river basin is strongly affected by large-scale temporal variabilities in precipitation that determine water runoff and discharge. Indeed, the Pinios basin is included in the wider region of Greece and the southern Balkans, which is characterized by large interannual and interdecadal variabilities in precipitation during the last decades, as noted by previous studies [4,19,22,81]. The interannual and interdecadal variabilities in precipitation are mostly determined by the large-scale variabilities of the atmospheric circulation bringing precipitation systems to the area and, thus, influencing the alteration between dry and wet periods [20,21,82–84].

The 37-year average discharges in the three sub-basins (PINUP, TITAR, and MREMA) are 11, 9.4, and 4 $m^3 \; s^{-1}$, respectively (Figure 2a–c). The discharge differences among the sub-basins are mostly attributed to the respective differences in their total area, altitude distribution, and location (Figure 1 and Table 1) that impact the precipitation amount received [4,26,44] and, thus, the discharge. The PINUP sub-basin has the largest average discharge because its western part is well located in the Pindus mountain range, which receives the most precipitation in Greece compared with other areas [85].

The average seasonal discharges in the three sub-basins (PINUP, TITAR, and MREMA) present high seasonality, as shown in Figure 2d–o. They are characterized by high variabilities between the seasons, presenting maxima in winter (December, January, and February—DJF), smaller values in spring (March, April, and May—MAM), even smaller values in autumn (September, October, and November—SON) and minima in summer (June, July, and August—JJA). For example, the maximum average discharge in the PINUP sub-basin for winter was estimated to be 40.7 $m^3 \; s^{-1}$ in 1998 (Figure 2d). However, the minimum one for summer reached about 0.1 $m^3 \; s^{-1}$ in 1990 (Figure 2j). Similar seasonal variabilities are demonstrated for all the sub-basins, mainly driven by the respective precipitation variabilities (Figure 2d–o). It is noteworthy that overall, 1990 and 1992 can be characterized as the driest years of the 37-year period studied (Figure 2). This is explained by the large-scale decrease in precipitation (depicted in Figure 2) and drought that influenced this period not only Pinios basin [23,24] but also in the eastern Mediterranean region [17,18,20].

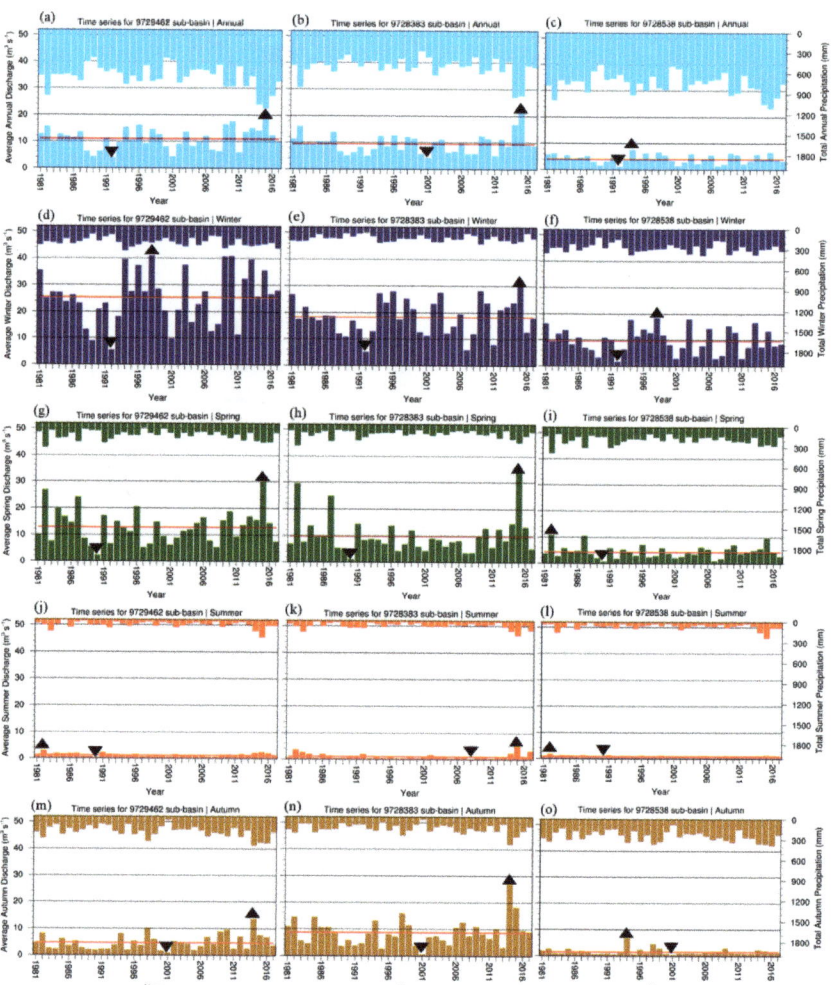

Figure 2. (**a–c**) Time series of average annual discharge (m^3 s^{-1}) in the PINUP, TITAR, and MREMA sub-basins from 1981 to 2017; (**d–f**) Similar to (**a–c**) but on a seasonal basis considering winter (December, January, and February—DJF); (**g–i**) Similarly for spring (March, April and May—MAM), (**j–l**) for summer (June, July, and August—JJA) and (**m–o**) for autumn (September, October, and November—SON). Triangles and inverted triangles depict maxima and minima, respectively, while red lines represent 37-year averages. Total annual and seasonal precipitation (mm) is also depicted for the three sub-basins, respectively, using the inverted y-axes on the right.

3.3. Generalized Additive Model Results

This sub-section presents a fit of the interannual trends of discharges of the three studied sub-basins using GAMs. Statistically significant non-linear trends of the monthly discharges were found for the sub-basins PINUP and TITAR (Table 4). In both sub-basins, discharge appears to decline from approximately 12 m^3 s^{-1} in 1981 to below 10 and 8 m^3 s^{-1}, respectively, in 1997, and then it increases to 14 m^3 s^{-1} and a little above 12 m^3 s^{-1} in 2017 (Figure 3a,c). This pattern is also shown in Figure 3b,d, where the negative rates of change for sub-basins PINUP and TITAR decrease until 1997 and then rise until 2017. On the other hand, although there are no statistically significant trends in the MREMA sub-basin

(Figure 3e,f), a declining linear trend is presented, probably due to its relatively increased agriculture (Table 1) that implies irrigation effects, thus complicating the investigation of pure climatic trends. When examining the trends separately for each month, significant interannual trends were noted only for September for sub-basins PINUP and TITAR, following the same pattern as described previously (Table 4, Figure 4), although it is obvious that the discharges were relatively low in 1981 (around 1 $m^3 s^{-1}$ for both sub-basins). It is noteworthy that sub-basin TITAR showed a sharp increase in the discharge, compared to the sub-basin PINUP, reaching 8 $m^3 s^{-1}$ in 2017 with a large rate of change. Therefore, September was wetter in the 2000s and 2010s compared to the 1980s and 1990s. This increasing interannual trend in the discharge of September is an interesting finding because the rivers in Greece are in their driest condition in September.

3.4. Interannual Increasing Trends of Discharge Maxima

Maximum analyses were conducted for every month of the year as well as for the selected period from July to October, aiming at untangling the interannual effects on discharge maxima. Only the statistically significant results with over 95% confidence level and some additional interesting results are presented here (Figures 5 and 6). The graphical depiction of the history of high discharge statistics provided in Figures 5 and 6 is an indication of the interannual changes in the magnitude of discharge events in the sub-basins PINUP, TITAR, and MREMA.

Results from the 4-month (July to October) analysis for the three sub-basins are presented side by side to facilitate the comparison (Figure 5). Increasing trends (represented by slopes) reaching 2.4% per year are apparent for the TITAR and PINUP sub-basins (Figure 5a–d). However, the trends are statistically significant in the 95% confidence level (p-value < 0.05) for both maximum 7-day and 30-day average discharges only for the TITAR sub-basin (Figure 5c,d). On the other hand, there is no significant trend (p-value > 0.1) at the sub-basin MREMA for both hydrological parameters (Figure 5e,f).

Based on the above results, further analysis was made to investigate changes in high flows employing a more detailed temporal resolution in the calculation of hydrological indices for each month. The analysis for maximum 1-day and 7-day average discharges resulted in statistically significant (p-value < 0.05) interannual trends only for September, while the results for October are interesting despite the fact that they are not statistically significant (Figures 6, A2 and A3). Therefore, the results only for these two months are presented in this sub-section. Figures 6, A2 and A3 consist of two hydrological variables (maximum 1-day and 7-day average discharges) describing discharge history relative to high flows that provide valuable information regarding long-term discharge modifications of the studied sub-basins throughout September and October.

Table 4. Results of GAMs used as response variables for the monthly discharge and the discharge of September.

Response Variable	Sub-Basin ID	Adj. R^2	Trend Sig. p Value	Month Sig. p Value
Monthly discharge	9729462 (PINUP)	0.564	0.019	≤0.001
	9728383 (TITAR)	0.433	0.002	≤0.001
	9728538 (MREMA)	0.441	NS	≤0.001
September's discharge	9729462 (PINUP)	0.299	0.002	NA
	9728383 (TITAR)	0.384	≤0.001	NA
	9728538 (MREMA)	0.099	NS	NA

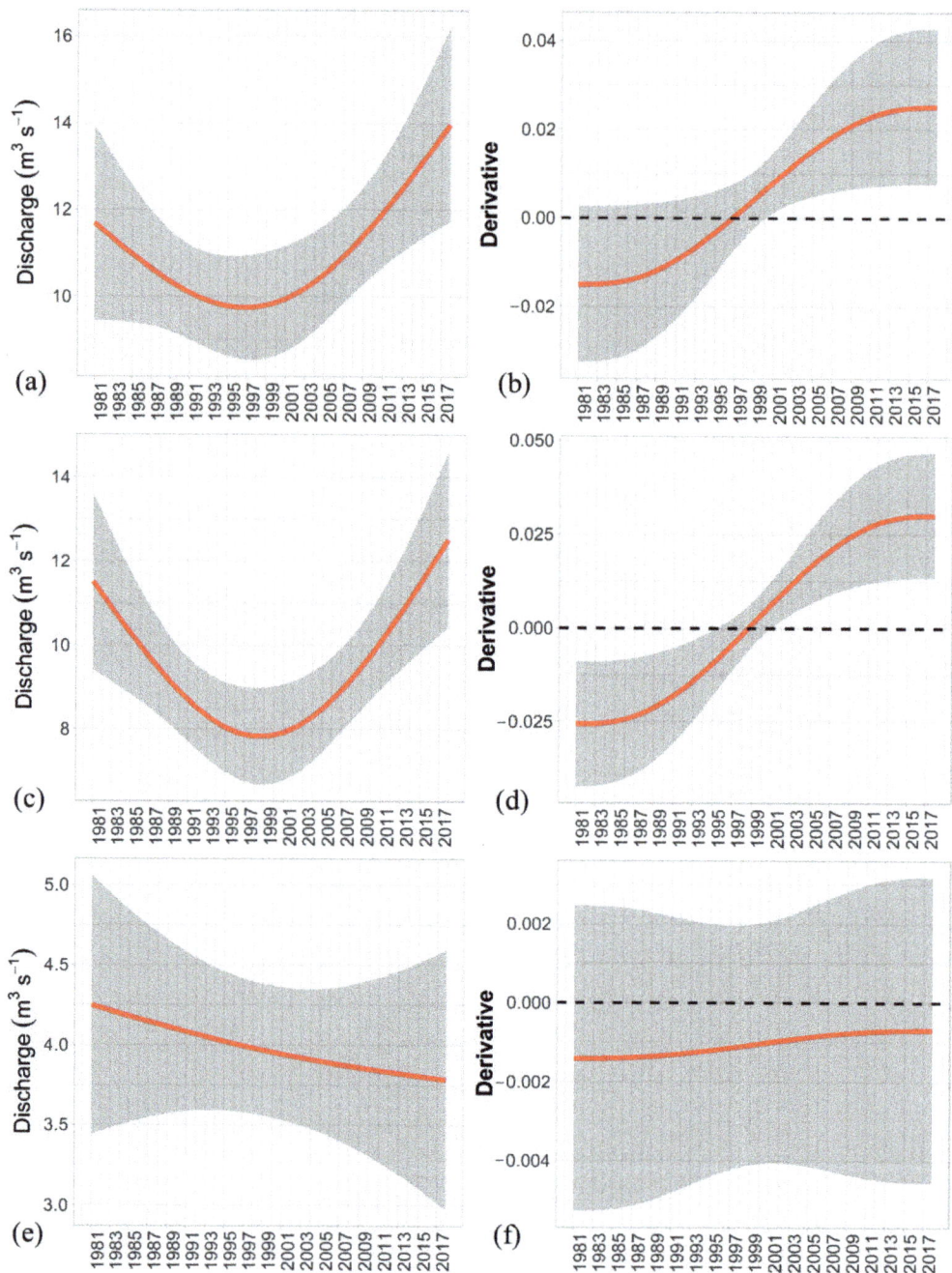

Figure 3. Generalized additive modeled fits of monthly discharge for sub-basins (**a**) PINUP, (**c**) TITAR, and (**e**) MREMA. Subplots (**b**,**d**,**f**) show the rate of change for sub-basins PINUP, TITAR, and MREMA, respectively. The shaded area represents the 95% confidence interval.

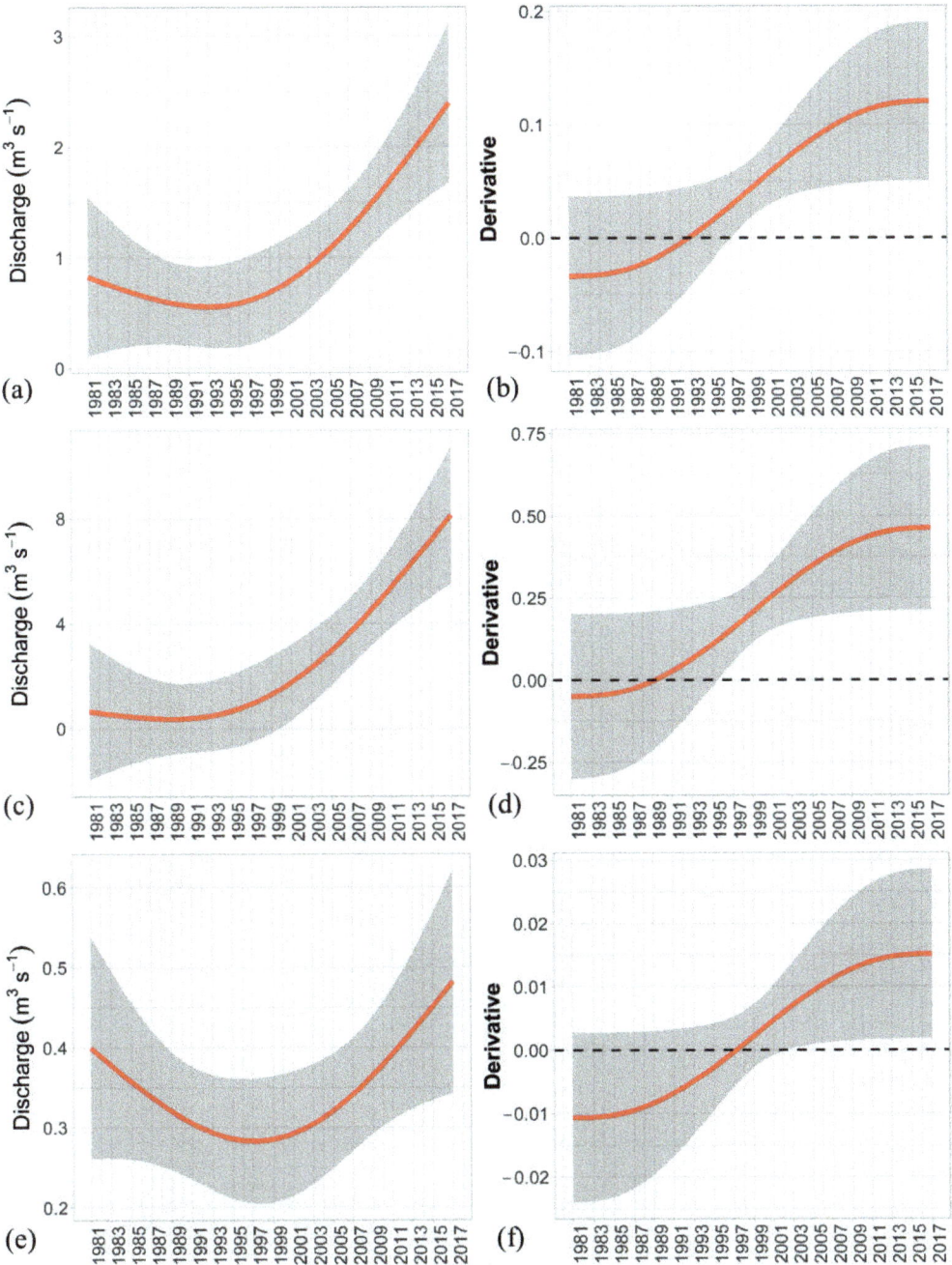

Figure 4. Generalized additive modeled fits of September's discharge for sub-basins (**a**) PINUP, (**c**) TITAR, and (**e**) MREMA. Subplots (**b**,**d**,**f**) show the rate of change for sub-basins PINUP, TITAR, and MREMA, respectively. The shaded area represents the 95% confidence interval.

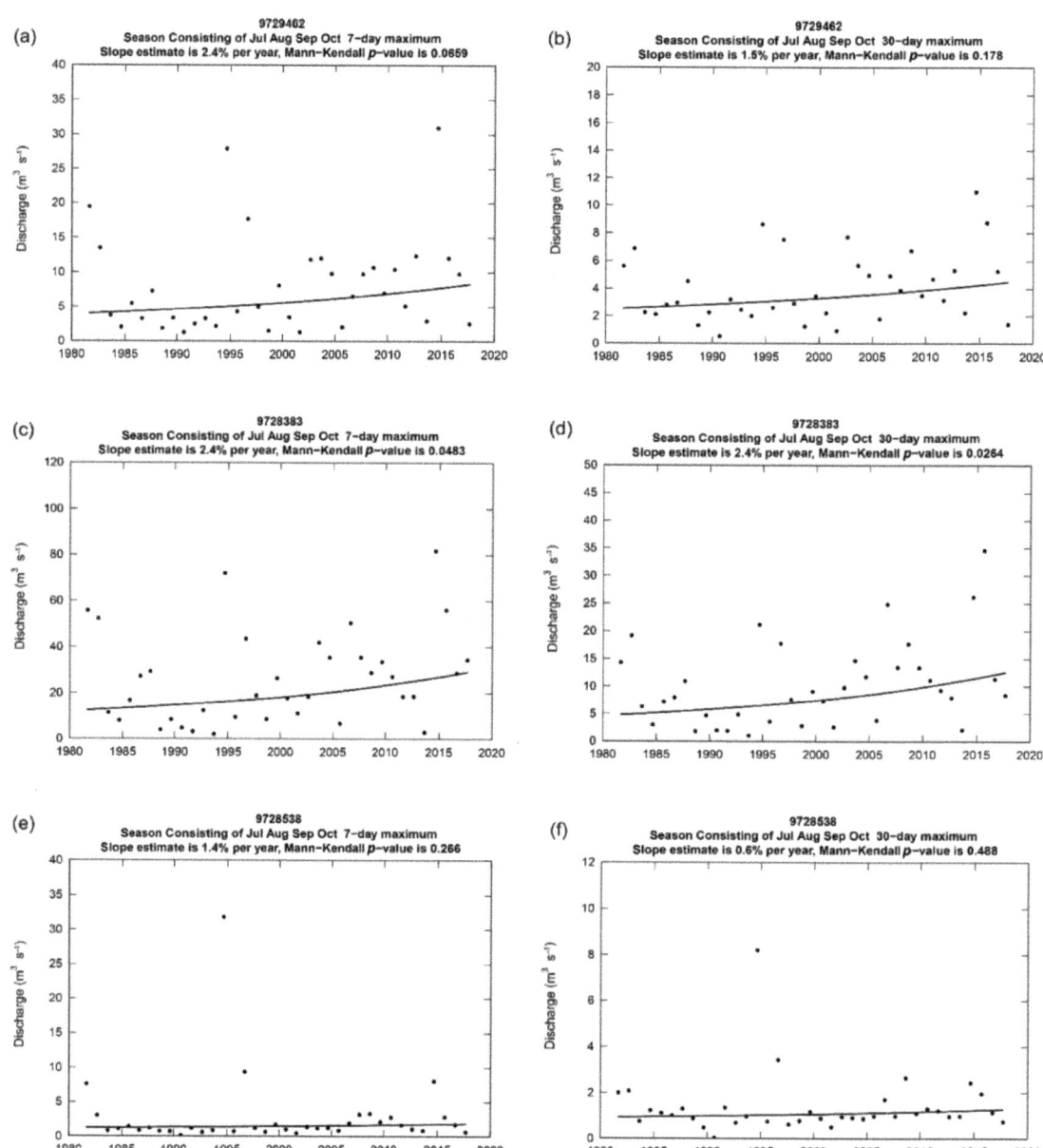

Figure 5. The left column shows trends of maximum 7-day average discharges in the (**a**) PINUP, (**c**) TITAR, and (**e**) MREMA sub-basins, whereas the right column shows trends of maximum 30-day average discharges in the (**b**) PINUP, (**d**) TITAR and (**f**) MREMA sub-basins, respectively, during the period 1981–2017 from July to October. A Thiel-Sen slope and a two-sided p-value for the Mann-Kendall trend test are also presented in the graphs.

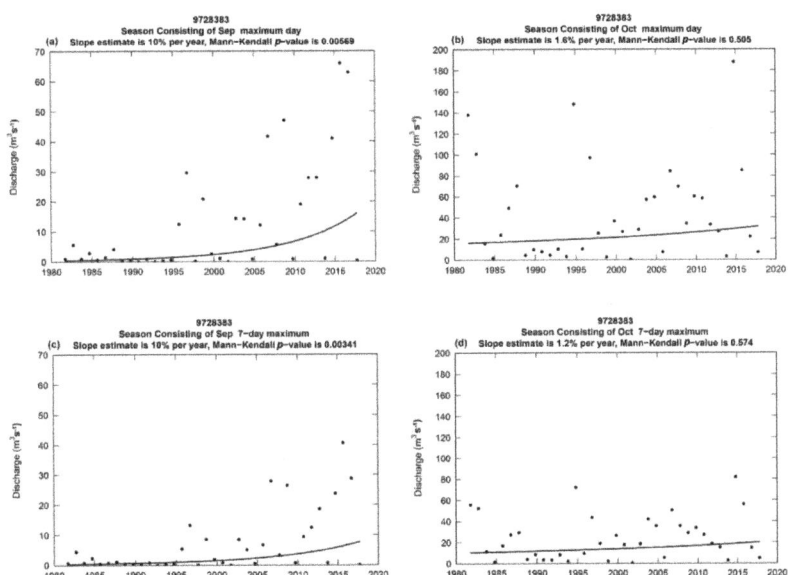

Figure 6. Trends of maximum 1-day and 7-day average discharges in the TITAR sub-basin during the period 1981–2017 for (**a,c**) September as well as (**b,d**) October, respectively. A Thiel-Sen slope and a two-sided *p*-value for the Mann-Kendall trend test are also presented in the graphs.

One of the most substantial results of this study is that positive statistically significant trends of 10% per year in the slope of the maximum 1-day and 7-day average discharge were estimated for September in the TITAR sub-basin (Figure 6a,c). The PINUP sub-basin is characterized by smaller positive significant trends of 4.6% and 4.3% per year, respectively (Figure A2a,c). Given this information, it is reasonable to assume that the trend in discharge across September's discharge distribution is positive and of substantial magnitude, particularly for the upper extremes of the discharge distribution. This result indicates that an interannual increasing trend in September's precipitation maxima hides behind the corresponding increase of discharge maxima since the sub-basins were selected to be upstream and anthropogenically low affected. Positive trends in September are critical because this month is one of the driest months in Pinios river, and thus, the estimated interannual increase in discharge maxima has significant implications. Regarding October, increasing trends are shown, but they are not statistically significant (Figure 6b,d and Figure A2b,d). Nevertheless, it should be noted that the highest maximum discharge value (188 m^3 s^{-1}) among the three sub-basins occurred in TITAR sub-basin in October 2014 (Figure 6b).

As far as the sub-basin MREMA is concerned, there are no significant trends for the examined hydrological indices in both September and October (Figure A3a–d). This may be attributed to the fact that in contrast to the other two sub-basins, PINUP and TITAR, the MREMA sub-basin is not characterized by high altitudes (Table 1) that favor precipitation. In addition, MREMA also includes a considerable agricultural area that impacts the simulated discharges through the introduction of parameterized irrigation demands in the model. These two factors have direct and indirect effects on the discharge regime, and thus, they cause inhomogeneities in the prevailing climatic trends of discharge in the upstream region of the Pinios river basin.

The magnitude of the trends across the range from low to high flows is demonstrated in Quantile-Kendall plots for the three sub-basins for September and October (Figure 7). The overall variability of September's discharge in all sub-basins has generally been increasing over time. Trends near the medians of the discharge distributions tend to be smaller positive values than the trends at the higher discharges for September, while the opposite occurs in October. For the highest flow days of September, across the probability distribution, the changes are statistically significant, especially for the sub-basins PINUP (Figure 7a,b) and TITAR (Figure 7c,d), whereas the upper quartile of the daily discharge distribution shows very likely to highly likely positive trends, typically above the range of 4% per year. Between the median and the 99th percentile of the distribution, discharges show a decline, least substantial (nevertheless, the trends are classified as likely downwards), for October. Trends in the range of low flows of the discharge distributions tend to be smaller positive values than the trends at the higher discharges with no significant increasing trends, except the PINUP sub-basin in October.

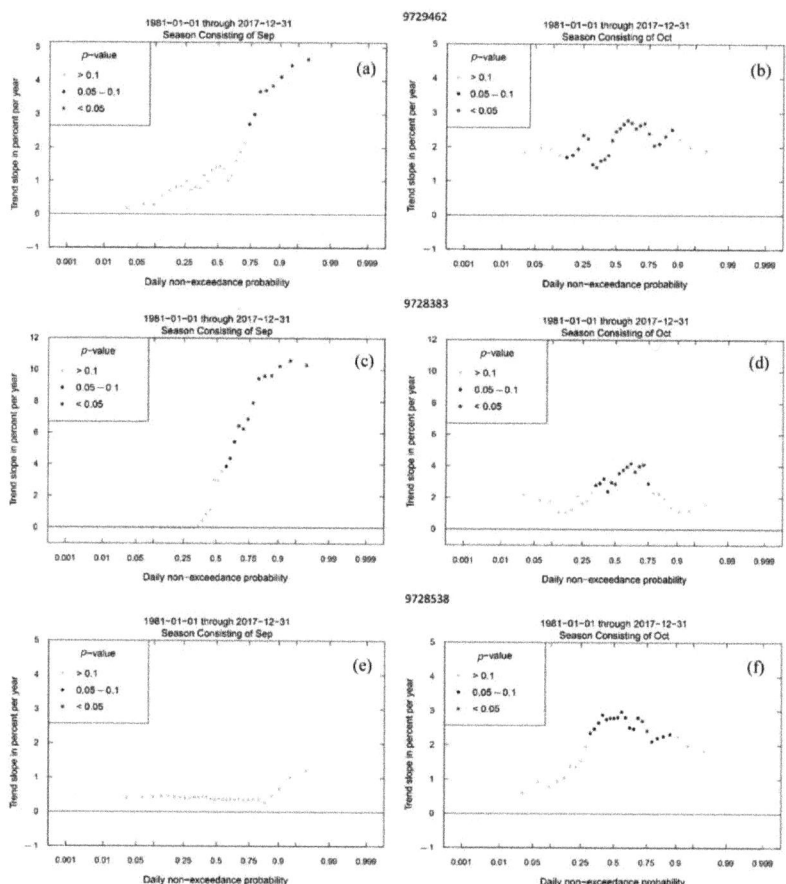

Figure 7. Quantile-Kendall plots showing 1981–2017 trends in simulated discharge for September (left column) and October (right column) in the (**a**,**b**) PINUP, (**c**,**d**) TITAR, as well as (**e**,**f**) MREMA sub-basins. The plots show the magnitude of the trends across the range from low to high flows (left side to right side, respectively) for each sub-basin, based on daily discharge data of (**a**,**c**,**e**) September and (**b**,**d**,**f**) October. The color represents the p-value for the Mann-Kendall test. Red indicates a trend that is significant at 0.05 level (95% confidence). Black indicates an attained significance between 0.05 and 0.1 (95% and 90% confidence). Grey dots indicate trends that are not significant at the 0.1 level (90% confidence).

4. Discussion
4.1. Climate Change Effects on Discharge Maxima in Early Autumn

Climate change should be considered as a main driver of change in the temporal distribution of water availability, affecting the magnitude and frequency of both precipitation and discharge maxima. The results of the present study highlighted an important aspect of rapid climate change during the last four decades regarding water resources by reporting significantly increasing trends in early-autumn discharge maxima in the TITAR and PINUP sub-basins during the period 1981–2017. Our study agrees, especially regarding September, with a previous study [36] mentioning that the total annual number of flood occurrences in Pinios presented an increasing trend from 1990 to 2010. Our findings also corroborate the latest knowledge of the scientific community, such as that presented in the IPCC Sixth Assessment Report [7], showcasing the triggering role of climate change on extreme hydrometeorological events that are frequently associated with discharge extremes. The substantial increasing trends in discharge maxima presented here are strongly dependent on climate change effects on the water cycle that bridges rivers with climate-related increases in precipitation maxima [7]. Several studies have shown that heavy precipitation events are influenced by various climate change-induced atmospheric parameters such as temperature, due point temperature, moisture, and wind but also sea parameters such as sea surface temperature and sea surface roughness [38,86–88]. Albeit there are large uncertainties in climate change effects on precipitation, it is important to note that atmospheric moisture is considered the most significant factor for the increase in short-duration precipitation extreme events. This is attributed to the fact that climate change strongly affects atmospheric moisture content with an increase of 7% per 1 °C of air temperature increase [89].

Our findings indicate that modeled discharge data can facilitate the scope of this study, as they are generally in good agreement with measurements. More specifically, average seasonal discharges in the three sub-basins (PINUP, TITAR, and MREMA) present high seasonality, whereas statistically significant non-linear trends of the monthly discharges were found for the sub-basins PINUP and TITAR. Based on the above results, further analysis was made to investigate changes in high flows employing a more detailed temporal resolution. One of the most substantial results of this study is that in comparison with other months, September has positive statistically significant interannual trends for the TITAR and PINUP sub-basins. September's increase in discharge maxima could imply a climate-induced increase in precipitation maxima in the Pinios river basin, as the two abovementioned sub-basins were selected to be upstream, partially mountainous, and anthropogenically low impacted. Moreover, given the fact that global warming enhances evapotranspiration, thus reducing water on the ground, the role of precipitation in increasing September's discharge maxima is critical. It is well-known that heavy precipitation events are largely dependent on energy and moisture content in the atmosphere. For example, high moisture amounts favoring such precipitation events in the Pinios basin and generally in Greece, are usually originated from the Aegean and Ionian Seas [32,34,38]. A warmer Mediterranean Sea during the last decades could positively contribute to the formation of high-energy and high-moisture atmospheric systems such as "medicanes". Ianos medicane, which occurred in September 2020, is a typical example of the abovementioned processes. Ianos formed in September when the Mediterranean Sea was still warm enough. However, the atmosphere began to cool down by southward movements of relatively colder air masses from northern-central Europe. This enhancement in the sea-air temperature contrast is a significant factor, and it hides behind the formation of Ianos. Thus, Ianos brought vast amounts of energy and moisture from the sea that triggered heavy rainfall that caused flash floods in southwestern parts of the Pinios river basin [35,90]. An interpretation of our findings reveals a connection between such phenomena, which could be more frequent in a warmer future climate, and, thus, more research in their hydrometeorological aspects and implications is needed, additionally considering potential climate change effects [91]. In this context, co-analyses between discharge, precipitation, and evapotranspiration during the last decades as well as in-depth analyses of extreme precipitation events in the Pinios basin driving to flash floods (e.g., that caused by the Ianos medicane in 2020), are interesting proposals for future research.

4.2. Necessity for Untangling Climate Effects and Direct Anthropogenic Pressures

This study focused on the unraveling of interannual variabilities and trends of annual average discharge and discharge maxima during the period from 1981 to 2017, avoiding parts of the Pinios river basin that are characterized by increased agricultural activity. However, the Pinios river basin is one of the most productive areas of the country regarding the agricultural sector, also including multiple anthropogenic pressures [45] such as big cities (e.g., Larissa city with more than 140,000 residents), irrigation processes, bridges, technical works, industries, buildings, pollution, etc. Thus, the results of the present study regarding the upstream of the river may deviate from low-lying areas. For example, the analyses showed an increase in an annual average discharge after the middle 1990s in the upstream sub-basins. However, this increase could be eliminated in lowlands if considering a probable increase in irrigation, removing water from the river and transferring it to the fields, or changes in land use. These factors are significant because agriculture had an increasing trend, especially in the 1990s in the Pinios basin, also presenting land use changes. Additionally, the increase in temperature since the 1990s, especially during summer, could have impacts on the increasing irrigation demands [25]. E-HYPE has sufficiently considered irrigation [55,56] and crop water demand [57]. Nevertheless, the combination of the abovementioned factors poses the necessity to dynamically consider the changes in irrigation and land use in low-lying sub-basins as well as the installation of small dams if the aim is to investigate the effects of both climate change and direct anthropogenic pressures on discharge. Such an effort presupposes a lot of detailed spatiotemporally distributed information about factors such as irrigation, land use, and dams, which is very difficult to find and introduce every year in the estimation of the hydrological cycle by the hydrological model. It is nevertheless a suggestion for future work in the Pinios river basin aiming for a more holistic approach to the issue of simulated long-term discharge trends in the whole river.

5. Conclusions

The scope of the present study was to investigate variabilities and significant interannual trends in the annual average discharge of three upstream sub-basins of the Pinios river while also unraveling significant trends in discharge maxima by examining each month. To this aim, daily discharges from the E-HYPE model from 1981 to 2017 were used. The most critical conclusions of this study are summarized below:

- The analysis of data revealed noteworthy findings for the upstream Pinios river. The average annual discharge in the three sub-basins decreased in the 1980s, reaching a minimum in the early 1990s when extensive droughts influenced not only Pinios river but also several areas of the Mediterranean region. Afterward, the average annual discharge gradually increased from the middle 1990s to 2017, reaching approximately the discharge levels of the early 1980s.
- The interannual and interdecadal discharge variabilities presented in this study were characterized by high amplitudes. These variabilities were induced by large-scale climatic forcings that largely determined the water resources of the Pinios river basin.
- The most striking finding of this study is that significantly increasing interannual trends in discharge maxima were identified for September, which is one of the driest months in the Pinios river basin. The performed discharge history analyses indicated the presence of large positive (10% and approximately 4.5% per year for TITAR and PINUP sub-basins, respectively) statistically significant (p-value < 0.05) interannual trends in maximum 1-day and 7-day average discharges for September.
- The increases in high flows were also much more substantial than those near the median flows for September, as evidenced by Quantile-Kendall plots. Significant

increasing trends up to 2.4 per year for the TITAR sub-basin were also resulted by analyzing maximum 7-day and 30-day average discharges considering the period from July to October.

The findings of this study regarding the interannual trends of both the annual average discharge and discharge maxima can be exploited by the scientific community to conduct climate studies as well as interdisciplinary projects. The implications of large variabilities in discharge over the years are related to multi-year wet and dry periods that influence the society, economy, and ecosystems of the Pinios basin. The increase of discharge maxima in early autumn, when the rivers in Greece are in their driest condition, is a clear signal of climate change that must be seriously considered in the future. The increase in maxima may potentially determine suitable hydrological conditions for the increased probability of floods during late summer and especially early autumn periods. This study presents a paradigm for probable discharge variabilities in the future, and thus, decision-makers and civil protection may consider the results when they make basin management strategies as well as climate change adaptation and mitigation plans.

Author Contributions: Conceptualization, G.V., C.P., K.S. and A.M.; methodology, G.V., C.P., K.S. and A.M.; software, G.V., C.P., K.S. and A.M.; validation, G.V. and A.M.; formal analysis, G.V., C.P., K.S., A.M., I.P., A.P. and E.D.; investigation, G.V., C.P., K.S., A.M., I.P., A.P. and E.D.; resources, I.P., A.P. and E.D.; data curation, G.V., C.P., K.S., A.M. and I.P.; writing—original draft preparation, G.V., C.P., K.S., A.M. and I.P.; writing—review and editing, G.V., C.P., K.S., A.M., I.P., A.P. and E.D.; visualization, G.V., C.P., K.S. and A.M.; supervision, A.P. and E.D.; project administration, G.V. All authors have read and agreed to the published version of the manuscript.

Funding: This research received no external funding.

Institutional Review Board Statement: Not applicable.

Informed Consent Statement: Not applicable.

Data Availability Statement: The data presented in this study are available on request from the corresponding author.

Acknowledgments: The European Hydrological Predictions for the Environment (E-HYPE) model climate dataset, including average daily discharge values from 1 January 1981 to 31 December 2017, was kindly provided by the Hydrological Research Unit at the Swedish Meteorological and Hydrological Institute (SMHI). Time series and maps from the E-HYPE model are available for inspection at http://hypeweb.smhi.se, accessed on 5 March 2023.

Conflicts of Interest: The authors declare no conflict of interest.

Appendix A

Statistical criteria of the E-HYPE model performance by comparing observed (o) and modeled (m) data of a sample size (n):

Pearson's correlation coefficient (R).

$$R = \frac{\sum_{i=1}^{n}(m_i - \overline{m})(o_i - \overline{o})}{\sqrt{\sum_{i=1}^{n}(m_i - \overline{m})^2}\sqrt{\sum_{i=1}^{n}(o_i - \overline{o})^2}} \tag{A1}$$

Percent bias ($PBIAS$).

$$PBIAS = 100 \times \frac{\sum_{i=1}^{n}(o_i - m_i)}{\sum_{i=1}^{n} o_i} \tag{A2}$$

Root mean square error ($RMSE$).

$$RMSE = \sqrt{\frac{\sum_{i=1}^{n}(o_i - m_i)^2}{n}} \tag{A3}$$

Ratio of the $RMSE$ to the Standard Deviation of observations (RSR).

$$RSR = \frac{\sqrt{\sum_{i=1}^{n}(o_i - m_i)^2}}{\sqrt{\sum_{i=1}^{n}(o_i - \overline{o})^2}} \quad (A4)$$

Nash-Sutcliffe efficiency (NSE).

$$NSE = 1 - \frac{\sum_{i=1}^{n}(o_i - m_i)^2}{\sum_{i=1}^{n}(o_i - \overline{o})^2} \quad (A5)$$

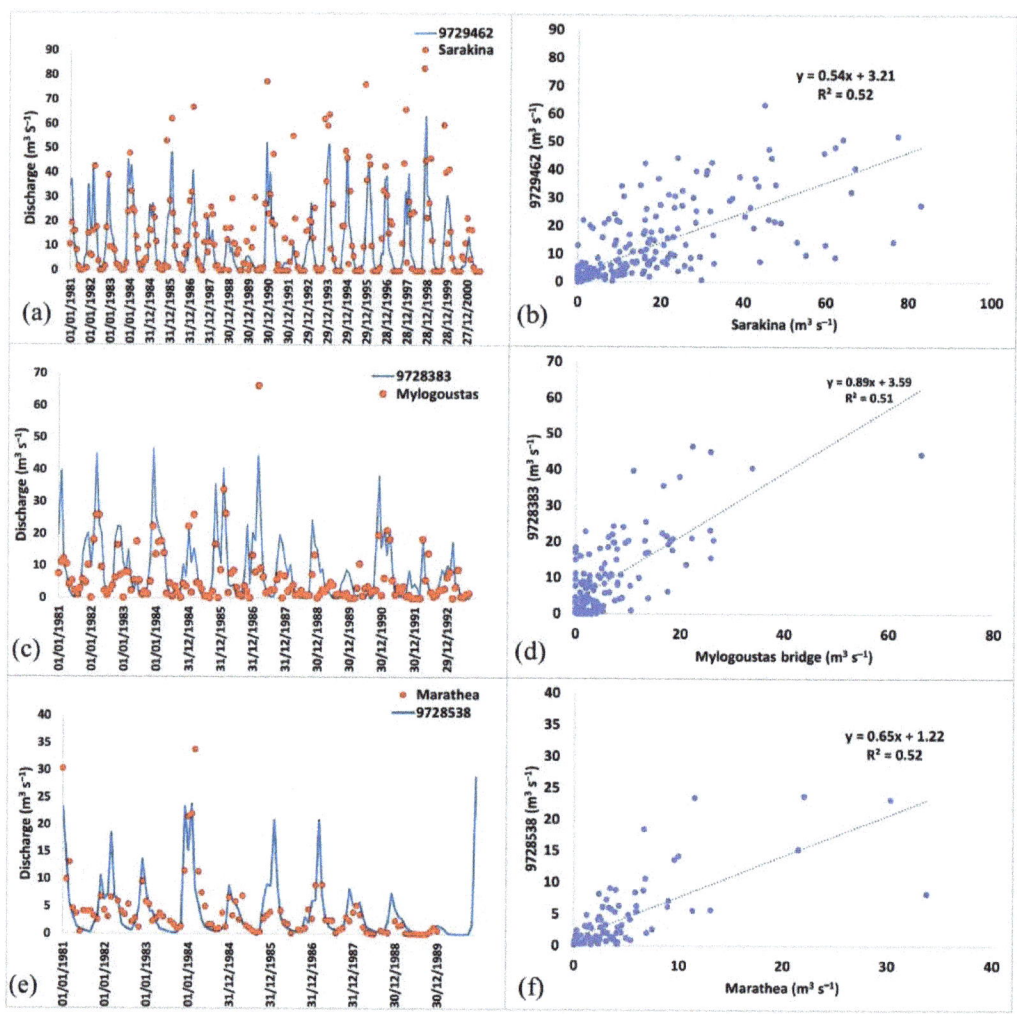

Figure A1. Simulated and measured discharge (m^3 s^{-1}) values and correlations of (**a**,**b**) PINUP—Sarakinas bridge, (**c**,**d**) TITAR—Mylogoustas bridge, and (**e**,**f**) MREMA—Marathea station.

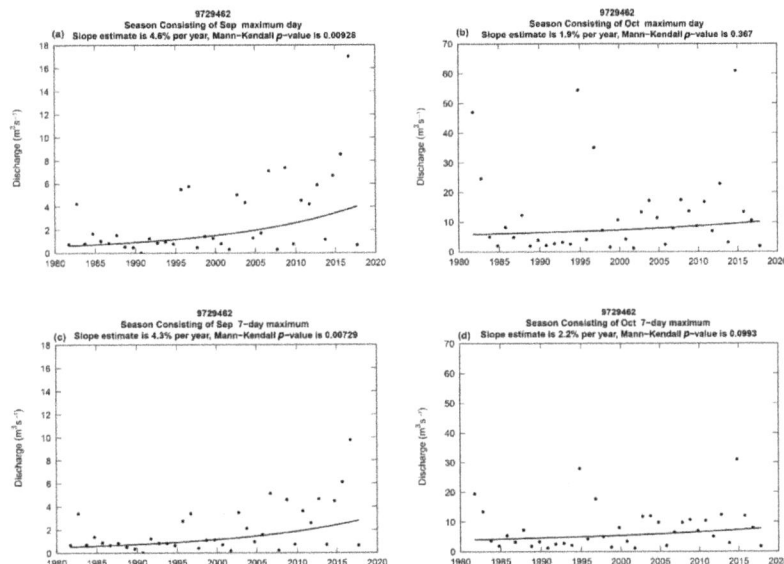

Figure A2. Trends of maximum 1-day and 7-day average discharges in the PINUP sub-basin during the period 1981–2017 for September (**a**,**c**) and October (**b**,**d**), respectively. A Thiel-Sen slope and a two-sided p-value for the Mann-Kendall trend test are also presented in the graphs.

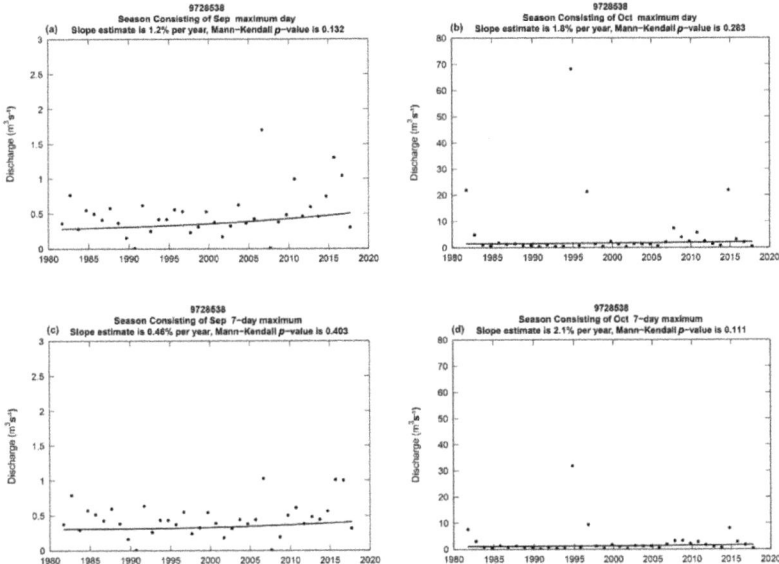

Figure A3. Trends of maximum 1-day and 7-day average discharges in the MREMA sub-basin during the period 1981–2017 for September (**a**,**c**) and October (**b**,**d**), respectively. A Thiel-Sen slope and a two-sided p-value for the Mann-Kendall trend test are also presented in the graphs.

References

1. Gudmundsson, L.; Boulange, J.; Do, H.X.; Gosling, S.N.; Grillakis, M.G.; Koutroulis, A.G.; Leonard, M.; Liu, J.; Schmied, H.M.; Papadimitriou, L.; et al. Globally Observed Trends in Mean and Extreme River Flow Attributed to Climate Change. *Science* **2021**, *371*, 1159–1162. [CrossRef]
2. Arnell, N.W.; Gosling, S.N. The Impacts of Climate Change on River Flood Risk at the Global Scale. *Clim. Chang.* **2016**, *134*, 387–401. [CrossRef]
3. Giorgi, F. Climate Change Hot-Spots. *Geophys. Res. Lett.* **2006**, *33*, 8707. [CrossRef]
4. Varlas, G.; Stefanidis, K.; Papaioannou, G.; Panagopoulos, Y.; Pytharoulis, I.; Katsafados, P.; Papadopoulos, A.; Dimitriou, E. Unravelling Precipitation Trends in Greece since 1950s Using ERA5 Climate Reanalysis Data. *Climate* **2022**, *10*, 12. [CrossRef]
5. Mentzafou, A.; Varlas, G.; Dimitriou, E.; Papadopoulos, A.; Pytharoulis, I.; Katsafados, P. Modeling the Effects of Anthropogenic Land Cover Changes to the Main Hydrometeorological Factors in a Regional Watershed, Central Greece. *Climate* **2019**, *7*, 129. [CrossRef]
6. Mentzafou, A.; Vamvakaki, C.; Zacharias, I.; Gianni, A.; Dimitriou, E. Climate Change Impacts on a Mediterranean River and the Associated Interactions with the Adjacent Coastal Area. *Environ. Earth Sci.* **2017**, *76*, 259. [CrossRef]
7. Masson-Delmotte, V.; Zhai, P.; Chen, Y.; Goldfarb, L.; Gomis, M.I.; Matthews, J.B.R.; Berger, S.; Huang, M.; Yelekçi, O.; Yu, R.; et al. *Contribution to the Sixth Assessment Report of the Intergovernmental Panel on Climate Change*; Cambridge University Press: Cambridge, UK; New York, NY, USA, 2021.
8. Masseroni, D.; Camici, S.; Cislaghi, A.; Vacchiano, G.; Massari, C.; Brocca, L. The 63-Year Changes in Annual Streamflow Volumes across Europe with a Focus on the Mediterranean Basin. *Hydrol. Earth Syst. Sci.* **2021**, *25*, 5589–5601. [CrossRef]
9. Blöschl, G.; Hall, J.; Viglione, A.; Perdigão, R.A.P.; Parajka, J.; Merz, B.; Lun, D.; Arheimer, B.; Aronica, G.T.; Bilibashi, A.; et al. Changing Climate Both Increases and Decreases European River Floods. *Nature* **2019**, *573*, 108–111. [CrossRef]
10. Montaldo, N.; Oren, R. Changing Seasonal Rainfall Distribution With Climate Directs Contrasting Impacts at Evapotranspiration and Water Yield in the Western Mediterranean Region. *Earth's Future* **2018**, *6*, 841–856. [CrossRef]
11. Brogli, R.; Sørland, S.L.; Kröner, N.; Schär, C. Causes of Future Mediterranean Precipitation Decline Depend on the Season. *Environ. Res. Lett.* **2019**, *14*, 114017. [CrossRef]
12. Giannakopoulos, C.; Kostopoulou, E.; Varotsos, K.V.; Tziotziou, K.; Plitharas, A. An Integrated Assessment of Climate Change Impacts for Greece in the near Future. *Reg. Environ. Chang.* **2011**, *11*, 829–843. [CrossRef]
13. Sellami, H.; Benabdallah, S.; La Jeunesse, I.; Vanclooster, M. Quantifying Hydrological Responses of Small Mediterranean Catchments under Climate Change Projections. *Sci. Total Environ.* **2016**, *543*, 924–936. [CrossRef] [PubMed]
14. Stefanidis, K.; Panagopoulos, Y.; Mimikou, M. Response of a Multi-Stressed Mediterranean River to Future Climate and Socio-Economic Scenarios. *Sci. Total Environ.* **2018**, *627*, 756–769. [CrossRef]
15. Papadaki, C.; Soulis, K.; Muñoz-Mas, R.; Martinez-Capel, F.; Zogaris, S.; Ntoanidis, L.; Dimitriou, E. Potential Impacts of Climate Change on Flow Regime and Fish Habitat in Mountain Rivers of the South-Western Balkans. *Sci. Total Environ.* **2016**, *540*, 418–428. [CrossRef]
16. Panagopoulos, A.; Arampatzis, G.; Tziritis, E.; Pisinaras, V.; Herrmann, F.; Kunkel, R.; Wendland, F. Assessment of Climate Change Impact in the Hydrological Regime of River Pinios Basin, Central Greece. *Desalination Water Treat.* **2016**, *57*, 2256–2267. [CrossRef]
17. Feidas, H.; Noulopoulou, C.; Makrogiannis, T.; Bora-Senta, E. Trend Analysis of Precipitation Time Series in Greece and Their Relationship with Circulation Using Surface and Satellite Data: 1955–2001. *Theor. Appl. Climatol.* **2007**, *87*, 155–177. [CrossRef]
18. Hatzianastassiou, N.; Katsoulis, B.; Pnevmatikos, J.; Antakis, V. Spatial and Temporal Variation of Precipitation in Greece and Surrounding Regions Based on Global Precipitation Climatology Project Data. *J. Clim.* **2008**, *21*, 1349–1370. [CrossRef]
19. Nastos, P.T.; Zerefos, C.S. Spatial and Temporal Variability of Consecutive Dry and Wet Days in Greece. *Atmos. Res.* **2009**, *94*, 616–628. [CrossRef]
20. Hurrell, J.W. Decadal Trends in the North Atlantic Oscillation: Regional Temperatures and Precipitation. *Science* **1995**, *269*, 676–679. [CrossRef]
21. Xoplaki, E.; González-Rouco, J.F.; Luterbacher, J.; Wanner, H. Wet Season Mediterranean Precipitation Variability: Influence of Large-Scale Dynamics and Trends. *Clim. Dyn.* **2004**, *23*, 63–78. [CrossRef]
22. Philandras, C.M.; Nastos, P.T.; Kapsomenakis, J.; Douvis, K.C.; Tselioudis, G.; Zerefos, C.S. Long Term Precipitation Trends and Variability within the Mediterranean Region. *Nat. Hazards Earth Syst. Sci.* **2011**, *11*, 3235–3250. [CrossRef]
23. Vasiliades, L.; Loukas, A. Hydrological Response to Meteorological Drought Using the Palmer Drought Indices in Thessaly, Greece. *Desalination* **2009**, *237*, 3–21. [CrossRef]
24. Loukas, A.; Vasiliades, L. Probabilistic Analysis of Drought Spatiotemporal Characteristics InThessaly Region, Greece. *Nat. Hazards Earth Syst. Sci.* **2004**, *4*, 719–731. [CrossRef]
25. Mentzafou, A.; Dimitriou, E.; Papadopoulos, A. Long-Term Hydrologic Trends in the Main Greek Rivers: A Statistical Approach. In *Handbook of Environmental Chemistry*; Springer: Berlin, Heidelberg, 2015; Volume 59, pp. 129–165.
26. Loukas, A. Surface Water Quantity and Quality Assessment in Pinios River, Thessaly, Greece. *Desalination* **2010**, *250*, 266–273. [CrossRef]
27. Zittis, G.; Bruggeman, A.; Lelieveld, J. Revisiting Future Extreme Precipitation Trends in the Mediterranean. *Weather Clim. Extrem.* **2021**, *34*, 100380. [CrossRef] [PubMed]

28. Camera, C.; Bruggeman, A.; Zittis, G.; Sofokleous, I.; Arnault, J. Simulation of Extreme Rainfall and Streamflow Events in Small Mediterranean Watersheds with a One-Way-Coupled Atmospheric-Hydrologic Modelling System. *Nat. Hazards Earth Syst. Sci.* **2020**, *20*, 2791–2810. [CrossRef]
29. Spyrou, C.; Varlas, G.; Pappa, A.; Mentzafou, A.; Katsafados, P.; Papadopoulos, A.; Anagnostou, M.N.; Kalogiros, J. Implementation of a Nowcasting Hydrometeorological System for Studying Flash Flood Events: The Case of Mandra, Greece. *Remote Sens.* **2020**, *12*, 2784. [CrossRef]
30. Varlas, G.; Anagnostou, M.N.; Spyrou, C.; Papadopoulos, A.; Kalogiros, J.; Mentzafou, A.; Michaelides, S.; Baltas, E.; Karymbalis, E.; Katsafados, P. A Multi-Platform Hydrometeorological Analysis of the Flash Flood Event of 15 November 2017 in Attica, Greece. *Remote Sens.* **2019**, *11*, 45. [CrossRef]
31. Giannaros, C.; Dafis, S.; Stefanidis, S.; Giannaros, T.M.; Koletsis, I.; Oikonomou, C. Hydrometeorological Analysis of a Flash Flood Event in an Ungauged Mediterranean Watershed under an Operational Forecasting and Monitoring Context. *Meteorol. Appl.* **2022**, *29*, e2079. [CrossRef]
32. Papaioannou, G.; Varlas, G.; Papadopoulos, A.; Loukas, A.; Katsafados, P.; Dimitriou, E. Investigating Sea-state Effects on Flash Flood Hydrograph and Inundation Forecasting. *Hydrol. Process.* **2021**, *35*, e14151. [CrossRef]
33. Fowler, H.J.; Lenderink, G.; Prein, A.F.; Westra, S.; Allan, R.P.; Ban, N.; Barbero, R.; Berg, P.; Blenkinsop, S.; Do, H.X.; et al. Anthropogenic Intensification of Short-Duration Rainfall Extremes. *Nat. Rev. Earth Environ.* **2021**, *2*, 107–122. [CrossRef]
34. Papaioannou, G.; Varlas, G.; Terti, G.; Papadopoulos, A.; Loukas, A.; Panagopoulos, Y.; Dimitriou, E. Flood Inundation Mapping at Ungauged Basins Using Coupled Hydrometeorological-Hydraulic Modelling: The Catastrophic Case of the 2006 Flash Flood in Volos City, Greece. *Water* **2019**, *11*, 2328. [CrossRef]
35. Lagouvardos, K.; Karagiannidis, A.; Dafis, S.; Kalimeris, A.; Kotroni, V. Ianos—A Hurricane in the Mediterranean. *Bull. Am. Meteorol. Soc.* **2022**, *103*, E1621–E1636. [CrossRef]
36. Bathrellos, G.D.; Skilodimou, H.D.; Soukis, K.; Koskeridou, E. Temporal and Spatial Analysis of Flood Occurrences in the Drainage Basin of Pinios River (Thessaly, Central Greece). *Land* **2018**, *7*, 106. [CrossRef]
37. Romero, R.; Emanuel, K. Medicane Risk in a Changing Climate. *J. Geophys. Res. Atmos.* **2013**, *118*, 5992–6001. [CrossRef]
38. Varlas, G.; Vervatis, V.; Spyrou, C.; Papadopoulou, E.; Papadopoulos, A.; Katsafados, P. Investigating the Impact of Atmosphere–Wave–Ocean Interactions on a Mediterranean Tropical-like Cyclone. *Ocean Model.* **2020**, *153*, 101675. [CrossRef]
39. European Environment Agency Copernicus Land Monitoring Service 2018. CORINE Land Cover CLC2018 Version 2020_20u1. Available online: https://land.copernicus.eu/pan-european/corine-land-cover/clc2018 (accessed on 5 March 2023).
40. Hellenic Statistical Authority of Greece Statistics-Gross Value Added by Industry. Available online: https://www.statistics.gr/en/statistics/-/publication/SEL45/2019 (accessed on 10 October 2022).
41. Migiros, G.; Bathrellos, G.D.; Skilodimou, H.D.; Karamousalis, T. Pinios (Peneus) River (Central Greece): Hydrological—Geomorphological Elements and Changes during the Quaternary. *Cent. Eur. J. Geosci.* **2011**, *3*, 215–228. [CrossRef]
42. Caputo, R.; Helly, B.; Rapti, D.; Valkaniotis, S. Late Quaternary Hydrographic Evolution in Thessaly (Central Greece): The Crucial Role of the Piniada Valley. *Quat. Int.* **2022**, *635*, 3–19. [CrossRef]
43. Kottek, M.; Grieser, J.; Beck, C.; Rudolf, B.; Rubel, F. World Map of the Köppen-Geiger Climate Classification Updated. *Meteorol. Z.* **2006**, *15*, 259–263. [CrossRef]
44. Mylopoulos, N.; Kolokytha, E.; Loukas, A.; Mylopoulos, Y. Agricultural and Water Resources Development in Thessaly, Greece in the Framework of New European Union Policies. *Int. J. River Basin Manag.* **2009**, *7*, 73–89. [CrossRef]
45. Mentzafou, A.; Varlas, G.; Papadopoulos, A.; Poulis, G.; Dimitriou, E. Assessment of Automatically Monitored Water Levels and Water Quality Indicators in Rivers with Different Hydromorphological Conditions and Pollution Levels in Greece. *Hydrology* **2021**, *8*, 86. [CrossRef]
46. U.S. Geological Survey. *Shuttle Radar Topography Mission 1 Arc-Second Global*; U.S. Geological Survey: Reston, VA, USA, 2017.
47. Lindström, G.; Pers, C.; Rosberg, J.; Strömqvist, J.; Arheimer, B. Development and Testing of the HYPE (Hydrological Predictions for the Environment) Water Quality Model for Different Spatial Scales. *Hydrol. Res.* **2010**, *41*, 295–319. [CrossRef]
48. Hundecha, Y.; Arheimer, B.; Donnelly, C.; Pechlivanidis, I. A Regional Parameter Estimation Scheme for a Pan-European Multi-Basin Model. *J. Hydrol. Reg. Stud.* **2016**, *6*, 90–111. [CrossRef]
49. Donnelly, C.; Andersson, J.C.M.; Arheimer, B. Using Flow Signatures and Catchment Similarities to Evaluate the E-HYPE Multi-Basin Model across Europe. *Hydrol. Sci. J.* **2016**, *61*, 255–273. [CrossRef]
50. Krysanova, V.; Vetter, T.; Eisner, S.; Huang, S.; Pechlivanidis, I.; Strauch, M.; Gelfan, A.; Kumar, R.; Aich, V.; Arheimer, B.; et al. Intercomparison of Regional-Scale Hydrological Models and Climate Change Impacts Projected for 12 Large River Basins Worldwide—A Synthesis. *Environ. Res. Lett.* **2017**, *12*, 105002. [CrossRef]
51. Vetter, T.; Reinhardt, J.; Flörke, M.; van Griensven, A.; Hattermann, F.; Huang, S.; Koch, H.; Pechlivanidis, I.G.; Plötner, S.; Seidou, O.; et al. Evaluation of Sources of Uncertainty in Projected Hydrological Changes under Climate Change in 12 Large-Scale River Basins. *Clim. Chang.* **2017**, *141*, 419–433. [CrossRef]
52. Pechlivanidis, I.G.; Arheimer, B.; Donnelly, C.; Hundecha, Y.; Huang, S.; Aich, V.; Samaniego, L.; Eisner, S.; Shi, P. Analysis of Hydrological Extremes at Different Hydro-Climatic Regimes under Present and Future Conditions. *Clim. Chang.* **2017**, *141*, 467–481. [CrossRef]

53. Samaniego, L.; Kumar, R.; Breuer, L.; Chamorro, A.; Flörke, M.; Pechlivanidis, I.G.; Schäfer, D.; Shah, H.; Vetter, T.; Wortmann, M.; et al. Propagation of Forcing and Model Uncertainties on to Hydrological Drought Characteristics in a Multi-Model Century-Long Experiment in Large River Basins. *Clim. Chang.* **2017**, *141*, 435–449. [CrossRef]
54. Berg, P.; Donnelly, C.; Gustafsson, D. Near-Real-Time Adjusted Reanalysis Forcing Data for Hydrology. *Hydrol. Earth Syst. Sci.* **2018**, *22*, 989–1000. [CrossRef]
55. Wriedt, G.; van der Velde, M.; Aloe, A.; Bouraoui, F. A European Irrigation Map for Spatially Distributed Agricultural Modelling. *Agric. Water Manag.* **2009**, *96*, 771–789. [CrossRef]
56. Siebert, S.; Döll, P.; Hoogeveen, J.; Faures, J.M.; Frenken, K.; Feick, S. Development and Validation of the Global Map of Irrigation Areas. *Hydrol. Earth Syst. Sci.* **2005**, *9*, 535–547. [CrossRef]
57. Portmann, F.T.; Siebert, S.; Döll, P. MIRCA2000-Global Monthly Irrigated and Rainfed Crop Areas around the Year 2000: A New High-Resolution Data Set for Agricultural and Hydrological Modeling. *Glob. Biogeochem. Cycles* **2010**, *24*, 1–24. [CrossRef]
58. European Environment Agency Corine Land Cover 2000 Coastline. Available online: http://www.eea.europa.eu/data-and-maps/data/corine-land-cover-2000-coastline/#tab-gis-data (accessed on 5 March 2023).
59. Hundecha, Y.; Arheimer, B.; Berg, P.; Capell, R.; Musuuza, J.; Pechlivanidis, I.; Photiadou, C. Effect of Model Calibration Strategy on Climate Projections of Hydrological Indicators at a Continental Scale. *Clim. Chang.* **2020**, *163*, 1287–1306. [CrossRef]
60. Kumar, R.; Livneh, B.; Samaniego, L. Toward Computationally Efficient Large-Scale Hydrologic Predictions with a Multiscale Regionalization Scheme. *Water Resour. Res.* **2013**, *49*, 5700–5714. [CrossRef]
61. Pechlivanidis, I.G.; Arheimer, B. Large-Scale Hydrological Modelling by Using Modified PUB Recommendations: The India-HYPE Case. *Hydrol. Earth Syst. Sci.* **2015**, *19*, 4559–4579. [CrossRef]
62. Samaniego, L.; Kumar, R.; Jackisch, C. Predictions in a Data-Sparse Region Using a Regionalized Grid-Based Hydrologic Model Driven by Remotely Sensed Data. *Hydrol. Res.* **2011**, *42*, 338–355. [CrossRef]
63. Beven, K. *Rainfall-Runoff Modelling: The Primer*, 2nd ed.; John Wiley & Sons, Ltd.: Chichester, UK, 2012; ISBN 9780470714591.
64. Pechlivanidis, I.G.; Jackson, B.M.; Mcintyre, N.R.; Wheater, H.S. Catchment Scale Hydrological Modelling: A Review of Model Types, Calibration Approaches and Uncertainty Analysis Methods in the Context of Recent Developments in Technology and Applications. *Glob. NEST J.* **2011**, *13*, 193–214.
65. Moriasi, D.; Arnold, J.; van Liew, M.; Bingner, R.; Harmel, R.; Veith, T. Model Evaluation Guidelines for Systematic Quantification of Accuracy in Watershed Simulations. *Trans. ASABE* **2007**, *50*, 885–900. [CrossRef]
66. Ministry for the Environment Physical Planning and Public Works. *Water Management Study of Pinios River Watershed. Part A: Physical System, Acheloos Diversion and Land Reclamation Works of Thessalian Plain*; Ministry for the Environment Physical Planning and Public Works: Athens, Greece, 2006.
67. Koutsoyiannis, D.; Efstratiadis, A.; Mamassis, N. *Appraisal of the Surface Water Potential and Its Exploitation in the Acheloos River Basin and in Thessaly, Ch. 5 of Study of Hydrosystems, Complementary Study of Environmental Impacts from the Diversion of Acheloos to Thessaly*; Ministry of Environment, Planning and Public Works: Athens, Greece, 2001.
68. Mylopoulos, Y. *Database Development of Surface Water and Groundwater Measurements and Evaluation of Reclamation Works in Thessaly*; Regional Development Fund of Thessaly: Volos, Greece, 2005.
69. Pedersen, E.J.; Miller, D.L.; Simpson, G.L.; Ross, N. Hierarchical Generalized Additive Models in Ecology: An Introduction with Mgcv. *PeerJ* **2019**, *7*, e6876. [CrossRef]
70. Stefanidis, K.; Varlas, G.; Papaioannou, G.; Papadopoulos, A.; Dimitriou, E. Trends of Lake Temperature, Mixing Depth and Ice Cover Thickness of European Lakes during the Last Four Decades. *Sci. Total Environ.* **2022**, *830*, 154709. [CrossRef]
71. Zuur, A.F.; Ieno, E.N.; Walker, N.; Saveliev, A.A.; Smith, G.M. *Mixed Effects Models and Extensions in Ecology with R*; Springer: New York, NY, USA, 2009. [CrossRef]
72. Hastie, T.J.; Tibshirani, R.J. *Generalized Additive Models*; Chapman and Hall: London, UK, 2017; ISBN 9781351445979.
73. de Souza, S.A.; Reis, D.S. Trend Detection in Annual Streamflow Extremes in Brazil. *Water* **2022**, *14*, 1805. [CrossRef]
74. Olden, J.D.; Poff, N.L. Redundancy and the Choice of Hydrologic Indices for Characterizing Streamflow Regimes. *River Res. Appl.* **2003**, *19*, 101–121. [CrossRef]
75. Hirsch, R.M.; de Cicco, L.A. User Guide to Exploration and Graphics for RivEr Trends (EGRET) and DataRetrieval: R Packages for Hydrologic Data. In *Techniques and Methods*; US Geological Survey: Reston, VA, USA, 2015. [CrossRef]
76. Papadaki, C.; Dimitriou, E. River Flow Alterations Caused by Intense Anthropogenic Uses and Future Climate Variability Implications in the Balkans. *Hydrology* **2021**, *8*, 7. [CrossRef]
77. Yue, S.; Pilon, P.; Phinney, B.; Cavadias, G. The Influence of Autocorrelation on the Ability to Detect Trend in Hydrological Series. *Hydrol. Process.* **2002**, *16*, 1807–1829. [CrossRef]
78. Grandry, M.; Gailliez, S.; Brostaux, Y.; Degré, A. Looking at Trends in High Flows at a Local Scale: The Case Study of Wallonia (Belgium). *J. Hydrol. Reg. Stud.* **2020**, *31*, 100729. [CrossRef]
79. Hirsch, R.M. *Daily Streamflow Trend Analysis*; U.S. Geological Survey Office of Water Information Blog: Reston, VA, USA, 2018.
80. Hinkle, D.; Wiersma, W.; Jurs, S. *Applied Statistics for the Behavioral Sciences*, 5th ed.; Houghton Mifflin College Division: Boston, MA, USA, 2003; Volume 663.
81. Markonis, Y.; Batelis, S.C.; Dimakos, Y.; Moschou, E.; Koutsoyiannis, D. Temporal and Spatial Variability of Rainfall over Greece. *Theor. Appl. Climatol.* **2017**, *130*, 217–232. [CrossRef]

82. Mastrantonas, N.; Herrera-Lormendez, P.; Magnusson, L.; Pappenberger, F.; Matschullat, J. Extreme Precipitation Events in the Mediterranean: Spatiotemporal Characteristics and Connection to Large-Scale Atmospheric Flow Patterns. *Int. J. Climatol.* **2021**, *41*, 2710–2728. [CrossRef]
83. Luterbacher, J.; Xoplaki, E.; Casty, C.; Wanner, H.; Pauling, A.; Küttel, M.; Rutishauser, T.; Brönnimann, S.; Fischer, E.; Fleitmann, D.; et al. Chapter 1 Mediterranean Climate Variability over the Last Centuries: A Review. *Dev. Earth Environ. Sci.* **2006**, *4*, 27–148.
84. Luterbacher, J.; Xoplaki, E. 500-Year Winter Temperature and Precipitation Variability over the Mediterranean Area and Its Connection to the Large-Scale Atmospheric Circulation. In *Mediterranean Climate*; Springer: Berlin, Heidelberg, 2003; pp. 133–153.
85. Nastos, P.T.; Kapsomenakis, J.; Philandras, K.M. Evaluation of the TRMM 3B43 Gridded Precipitation Estimates over Greece. *Atmos. Res.* **2016**, *169*, 497–514. [CrossRef]
86. Koseki, S.; Mooney, P.A.; Cabos, W.; Angel Gaertner, M.; de La Vara, A.; Jesus Gonzalez-Aleman, J. Modelling a Tropical-like Cyclone in the Mediterranean Sea under Present and Warmer Climate. *Nat. Hazards Earth Syst. Sci.* **2021**, *21*, 53–71. [CrossRef]
87. Varlas, G.; Katsafados, P.; Papadopoulos, A.; Korres, G. Implementation of a Two-Way Coupled Atmosphere-Ocean Wave Modeling System for Assessing Air-Sea Interaction over the Mediterranean Sea. *Atmos. Res.* **2018**, *208*, 201–217. [CrossRef]
88. Manola, I.; van den Hurk, B.; de Moel, H.; Aerts, J.C.J.H. Future Extreme Precipitation Intensities Based on a Historic Event. *Hydrol. Earth Syst. Sci.* **2018**, *22*, 3777–3788. [CrossRef]
89. Karl, T.R.; Trenberth, K.E. Modern Global Climate Change. *Science* **2003**, *302*, 1719–1723. [CrossRef]
90. Tegos, A.; Ziogas, A.; Bellos, V.; Tzimas, A. Forensic Hydrology: A Complete Reconstruction of an Extreme Flood Event in Data-Scarce Area. *Hydrology* **2022**, *9*, 93. [CrossRef]
91. de Girolamo, A.M.; Barca, E.; Leone, M.; lo Porto, A. Impact of Long-Term Climate Change on Flow Regime in a Mediterranean Basin. *J. Hydrol. Reg. Stud.* **2022**, *41*, 101061. [CrossRef]

Disclaimer/Publisher's Note: The statements, opinions and data contained in all publications are solely those of the individual author(s) and contributor(s) and not of MDPI and/or the editor(s). MDPI and/or the editor(s) disclaim responsibility for any injury to people or property resulting from any ideas, methods, instructions or products referred to in the content.

Article

Integrated and Individual Impacts of Land Use Land Cover and Climate Changes on Hydrological Flows over Birr River Watershed, Abbay Basin, Ethiopia

Demelash Ademe Malede [1,2,*], Tena Alamirew [3] and Tesfa Gebrie Andualem [4,5]

[1] Department of Hydrology and Water Resource Management at Africa Center of Excellence for Water Management, Addis Ababa University, Addis Ababa P.O. Box 1176, Ethiopia
[2] Department of Natural Resources Management, Debre Markos University, Debre Markos P.O. Box 269, Ethiopia
[3] Water and Land Resource Center, Ethiopian Institute of Water Resource, Addis Ababa University, Addis Ababa P.O. Box 1176, Ethiopia
[4] Department of Hydraulic and Water Resources Engineering, Debre Tabor University, Debre Tabor 272, Ethiopia
[5] UniSA-STEM, University of South Australia, Adelaide, SA 5095, Australia
* Correspondence: demelashade@gmail.com or demelash.ademe@aau.edu.et

Citation: Malede, D.A.; Alamirew, T.; Andualem, T.G. Integrated and Individual Impacts of Land Use Land Cover and Climate Changes on Hydrological Flows over Birr River Watershed, Abbay Basin, Ethiopia. *Water* 2023, 15, 166. https://doi.org/10.3390/w15010166

Academic Editors: Sonia Raquel Gámiz-Fortis and Matilde García-Valdecasas Ojeda

Received: 10 November 2022
Revised: 26 December 2022
Accepted: 28 December 2022
Published: 31 December 2022

Copyright: © 2022 by the authors. Licensee MDPI, Basel, Switzerland. This article is an open access article distributed under the terms and conditions of the Creative Commons Attribution (CC BY) license (https://creativecommons.org/licenses/by/4.0/).

Abstract: Land use/land cover (LULC) and climate change are the two major environmental factors that affect water resource planning and management at different scales. This study aims to investigate the effects of LULC and climate change patterns for a better understanding of the hydrological processes of the Birr River watershed. To examine the effects of LULC and climate change patterns on hydrology, three periods of climate data (1986–1996, 1997–2007 and 2008–2018) and three sets of LULC maps (1986, 2001 and 2018) were established. The changes in hydrological flow caused by climate and LULC changes were estimated using the soil and water assessment tool (SWAT) and indicator of hydrological alteration (IHA) method. Results showed that the SWAT model performed well during the calibration and validation period at monthly timestep, with R^2 and NSE values of (0.83 and 0.81) and (0.80 and 0.71), respectively. The LULC change increased surface runoff while decreasing baseflow, water yield, and evapotranspiration. This was due to increased agriculture and settlements, and a reduction in bushland, forest, and grassland. Climate change increased surface runoff and water yield while decreasing baseflow and evapotranspiration during 1996–2006. The combined effect of LULC and climate reveals increased surface runoff and a decreased trend of evapotranspiration, whereas baseflow and water yield showed inconsistency. In addition, the IHA found no statistically significant increasing trend for one-day, three-days, seven-day, and thirty-day minimum and maximum daily streamflow in the Birr River watershed. These findings will be useful to authorities, water engineers, and managers concerned with hydrology, LULC, and climate.

Keywords: Birr River watershed; climate change; land use/land cover; streamflow; SWAT

1. Introduction

Lan use/land cover (LULC) and climate change are the two major environmental components that had a significant effect on the hydrological process and socioeconomic activities, which directly affects water resource management and development [1–5]. The patterns of LULC and climate change are dynamic, with the effects of natural phenomena and anthropogenic activities primarily governing their prevalent processes [6,7]. Various anthropogenic activities of LULC changes, such as rapid urbanization, agricultural expansion, deforestation, industrialization, and other human activities can have a significant effect on streamflow, surface runoff, groundwater flow, water yield, and evapotranspiration of a watershed [8–10]. Climate change also has a wide range of effects on the hydrological cycle in a variety of ways, including changes in peak flow and volume [11–14]. Moreover, both the quantity and quality of water are constantly deteriorating as a result of mining activities and the conversion of forests into agricultural areas. Mining increases land degradation and soil erosion, whereas deforestation increases surface runoff and decreases baseflow on agricultural and urbanization fields [15] As a result, LULC and climate change had a detrimental effect on long-term water resource management, and development [16,17]. Hence, to effectively manage the available water resource for various uses, the watershed's water should be adequately quantified and estimated.

Analyzing and quantifying the impact of LULC and climate change on the hydrological flows of a watershed is a challenge because of the complex relationship between landscape, LULC, climate, and hydrology [6]. Thus, understanding these impacts at a watershed scale is essential for land use planners, policymakers, stakeholders, and water planners and managers. Studies of LULC and climate change patterns, and implications can also pave the way for the development and implementation of appropriate land management strategies [18,19]. Local and regional knowledge is especially important in the Birr watershed, which contains many small rivers and streams that supply water to the Birr River [20]. Several studies have been conducted to better understand the separate impact of LULC or climate change on hydrological flow in a watershed [19,21–23]. However, it is critical to recognize the cumulative effect of these LULC and climate change factors. [24–26]. The integrated and individual effects of climate and land cover change influence the hydrological processes of a watershed [27–29]. Although, the magnitude of the change varies among watersheds depending on the watershed's characteristics such as vegetation cover, climate condition, land cover, and topography. Therefore, studying the effects of changes in climate and land cover drivers on hydrological response has become a major research topic in recent decades [30]. Many scholars studied LULC and climate change effects on hydrological processes all over the world [24,31–36], however, it is complicated to quantify the hydrological process [37]. Some researchers recently also concluded that the effects of LULC and climate change on streamflow varied over several regions as a result of the difference in soil type, terrain, human activities, and climate conditions [1,38–40]. For instance, Kuma et al. and Chen et al. [28,29] reported that climate change has a greater impact on hydrologic response than land cover change. On the other hand, land cover changes are more sensitive to hydrologic responses than climate change [5,7,41]. Patel and Verma [42] stated that changes in LULC contribute to an increase in streamflow and a decrease in evapotranspiration, primarily as a result of increased urbanization and decreased water bodies and forest cover. Thus, it is important to consider the effect of both climate and land cover changes on hydrological processes. This study used soil and water assessment tools (SWAT) and indicators of hydrological alteration (IHA) to investigate how changes in hydrological processes within the Birr River watershed. Because of the limited data, there are fewer studies in which SWAT has been applied in the Birr River watershed. IHA is used to estimate the magnitude of changes in hydrological flow fluctuation caused by climate and anthropogenic changes Additionally, this study raised concerns about resource degradation, particularly the loss of vegetation and its transformation in other land LULC types. The transformation of land use is the result of human-induced systems, which are primarily disrupting watershed streamflow regimes.

Many types of research have been conducted to investigate the effect of climate and LULC change on hydrological flows using a hydrological model due to the enormous economic and social importance of these climate and LULC changes [25,43–48]. Regional and local studies have been conducted using the SWAT model to investigate the response of hydrological variables to LULC and climate change in a watershed [49–55]. The SWAT model is a physically-based semi-distributed hydrological model that is used worldwide. It is user-friendly and freely available, and it estimates surface runoff using a modified Soil Conservation Service Curve Number (SCS–CN) technique. The model has been used to monitor and control, changes in LULC and climate changes in the Birr River watershed.

The Birr River watershed mainly experienced rapid LULC and climate change, high population growth, and decreased surface water availability, due to high agricultural water demand [56]. Most local people in the watershed depend on rainfed agriculture and small-scale irrigation schemes for their livelihoods, which have been severely impacted by LULC and climate change. Land degradation from soil erosion, deforestation, and uncontrolled hillside farming is a serious problem in the watershed [20]. The study area has high deforestation and steep slopes employed by farmland for crop production, resulting in severe land degradation. Vegetations are becoming scarce as a result of increased cultivation, settlements, and land degradation. Thus, quantifying the individual and integrated effects of land use and climate change on various hydrological processes on a local scale (watershed level) and identifying the relative contribution of these changes is the novelty of the study. The study provides more information on the integrated and individual effects of land use and climate change drivers for a better understanding of the hydrological processes in the Birr River watershed. Furthermore, the study would have the highest impact and be useful in developing policies and strategies for sustainable land and water resource management practices in the study area. Therefore, the main goal of this study was (i) to determine the impact of integrated climate and LULC change on hydrological processes (surface runoff, baseflow, water yield, and evapotranspiration) over 32 years (1986–2018) (ii) to analyze the relative contribution of individual climate and LULC cover change on hydrological processes (iii) to model and understand the availability of streamflow in the Birr River watershed for the effect of climate and land use changes using the IHA.

2. Materials and Method

2.1. Study Area

Birr River watershed is situated in the northwestern highlands of Ethiopia. Geographically, the watershed is located between the longitude of $37°10'$ and $38°50'$ E and the latitude of $10°30'$ and $11°10'$ N (Figure 1). The watershed is characterized by rough topography and a wide range of elevations ranging from 1691 to 4084 m above sea level. The total drainage area of the watershed is 3062 km^2. Many small tributary streams contribute to high discharge to the Birr River watershed during the summer season and are distinguished by substantial spatial and seasonal variability in rainfall. Based on the rainfall patterns, there are three distinct seasons: the main rainy season, which runs from June to September, the minor rainy season, which runs from February to Ma, and the dry season, which runs from October to February [57,58]. The estimated mean annual rainfall for the watershed is 1389 mm based on 32 years of recorded data (1986–2018) obtained from nearby representative meteorological stations, (Figure 2). The northwestern region of the watershed has an estimated annual rainfall was 1391 mm, while the southwestern tip near the mouth of the basin was 1026 mm. The estimated maximum temperature also ranges between 23 and 30 °C, while the lowest temperature varied between 8 to 12 °C, with an average temperature of 18 °C (Figure 2). Figure 2 depicts the average maximum and minimum temperatures, as well as rainfall, in the Birr River watershed. The estimated gauged area of the Birr River watershed covers 1500 km^2 (Figure 1). The watershed is significant on local and national magnitude. It includes a high irrigation potential, a high value of cash crops, livestock production as well as tourism, because of the existence of an impressive landscape, and a unique source of biological diversity [20,57].

2.2. Data Collection and Quality Control

2.2.1. Spatial Data

A spatial dataset of digital elevation model (DEM), soil, and LULC was employed to simulate the SWAT hydrological model. DEM data is required to define watershed and sub-watershed boundaries, as well as the delineation of the hydrological response unit, and slope reclassification. DEM data with the 30-m spatial resolution was obtained from the Shuttle Radar Topographic Mission (SRTM) USGS website http://earthexplorer.usgs.gov/ (accessed on 1 August 2022) [59]. The soil and LULC map along with the watershed and sub-watersheds delineated from DEM was used to determine the hydrological parameters and the Hydrological Response Units (HRUs). The soil map was collected from the Ethiopian Ministry of Water, Irrigation, and Energy (MoWIE). There are six major soil types found in the Birr River watershed, which include Haplic Alisols, Eutric Fluvisols, Haplic Luvisols, Eutrc Leptosols, Haplic Nitisols, and Eutric Vertisols (Figure 3). Among all soil types Haplic Alisols is most dominated soil types, which covers about 816 km^2 (59.78%) followed by Eutric Fluvisols 236 km^2 (17.29%), Eutrc Leptosols 125km^2 (9.16%), Haplic Luvisols 96km^2 (7.03%), Haplic Nitisols 65km^2 (4.76%), and Eutric Vertisols 5km^2 (0.37%). The Birr River watershed's detailed soil physical and chemical property parameters, such as soil texture, bulk density, available water content, hydraulic conductivity, soil depth, and organic carbon content, were derived from the world's digital soil map [60].

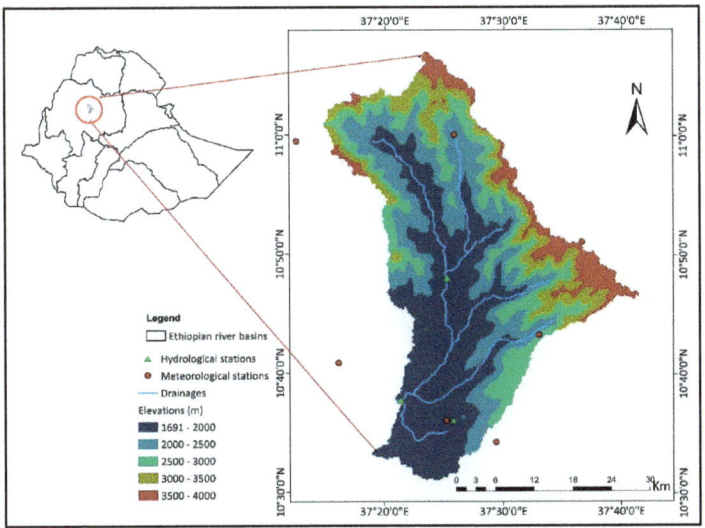

Figure 1. Location map of the study area, meteorological and hydrological gauging stations.

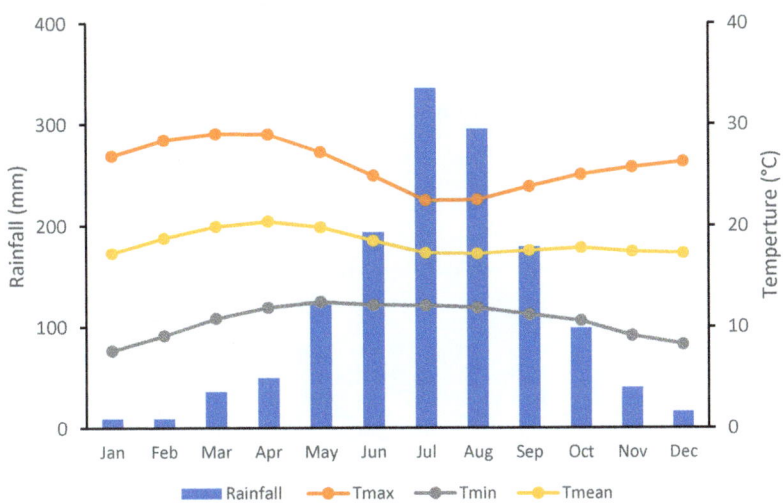

Figure 2. Monthly maximum temperature (Tmax), minimum temperature (Tmin), mean temperature (Tmean), and mean rainfall in the Birr River watershed (1986–2018).

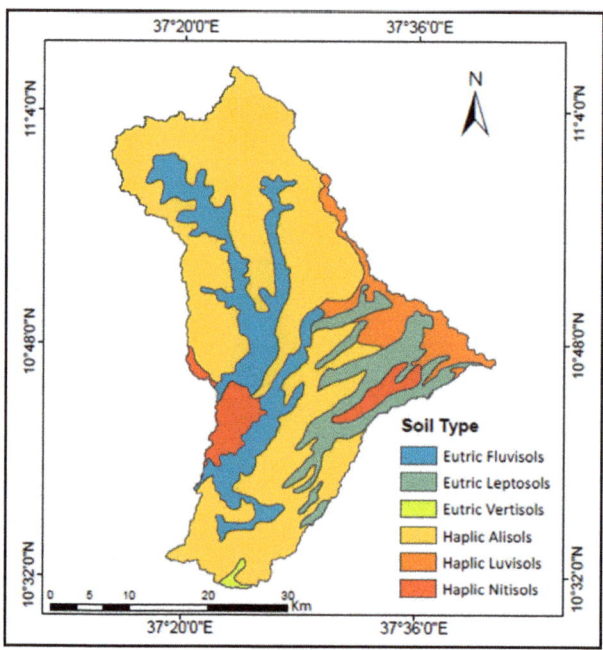

Figure 3. Major soil types in the study Birr River watershed.

Landsat imagery from the years 1986, 2001, and 2018 was employed to determine how LULC was changing over time. These satellite images were downloaded from the USGS website https://earthexplorer.usgs.gov/ (accessed on 1 August 2022). Table 1 summarizes the sensors, path/raw, spatial resolution, and acquisition dates that were used in the study area. Each Landsat image was georeferenced to World Geodetic System 1984 (WGS84) Universal Transverse Mercator (UTM) zone 37 N. Image preprocessing such as band composite, layer stacking, mosaic, sub-setting, noise, and haze correction was performed. Then based on the prevalent land covers, the spectral responses of features on Landsat images, extensive field observation, and a literature review, the Birr River watershed was divided into five LULC classes or types that were generated namely; agricultural, bushland, forest, grassland, and settlement (Table 2). The maximum likelihood classification method was used to process the image classification using the ERDAS imagine 2015 software (version 15.0). Then, the SWAT model includes these five LULC classes (agricultural, forest, grassland, shrub/bushland, and settlements), with the LULC maps beginning in 1986, 2001, and 2018.

Table 1. Satellite data acquisition in the study area.

Source	Sensors	Path/Row	Spatial Resolution	Acquisition Date
Earthexplorer.usgs.gov (accessed on 1 August 2022)	Landsat TM	169/053	30 m × 30 m	January 1986
	Landsat TM	169/053	30 m × 30 m	January 2001
	Landsat OLI/TIRS	169/053	30 m × 30 m	February 2018

Table 2. General description of LULC classes.

LULC Classes	Description of LULC Classes
Agricultural land	Enclosed with permanent crops, following land, and irrigated cultivation
Bushlands	Covered with small to medium-sized perennial woody or natural vegetation
Forest land	Trees taller than 5 m and covering more than 0.5 hectares of land
Grasslands	Terrestrial vegetation dominated by grass, suitable for grazing by livestock
Settlements	Built-up areas and roads, the establishment of a person in a new region

Accuracy Assessment of the Classified LULC Types

Accuracy assessment is an important step for LULC classification. It compares the classified image to another data source that is considered to be accurate or ground truth data. The ground truth should be chosen in such a way that it appears in both the Landsat image and the google earth map. The overall accuracy demonstrates the ability of the classifier to preview the classes. It was calculated by dividing the total number of correctly classified pixels (diagonal) by the total number of reference pixels. The recommended one should be between 85–95% [61]. For accuracy evaluations, the Kappa coefficient (K), which represents the agreement between the classified image and the reference or ground truth, was used. The Kappa coefficient is calculated using Equation (1).

$$Kappa\ Coefficient\ (K) = \frac{(Total\ sample \times Total\ corectely\ classified\ sample) - \sum(Column\ total * Row\ total)}{(Total\ sample)^2 - \sum(Column\ total - Row\ total)} * 100 \quad (1)$$

Kappa coefficient statistics criteria agreements are as follows: poor when *Kappa* < 0.4, good when 0.4 < *kappa* < 0.7, and excellent when *k* > 0.75 [62].

2.2.2. Temporal Data

Climatological data is the main requirement of the SWAT hydrological model. Daily climate data (rainfall, maximum and minimum temperature data, relative humidity, sunshine hours, and wind speed) were obtained from the Ethiopian National Meteorological Agency (NMA). These climate data were used between 1986 and 2018 at eight meteorological stations (Adet, Dembecha, Dengayber, Feresebet, Finoteselam, Qaurit, Sekela, and Yechereka). SWAT weather generator was used to simulate relative humidity, solar radiation, and wind speed data from the Adet station [63]. The weather generator was also used to fill the missing rainfall, and temperature data. The double mass curve method was also used to assess the consistency of data elements [64]. The daily evapotranspiration was estimated using the Penman-Monteith method, which is the only accepted method of calculating evaporation [65,66]. The consistent hydrological streamflow data were also collected from MoWIE. The missing and its homogeneity test were investigated using indicators of hydrological alterations (IHA), which employed long-term daily streamflow data [67].

2.3. The SWAT Model

SWAT predicts the effect of human activities on the quality and quantity of hydrological processes at different scales [26,43,68–70]. It is a physically-based, semi-distributed, computationally efficient, robust, process-based hydrological model, that was created to estimate the long-term effect of land use management practices on water, sediment, and agricultural chemical yields [65,68,71]. SWAT also simulates the water balance components of surface runoff, groundwater flow, lateral flow, total water yield, and evapotranspiration [68,72]. The water balance equation is given (Equation (2)).

$$SW_t = SW_O + \sum \left(R_{day} - SURQ - E_a - W_{seep} - GWQ \right) \quad (2)$$

where SW_t is the final soil water content (mm), SW_O is the initial soil water content (mm), t is time in days, R_{day} is the amount of precipitation (mm), $SURQ$ is the amount of surface runoff (mm), E_a is the amount of evapotranspiration (mm), W_{seep} is the amount of water entering the vadose zone from the soil profile (mm), and GWQ is the amount of groundwater flow (mm).

ArcSWAT 2012.10.4.21 was used with ArcGIS 10.4.1 to delineate watershed boundaries and stream networks. The model simulates a watershed by separating it into sub-watersheds, which are then subdivided into hydrologic response units (HRUs), which have smaller units and specific land use, soil, and slope combinations [69,71].

2.4. Calibration and Validation

A two-year warm-up period was used to simulate the Birr River watershed [73]. The sequential uncertainty fitting version 2 (SUFI-2) algorithm was used to achieve an acceptable satisfactory agreement between simulated and observed streamflow data. SWAT model parameters were calibrated using monthly observed streamflow data from the Birr River gauging stations between 1994–2001. The validation process of a model was used to examine simulation consistency and validated using monthly streamflow data from 2002 to 2005 period.

2.5. Model Performance Evaluation

SWAT model simulation was evaluated using monthly timescale streamflow data from the Birr River gauging station because daily timescale simulations may not clearly show the effects of LULC and climate change. Various statistics can be used to calculate the degree of agreement between observed and simulated data [74–77]. The model was evaluated using the coefficient of determination (R^2), Nash Sutcliffe Efficiency (NSE), and percent of bias (PBIAS) as described (Equations (3)–(5)). R^2 is the correlation between observed and simulated streamflow data, and it ranges from 0 to 1. $R^2 > 0.5$ is regarded as acceptable, as is the model's ability to predict observed values reliability [78,79]. NSE calculates the relative magnitude of the residual variance in comparison to the measured data variance, and how well the observed versus simulated data plot fits [80]. NSE value ranges from $-\infty$ to 1. As the NSE is closer to 1, the more accurate the model is. NSE = 1, a perfect match between the observed and simulated streamflow data. NSE = 0, showed that the model predictions are as accurate as the mean of the observed data. $-\infty < NSE < 0$, indicates that the observed mean is a better predictor than the model. The PBIAS compares the average tendency of the simulated to the observed data [81]. The best value of BPIAS is 0. The positive value of PBIAS indicates an underestimation and the negative value indicates an overestimation of the model [81]. The statistical indices were estimated using the equations listed below.

$$R^2 = \left[\frac{\sum_{i=1}^{n}(O_i - \overline{O})(S_i - \overline{S})}{\sqrt{\sum_{i=1}^{n}(O_i - \overline{O})^2} \sqrt{\sum_{i=1}^{n}(S_i - \overline{S})^2}} \right] \quad (3)$$

$$NSE = \frac{\sum_{i=1}^{n}(O_i - \overline{O})^2 - \sum_{i=1}^{n}(S_i - O_i)^2}{\sum_{i=1}^{n}(O_i - \overline{O})^2} \quad (4)$$

$$PBIAS = \left[\frac{(O_i - S_i) * 100}{\sum_{i=1}^{n}(O_i)} \right] \quad (5)$$

where the total number of observations, O_i is the ith observed value, $\overline{O}$ is the mean observed value, S_i is the ith model simulated value, and $\overline{S}$ is the mean model simulated value.

2.6. Simulation of LULC and Climate Change Impacts

A fixing-changing method was used to assess the effect of LULC and climate change on the hydrological process [7,25,82–84]. Climate data from 1986 to 2018 were divided into three time periods (1986–2000, 2001–2010, and 2011–2018) for the LULC map of 1986,

2001, and 2018. Based on these climate data and a set of LULC maps, nine scenarios were established (Table 3). If the LULC map and climate data are from the same period, it is referred to as a baseline scenario, and if they are from different periods, it is referred to as an assumed scenario [6]. For instance, in scenario 1, the LULC map from 1986 and climate data from 1986–2000 was used, and this is known as the baseline scenario. The LULC map from 2001 and climate data from 1986–2000 were used in scenario 2, which is also known as an assumed scenario. These scenarios would provide more detailed information about the effects of LULC and climate change on the Birr River watershed.

Table 3. Different simulation scenarios for evaluating the effect of LULC and climate change on hydrological processes from 1986 to 2018.

Scenarios Considered	LULC map	Climate Data	Remarks
S1	1986	1986–1996	Bassline
S2	2001	1986–1996	Assumed
S3	2018	1986–1996	Assumed
S4	1986	1997–2007	Assumed
S5	2001	1997–2007	Baseline
S6	2018	1997–2007	Assumed
S7	1986	2008–2018	Assumed
S8	2001	2008–2018	Assumed
S9	2018	2008–2018	Baseline

The difference in the hydrological process obtained from scenarios S2 and S1 represents the separated effect of LULC from 1986 to 2001. The main goal of this evaluation is to determine whether LULC change is a driver for changes in hydrological processes in the Birr River watershed or not while keeping the DEM and soil data constant [6,25,83,84]. The difference between S4 and S1, on the other hand, would indicate the effect of climate change between 1986–2010. Furthermore, the difference between S5 and S1 (baseline scenarios) represents the combined effect of LULC and climate change between 1986 and 2010. Equations (6)–(8) provided the percentage changes resulting from the contributions of LULC, climate, and combined LULC and climate change on the hydrological flows [6].

$$\Delta H_{LULC} = (\tfrac{S2-S1}{S1}) \times 100 \tag{6}$$

$$\Delta H_{Climate} = (\tfrac{S4-S1}{S1}) \times 100 \tag{7}$$

$$\Delta H_{Combined} = (\tfrac{S5-S1}{S1}) \times 100 \tag{8}$$

where ΔH is a change in percentages for the effect of LULC, climate, and combined on hydrological processes, and S1, S2, S4, and S5 are scenarios considered. For other periods, between 1986 to 2018 and 2010 to 2018 a similar analogy is used to analyze the effect of LULC and climate changes.

2.7. Indicator of Hydrological Alteration Method

The indicator of hydrological alteration (IHA) method is used to estimate the magnitude of changes in hydrological flow fluctuation caused by climate and anthropogenic changes. The IHA is a software program that was created in the 1990s by the US Nature Conservancy to process hydrological records [85]. The IHA parameters of one day, three days, seven days, thirty days, and ninety days were investigated for the minimum and maximum magnitudes and durations of streamflow conditions. The parameters were evaluated and compared using a p-value of 5% significance. Modeling water resources with IHA provides useful information and identification of hydrological regimes in the watershed that is influenced by climate and anthropogenic factors in the watershed [86].

3. Results

3.1. Land Use Land Cover Change Detections

There are five major land use classes identified in the study area: agricultural land, bushland, forest, grassland, and settlements. Agriculture covered the largest area in the watershed than the other LULC types in all three years (1986, 2001, and 2018), whereas forests and settlements covered less area [87]. This indicates agriculture is critical to the socioeconomic development of the study watershed. Table 4 depicts the LULC classification in the watershed over 32 years (1986, 2001, and 2018). Agricultural land increased from 56.39–70.19% between (1986–2018) because increased population density leads to an increase in cultivated area and settlements. This finding is in agreement with another finding [30,88].

Table 4. The proportional area coverage in kilometer squares (km^2) and percentage (%) of LULC classes in the Birr River watershed.

LULC Classes	LULC Area in Kilometer Squares (km^2) and Percentage (%)					
	1986		2001		2018	
	km^2	%	km^2	%	km^2	%
Agriculture	773.04	56.39	849.90	61.99	962.34	70.19
Bushland	358.91	26.18	326.35	23.80	264.64	19.30
Forest	67.69	4.94	24.23	1.77	26.39	1.92
Grassland	161.38	11.77	154.26	11.25	98.22	7.16
Settlements	9.96	0.73	16.25	1.19	19.40	1.42

Accuracy Assessment of Classified LULC

According to the results of three classified LULCs, the overall accuracy of the maps from 1986, 2001, and 2018 was 90.69%, 91.01%, and 92.22%, respectively. The classification performed in this study produces an overall accuracy that meets the minimum accuracy level of 85 defined by Anderson et al. [61]. The Kappa coefficients for the 1986, 2001, and 2018 maps were also 0.84, 0.86, and 0.89, respectively. Therefore, the classification used in this study has an almost perfect agreement for the years 1986, 2001, and 2018 (Tables 5–7).

Table 5. Accuracy assessment of LULC map classification, 1986.

Land Use Classes	Agriculture	Bushland	Forest	Grassland	Settlement	Row Total	Users Accuracy
Agriculture	27	0	0	0	0	27	100%
Bushland	1	22	0	2	0	25	88%
Forest	1	0	9	0	0	10	85%
Grassland	2	1	0	17	0	20	90%
Settlement	1	0	0	0	3	4	75%
Column total	32	23	9	20	3	86	
Producers accuracy	84%	95%	100%	90%	100%		
Overall classification accuracy = 90.69%				Kappa Coefficient = 0.87			

Table 6. Accuracy assessment of LULC map classification, 2001.

Land Use Classes	Agriculture	Bushland	Forest	Grassland	Settlement	Row Total	Users Accuracy
Agriculture	25	3	1	1	0	30	83.33%
Bushland	1	23	0	1	0	25	92%
Forest	0	0	10	0	0	10	100%
Grassland	1	0	0	19	0	20	95%
Settlement	0	0	0	0	4	4	100%
Column total	27	26	11	21	4	89	
Producers accuracy	92%	88%	90%	90%	100%		
Overall classification accuracy = 91.01%				Kappa Coefficient = 0.88			

Table 7. Accuracy assessment of LULC map classification, 2018.

Land Use Classes	Agriculture	Bushland	Forest	Grassland	Settlement	Row Total	Users Accuracy
Agriculture	28	1	0	1	0	30	93%
Bushland	2	22	1	0	0	25	88%
Forest	0	0	10	0	0	10	100%
Grassland	1	1	0	18	0	20	90%
Settlement	0	0	0	0	5	5	100%
Column total	31	26	12	19	5	90	
Producers accuracy	90%	85%	83%	95%	100%		
Overall classification accuracy = 92.22%				Kappa Coefficient = 0.89			

The extent and rate of LULC change patterns from 1986 to 2018 are also presented in (Table 8). The result showed that agricultural land and settlements increased by 24.49%, and 54.78%, respectively, whereas bushland, forest, and grasslands had a decreasing trend. The rate of change in agricultural land and settlements was also raised by 0.77% and 2.96%, respectively. Bushland, forest, and grasslands had also dropped by −0.82%), −1.91%, and −1.22% respectively. Similarly, Ewunetu et al. [89] showed the highest gain in agricultural land was obtained from grassland and bushland in the North Gojjam sub-basin from 1986 to 2017.

Table 8. Rate of LULC changes between 1986 and 2018.

LULC Classes	1986	2018	Change in 1986 and 2018		Rate of Changes	
	km^2	km^2	km^2	%	km^2/year	%
Agricultural land	773.04	962.34	189.3	24.49	5.92	0.77
Bushland	358.91	264.64	−94.27	−26.27	−2.95	−0.82
Forest	67.69	26.39	−41.30	−61.01	−1.29	−1.91
Grassland	161.38	98.22	−63.16	−39.14	−1.97	−1.22
Settlements	9.96	19.40	9.44	54.78	0.31	2.96

3.2. SWAT Model Calibration and Validation

The SWAT model had been calibrated using the monthly observed streamflow covering from 1994 to 2001 over the Birr River watershed. Streamflow data from the previous years was used for the warm-up period from 1992 to 1993. SWAT-CUP automatic calibration with the sequential uncertainty fitting version 2 (SUFI-2) algorithm at the Birr River gauging station from 1994 to 2001 was used. Before beginning to calibrate and validate the SWAT model, the model's developer and users provided detailed readings and observations, which aided in determining the calibration and validation parameters that needed to be adjusted. The parameters such as SCS runoff curve number for moisture conditions II (CN2), soil evaporation compensation factor (ESCO), The threshold water level shallow aquifer baseflow (GWQMN), maximum canopy index (Canmx), an available water capacity of the soil layer (SOL_AWS), and soil depth (SOL_Z) were employed for the model calibration. The model was calibrated by varying the parameter range values between the lower and upper limits. These SWAT parameters, range values, fitted values, and parameter descriptions used in the SWAT model simulation are shown in (Table 9).

Figures 4 and 5 depict the comparison and relationship between the simulated and observed monthly streamflow at the Jiga gauging station. To calibrate the SWAT model, climate data from 1997 to 2007 were used, as well as the 2001 LULC map. The peak value of the simulated streamflow closely matches those of the observed one, but at different magnitudes (Figure 4). According to the simulation results, the SWAT model demonstrated that monthly streamflow has a better agreement in the Birr River watershed. The SWAT model calibration and validation confirmed that it could be used to assess the effects of LULC and climate change on water balance components. Figure 5 shows that for the lower values of observed streamflow, the simulated streamflow values are distributed uniformly along a one-to-one line. However, at higher discharge values, the model simulation values showed slight underestimation.

Table 9. The parameter values used to simulate the SWAT model.

Parameters	Range Value	Fitted Value	Parameter Description
CN2	−50 to 50	23.33	SCS curve number for moisture conditions II
ESCO	0 to 1	0.23	Soil evaporation compensation factor
GWQMN	0 to 5000	166.66	The threshold depth of water in the shallow aquifer required for return flow occurs (mm)
Canmax	0 to 10	7.66	Maximum canopy index
SOL_AWS	−50 to 50	13.33	Available water capacity of the soil layer (mm mm^{-1})
SOL_Z	−50 to 50	9.99	Depth from the soil surface to the bottom of the layer (mm)

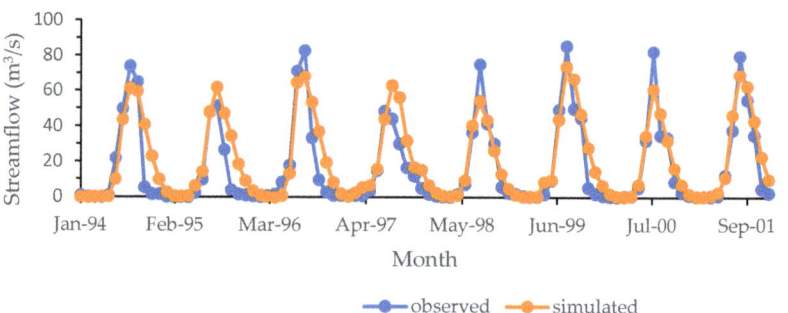

Figure 4. Comparison of observed and simulated streamflow for model calibrated value (1994–2001).

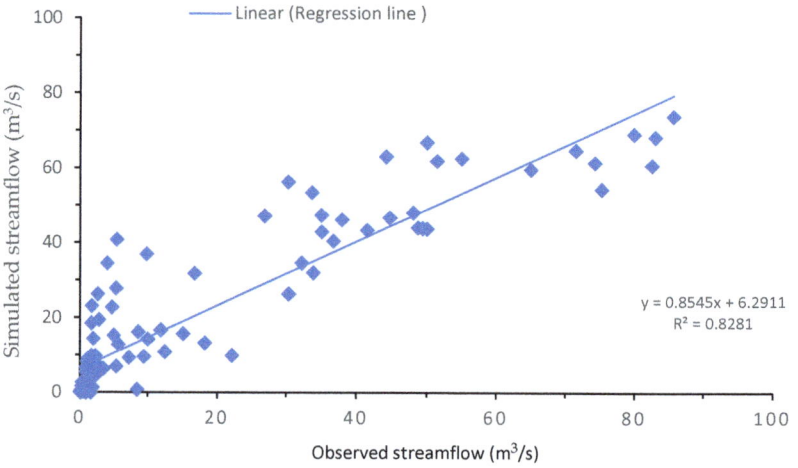

Figure 5. Scatter plot between observed and simulated streamflow for model calibration period (1994–2001).

For further investigation of the calibrated SWAT model, the simulated and observed monthly streamflow at Jiga gauging stations validation period was also compared as shown in (Figures 6 and 7). The Birr River watershed gauging site has a good agreement between simulated and observed monthly streamflow.

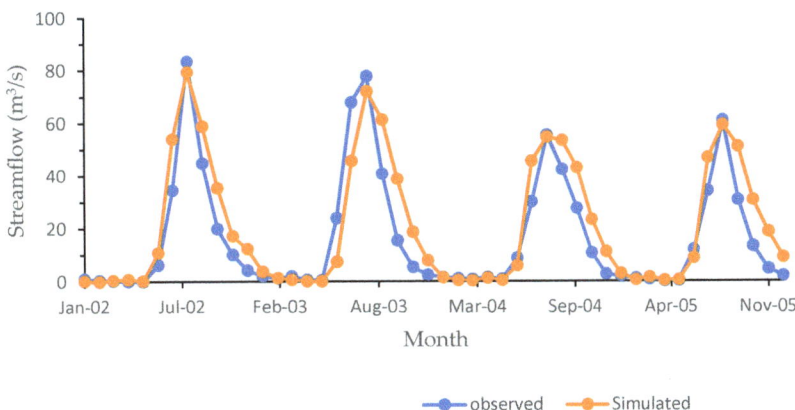

Figure 6. Observed and simulated streamflow for a model validation value (2002–2005).

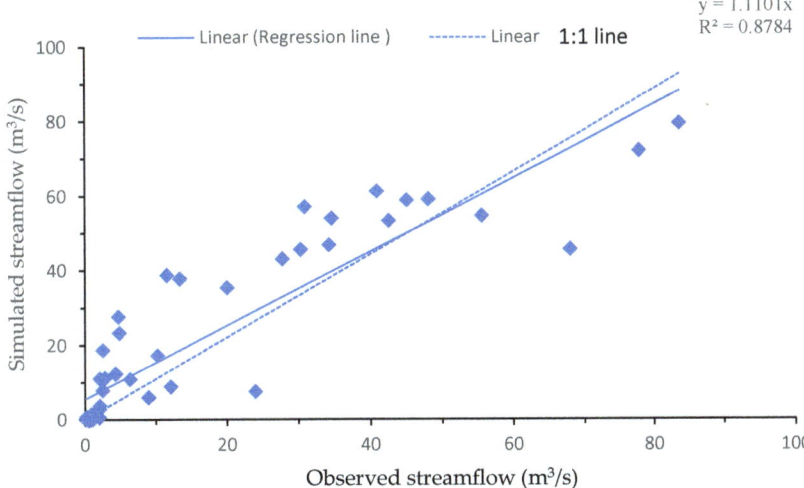

Figure 7. Scatter plot between observed and simulated streamflow for model validation period (2002–2005).

The statistical parameters of R^2, NSE, and PBIAS all indicate a satisfactory relationship between observed and simulated streamflow (Table 10). It has been observed that for the period 1994–2001, the model performed well in terms of R^2 and NSE, which were found to be 0.83 and 0.80, respectively during calibration. For the model validation period the R^2 and NSE were also found to be 0.81 and 0.71, respectively. The model performed better during the calibration period than during the validation period, which could be attributed to the poor quality of streamflow data recorded during the validation period. Furthermore, a lack of consistent hydroclimate and spatial data (LULC and soil data), may result in a slight discrepancy in the model simulation, however, the graphical interpretation of the simulated and observed monthly streamflow hydrographs, as well as the performance of the statistical indices, meet the criteria suggested by Moriasi et al. [90]. As a result, the SWAT model results showed that the overall prediction of monthly streamflow during the calibration and validation period was satisfactory and acceptable for further investigation.

Table 10. Statistical analysis of the observed and simulated monthly streamflow during calibration (1997–2007) and validation (2008–2013).

Period	Statistical Parameters	Value
Calibrations	R^2	0.83
	NSE	0.80
	PBIAS	−15.23
Validations	R^2	0.81
	NSE	0.71
	PBIAS	−14.45

3.3. The Effects of LULC and Climate Change on Hydrological Flows

To assess the effect of LULC and climate change on hydrological flows, different land use scenarios are compared with various climatic settings. The SWAT model simulated various hydrological water balance components, namely, streamflow, surface runoff, baseflow, water yield, and evapotranspiration under various LULC and climatic conditions (Tables 11–13).

Table 11. Average monthly hydrological components for the effect of various land use types with fixed climate data.

Scenarios	LULC	Climate	Surface Runoff (mm)	Baseflow (mm)	Water Yield (mm)	Evapotranspiration (mm)
S1	1986	1986–1996	8.26	32.76	50.33	48.83
S2	2001	1986–1996	8.65	31.01	50.30	48.78
S3	2018	1986–1996	9.14	30.55	49.45	48.75

Table 12. Average monthly hydrological components for the effect of various land use types with fixed climate data.

Scenarios	LULC	Climate	Surface Runoff (mm)	Baseflow (mm)	Water Yield (mm)	Evapotranspiration (mm)
S4	1986	1997–2007	15.15	39.02	65.03	44.27
S5	2001	1997–2007	15.70	37.55	64.54	43.12
S6	2018	1997–2007	16.09	35.85	63.07	42.14

Table 13. Average monthly hydrological components for the effect of various land use types with fixed climate data.

Scenarios	LULC	Climate	Surface Runoff (mm)	Baseflow (mm)	Water Yield (mm)	Evapotranspiration (mm)
S7	1986	2008–2018	16.89	37.96	65.62	44.98
S8	2001	2008–2018	17.00	36.63	64.23	43.62
S9	2018	2008–2018	17.80	34.54	63.36	42.30

3.3.1. Effects of LULC Change on Hydrological Flows

Table 11 depicts the three land use scenarios evaluated (S1, S2, and S3) from 1986 to 1996, illustrating the predominant influence of LULC changes with constant climatic circumstances. Similarly, Table 12 shows the different scenarios considered in (S4, S5, and S6), which is the sole effect of land use change with constant climatic settings between 1997–2007, as well as Table 13 depicts the various scenarios considered in (S7, S8, and S9), which is the sole effect of land uses change with constant climatic settings between 2008–2018. Surface runoff increased by 0.39 mm (4.72%), and 0.88 mm (10.65%) in 2001 and 2018, respectively, when compared to the baseline scenario in S1 with S2 and S3, whereas baseflow decreased by −1.75 mm (−5.34%), and −2.21 mm (−6.75%). In different assumed scenarios considered from S4 to S9, similar results of increased surface runoff and decreased baseflow patterns were observed. However, the magnitudes of surface runoff and baseflow have differed (Tables 12 and 13). The changes in LULC play a significant role in runoff variations, particularly in tropical areas [1]. Baker and Miller [91] reported that the dramatic changes in LULC have resulted in increased surface runoff and decreased groundwater recharge. Similarly, water yield and evapotranspiration decreased in all considered assumed scenarios between S1–S9 during the study period (1986–2018).

Agricultural land increased from 56.39% in 1986 to 70.19% in 2018, while settlements increased from 0.73% in 1986 to 1.42% in 2018, in the study watershed. On the other hand, bushland, forest, and grassland decreased from 26.18%, 4.94%, and 11.77% in 1986 to 19.30%, 1.92%, and 7.16% in 2018 respectively. The reduction of baseflow increases surface runoff and results in more frequent and severe flooding [92,93]. As a result, changes in LULC may be the reason for increasing surface runoff and decreasing baseflow and evapotranspiration. These findings imply the separate effect of LULC change in the Birr River watershed between the assumed scenarios.

3.3.2. Effects of Climate Change on Hydrological Flows

The SWAT model was used to assess and address the effects of climate change (i.e., precipitation and temperature) on hydrological flows in the Birr River watershed. The comparisons were carried out using scenarios S1, S4, and S7, in which the constant LULC map of 1986 was compared to the various climatic settings from 1986–1996, 1997–2007, and 2008–2018 (Tables 11–13). The results reveal that surface runoff increased by 6.89 mm from S1 to S4, whereas in S4 to S7 surface runoff also increased by 0.9 mm although the magnitude is different. Furthermore, baseflow increased by 6.26 mm from S1 to S4, but decreased by −1.06 mm from S4 to S7. The variation between S1, S4, and S7 scenarios noticeably indicates separate climate variability has a distinct effect on the study River watershed. Repeated trials on (S2, S5, S8), and (S3, S6, S9) scenarios revealed an increase and decrease in the magnitude of surface runoff and baseflow patterns. These study findings revealed that the impact of climate change is much greater than the impact of LULC change on surface runoff in the Birr River watershed. This study was in agreement with the previous finding [1,7,25,94]. Furthermore, water yield increased from S1–S7, whereas evapotranspiration decreased from S1 to S4 and increased from S4 to S7 for the individual effect of climate change. Evapotranspiration is more sensitive to LULC than to climate change [5].

3.3.3. Integrated Effects of LULC and Climate Change on Hydrological Flows

To better understand the hydrological flow of the Birr River watershed, baseline scenarios with the combined effect of LULC and climate change were presented. To determine the relative contribution of the combined effect for climate and LULC, climate data from 1986–1996 were selected with the LULC map of 1986 (S1), climate data from 1997–2007 were selected with the LULC map of 2001 (S5), and climate data from 2008–2018 were selected with LULC map of 2018 (S9) (Tables 11–13). The results show that surface runoff increased by 9.54 mm from S1 to S9, baseflow increased by 1.78 mm, water yield increased by 13.03 mm, and evapotranspiration decreased by 3.85 mm. It has been also observed that from S5 to S9, surface runoff increased by 2.65 mm, baseflow decreased by 4.48 mm, water yield decreased by 1.67 mm, and evapotranspiration increased by 2.84 mm. This indicates that the combined effects of LULC and climate change have a significant impact on the changing pattern of hydrological components in the Birr River watershed. The combined effect of climate and LULC, however, did not clearly show a one-dimensional pattern from S1–S9. Overall, the combined effect of LULC and climate change increases surface runoff in the Birr River watershed, but the magnitude of baseflow, water yield, and evapotranspiration varies from 1986 to 2018. The relative contribution of the combined effects of climate and LULC changes to the hydrological flow is not consistent in the study area. Previous studies have also reported similar findings [88,95,96].

3.4. Indicator of Hydrological Alteration

The IHA findings for one-day, three-day, seven-day, and thirty-day minimum and maximum daily streamflow in the Birr River watershed revealed no statistically significant increasing trend (Table 14). To assess the trend results, the Z value and computed two tailed probability (P) were compared at 5% confidence level. A small amount of positive increasing patterns were shown in both minimum and maximum daily streamflow, however, after ninety days the minimum and maximum streamflow showed a statistically significant positive increasing pattern. Streamflow regimes in the watershed were also investigated using rise and fall rate parameters. The result showed that there was no statistically significant trend. The rising rate was positive, while the falling rates showed negative streamflow patterns in the Birr River watershed.

Table 14. Results in Indicators of hydrological alteration parameters.

IHA Parameters	1-Day min	3-Day min	7-Day min	30-Day min	90-Day min	1-Day max	3-Day max	7-Day max	30-Day max	90-Day max	Rise Rate	Fall Rate
p value	0.35	0.31	0.07	0.01	0.00	0.48	0.35	0.14	0.29	0.03	0.01	0.00
Z value	0.93	1.02	1.79	0.82	2.21	0.70	0.94	1.48	1.06	2.20	0.78	−1.78

Note: min = minimum, and max = maximum.

4. Discussion

The SWAT model was found to be suitable for investigating the impact of climate and LULC change on hydrologic processes in the Birr River watershed. Overall, the SWAT model performance classification for the watershed was very good [90]. As the SWAT model results revealed that LULC and climate changes had effect on the hydrologic process (i.e., streamflow, surface runoff, baseflow, water yield, and ET) of the Birr River watershed. The observed changes in hydrological processes were attributed to LULC and climate change for this study. Substantial changes in LULC have been observed in the Birr River watershed over the last 32 years. Agriculture and settlement, for example, increased between 1986–2018, while bushland, forest, and grassland decreased (Table 8). The changes in LULC were mainly driven by anthropogenic activities (population pressure). For example, the population size of Quarit district (including urbanization), which is entirely within the Birr watershed is 114,771 in 2007, and this population number increased to 142,675 in 2022 with a 1.4% annual population change [97]. Natural vegetation is being converted into agricultural areas in the watershed. This could increase surface runoff and reduce baseflow, resulting increase in land degradation, soil erosion, and shortage of water resources. These findings were in line with previous scholars [30,88,95,98] For example, Wedajo et al. [30] indicated the transformation of natural vegetation into agricultural land in the Dhidhessa River basin. These could be increased surface runoff and decreased baseflow. Similarly, Gashaw et al. [88] reported that during 1985–2015, there was a continuous expansion of cultivated land and settlements, and a withdrawal of forest, shrubland, and grassland. Similarly, malede et al. [87], Demeke and Andualem [99], and Andualem et al. [100] indicated there is a significant land use change in the highlands of Ethiopia.

Individual LULC changes in the Birr River watershed also showed a positive increase in surface runoff while decreasing baseflow, however, the amount of increment is small (Tables 11–13). From 1997–2007 and 2008–2018, surface runoff increased by 3.63% and 4.71%, while baseflow changes by −3.77% and −9.34%, respectively (Figure 8a,b). The primary cause of the growth in surface runoff and decline in baseflow is agricultural land and small-extent settlement expansion, which reduces the water infiltration rate in the watershed. Thus, surface runoff increases and reduces baseflow characteristics in the study Birr River watershed. Between 1986 and 2018, the area of agricultural land and settlements increased by 24.49% and 54.78%, respectively, resulting in more surface runoff and reduced baseflow (Table 8). Agriculture and settlement areas increased to 9.4% and 43.15%, respectively, during 1986–2001, resulting in increased surface runoff.

Agriculture and settlement areas also further increased to 13.23% and 19.38%, respectively, during 2001–2018, which showed increased surface runoff and reduced baseflow. This result is consistent with previous studies by [6,7,75,91,101]. Agricultural and settlement areas increase impervious areas, which reduces soil infiltrations. Flooding becomes more frequent and severe as baseflow decreases and surface runoff increases [92,93,102]. The increase in surface runoff and decrease in baseflow is also attributed to due to a reduction in water bodies, forests, and bushlands.

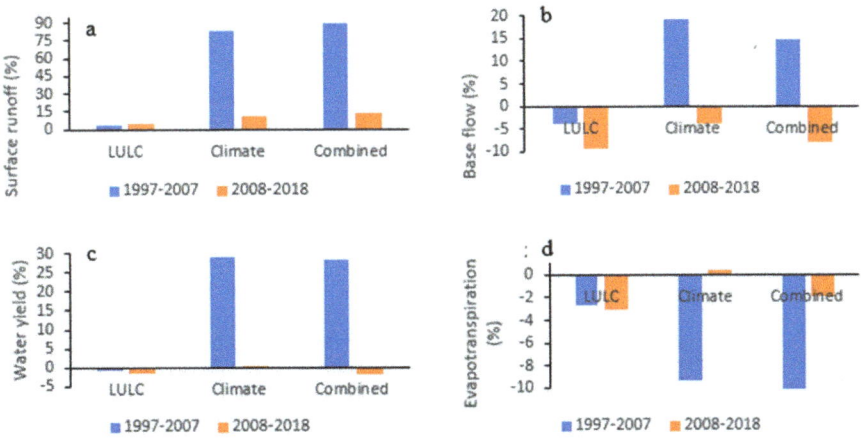

Figure 8. Separate and combined effects of LULC and climate change on: (**a**). surface runoff; (**b**). baseflow; (**c**). water yield, and (**d**). evapotranspiration.

Water yield and evapotranspiration (ET) revealed a decreased trend due to the effect of LULC change. During 19,997–2007 and 2008–2018, ET decreased by −2.61% and −3.03%, whereas water yield decreased by −0.75% and −1.35%, respectively (Figure 8d). The decrease in evapotranspiration and water yield is primarily caused by a reduction in bushland, forest cover, and shrublands. Between 1986–2018, bushland, forest cover, and grasslands decreased by −26.27%, −61.01%. and −3914%, respectively (Table 8). Liu et al. [103] reported that for all cover types, forest area contributes the greatest proportion of total evapotranspiration.

The spatiotemporal change in rainfall and temperature (climate change) was observed in the last three decades over the Birr River watershed [17,87]. This climate change increased surface runoff by 83.41% and 10.63% during 1997–2007, and 2008–2018, respectively, whereas baseflow increased by 19.11% from 1997–2007 and decrease by −3.65% during the 2008–2018 period (Figure 8). Climate change also resulted in a decrease in ET during 1997–2007, and it increases in 2008–2018. Individual climate change has a greater impact on surface runoff than individual LULC change over the Birr River watershed. The rising trend of rainfall and temperature causes more surface runoff [17,104]. Furthermore, increased rainfall in the Didessa River basin during the analysis study period have been contributed to increasing surface runoff [30].

Surface runoff was found to be more sensitive to the integrated effects of LULC and climate change than to the individual effects of LULC and climate change. Surface runoff increased by 90.07% from 1997–2007, and 13.38% from 2008–2018 (Figure 8a). Surface runoff increased significantly from 1997 to 2007, compared to 2008–2018. The decrease in surface runoff from 2008–2018 could be attributed to the Ethiopian government's planned afforestation and reforestation package program, which began in 2010 [105]. The cumulative impact of climate and LULC change on baseflow also showed a 14.62% increase from 1997–2007 but it decreased by −8.02 from 2008–2018. The combined effect of LULC and climate change on ET indicated a small decrease in ET in the period between 2008–2018 as compared to a decrease from 1997–2007, ET decreased by −1.9%. In general, proper implementation of integrated watershed management such as soil and water conversion practices, afforestation, and reforestation could play important roles in mitigating the impact of the hydrological process in the Birr River watershed [30]. According to Wedajo et al. [30] integrated watershed management reduced surface runoff by reducing peak runoff and increasing infiltration. Conservation measures also improve soil fertility, healthy vegetation growth, improved agricultural yield, increased water resource availability and combating climate change and watershed degradation in the Birr River watershed.

In addition to the SWAT model, the IHA is appropriate for evaluating the variations of daily streamflow due to climate and anthropogenic effects. The IHA model found to be there is no statistically significant increasing trend in the Birr River watershed for one-day, three-days, seven-day, and thirty-day minimum and maximum daily streamflow. However, all showed a positive small increment pattern in both minimum and maximum streamflow. In contrast, after ninety days, the minimum and maximum streamflow showed a statistically significant positive increasing pattern. Moreover, the streamflow regimes in the watershed were also investigated using the rise and fall rate parameters. There was no statistically significant trend observed in either the rise or fall rates; however, the rising rate was positive, whereas the falling rate was negative streamflow patterns in the Birr River watershed. These findings were consistent with those previously studied by [86,106]. For example, Gebremicael et al. [86], stated that the daily streamflow rising rate has remained constant, while the falling rate has significantly increased. Moreover, the one-day and seven-day maximum flows remain unchanged at Embamadre station. The one day and seven-day maximum flows did not change significantly could be the attributed to the homogenization of the low flow and peak flow hydrographs following the construction of the hydropower dam.

According to the integrated SWAT and IHA modeling, the Birr River watershed revealed a small amount of streamflow. However, the demand for water resources is increasing for different purposes such as irrigation, and water supply. Therefore, studying the important hydrological processes (streamflow, surface runoff, baseflow, water yield, and evapotranspiration) is serious for precise water resource planning, management, and development of this scarce water resource in the watershed.

5. Conclusions

An integrated study that quantifies the combined and separate effects of climate and land use change on the hydrological flow is ideal and necessary for effective water resource planning and management. Most previous studies concentrated on the separate effects of LULC and climate change on hydrological responses. This study analyzed the separate and combined effects of LULC and climate change over the Birr River watershed during the last 32 years using the SWAT hydrological model. In the Birr River watershed, it is important to investigate and identify the combined and separate effects of LULC and climate change on the hydrology at the watershed level (local) as well as the relative contribution of their changes. The SWAT model, which is GIS-enabled, was used to investigate the impact of LULC and climate change on hydrological flows in the Birr River watershed. The calibration results for the SWAT model indicate that it is a reliable and effective model for analyzing the hydrological process in the Birr River watershed with the effects of LULC and climate change.

The study results indicate that the sole effect of LULC change on surface runoff increased by 3.64% and 4.71%, during 1997–2007, and 2008–2018, respectively, whereas baseflow, water yield, and evapotranspiration decreased by −3.77%, and −9.34%; −0.75% and −1.35%; −2.61%, and −3.03% during the same period. Increased agricultural and settlement expansion was attributed to the increase in surface runoff. The decreasing trend in evapotranspiration can also be attributed to the reduction of bushland, forest, and grassland, while increasing agriculture and settlements. A decrease in baseflow and water yield, on the other hand, could be due to decreased groundwater recharge as a result of the transformations of vegetation cover to agricultural land. Streamflow increases during the wet season but declines during the dry season, affecting agricultural activities and water availability in the watershed. Individual climate change has a much greater relative contribution to surface runoff in the Birr River watershed than LULC change. Surface runoff and water yield showed positive values for the effect of climate change, whereas baseflow and evapotranspiration were revealed as uneven patterns. Surface runoff, baseflow, and water yield were more affected by climate change than LULC changes. In the Birr River watershed, the cumulative effect of climate and LULC change on hydrological flow is greater than the individual effect of climate and LULC change. The magnitude of surface runoff showed much-increased while decreasing evapotranspiration. Moreover, climate and LULC change caused an increase in baseflow and water yield between 1997–2007, and a decrease between 2008–2018. Overall, the results of the hydrological response to the effect of LULC and climate change showed a negative effect on the availability of water resources for agricultural production and others. Therefore, to reduce the impact of environmental changes on the hydrological processes of the Birr River watershed, appropriate integrated watershed management strategies, such as soil and water conservation practices, afforestation, and climate change strategies, must be implemented.

Author Contributions: D.A.M.: idea conceptualization, methodology setup, data curation and analysis, investigation, original draft, writing, and revision of the manuscript, editing, visualizing. T.A.: idea conceptualization, methodology, editing, review, visualization, and supervision. T.G.A.: review, editing, visualization, writing, and revision of the manuscript. All authors have read and agreed to the published version of the manuscript.

Funding: This study was carried out with the support of the Africa Center of Excellent for Water Management, Addis Ababa University.

Institutional Review Board Statement: Not applicable.

Informed Consent Statement: Not applicable.

Data Availability Statement: The data that supports the funding of this study is available from the corresponding author upon reasonable request.

Acknowledgments: The authors would like to thank the Ethiopian National Meteorology Agency and the Ministry of Water Irrigation and Energy for providing climate and streamflow data. Africa Center of excellent for water management, Addis Ababa University for the support to conduct this research.

Conflicts of Interest: The authors declare no conflict of interest.

References

1. Ahmed, N.; Wang, G.; Booij, M.J.; Xiangyang, S.; Hussain, F.; Nabi, G. Separation of the Impact of Landuse/Landcover Change and Climate Change on Runoff in the Upstream Area of the Yangtze River, China. *Water Resour. Manag.* **2022**, *36*, 181–201. [CrossRef]
2. Petrovic, F. Hydrological Impacts of Climate Change and Land Use. *Water* **2021**, *13*, 799. [CrossRef]
3. Talib, A.; Randhir, T.O. Climate Change and Land Use Impacts on Hydrologic Processes of Watershed Systems. *J. Water Clim. Chang.* **2017**, *8*, 363–374. [CrossRef]
4. Kirby, J.M.; Mainuddin, M.; Mpelasoka, F.; Ahmad, M.D.; Palash, W.; Quadir, M.E.; Shah-Newaz, S.M.; Hossain, M.M. The Impact of Climate Change on Regional Water Balances in Bangladesh. *Clim. Chang.* **2016**, *135*, 481–491. [CrossRef]
5. Berihun, M.L.; Tsunekawa, A.; Haregeweyn, N.; Meshesha, D.T.; Adgo, E.; Tsubo, M.; Masunaga, T.; Fenta, A.A.; Sultan, D.; Yibeltal, M.; et al. Hydrological Responses to Land Use/Land Cover Change and Climate Variability in Contrasting Agro-Ecological Environments of the Upper Blue Nile Basin, Ethiopia. *Sci. Total Environ.* **2019**, *689*, 347–365. [CrossRef]
6. Kumar, M.; Denis, D.M.; Kundu, A.; Joshi, N.; Suryavanshi, S. Understanding Land Use/Land Cover and Climate Change Impacts on Hydrological Components of Usri Watershed, India. *Appl. Water Sci.* **2022**, *12*, 39. [CrossRef]
7. Yang, L.; Feng, Q.; Yin, Z.; Wen, X.; Si, J.; Li, C.; Deo, R.C. Identifying Separate Impacts of Climate and Land Use/Cover Change on Hydrological Processes in Upper Stream of Heihe River, Northwest China. *Hydrol Process.* **2017**, *31*, 1100–1112. [CrossRef]
8. Schäfer, M.P.; Dietrich, O.; Mbilinyi, B. Streamflow and Lake Water Level Changes and Their Attributed Causes in Eastern and Southern Africa: State of the Art Review. *Int. J. Water Resour. Dev.* **2015**, *32*, 853–880. [CrossRef]

9. Getachew, B.; Manjunatha, B.R.; Bhat, H.G. Modeling Projected Impacts of Climate and Land Use/Land Cover Changes on Hydrological Responses in the Lake Tana Basin, Upper Blue Nile River Basin, Ethiopia. *J. Hydrol.* **2021**, *595*, 125974. [CrossRef]
10. Zhou, F.; Xu, Y.; Chen, Y.; Xu, C.Y.; Gao, Y.; Du, J. Hydrological Response to Urbanization at Different Spatio-Temporal Scales Simulated by Coupling of CLUE-S and the SWAT Model in the Yangtze River Delta Region. *J. Hydrol.* **2013**, *485*, 113–125. [CrossRef]
11. Tekleab, S.; Mohamed, Y.; Uhlenbrook, S. Hydro-Climatic Trends in the Abay/Upper Blue Nile Basin, Ethiopia. *Phys. Chem. Earth Parts A/B/C* **2013**, *61–62*, 32–42. [CrossRef]
12. Hassan, M.M. Monitoring Land Use/Land Cover Change, Urban Growth Dynamics and Landscape Pattern Analysis in Five Fastest Urbanized Cities in Bangladesh. *Remote Sens. Appl. Soc. Environ.* **2017**, *7*, 69–83. [CrossRef]
13. Osei, M.A.; Amekudzi, L.K.; Wemegah, D.D.; Preko, K.; Gyawu, E.S.; Obiri-Danso, K. The Impact of Climate and Land-Use Changes on the Hydrological Processes of Owabi Catchment from SWAT Analysis. *J. Hydrol. Reg. Stud.* **2019**, *25*, 100620. [CrossRef]
14. Wang, S.; Kang, S.; Zhang, L.; Li, F. Modelling Hydrological Response to Different Land-Use and Climate Change Scenarios in the Zamu River Basin of Northwest China. *Hydrol. Process.* **2008**, *22*, 2502–2510. [CrossRef]
15. Venkatesh, K.; Ramesh, H.; Das, P. Modelling Stream Flow and Soil Erosion Response Considering Varied Land Practices in a Cascading River Basin. *J. Env. Manag.* **2020**, *264*, 110448. [CrossRef] [PubMed]
16. Duveiller, G.; Caporaso, L.; Abad-Viñas, R.; Perugini, L.; Grassi, G.; Arneth, A.; Cescatti, A. Local Biophysical Effects of Land Use and Land Cover Change: Towards an Assessment Tool for Policy Makers. *Land Use Policy* **2020**, *91*, 104382. [CrossRef]
17. Malede, D.A.; Agumassie, T.A.; Kosgei, J.R.; Andualem, T.G.; Diallo, I. Recent Approaches to Climate Change Impacts on Hydrological Extremes in the Upper Blue Nile Basin, Ethiopia. *Earth Syst. Environ.* **2022**, *6*, 669–679. [CrossRef]
18. Winkler, K.; Fuchs, R.; Rounsevell, M.; Herold, M. Global Land Use Changes Are Four Times Greater than Previously Estimated. *Nat. Com.* **2021**, *12*, 2501. [CrossRef]
19. Getachew, B.; Rachotappa, B.; Getachew, M.B.; Programme, G.; Getachew, B.; Manjunatha, B.R. Impacts of Land-Use Change on the Hydrology of Lake Tana Basin, Upper Blue Nile River Basin, Ethiopia. *Wiley Online Libr.* **2022**, *6*, 2200041. [CrossRef]
20. Minwyelet, M. *Birr Watershed Integrated Natural Resource Management*; Bahir Dar University: Bahir Dar, Ethiopia, 2014.
21. Wang, Z.; Tian, J.; Feng, K. Response of Runoff towards Land Use Changes in the Yellow River Basin in Ningxia, China. *PLoS One* **2022**, *17*, e0265931. [CrossRef]
22. Emiru, N.C.; Recha, J.W.; Thompson, J.R.; Belay, A.; Aynekulu, E.; Manyevere, A.; Demissie, T.D.; Osano, P.M.; Hussein, J.; Molla, M.B.; et al. Impact of Climate Change on the Hydrology of the Upper Awash River Basin, Ethiopia. *Hydrology* **2021**, *9*, 3. [CrossRef]
23. Getu Engida, T.; Nigussie, T.A.; Aneseyee, A.B.; Barnabas, J. Land Use/Land Cover Change Impact on Hydrological Process in the Upper Baro Basin, Ethiopia. *Appl. Env. Soil Sci.* **2021**, *2021*, 6617541. [CrossRef]
24. Chawla, I.; Mujumdar, P.P. Isolating the Impacts of Land Use and Climate Change on Streamflow. *Hydrol. Earth Syst. Sci.* **2015**, *19*, 3633–3651. [CrossRef]
25. Yin, J.; He, F.; Xiong, Y.J.; Qiu, G.Y. Effects of Land Use / Land Cover and Climate Changes on Surface Runoff in a Semi-Humid and Semi-Arid Transition Zone in Northwest China. *Hydrol. Earth Syst. Sci.* **2017**, *21*, 183–196. [CrossRef]
26. Kim, J.; Choi, J.; Choi, C.; Park, S. Impacts of Changes in Climate and Land Use/Land Cover under IPCC RCP Scenarios on Streamflow in the Hoeya River Basin, Korea. *Sci. Total Environ.* **2013**, *452–453*, 181–195. [CrossRef]
27. Gurara, M.A.; Jilo, N.B.; Tolche, A.D. Modelling Climate Change Impact on the Streamflow in the Upper Wabe Bridge Watershed in Wabe Shebele River Basin, Ethiopia. *Int. J. River Basin Manag.* **2021**, 1–13. [CrossRef]
28. Chen, Q.; Chen, H.; Wang, J.; Zhao, Y.; Chen, J.; Xu, C. Impacts of Climate Change and Land-Use Change on Hydrological Extremes in the Jinsha River Basin. *Water* **2019**, *11*, 1398. [CrossRef]
29. Kuma, H.G.; Feyessa, F.F.; Demissie, T.A. Hydrologic Responses to Climate and Land-Use/Land-Cover Changes in the Bilate Catchment, Southern Ethiopia. *J. Water Clim. Chang.* **2021**, *12*, 3750–3769. [CrossRef]
30. Wedajo, G.K.; Muleta, M.K.; Awoke, B.G. Impacts of Combined and Separate Land Cover and Climate Changes on Impacts of Combined and Separate Land Cover and Climate Changes on Hydrologic Responses of Dhidhessa River Basin, Ethiopia Gizachew Kabite Wedajo, Misgana Kebede Muleta & Berhan Gessesse. *Int. J. River Basin Manag.* **2022**, 1–14. [CrossRef]
31. Chanapathi, T.; Thatikonda, S. Investigating the Impact of Climate and Land-Use Land Cover Changes on Hydrological Predictions over the Krishna River Basin under Present and Future Scenarios. *Sci. Total Environ.* **2020**, *721*, 137736. [CrossRef]
32. Idrissou, M.; Diekkrüger, B.; Tischbein, B.; de Hipt, F.O.; Näschen, K.; Poméon, T.; Yira, Y.; Ibrahim, B. Modeling the Impact of Climate and Land Use/Land Cover Change on Water Availability in an Inland Valley Catchment in Burkina Faso. *Hydrology* **2022**, *9*, 12. [CrossRef]
33. Astuti, I.S.; Sahoo, K.; Milewski, A.; Mishra, D.R. Impact of Land Use Land Cover (LULC) Change on Surface Runoff in an Increasingly Urbanized Tropical Watershed. *Water Resour. Manag.* **2019**, *33*, 4087–4103. [CrossRef]
34. Khare, D.; Patra, D.; Mondal, A.; Kundu, S. Impact of Landuse/Land Cover Change on Run-off in the Catchment of a Hydro Power Project. *Appl. Water Sci.* **2017**, *7*, 787–800. [CrossRef]
35. Bronstert, A.; Niehoff, D.; Brger, G. Effects of Climate and Land-Use Change on Storm Runoff Generation: Present Knowledge and Modelling Capabilities. *Hydrol. Process.* **2002**, *16*, 509–529. [CrossRef]

36. Pfister, L.; Kwadijk, J.; Musy, A.; Bronstert, A.; Hoffmann, L. Climate Change, Land Use Change and Runoff Prediction in the Rhine-Meuse Basins. *River Res. Appl.* **2004**, *20*, 229–241. [CrossRef]
37. Suryavanshi, S.; Pandey, A.; Chaube, U.C. Hydrological Simulation of the Betwa River Basin (India) Using the SWAT Model. *Hydrol. Sci. J.* **2017**, *62*, 960–978. [CrossRef]
38. Wang, Q.; Xu, Y.; Wang, Y.; Zhang, Y.; Xiang, J.; Xu, Y.; Wang, J. Individual and Combined Impacts of Future Land-Use and Climate Conditions on Extreme Hydrological Events in a Representative Basin of the Yangtze River Delta, China. *Atmos Res.* **2020**, *236*, 104805. [CrossRef]
39. Chang, H.; Johnson, G.; Hinkley, T.; Jung, I.W. Spatial Analysis of Annual Runoff Ratios and Their Variability across the Contiguous U.S. *J. Hydrol.* **2014**, *511*, 387–402. [CrossRef]
40. Zuo, D.; Xu, Z.; Yao, W.; Jin, S.; Xiao, P.; Ran, D. Assessing the Effects of Changes in Land Use and Climate on Runoff and Sediment Yields from a Watershed in the Loess Plateau of China. *Sci. Total Environ.* **2016**, *544*, 238–250. [CrossRef]
41. Kiprotich, P.; Wei, X.; Zhang, Z.; Ngigi, T.; Qiu, F.; Wang, L. Assessing the Impact of Land Use and Climate Change on Surface Runoff Response Using Gridded Observations and SWAT+. *Hydrology* **2021**, *8*, 48. [CrossRef]
42. Patel, S.K.; Verma, P. Agricultural Growth and Land Use Land Cover Change in Peri-Urban India. *Env. Monit. Assess.* **2019**, *191*, 1–17. [CrossRef] [PubMed]
43. Zhang, L.; Podlasly, C.; Ren, Y.; Feger, K.H.; Wang, Y.; Schwärzel, K. Separating the Effects of Changes in Land Management and Climatic Conditions on Long-Term Streamflow Trends Analyzed for a Small Catchment in the Loess Plateau Region, NW China. *Hydrol. Process.* **2014**, *28*, 1284–1293. [CrossRef]
44. Bogale, S. Hydrological Response to Land Use and Land Cover Changes of Ribb Watershed, Ethiopia. *Hydrology* **2021**, *9*, 1. [CrossRef]
45. Dibaba, W.T.; Demissie, T.A.; Miegel, K. Watershed Hydrological Response to Combined Land Use/Land Cover and Climate Change in Highland Ethiopia: Finchaa Catchment. *Water* **2020**, *12*, 1801. [CrossRef]
46. Getachew, B.; Manjunatha, B.R. Potential Climate Change Impact Assessment on the Hydrology of the Lake Tana Basin, Upper Blue Nile River Basin, Ethiopia. *Phys. Chem. Earth Parts A/B/C* **2022**, *127*, 103162. [CrossRef]
47. Woldesenbet, T.A.; Elagib, N.A.; Ribbe, L.; Heinrich, J. Catchment Response to Climate and Land Use Changes in the Upper Blue Nile Sub-Basins, Ethiopia. *Sci. Total Environ.* **2018**, *644*, 193–206. [CrossRef]
48. Gebremicael, T.G.; Mohamed, Y.A.; van der Zaag, P. Attributing the Hydrological Impact of Different Land Use Types and Their Long-Term Dynamics through Combining Parsimonious Hydrological Modelling, Alteration Analysis and PLSR Analysis. *Sci. Total Environ.* **2019**, *660*, 1155–1167. [CrossRef]
49. Bekele, D.; Alamirew, T.; Kebede, A.; Zeleke, G.; Melesse, A.M. Modeling the Impacts of Land Use and Land Cover Dynamics on Hydrological Processes of the Keleta Watershed, Ethiopia. *Sustain. Environ.* **2021**, *7*, 1909860. [CrossRef]
50. Mehdi, B.; Lehner, B.; Gombault, C.; Michaud, A.; Beaudin, I.; Sottile, M.F.; Blondlot, A. Simulated Impacts of Climate Change and Agricultural Land Use Change on Surface Water Quality with and without Adaptation Management Strategies. *Agric. Ecosyst. Env.* **2015**, *213*, 47–60. [CrossRef]
51. El-Khoury, A.; Seidou, O.; Lapen, D.R.L.; Que, Z.; Mohammadian, M.; Sunohara, M.; Bahram, D. Combined Impacts of Future Climate and Land Use Changes on Discharge, Nitrogen and Phosphorus Loads for a Canadian River Basin. *J. Env. Manag.* **2015**, *151*, 76–86. [CrossRef]
52. Wang, K.; Qian, M.; Xu, S.; Liang, S.; Chen, H.; Hu, Y.; Su, C.; Zhao, M.; Li, W.; Wang, J. Impacts of Climate Change on Water Resources in the Huaihe River Basin. *MATEC Web Conf.* **2018**, *246*, 01090. [CrossRef]
53. Mishra, H.; Denis, D.M.; Suryavanshi, S.; Kumar, M.; Srivastava, S.K.; Denis, A.F.; Kumar, R. Hydrological Simulation of a Small Ungauged Agricultural Watershed Semrakalwana of Northern India. *Appl. Water Sci.* **2017**, *7*, 2803–2815. [CrossRef]
54. Lopes, T.R.; Zolin, C.A.; Mingoti, R.; Vendrusculo, L.G.; de Almeida, F.T.; de Souza, A.P.; de Oliveira, R.F.; Paulino, J.; Uliana, E.M. Hydrological Regime, Water Availability and Land Use/Land Cover Change Impact on the Water Balance in a Large Agriculture Basin in the Southern Brazilian Amazon. *J. South Am. Earth Sci.* **2021**, *108*, 103224. [CrossRef]
55. Bal, M.; Dandpat, A.K.; Naik, B. Hydrological Modeling with Respect to Impact of Land-Use and Land-Cover Change on the Runoff Dynamics in Budhabalanga River Basing Using ArcGIS and SWAT Model. *Remote Sens. Appl. Soc. Environ.* **2021**, *23*, 100527. [CrossRef]
56. Melesse, A.M.; Loukas, A.G.; Senay, G.; Yitayew, M. Advanced Bash-Scripting Guide An In-Depth Exploration of the Art of Shell Scripting Table of Contents. *Hydrol. Process.* **2010**, *2274*, 2267–2274. [CrossRef]
57. Bekele, A.A.; Pingale, S.M.; Hatiye, S.D.; Tilahun, A.K. Impact of Climate Change on Surface Water Availability and Crop Water Demand for the Sub-Watershed of Abbay Basin, Ethiopia. *Sustain. Water Resour. Manag.* **2019**, *5*, 1859–1875. [CrossRef]
58. Gebere, S.B.; Alamirew, T.; Merkel, B.J.; Melesse, A.M. Performance of High-Resolution Satellite Rainfall Products over Data-Scarce Parts of Eastern Ethiopia. *Remote Sens.* **2015**, *7*, 11639–11663. [CrossRef]
59. SRTM Source of the 30 M Resolution. Available online: https://earthexplorer.usgs.gov/ (accessed on 1 October 2022).
60. FAO. *World Soil Resources: An Explanatory Note on the FAO (Food and Agriculture Organization) World Soil Resources Map at 1:25 000 00 Scale*; FAO: Rome, Italy, 1995.
61. Anderson, J. *A Land Use and Land Cover Classification System for Use with Remote Sensor Data*; USGS: Reston, VA, USA, 1976.
62. Ismail, M.; Jusoff, K. Satellite Data Classification Accuracy Assessment Based from Reference Dataset. *Int. J. Geol. Environ. Eng.* **2008**, *2*, 23–29.

63. Neitsch, S.L.; Arnold, J.G.; Kiniry, J.R.; Williams, J.R. Soil & Water Assessment Tool Theoretical Documentation Version 2009; Technical Report; Texas Water Resources Institute: College Station, TX, USA, 2011; pp. 1–647. [CrossRef]
64. Qiu, B.L.; Peng, D.; Fang, J.; Zhang, Z. Estimation of Hydrological Responses to Climate Changes and Human Activities Estimation of Hydrological Responses to Climate Changes and Human Activities in the Xitiaoxi River Basin. In Proceedings of the 3rd IMA International Conference on Flood Risk, Swansea University, Wales, UK, 30–31 March 2015.
65. Dile, Y.T.; Daggupati, P.; George, C.; Srinivasan, R.; Arnold, J. Introducing a New Open-Source GIS User Interface for the SWAT Model. *Environ. Model. Softw.* **2016**, *85*, 129–139. [CrossRef]
66. Zotarelli, L.; Dukes, M.D.; Romero, C.C.; Migliaccio, K.W.; Kelly, T. Step by Step Calculation of the Penman-Monteith Evapotranspiration (FAO-56 Method). *Inst. Food Agric. Sciences. Univ. Fla.* **2020**, *8*, 1–10.
67. Masih, I.; Uhlenbrook, S.; Maskey, S.; Smakhtin, V. Streamflow Trends and Climate Linkages in the Zagros Mountains, Iran. *Clim. Chang.* **2011**, *104*, 317–338. [CrossRef]
68. Arnold, J.G.; Srinivasan, R.; Muttiah, R.S.; Williams, J.R. LARGE AREA HYDROLOGIC MODELING AND ASSESSMENT PART I: MODEL DEVELOPMENT1. *JAWRA J. Am. Water Resour. Assoc.* **1998**, *34*, 73–89. [CrossRef]
69. Arnold, J.G.; Moriasi, D.N.; Gassman, P.W.; Abbaspour, K.C.; White, M.J.; Srinivasan, R.; Santhi, C.; Harmel, R.D.; van Griensven, A.; Van Liew, M.W.; et al. SWAT: Model Use, Calibration, and Validation. *Trans. ASABE* **2012**, *55*, 1491–1508. [CrossRef]
70. Neitsch, S.L.; Arnold, J.G.; Srinivasan, R. Pesticides Fate and Transport Predicted by the Soil and Water Assessment Tool (SWAT) Atrazine, Metolachlor and Trifluralin in the Sugar Creek Watershed. *BRC Rep.* **2002**, *3*, 1–100.
71. Gassman, P.W.; Sadeghi, A.M.; Srinivasan, R. Applications of the SWAT Model Special Section: Overview and Insights. *J. Env. Qual.* **2014**, *43*, 1–8. [CrossRef] [PubMed]
72. Vazquez-Amábile, G.G.; Engel, B.A. Use of SWAT to Compute Groundwater Table Depth and Streamflow in the Muscatatuck River Watershed. *Trans. Am. Soc. Agric. Eng.* **2005**, *48*, 991–1003. [CrossRef]
73. Daggupati, P.; Pai, N.; Ale, S.; Douglas-Mankin, K.R.; Zeckoski, R.W.; Jeong, J.; Parajuli, P.B.; Saraswat, D.; Youssef, M.A. A Recommended Calibration and Validation Strategy for Hydrologic and Water Quality Models. *Trans. ASABE* **2015**, *58*, 1705–1719. [CrossRef]
74. Murty, P.S.; Pandey, A.; Suryavanshi, S. Application of Semi-distributed Hydrological Model for Basin Level Water Balance of the Ken Basin of Central India. *Wiley Online Libr.* **2013**, *28*, 4119–4129. [CrossRef]
75. Andualem, T.; Eng, B.G. Impact of Land Use Land Cover Change on Stream Flow and Sediment Yield: A Case Study of Gilgel Abay Watershed, Lake Tana Sub-Basin, Ethiopia. *Int. J. Technol. Enhanc. Merg.* **2015**, *3*, 28.
76. Tenagashaw, D.Y.; Andualem, T.G. Analysis and Characterization of Hydrological Drought under Future Climate Change Using the SWAT Model in Tana Sub-Basin, Ethiopia. *Water Conserv. Sci. Eng.* **2022**, *7*, 131–142. [CrossRef]
77. Andualem, T.G.; Guadie, A.; Belay, G.; Ahmad, I.; Dar, M.A. Hydrological Modeling of Upper Ribb Watershed, Abbay Basin, Ethiopia. *Glob. Nest J.* **2020**, *22*, 158–164. [CrossRef]
78. Santhi, C.; Arnold, J.G.; Williams, J.R.; Dugas, W.A.; Srinivasan, R.; Hauck, L.M. Validation of the Swat Model on A Large Rwer Basin with Point and Nonpoint Sources. *JAWRA J. Am. Water Resour. Assoc.* **2001**, *37*, 1169–1188. [CrossRef]
79. Van Liew, M.W.; Arnold, J.G.; Garbrecht, J.D. Hydrologic Simulation on Agricultural Watersheds: Choosing between Two Models. *Trans. Am. Soc. Agric. Eng.* **2003**, *46*, 1539–1551. [CrossRef]
80. Nash, J.E.; Sutcliffe, J.V. River Flow Forecasting through Conceptual Models Part I—A Discussion of Principles. *J. Hydrol.* **1970**, *10*, 282–290. [CrossRef]
81. Gupta, H.V.; Sorooshian, S.; Yapo, P.O. Status of Automatic Calibration for Hydrologic Models: Comparison with Multilevel Expert Calibration. *J. Hydrol.Eng.* **1999**, *4*, 135–143. [CrossRef]
82. Li, Z.; Xu, Z.; Shao, Q.; An, J.Y.-H.P. Parameter Estimation and Uncertainty Analysis of SWAT Model in Upper Reaches of the Heihe River Basin. *Hydrol. Process. Int. J.* **2009**, *23*, 2744–2753. [CrossRef]
83. Mekonnen, D.; Duan, Z.; Rientjes, T.; Disse, M. Analysis of the Combined and Single Effects of LULC and Climate Change on the Streamflow of the Upper Blue Nile River Basin (UBNRB): Using Statistical Trend Tests, Remote Sensing Landcover Maps and the SWAT Model. *Hydrol. Earth Syst. Sci. Discuss.* **2017**, 1–26. [CrossRef]
84. Woldesenbet, T.A.; Elagib, N.A.; Ribbe, L.; Heinrich, J. Hydrological Responses to Land Use/Cover Changes in the Source Region of the upper Blue Nile Basin, Ethiopia. *Sci. Total Environ.* **2017**, *575*, 724–741. [CrossRef]
85. Mathews, R.; Richter, B.D. Application of the Indicators of Hydrologic Alteration Software in Environmental Flow Setting. *JAWRA J. Am. Water Resour. Assoc.* **2007**, *43*, 1400–1413. [CrossRef]
86. Gebremicael, T.G.; Mohamed, Y.A.; Zaag, P.V.; Hagos, E.Y. Temporal and Spatial Changes of Rainfall and Streamflow in the Upper Tekezē-Atbara River Basin, Ethiopia. *Hydrol. Earth Syst. Sci.* **2016**, *21*, 2127–2142. [CrossRef]
87. Malede, D.A.; Alamirew, T.; Kosgie, J.R.; Andualem, T.G. Analysis of Land Use/Land Cover Change Trends over Birr River Watershed, Abbay Basin, Ethiopia. *Environ. Sustain. Indic.* **2023**, *17*, 100222. [CrossRef]
88. Gashaw, T.; Tulu, T.; Argaw, M.; Worqlul, A.W. Modeling the Hydrological Impacts of Land Use/Land Cover Changes in the Andassa Watershed, Blue Nile Basin, Ethiopia. *Sci. Total Environ.* **2018**, *619–620*, 1394–1408. [CrossRef] [PubMed]
89. Ewunetu, A.; Simane, B.; Teferi, E.; Zaitchik, B.F. Land Cover Change in the Blue Nile River Headwaters: Farmers' Perceptions, Pressures, and Satellite-Based Mapping. *Land* **2021**, *10*, 68. [CrossRef]
90. Moriasi, D.N.; Arnold, J.G.; van Liew, M.W.; Bingner, R.L.; Harmel, R.D.; Veith, T.L. Model Evaluation Guidelines for Systematic Quantification of Accuracy in Watershed Simulations. *Trans. ASABE* **2007**, *50*, 885–900. [CrossRef]

91. Baker, T.J.; Miller, S.N. Using the Soil and Water Assessment Tool (SWAT) to Assess Land Use Impact on Water Resources in an East African Watershed. *J. Hydrol.* **2013**, *486*, 100–111. [CrossRef]
92. Huang, H.J.; Cheng, S.J.; Wen, J.C.; Lee, J.H. Effect of Growing Watershed Imperviousness on Hydrograph Parameters and Peak Discharge. *Hydrol. Process.* **2008**, *22*, 2075–2085. [CrossRef]
93. Wang, R.; Kalin, L.; Kuang, W.; Tian, H. Individual and Combined Effects of Land Use/Cover and Climate Change on Wolf Bay Watershed Streamflow in Southern Alabama. *Hydrol. Process.* **2014**, *28*, 5530–5546. [CrossRef]
94. Chim, K.; Tunnicliffe, J.; Shamseldin, A.Y.; Bun, H. Assessment of Land Use and Climate Change Effects on Hydrology in the Upper Siem Reap River and Angkor Temple Complex, Cambodia. *Env. Dev* **2021**, *39*, 100615. [CrossRef]
95. Teklay, A.; Dile, Y.T.; Setegn, S.G.; Demissie, S.S.; Asfaw, D.H. Evaluation of Static and Dynamic Land Use Data for Watershed Hydrologic Process Simulation: A Case Study in Gummara Watershed, Ethiopia. *Catena* **2019**, *172*, 65–75. [CrossRef]
96. Worku, T.; Khare, D.; Tripathi, S.K. Modeling Runoff–Sediment Response to Land Use/Land Cover Changes Using Integrated GIS and SWAT Model in the Beressa Watershed. *Env. Earth Sci.* **2017**, *76*, 550. [CrossRef]
97. Central Statistical Agency (CSA). *Summary and Statistical Report of the 2007 Population and Housing Census Results*; Central Statistical Agency (CSA): Addis Ababa, Ethiopia, 2007.
98. Yang, X.L.; Ren, L.L.; Liu, Y.; Jiao, D.L.; Jiang, S.H. Hydrological Response to Land Use and Land Cover Changes in a Sub-Watershed of West Liaohe River Basin, China. *J. Arid. Land* **2014**, *6*, 678–689. [CrossRef]
99. Demeke, G.G.; Andualem, T.G. Application of Remote Sensing for Evaluation of Land Use Change Responses on Hydrology of Muga Watershed, Abbay River Basin, Ethiopia. *J. Earth Sci. Clim. Chang.* **2018**, *9*, 2. [CrossRef]
100. Andualem, T.G.; Belay, G.; Guadie, A. Land Use Change Detection Using Remote Sensing Technology. *J. Earth Sci. Clim. Chang.* **2018**, *9*, 1–6. [CrossRef]
101. Fenta Mekonnen, D.; Duan, Z.; Rientjes, T.; Disse, M. Analysis of Combined and Isolated Effects of Land-Use and Land-Cover Changes and Climate Change on the Upper Blue Nile River Basin's Streamflow. *Hydrol. Earth Syst. Sci.* **2018**, *22*, 6187–6207. [CrossRef]
102. Kim, Y.; Engel, B.A.; Lim, K.J.; Larson, V.; Duncan, B. Runoff Impacts of Land-Use Change in Indian River Lagoon Watershed. *J. Hydrol. Eng.* **2002**, *7*, 245–251. [CrossRef]
103. Liu, J.; Chen, J.M.; Cihlar, J. Mapping Evapotranspiration Based on Remote Sensing: An Application to Canada's Landmass. *Water Resour. Res.* **2003**, *39*, 1189. [CrossRef]
104. Malede, D.A.; Agumassie, T.A.; Kosgei, J.R.; Linh, N.T.T.; Andualem, T.G. Analysis of Rainfall and Streamflow Trend and Variability over Birr River Watershed, Abbay Basin, Ethiopia. *Environ. Chall.* **2022**, *7*, 100528. [CrossRef]
105. Takele, A.; Lakew, H.B.; Kabite, G. Does the Recent Afforestation Program in Ethiopia Influenced Vegetation Cover and Hydrology? A Case Study in the Upper Awash Basin, Ethiopia. *Heliyon* **2022**, *8*, e09589. [CrossRef]
106. Gao, B.; Li, J.; Wang, X. Analyzing Changes in the Flow Regime of the Yangtze River Using the Eco-Flow Metrics and IHA Metrics. *Water* **2018**, *10*, 1552. [CrossRef]

Disclaimer/Publisher's Note: The statements, opinions and data contained in all publications are solely those of the individual author(s) and contributor(s) and not of MDPI and/or the editor(s). MDPI and/or the editor(s) disclaim responsibility for any injury to people or property resulting from any ideas, methods, instructions or products referred to in the content.

Article

A Laboratory Study of the Role of Nature-Based Solutions in Improving Flash Flooding Resilience in Hilly Terrains

Shees Ur Rehman [1], Afzal Ahmed [1], Gordon Gilja [2,*], Manousos Valyrakis [3], Abdul Razzaq Ghumman [4], Ghufran Ahmed Pasha [1] and Rashid Farooq [5]

[1] Department of Civil Engineering, University of Engineering and Technology, Taxila 47050, Pakistan; sheesurrehman9@gmail.com (S.U.R.); afzal.ahmed@uettaxila.edu.pk (A.A.); ghufran.ahmed@uettaxila.edu.pk (G.A.P.)
[2] Department of Hydroscience and Engineering, Faculty of Civil Engineering, University of Zagreb, Fra Andrije Kacica Miosica 26, 10000 Zagreb, Croatia
[3] Department of Civil Engineering, Aristotle University of Thessaloniki, 54124 Thessaloniki, Greece; mvalyrak@gmail.com
[4] Department of Civil Engineering, College of Engineering, Qassim University, Buraydah 51452, Saudi Arabia; abdul.razzaq@qec.edu.sa
[5] Department of Civil Engineering, International Islamic University, Islamabad 44000, Pakistan; rashid.farooq@iiu.edu.pk
* Correspondence: gordon.gilja@grad.unizg.hr

Citation: Rehman, S.U.; Ahmed, A.; Gilja, G.; Valyrakis, M.; Ghumman, A.R.; Pasha, G.A.; Farooq, R. A Laboratory Study of the Role of Nature-Based Solutions in Improving Flash Flooding Resilience in Hilly Terrains. *Water* **2024**, *16*, 124. https://doi.org/10.3390/w16010124

Academic Editors: Sonia Raquel Gámiz-Fortis and Matilde García-Valdecasas Ojeda

Received: 23 October 2023
Revised: 24 December 2023
Accepted: 26 December 2023
Published: 29 December 2023

Copyright: © 2023 by the authors. Licensee MDPI, Basel, Switzerland. This article is an open access article distributed under the terms and conditions of the Creative Commons Attribution (CC BY) license (https://creativecommons.org/licenses/by/4.0/).

Abstract: Nature-based solutions (NBSs) always provide optimal opportunities for researchers and policymakers to develop sustainable and long-term solutions for mitigating the impacts of flooding. Computing the hydrological process in hilly areas is complex compared to plain areas. This study used a laboratory-scaled hillslope model to study rainfall-runoff responses considering the natural hillslope conditions prevailing in hill torrents creating flash floods. The objective of this study was to estimate the impact of nature-based solutions on time-to-peak for flash flooding events on hilly terrains under different scenarios. Many factors decide the peak of runoff generation due to rainfall, like land use conditions, e.g., soil porosity, vegetation cover, rainfall intensity, and terrain slope. To reduce these complexities, the model was designed with thermopore sheets made of impermeable material. A hillslope model using NBS was designed to evaluate flood hydrograph attenuation to minimize the peak discharge (Qp) and increase time-to-peak (Tp) under varying rainfall, land cover, and drainage channel slope conditions. A rainfall simulator was used to analyze the formation of hydrographs for different conditions, e.g., from barren to vegetation under three different slopes (S0, S1, S2) and three rainfall intensities (P1, P2, P3). Vegetation conditions used were no vegetation, rigid vegetation, flexible vegetation, and the combination of both rigid and flexible vegetation. The purpose of using all these conditions was to determine their mitigation effects on flash flooding. This experimental analysis shows that the most suitable case to attenuate a flood hydrograph was the mixed vegetation condition, which can reduce the peak discharge by 27% to 39% under different channel slopes. The mixed vegetation condition showed an increase of 49% in time-to-peak (Tp) compared to the no vegetation condition. Additionally, under P1 rainfall and a bed slope of 0°, it reduced the peak discharge by up to 35% in the simulated flood and effectively minimized its potentially destructive impacts.

Keywords: nature-based solutions; time-to-peak; efficiency; complex; flexible vegetation; rigid vegetation

1. Introduction

Nature-based solutions (NBSs) are widely adopted to minimize climate change impacts and enhance resilience in areas of meteorological risk to society, such as flooding and sustainable development [1,2]. In the area surrounding D.G. Khan, a total of 13 hill torrents, Kaura, SoriJanubi, Mithawan, Sanghar, Pitok, Vehova, SakhiSarwar, Kaha, Sorilund,

Chadhar, SoriShumali, Zangi, and Vidor, serve as conduits for floodwater from nearby catchments [3]. These torrents enter the Indus River from the right bank of the Chashma River, the D.G. Khan canal, and the Kachhi canal, originating from the Koh-e-Suleman Range. In 2021, a research-based study was conducted to recommend risk management solutions for the Koh-e-Suleiman hilly areas to protect the nearby community, reduce damage to infrastructure, and minimize damage to already standing crops in the event of a channel breach. Climate change has increased the flood frequency and magnitude with concentrated rainfall contributing to floods in river catchments. Flash flooding is difficult to predict in countries like Pakistan, which has varied topography, with steep slopes in hilly areas, and is severely impacted by climate change [4]. When heavy monsoon rainfall take place in the Koh e Suleman hills (D.G. Khan), the surface runoff from different hilly terrains starts flowing toward the connected plain areas in the form of stormwater, causing damage to standing crops and nearby populations. The flooding generated from these hill torrents, like the Mithwan hill torrents, Kaura hill torrents, Vehova, and Sakhi Sarwar in D.G. Khan, has high peaks within a very short time [5]. In the rainy season, due to intense rainfall, water flows from the hill torrents towards the lower plain areas. This heavy rainfall is the main cause of flash flooding in these areas every year [6].

The discharge at the outflow region highly depends on various factors, including the topography of the catchment area, total catchment area, and rainfall intensity and duration. Flash floods are amongst the most dangerous types of floods, occurring suddenly and not allowing for enough response time, resulting in the enormous loss of human lives, standing crops, and livestock [7–9]. People usually have less time to respond to these types of flash floods, resulting in an enormous loss of human lives, standing crops, and livestock. Many hilly areas in Pakistan are hill torrents. Among these, the hill torrents in Southern Punjab and Baluchistan have steep slopes, and barren mountainous regions are responsible for flash floods. Pakistan's constrained resources have contributed minimally to flash flood routing and management research [10].

There is a critical need for a floodwater management plan to mitigate the impacts of hill torrents, particularly during the monsoon season. Such a plan may involve structures capable of withstanding large water quantities and reducing the impact of hill torrents. While various models have quantified runoff from ungauged catchments, there has been a limited focus on developing mitigation strategies for flash floods [11]. Developing rainfall-runoff relations from green surfaces can also be used to calibrate traditional infiltration models in urban drainage engineering [12,13].

More knowledge about the potential application of nature-based solutions is needed to help engineers and practitioners to cope with flash floods. One of the main concerns is the protection of the nearby population, homes, and other assets from flash flooding in the rainy season. This research focuses on peak discharge, the time required to reach peak discharge, how the peak discharge can be decreased, and how the time to reach peak discharge can be increased.

Rainfall-runoff relation can be developed using physical, empirical, and conceptual models [14]. Based on the existing data, the empirical model can be applied to develop the relationship between rainfall and runoff. However, artificial neural networks [15] or fuzzy logic have been used by researchers as a conceptual modeling technique [16]. Moreover, rainfall simulators have been used by many researchers to generate runoff in a study area, which were further used to predict erosion along roads [17,18]. They have also been used to develop a relationship between sediment yield and runoff under variable rainfall intensities in a vineyard plantation in Spain [19]. Hence, researchers have indicated that a rainfall simulator could be a useful tool for representing natural rainfall conditions [20]. The runoff volume and peak discharge estimation are important measurements for designing hydraulic structures [21]. Various models can be used to simulate the rainfall-runoff relationship, resulting in input data used for the design of the structures. Among these models, synthetic hydrographs have been used to quantify runoff responses generated from hill torrents with a dense canopy [22,23]. In the current study, rainfall-runoff responses are analyzed by using a rainfall simulator at a laboratory scale. This study simulates natural conditions in hill torrents that are susceptible to flash floods using a lab-scaled hilly model. The complexity of computing hydrological processes in hilly terrains is addressed, in contrast to conventional studies in plain areas designed using a laboratory model, which uses impermeable thermopore sheets to reduce complexity and incorporates nature-based solutions. By considering a few variables, including land cover, rainfall intensity, and drainage channel slope, this study assesses flood hydrograph attenuation and offers insightful information about the mitigating effects of various vegetation conditions, similar to those used by [24–26]. Runoff was measured from the hilly model without any vegetation, and the results were compared with flexible vegetation (grass bed), rigid vegetation (tree branches), and mixed vegetation (a combination of both rigid and flexible). The role of vegetation is to create resistance on the surface flow path. Rigid vegetation, being less dense, is only a direct obstruction against rainfall impact on the land's surface, providing limited resistance. In contrast, flexible vegetation has shown more resistance, as surface runoff is continuously facing resistance in its path [27]. The combination of both rigid and flexible vegetation results in a collaborative resistance against surface runoff, making this mixed approach the most efficient.

2. Materials and Methods

In this study, a rainfall simulator FM-1849-45 by Infinit Technologies, Rosedale, MD, USA (https://infinit-technologies.com accessed on 11 November 2023) was utilized at the University of Engineering & Technology Taxila, Pakistan's water resources laboratory. It is a key tool for generating and measuring rainfall patterns and their impact on different hydrological processes. The equipment includes networks of pipes, 11 rainfall sprinklers, a discharge measuring weir, and control valves, as shown in (Figure 1a) with their components. These components were utilized to simulate different rainfall intensities at three different channel slopes of $0°$, $1°$, and $2°$ and to measure the corresponding runoff responses in the model.

The designed rigid vegetation model for the experimental analysis used in the rainfall simulator apparatus is shown in Figure 1b, representing the hilly area model with trees on a sloped surface. The dimensions of the catchment area of the rainfall simulator were 1 m (W) × 2 m (L) × 0.12 m (H). The hill model used in the experiment was constructed from polystyrene sheets with a length of 1.85 m and a width of 1 m. The width of inclined region A4 or A2 of the model represented by light blue arrows was 0.26 m, while the width of the rectangular channel A3 between the two inclined hilly areas A2 and A4 was 0.23 m. The model's remaining 0.25 m width (A1 and A5) was flat, as depicted in Figure 1c. The model's dimensions on each channel side are symmetrical, with areas A1 = A5 and A2 = A4.

After placing the model without vegetation (NV) (Figure 2a), the rainfall sprinkles were started, keeping uniform rainfall throughout the catchment area. The rigid vegetation (RV), i.e., tree branches, was placed on the polystyrene model (Figure 2b) to examine the effect of rigid vegetation on the runoff generation. The tree model was placed in a staggered arrangement, having two lanes on each side of A2 and A4. The average height of each tree was 25 cm. To represent flexible vegetation (FV) like grass and bushes, an artificial 9 mm tall grass carpet was used, as shown in Figure 2c. Finally, the combined effect of both flexible and rigid vegetation (MV) was observed by placing both RV and FV types of vegetation, as shown in Figure 2d. During the experimental work, there were three rainfall intensities, i.e., P1 = 0.3 cm/min, P2 = 0.4 cm/min, and P3 = 0.5 cm/min, used to examine the effect of rainfall intensity on hydrograph size and shape.

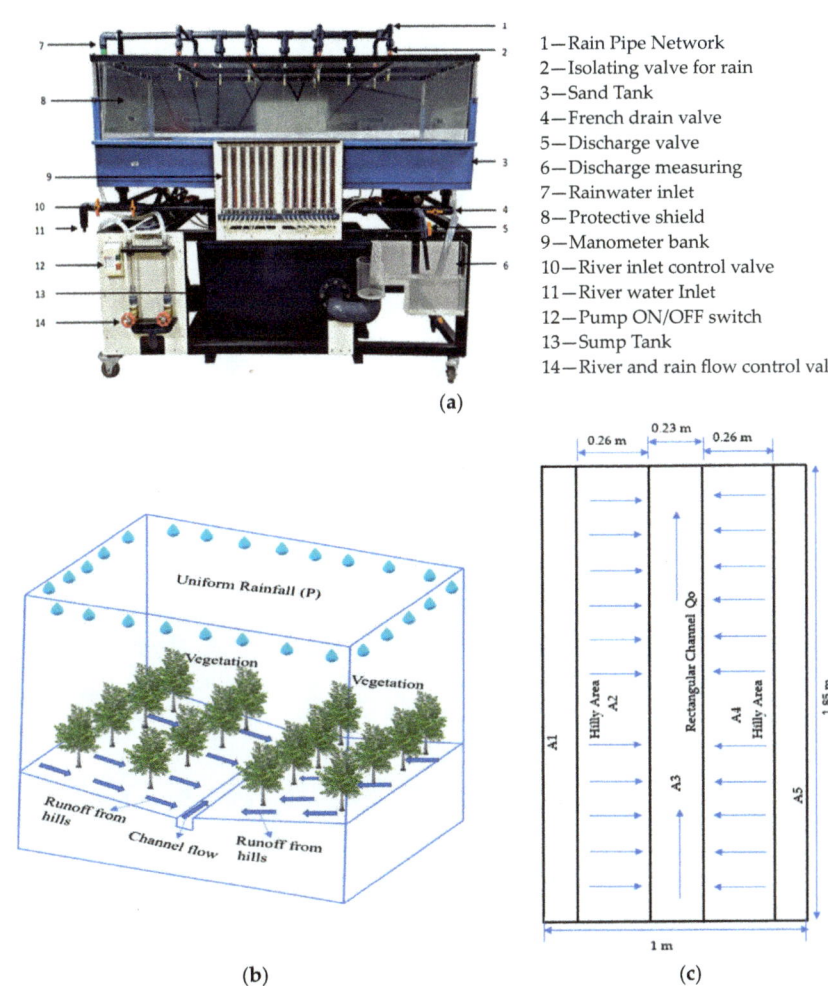

Figure 1. Photo of the rainfall simulator with featured elements—front view (**a**), scheme of a rigid vegetation model design (**b**), layout of the model topography—top view (**c**).

The runoff was measured using the weir installed on the downstream side of the rainfall simulator. The time-to-peak discharge (Tp) was calculated indirectly from the incrementally measured outflow data—for each unit liter, outflow time was measured with a stopwatch. The peak discharge (Qp) was measured by developing the hydrograph for each case. For each case, the rainfall simulator was run for 5 min and then switched off. After stopping the rainfall, the outflow continued through the catchment, producing different outflow hydrographs for each case, depending upon the land use conditions, the rainfall intensity, and the channel slope. The outflow hydrograph was simplified into three main components—rising limb, peak, and falling limb, as shown in Figure 3—and all the cases used in the experiments are presented in Table 1.

Figure 2. Photos of different model setups in the rainfall simulator: model without vegetation (**a**), model with rigid vegetation (**b**), model with flexible vegetation (**c**), and model with mixed vegetation (**d**).

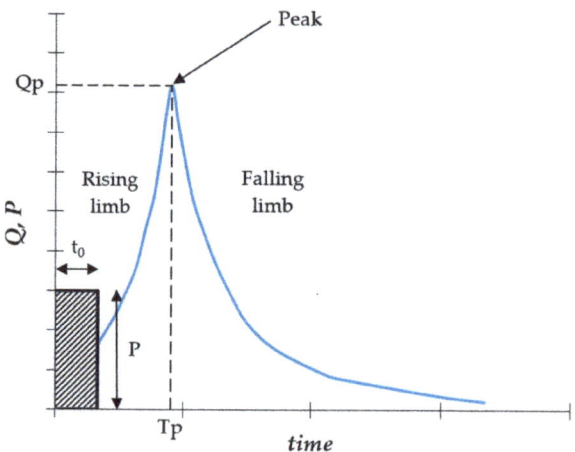

Figure 3. Breakdown of the direct runoff hydrograph components.

Table 1. The experimental matrix of conditions for the rainfall simulations.

Simulation No.	Vegetation Cover	Rainfall Intensity [cm/min]			Drainage Channel Slope	Total Number of Simulations
		P1	P2	P3		
1	NV	0.3	0.4	0.5	0°	3
2	NV	0.3	0.4	0.5	1°	3
3		0.3	0.4	0.5	2°	3
4		0.3	0.4	0.5	0°	3
5	RV	0.3	0.4	0.5	1°	3
6		0.3	0.4	0.5	2°	3
7		0.3	0.4	0.5	0°	3
8	FV	0.3	0.4	0.5	1°	3
9		0.3	0.4	0.5	2°	3
10		0.3	0.4	0.5	0°	3
11	MV	0.3	0.4	0.5	1°	3
12		0.3	0.4	0.5	2°	3

Peak discharge for each experimental case was measured directly from the respective hydrograph, and similarly, the time-to-peak was also calculated from the corresponding hydrograph peak, as shown in Figure 3. The relative reduction of peak discharges for rigid vegetation (RV), flexible vegetation (FV), and mixed vegetation (MV) are measured by comparing them with the benchmark values of barren (NV) conditions. For the percentage relative peak discharge reduction, the following equation was used:

$$Q_{p(i),red} = \frac{Q_{p(i)} - Q_{p(i,NV)}}{Q_{p(i,NV)}} \; [\%], \qquad (1)$$

For the percentage relative time-to-peak reduction, the following equation was used:

$$T_{p(i),red} = \frac{T_{p(i)} - T_{p(i,NV)}}{T_{p(i,NV)}} \; [\%], \qquad (2)$$

where i represents the type of vegetation to which the relative time-to-peak or the relative peak discharge is being calculated, and i,NV represents baseline condition for each respective case.

3. Results

3.1. Peak Discharge

It was observed that for different vegetation conditions, i.e., NV, RV, FV, and MV with rainfall intensities of P1, P2, and P3, a similar runoff pattern emerged for all bed slopes. Initially, the peak discharge was highest for the no vegetation condition, gradually declining for the case of rigid vegetation and further for the case of flexible vegetation, as given in Table 2. The mixed vegetation condition consistently exhibited the lowest peak discharge across these scenarios. The absolute Qp values were highest for the steepest channel slope and smallest for the flat channel bed (Figures 4–6), as expected.

Table 2. Peak discharge observed at the system outlet for all experiments.

Channel Slope	Rainfall Intensity [cm/min]	Qp(NV) [L/min]	Qp(RV) [L/min]	Qp(FV) [L/min]	Qp(MV) [L/min]
0°	0.3	62	53	45	41
	0.4	71	59	50	44
	0.5	80	65	54	49
1°	0.3	71	59	50	43
	0.4	77	71	67	56
	0.5	83	77	73	61
2°	0.3	71	63	47	44
	0.4	83	77	67	56
	0.5	91	83	77	67

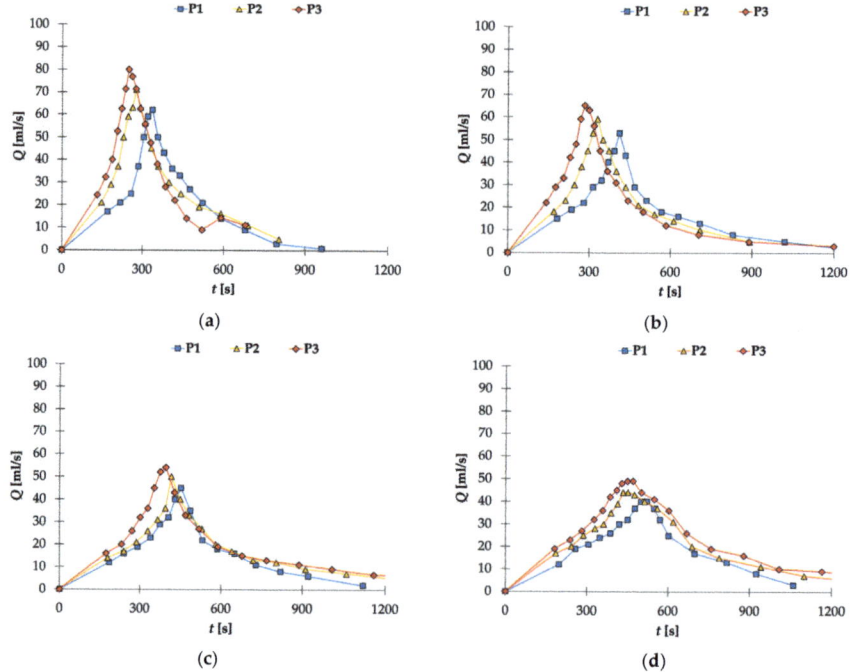

Figure 4. Outflow hydrographs for channel slope of 0° for: no vegetation condition (**a**), rigid vegetation condition (**b**), flexible vegetation condition (**c**), and mixed vegetation condition (**d**).

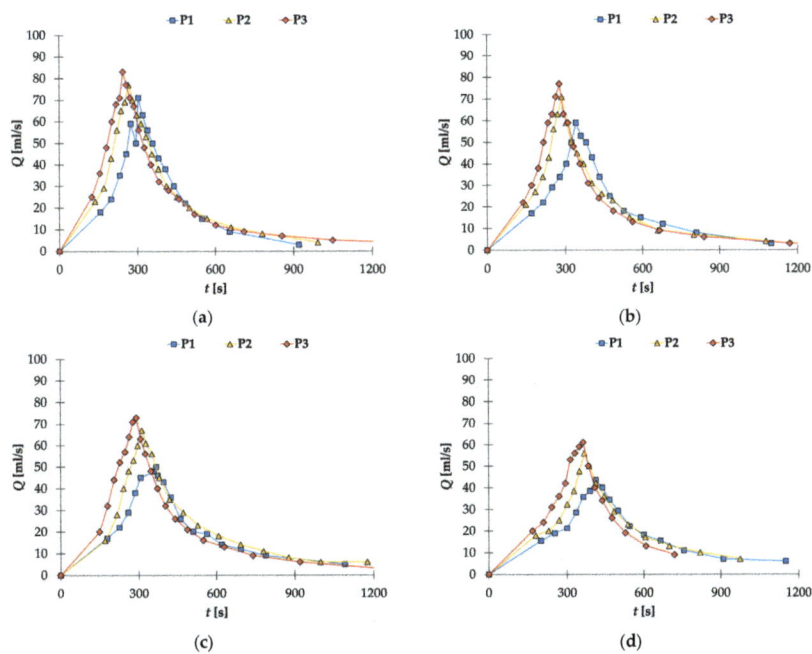

Figure 5. Outflow hydrographs for channel slope of 1° for: no vegetation condition (**a**), rigid vegetation condition (**b**), flexible vegetation condition (**c**), and mixed vegetation condition (**d**).

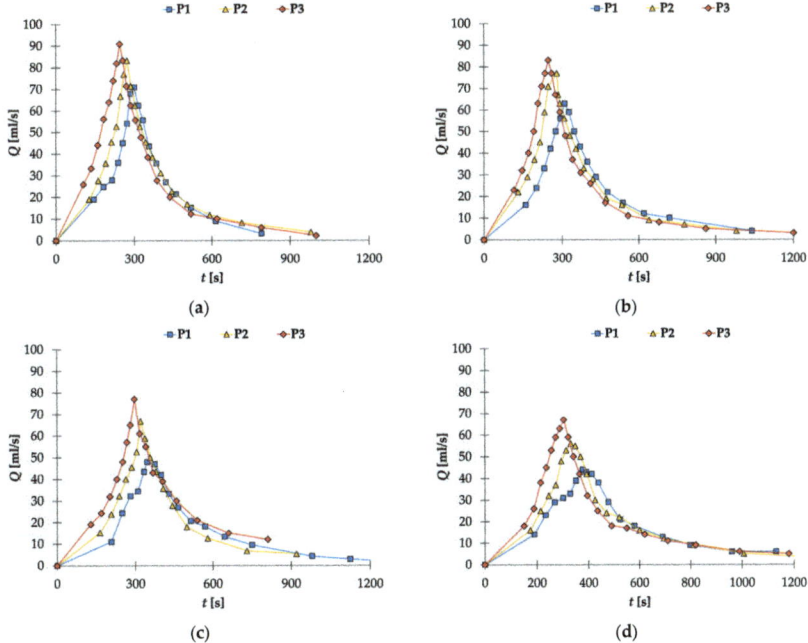

Figure 6. Outflow hydrographs for channel slope of 2° for: no vegetation condition (**a**), rigid vegetation condition (**b**), flexible vegetation condition (**c**), and mixed vegetation condition (**d**).

3.2. Relative Peak Discharge

It was found that for all the vegetation conditions used at the 0° slope, the rainfall intensity from P1 to P2 and P3 increases the percentage relative peak discharge reduction compared to the respective no vegetation condition [28]. When slopes of 1° and 2° were provided to the drainage channel, this trend changed. At the 1° and 2° channel slopes, the percentage relative peak discharge was highest for P1, and it decreased as rainfall intensity increased to P2 and P3 compared to the respective no vegetation conditions. The difference in Qp attenuation for the same vegetation type was negligible under different rainfall intensities at the 0° slope. When the channel slope increased to 1° and 2°, Qp attenuation decreased with the increase in rainfall intensity, with notable differences for different experiments. This analysis shows that both the channel slope and vegetative cover contribute to Qp attenuation, but vegetation is the main parameter which has a significant influence. The most efficient vegetation type found was mixed vegetation, which offered significant improvements over RV and FV [29]. Moreover, the experimental analysis showed that flexible vegetation significantly decreased the peaks compared to rigid vegetation, as shown in Figure 7, Table 3.

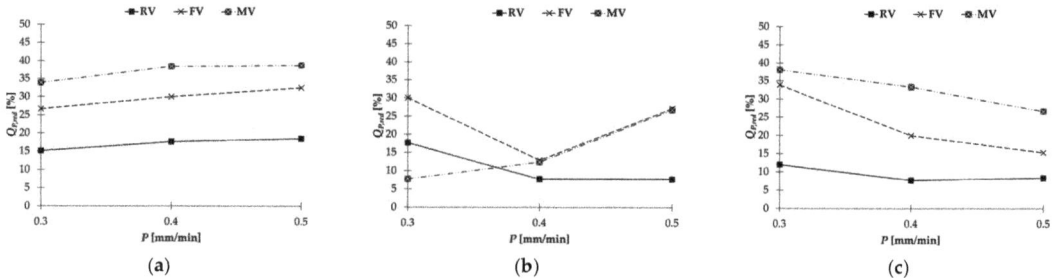

Figure 7. Change in relative peak discharge in response to variation of rainfall intensities for RV, FV, and MV: channel slope 0° (**a**), channel slope 1° (**b**), and channel slope 2° (**c**).

Table 3. Reduction in peak discharge of the hydrograph at the system outlet, expressed as percentage relative peak discharge.

Channel Slope	Rainfall Intensity [cm/min]	$Q_{p,red}(RV)$ [%]	$Q_{p,red}(FV)$ [%]	$Q_{p,red}(MV)$ [%]
0°	0.3	15	27	34
	0.4	18	30	38
	0.5	19	33	39
1°	0.3	18	30	39
	0.4	8	13	27
	0.5	8	12	27
2°	0.3	12	34	38
	0.4	8	20	33
	0.5	8	15	27

3.3. Time-to-Peak

Time-to-peak discharge is the key factor in defining the time for the community to respond to a flood [30,31]. The Tp for all experiments decreased with the increase in the rainfall intensity, emphasizing the fact that the rainfall duration was shorter than the catchment concentration time (Tables 4 and 5). Similar to the Qp analysis, the Tp increased from RV, over FV to MV, revealing that MV contributes the most to both the decrease in Qp and increase in Tp, which when combined, reduced the flood risk. The MV case at slope 0° under rainfall intensity P3 showed the maximum resistance to the flow, followed by the FV and RV cases.

Table 4. Time-to-peak for different land cover conditions.

Channel Slope	Rainfall Intensity [cm/min]	$Tp(NV)$ [s]	$Tp(RV)$ [s]	$Tp(FV)$ [s]	$Tp(MV)$ [s]
0°	0.3	337	411	452	501
	0.4	277	330	416	452
	0.5	249	283	396	450
1°	0.3	304	341	367	412
	0.4	266	286	312	367
	0.5	245	277	291	363
2°	0.3	300	310	375	376
	0.4	273	280	320	330
	0.5	245	247	297	303

Table 5. Relative increase in time-to-peak for different land cover conditions in comparison to no vegetation condition.

Channel Slope	Rainfall Intensity [cm/min]	$Tp,red(RV)$ [%]	$Tp,red(FV)$ [%]	$Tp,red(MV)$ [%]
0°	0.3	22	34	49
	0.4	19	50	63
	0.5	14	59	81
1°	0.3	18	21	36
	0.4	8	17	38
	0.5	13	19	48
2°	0.3	3	25	25
	0.4	3	17	21
	0.5	1	21	24

It was found that the time-to-peak (Tp) increased as the vegetation condition changed from barren to rigid, from rigid to flexible, and from flexible to mixed vegetation conditions. The impact of slope and rainfall intensity followed the same trend as the Tp duration. As we increased the slope from 0° to 1° and 2° and increased the rainfall intensity from P1 to P2 and P3, the time-to-peak (Tp) duration decreased with all other conditions remaining the same (Tables 4 and 5). The experimental analysis for Tp showed that the slope and rainfall intensity both had an inverse relation with Tp (Tables 4 and 5). It was analyzed that Tp follows a distinct pattern, where the MV condition consistently produces the longest Tp, followed by the FV condition, which exhibits a shorter duration. Moreover, the rigid vegetation condition, although more structured than the mixed and flexible vegetations, also demonstrated a relatively shorter Tp when compared with the mixed and flexible vegetation conditions. An increase in the channel slope resulted in a decrease in Tp, but the presence of the vegetation still significantly reduced the Tp. When the channel slope was the steepest, the difference between the MV and FV cases became negligible for all rainfall intensities. The results also show that the maximum relative percentage of the Tp reduction was achieved for the mixed vegetation and 0° slope (81% under P3 rainfall) (Figures 8 and 9). The minimum relative percentage of the Tp (1%) was observed in the case of the rigid vegetation condition at a 2° slope under P3 rainfall, showing that the effect of rigid vegetation is severely limited.

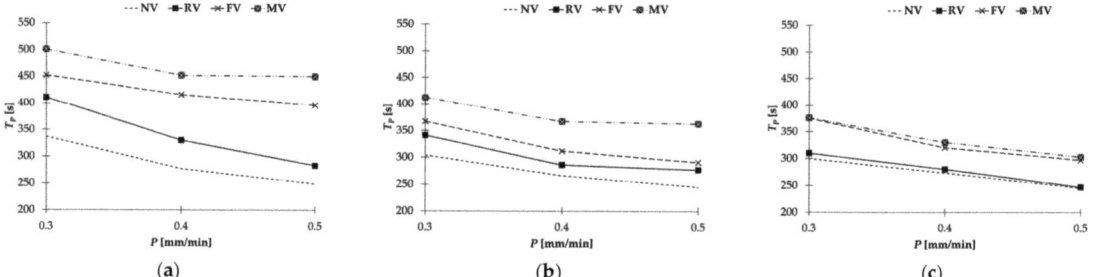

Figure 8. Change in time-to-peak in response to variation of rainfall intensities for RV, FV, and MV: channel slope 0° (a), channel slope 1° (b), and channel slope 2° (c).

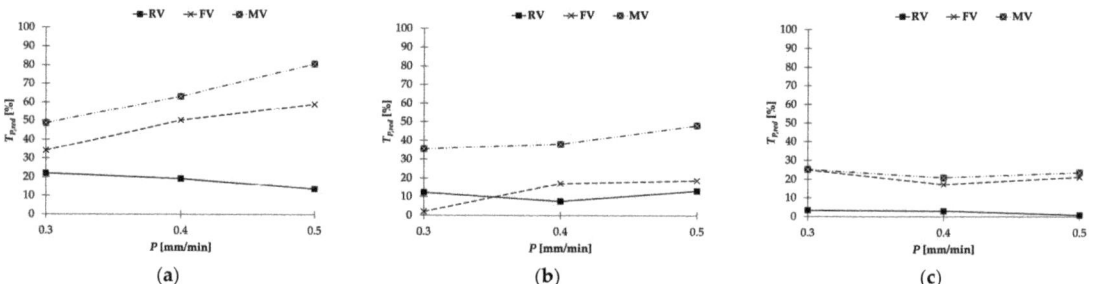

Figure 9. Change in relative time-to-peak in response to variations in rainfall intensities for RV, FV, and MV: channel slope 0° (a), channel slope 1° (b), and channel slope 2° (c).

4. Discussion

4.1. Comparison of the Findings to the Literature

Hilly areas subject to flash flooding produced by hill torrents were the focus of this [32] study, which concentrated on applying nature-based solutions to reduce the impact of flash floods. A laboratory-scale hill slope model was developed to examine the rainfall-runoff responses and evaluate the efficiency of various NBS configurations in lowering the peak discharge and increasing the time-to-peak [33]. The rigid vegetation used in this study showed a resistance in the range of 8 to 15% to reduce flood peaks due to rainfall. Flexible and a combination of both rigid and flexible vegetations reduced the flood peaks from 12 to 33% and 27 to 39%, respectively. There are numerous instances of governments using tree planting as a flood control measure throughout the world. To lessen flooding, the municipal authorities of Pickering, North Yorkshire, England, planted trees as part of the project "Slowing the Flow." According to a scheme analysis, the measures decreased peak river flow by 15–20%. The program was launched in 2009 following the town's four significant floods in a ten-year period, with the 2007 floods resulting in damage estimated at around £7 million.

Similar findings were reported in the laboratory-based study by Chouksey et al. [25], who also used a rainfall simulator over an experimental hillslope plot to investigate hydrological modeling, providing a better understanding of the effectiveness of nature-based solutions in mitigating flash flood impacts. The current findings follow the same trends as the field-based study by Flores et al. [26], who compared three daily rainfall-runoff hydrological models using four evapotranspiration models in four small, forested watersheds with different land covers in South-Central Chile. This demonstrates that the presented insights can directly find application in real-world situations.

Flexible vegetation throughout the catchment retained some amount of rainfall water and also effectively resisted surface runoff to reach the catchment outlet. This vegetation effectively increased the time to flow from the catchment to the outlet and decreased the flood peak by delaying the process.

The current experimental design utilized thermopore sheets made of impermeable material, which enables controlled simulations of various NBS conditions. These design considerations simplified the experimental setup and reduced the model complexity but may also have impacted its representativeness compared to real-life scenarios. Future studies may explore more realistic soil types and vegetation characteristics for the hillslope model to offer a more accurate assessment of the effectiveness of nature-based solutions. Also, the use of a laboratory-scaled hillslope model may not fully capture the complexities of real-world hillslope conditions. The conversion of scientific discoveries into useful applications is facilitated using laboratory-sized models that enable controlled experiments that can be replicated and scaled up to real-world situations [34]. For such cases, the potential impact of human activities, such as land use changes and urbanization on flash flooding in hilly terrains, can be better assessed to inform the development of more effective strategies for flash flood resilience.

This study's results emphasize the significant effect of vegetation and ground slope in reducing flash flooding impacts. The mixed vegetation condition with a channel slope of 0° was found to be the most effective for minimizing severe flash flooding, as was the case in other studies [35,36]. This study offers important insights into developing long-term solutions by demonstrating the efficiency of NBS in flood mitigation, particularly in mountainous regions. The peak discharge was simultaneously minimized, while the maximum Tp under P3 rainfall was produced. Compared to previous NBS cases, the mixed vegetation condition reduced the peak discharge up to 39% and increased the Tp by 81%. Practitioners should consider integrating nature-based solutions, such as mixed vegetation, into land use planning and development strategies for hilly terrains. This can help to reduce the risk of flash flooding and improve community resilience by increasing the Tp and reducing peak discharge, according to the findings herein. This could include providing financial incentives for landowners to adopt mixed vegetation or other nature-based solutions as well as incorporating these strategies into broader flood risk management plans and policies. Raising awareness about the benefits of nature-based solutions for flash flood mitigation and engaging local communities in their implementation can also help sustain their success.

4.2. Future Research Direction

The results of the conducted research show that mixed vegetation conditions with a 0° bed slope are particularly effective for increasing the Tp and lowering the Qp. This study offers important new information for applying improved flood management practices and aims to provide the groundwork for future investigations and NBS applications to lessen the negative effects of flash flooding and improve community resilience [13,35], building upon the already known benefits of nature-based solutions in reducing the impacts of floods in hilly areas subject to flash flooding [32,37]. Future research could focus on improving the laboratory-scale hillslope model to use better field conditions. This might involve incorporating more realistic soil types, vegetation characteristics, and hydrological processes to enhance the model's accuracy and applicability to real-world scenarios. Further, additional land use and climate change conditions can be explored in future studies. These, for example, could investigate the effectiveness of different land use conditions and vegetation types in mitigating flash flooding in hilly terrains. This could help identify the most effective nature-based solutions for specific regions and inform the development of targeted strategies for flash flood resilience. Future research could examine the performance of nature-based solutions, including to what extent vegetation is useful, under various climate change scenarios, such as increased rainfall intensity or more frequent extreme weather events. This would provide valuable insights into the long-term effectiveness of these solutions and help in the form of adaptation strategies for flash flood resilience in a changing climate.

5. Conclusions

Decreasing the peak discharge and increasing the time-to-peak can reduce the impact of flash floods by providing the community with more time to respond. Laboratory models can be used to evaluate the sustainability and long-term viability of nature-based solutions for flood mitigation in real conditions, providing information about land cover density and type and different rainfall patterns that might occur due to climate change. The experimental results of this study show that the mixed vegetation condition is the best one to reduce the peak discharge and increase the time-to-peak under different rainfall intensities and channel bed slopes. It was also observed that flexible vegetation contributed much more than rigid vegetation to increasing the time-to-peak and mitigating the flood peak:

- The peak discharge of a hydrograph was positively correlated with rainfall intensity and the channel bed slope. An increase in either of these factors led to a higher peak of the hydrograph and vice versa.
- The hydrograph formation for the no vegetation condition exhibited the maximum peak discharge, while the mixed vegetation condition, comprising both flexible and rigid vegetation, showed the minimum peak discharge.
- Flexible vegetation showed greater resistance to runoff than rigid vegetation.
- The order of resistance to flow for time-to-peak discharge increased from rigid to flexible and was the highest for the mixed vegetation condition.
- The time-to-peak discharge of the hydrograph was negatively correlated with rainfall intensity and channel bed slope.
- The mixed vegetation condition with the lowest bed slope and maximum rainfall intensity of P3 reduced the peak discharges by 39% and increased the time-to-peak by 81%. The same mixed vegetation condition reduced the peak discharge by almost 27% and increased the time-to-peak discharge by 24% under the same rainfall condition with the maximum channel slope.

Author Contributions: Conceptualization, S.U.R., A.A. and R.F.; methodology, S.U.R., A.A. and G.A.P.; validation, S.U.R., A.A. and A.R.G.; formal analysis, S.U.R., A.A., M.V. and G.G.; investigation, S.U.R. and A.A.; resources, S.U.R. and A.A.; data curation, S.U.R. and A.A.; writing—original draft preparation, S.U.R., A.A. and G.G.; writing—review and editing, S.U.R., A.A., M.V. and G.G.; visualization, S.U.R., A.A., M.V. and G.G.; supervision, A.A. and M.V. All authors have read and agreed to the published version of the manuscript.

Funding: The authors acknowledge the support of HEC Pakistan under the National Research Program for Universities. No: Ref No. 20-16414/NRPU/R&D/HEC/2021.

Data Availability Statement: Data available on request.

Conflicts of Interest: The authors declare no conflicts of interest.

References

1. Cohen-Shacham, E.; Walters, G.; Janzen, C.; Maginnis, S. *Nature-Based Solutions to Address Global Societal Challenges*; IUCN: Gland, Switzerland, 2016; p. 97.
2. Renaud, F.G.; Sudmeier-Rieux, K.; Estrella, M.; Nehren, U. *Ecosystem-Based Disaster Risk Reduction and Adaptation in Practice*; Springer: Berlin/Heidelberg, Germany; Cham, Switzerland, 2016; p. 598. [CrossRef]
3. Saleem, M.; Arfan, M.; Ansari, K.; Hassan, D. Analyzing the Impact of Ungauged Hill Torrents on the Riverine Floods of the River Indus: A Case Study of Koh E Suleiman Mountains in the DG Khan and Rajanpur Districts of Pakistan. *Resources* **2023**, *12*, 26. [CrossRef]
4. Munir, B.A.; Ahmad, S.R.; Rehan, R. Torrential Flood Water Management: Rainwater Harvesting through Relation Based Dam Suitability Analysis and Quantification of Erosion Potential. *ISPRS Int. J. Geo-Inf.* **2021**, *10*, 27. [CrossRef]
5. Ahmed, T.F.; Shah, S.-S.; Sheikh, A.A.; Hashmi, H.N.; Khan, M.A.; Afzal, M.A. Flood Water Management from Hill Torrents of Pakistan for Agriculture Livelihood Improvement. *Pak. J. Agric. Res.* **2021**, *34*, 3. [CrossRef]

6. Munir, B.A.; Iqbal, J. Flash flood water management practices in Dera Ghazi Khan City (Pakistan): A remote sensing and GIS prospective. *Nat. Hazards* **2016**, *81*, 1303–1321. [CrossRef]
7. Prokešová, R.; Horáčková, Š.; Snopková, Z. Surface runoff response to long-term land use changes: Spatial rearrangement of runoff-generating areas reveals a shift in flash flood drivers. *Sci. Total Environ.* **2022**, *815*, 151591. [CrossRef] [PubMed]
8. Kundzewicz, Z.W.; Kanae, S.; Seneviratne, S.I.; Handmer, J.; Nicholls, N.; Peduzzi, P.; Mechler, R.; Bouwer, L.M.; Arnell, N.; Mach, K.; et al. Flood risk and climate change: Global and regional perspectives. *Hydrol. Sci. J.* **2014**, *59*, 1–28. [CrossRef]
9. Jonkman, S.N.; Vrijling, J.K. Loss of life due to floods. *J. Flood Risk Manag.* **2008**, *1*, 43–56. [CrossRef]
10. Ahmad, N.; Khan, S.; Ehsan, M.; Rehman, F.U.; Al-Shuhail, A. Estimating the Total Volume of Running Water Bodies Using Geographic Information System (GIS): A Case Study of Peshawar Basin (Pakistan). *Sustainability* **2022**, *14*, 3754. [CrossRef]
11. Nakhaei, M.; Nakhaei, P.; Gheibi, M.; Chahkandi, B.; Wacławek, S.; Behzadian, K.; Chen, A.S.; Campos, L.C. Enhancing community resilience in arid regions: A smart framework for flash flood risk assessment. *Ecol. Indic.* **2023**, *153*, 110457. [CrossRef]
12. Rammal, M.; Berthier, E. Runoff Losses on Urban Surfaces during Frequent Rainfall Events: A Review of Observations and Modeling Attempts. *Water* **2020**, *12*, 2777. [CrossRef]
13. Redfern, T.W.; Macdonald, N.; Kjeldsen, T.R.; Miller, J.D.; Reynard, N. Current understanding of hydrological processes on common urban surfaces. *Prog. Phys. Geogr.* **2016**, *40*, 699–713. [CrossRef]
14. Aksoy, H.; Kavvas, M.L. A review of hillslope and watershed scale erosion and sediment transport models. *CATENA* **2005**, *64*, 247–271. [CrossRef]
15. Hundecha, Y.; Bardossy, A.; Werner, H.-W. Development of a fuzzy logic-based rainfall-runoff model. *Hydrol. Sci. J.* **2001**, *46*, 363–376. [CrossRef]
16. Adams, R.; Parkin, G.; Rutherford, J.C.; Ibbitt, R.P.; Elliott, A.H. Using a rainfall simulator and a physically based hydrological model to investigate runoff processes in a hillslope. *Hydrol. Process. Int. J.* **2005**, *19*, 2209–2223. [CrossRef]
17. Arnaez, J.; Lasanta, T.; Ruiz-Flaño, P.; Ortigosa, L. Factors affecting runoff and erosion under simulated rainfall in Mediterranean vineyards. *Soil Tillage Res.* **2007**, *93*, 324–334. [CrossRef]
18. Sheridan, G.J.; Noske, P.J.; Lane, P.N.J.; Sherwin, C.B. Using rainfall simulation and site measurements to predict annual interrill erodibility and phosphorus generation rates from unsealed forest roads: Validation against in-situ erosion measurements. *CATENA* **2008**, *73*, 49–62. [CrossRef]
19. Quinton, J.N.; Edwards, G.M.; Morgan, R.P.C. The influence of vegetation species and plant properties on runoff and soil erosion: Results from a rainfall simulation study in south eastsoutheast Spain. *Soil Use Manag.* **1997**, *13*, 143–148. [CrossRef]
20. Khaleghi, M.R.; Gholami, V.; Ghodusi, J.; Hosseini, H. Efficiency of the geomorphologic instantaneous unit hydrograph method in flood hydrograph simulation. *CATENA* **2011**, *87*, 163–171. [CrossRef]
21. Pradhananga, D.; Pomeroy, J.W. Diagnosing changes in glacier hydrology from physical principles using a hydrological model with snow redistribution, sublimation, firnificationfirnification, and energy balance ablation algorithms. *J. Hydrol.* **2022**, *608*, 127545. [CrossRef]
22. Liao, G.; He, P.; Gao, X.; Lin, Z.; Huang, C.; Zhou, W.; Deng, O.; Xu, C.; Deng, L. Land use optimization of rural production–living–ecological space at different scales based on the BP-ANN and CLUE-S models. *Ecol. Indic.* **2022**, *137*, 108710. [CrossRef]
23. Zhou, Y.; Cui, Z.; Lin, K.; Sheng, S.; Chen, H.; Guo, S.; Xu, C.-Y. Short-term flood probability density forecasting using a conceptual hydrological model with machine learning techniques. *J. Hydrol.* **2022**, *604*, 127255. [CrossRef]
24. Zhou, M.; Qu, S.; Chen, X.; Shi, P.; Xu, S.; Chen, H.; Zhou, H.; Gou, J. Impact Assessments of Rainfall–Runoff Characteristics Response Based on Land Use Change via Hydrological Simulation. *Water* **2019**, *11*, 866. [CrossRef]
25. Chouksey, A.; Lambey, V.; Nikam, B.R.; Aggarwal, S.P.; Dutta, S. Hydrological Modelling Using a Rainfall Simulator over an Experimental Hillslope Plot. *Hydrology* **2017**, *4*, 17. [CrossRef]
26. Flores, N.; Rodríguez, R.; Yépez, S.; Osores, V.; Rau, P.; Rivera, D.; Balocchi, F. Comparison of Three Daily Rainfall-Runoff Hydrological Models Using Four Evapotranspiration Models in Four Small Forested Watersheds with Different Land Cover in South-Central Chile. *Water* **2021**, *13*, 3191. [CrossRef]
27. Zhang, X.; Yu, Y.; Hu, C.; Ping, J. Study on the influence of vegetation change on runoff generation mechanism in the Loess Plateau, China. *Water Supply* **2020**, *21*, 683–695. [CrossRef]
28. Mu, W.; Yu, F.; Li, C.; Xie, Y.; Tian, J.; Liu, J.; Zhao, N. Effects of Rainfall Intensity and Slope Gradient on Runoff and Soil Moisture Content on Different Growing Stages of Spring Maize. *Water* **2015**, *7*, 2990–3008. [CrossRef]
29. Han, D.; Deng, J.; Gu, C.; Mu, X.; Gao, P.; Gao, J. Effect of shrub-grass vegetation coverage and slope gradient on runoff and sediment yield under simulated rainfall. *Int. J. Sediment Res.* **2021**, *36*, 29–37. [CrossRef]
30. Grimaldi, S.; Petroselli, A.; Tauro, F.; Porfiri, M. Time of concentration: A paradox in modern hydrology. *Hydrol. Sci. J.* **2012**, *57*, 217–228. [CrossRef]
31. Beven, K.J. A history of the concept of time of concentration. *Hydrol. Earth Syst. Sci.* **2020**, *24*, 2655–2670. [CrossRef]
32. Ben-Zvi, A. Laboratory examination of linearity in rainfall–Runoff relationships. *Hydrol. Sci. J.* **2020**, *65*, 1794–1801. [CrossRef]
33. Ramachandra Rao, A.; Tirtotjondro, W. Computation of unit hydrographs by a Bayesian method. *J. Hydrol.* **1995**, *164*, 325–344. [CrossRef]
34. Fang, X.; Thompson David, B.; Cleveland Theodore, G.; Pradhan, P.; Malla, R. Time of Concentration Estimated Using Watershed Parameters Determined by Automated and Manual Methods. *J. Irrig. Drain. Eng.* **2008**, *134*, 202–211. [CrossRef]

35. Lora, M.; Camporese, M.; Salandin, P. Design and performance of a nozzle-type rainfall simulator for landslide triggering experiments. *CATENA* **2016**, *140*, 77–89. [CrossRef]
36. Wong, T.S. Evolution of Kinematic Wave Time of Concentration Formulas for Overland Flow. *J. Hydrol. Eng.* **2009**, *14*, 739–744. [CrossRef]
37. Ben-Zvi, A. Runoff peaks from two-dimensional laboratory watersheds. *J. Hydrol.* **1984**, *68*, 115–139. [CrossRef]

Disclaimer/Publisher's Note: The statements, opinions and data contained in all publications are solely those of the individual author(s) and contributor(s) and not of MDPI and/or the editor(s). MDPI and/or the editor(s) disclaim responsibility for any injury to people or property resulting from any ideas, methods, instructions or products referred to in the content.

MDPI
St. Alban-Anlage 66
4052 Basel
Switzerland
www.mdpi.com

Water Editorial Office
E-mail: water@mdpi.com
www.mdpi.com/journal/water

Disclaimer/Publisher's Note: The statements, opinions and data contained in all publications are solely those of the individual author(s) and contributor(s) and not of MDPI and/or the editor(s). MDPI and/or the editor(s) disclaim responsibility for any injury to people or property resulting from any ideas, methods, instructions or products referred to in the content.